Energy, Environment, and Sustainability

Series editors

This books series publishes cutting edge monographs and professional books focused on all aspects of energy and environmental sustainability, especially as it relates to energy concerns. The Series is published in partnership with the International Society for Energy, Environment, and Sustainability. The books in these series are editor or authored by top researchers and professional across the globe. The series aims at publishing state-of-the-art research and development in areas including, but not limited to:

- Renewable Energy
- Alternative Fuels
- Engines and Locomotives
- Combustion and Propulsion
- Fossil Fuels
- Carbon Capture
- Control and Automation for Energy
- Environmental Pollution
- Waste Management
- Transportation Sustainability

More information about this series at http://www.springer.com/series/15901

Reeta Rani Singhania · Rashmi Avinash Agarwal
R. Praveen Kumar · Rajeev K. Sukumaran
Editors

Waste to Wealth

Springer

Editors
Reeta Rani Singhania
DBT-Energy Bioscience Overseas Fellow
DBT-IOC Center for Advanced Bio-Energy Research, Indian Oil Corporation Limited R & D Campus
Faridabad, Haryana
India

Rashmi Avinash Agarwal
Department of Civil Engineering
Indian Institute of Technology Kanpur
Kanpur, Uttar Pradesh
India

R. Praveen Kumar
Department of Biotechnology
Arunai Engineering College
Tiruvannamalai, Tamil Nadu
India

Rajeev K. Sukumaran
Biofuels and Biorefineries Section, Microbial Processes and Technology Division
CSIR-National Institute for Interdisciplinary Science and Technology
Thiruvananthapuram, Kerala
India

ISSN 2522-8366 ISSN 2522-8374 (electronic)
Energy, Environment, and Sustainability
ISBN 978-981-10-7430-1 ISBN 978-981-10-7431-8 (eBook)
https://doi.org/10.1007/978-981-10-7431-8

Library of Congress Control Number: 2017959151

Printed on acid-free paper

This Springer imprint is published by Springer Nature
The registered company is Springer Nature Singapore Pte Ltd.
The registered company address is: 152 Beach Road, #21-01/04 Gateway East, Singapore 189721, Singapore

Preface

Energy demand has been rising remarkably due to increasing population and urbanization. Global economy and society are significantly dependent on the energy availability because it touches every facet of human life and its activities. Transportation and power generation are major examples of energy. Without the transportation by millions of personalized and mass transport vehicles and availability of 24×7 power, human civilization would not have reached contemporary living standards.

The first international conference on "Sustainable Energy and Environmental Challenges" (SEEC-2017) was organized under the auspices of "International Society for Energy and Environmental Sustainability" (ISEES) by the "Center of Innovative and Applied Bioprocessing" (CIAB), Mohali, from February 26 to 28, 2017. ISEES was founded at the Indian Institute of Technology Kanpur in January 2014, with the aim of spreading knowledge in the fields of energy, environment, sustainability and combustion. The society's goal is to contribute to the development of clean, affordable and secure energy resources and a sustainable environment for the society and to spread knowledge in the above-mentioned areas and awareness about the environmental challenges, which the world is facing today. ISEES is involved in various activities such as conducting workshops, seminars and conferences in the domains of its interest. The society also recognizes the outstanding work done by the young scientists and engineers for their contributions in these fields by conferring them awards under various categories.

This conference provided a platform for discussions between eminent scientists and engineers from various countries including India, the USA, South Korea, Norway, Malaysia and Australia. In this conference, eminent speakers from all over the world presented their views related to different aspects of energy, combustion, emissions and alternative energy resource for sustainable development and cleaner environment. The conference started with four mini-symposiums on very topical themes, which included (i) New Fuels and Advanced Engine Combustion, (ii) Sustainable Energy, (iii) Experimental and Numerical Combustion and (iv) Environmental Remediation and Rail Road Transport. The conference had 14 technical sessions on topics related to energy and environmental sustainability and a panel discussion on "Challenges, Opportunities and Directions of Technical Education & Research in the Area of Energy, Environment and Sustainability" to wrap up the three-day technical extravaganza. The conference included 2 plenary talks, 12 keynote talks, 42 invited talks from prominent scientists, 49 contributed talks and 120 posters. A total of 234 participants and speakers attended this three-day conference, which hosted Dr. V. K. Saraswat, Member, NITI Aayog, India, as a chief guest for the award ceremony of ISEES. This conference laid out the road map for technology development, opportunities and challenges in this technology domain. The technical sessions in the conference included Advances in IC Engines and Fuels; Conversion of Biomass to Biofuels; Combustion Processes; Renewable Energy: Prospects and Technologies; Waste to Wealth—Chemicals and Fuels; Energy Conversion Systems; Numerical Simulation of Combustion Processes; Alternate Fuels for IC Engines; Sprays and Heterogeneous Combustion of Coal/Biomass; Biomass Conversion to Fuels and Chemicals—Thermochemical Processes; Utilization of Biofuels; and Environmental Protection and Health. All these topics are very relevant to the country and the world in the present context. The society is grateful to Prof. Ashok Pandey for organizing and hosting this conference, which led to the germination of this series of monographs, which included 16 books related to different aspects of energy, environment and sustainability. This is the first time that such a voluminous and high-quality outcome has been achieved by any society in India from one conference.

The editors would like to express their sincere gratitude to the authors for submitting their work in a timely manner and revising it appropriately at short notice. We would like to express our special thanks to Prof. Ashok Pandey, Dr. Jitendra Saini, Dr. Ruchi Gaur, Dr. Manali Kapoor, Dr. Tirath Raj, Dr. Sandeep Sharma, Dr. Anil Patel, Dr. Nisha Singh, Dr. Mukund Adsul, Dr. Sindhu Raveendran, Dr. A. Sabu, Prof. K. Jayachandran, Dr. Suresh P. V., Dr. Bharatiraja B., Mr. Chozhavendran, Dr. Kartik Rajendran, Mr. Sivarathanakumar S., Mr. Raja Sathendra E., Mr. Vinoth Arulraj J. and Dr. Shanmugaprakash, who reviewed various chapters of this monograph and provided their valuable suggestions to improve the manuscripts. We acknowledge the support received from various funding agencies and organizations for successfully conducting the first ISEES conference SEEC-2017, where these monographs germinated. These include Department of Science and Technology, Government of India (special thanks to Dr. Sanjay Bajpai); TSI, India (special thanks to Dr. Deepak Sharma); Tesscorn, India (special thanks to Sh. Satyanarayana); AVL, India; Horiba, India; Springer (special thanks to Swati Mehershi); CIAB (special thanks to Dr. Sangwan).

In this era of technology development, it is inevitable to find eco-friendly and sustainable solutions for waste-to-value-added product conversion. With the advent of biotechnology, it is possible to utilize waste for the benefits of mankind. Bioenergy from biomass or algae lipids, bioelectricity from municipal waste, biochar application for agriculture and various agricultural residues' utilization for getting valuable products have been made successful which helps in handling the waste as well as getting value out of it.

This monograph is intended for biotechnologists who practice in the area of waste conversion, and we hope that the book would be of great interest to the professionals and postgraduate students involved in finding advanced eco-friendly sustainable solutions for waste handling and its value addition. The main objective of this monograph is to promote a better and more accurate understanding of the possible ways to tackle waste generated from various streams, its possible utilization for the benefit of mankind and the challenges that need to be tackled.

Faridabad, India — Reeta Rani Singhania
Kanpur, India — Rashmi Avinash Agarwal
Tiruvannamalai, India — R. Praveen Kumar
Thiruvananthapuram, India — Rajeev K. Sukumaran

Contents

About the Editors

Dr. Reeta Rani Singhania, M.Sc., Ph.D. is DBT—Energy Bioscience Overseas Fellow working at the Center for Advanced Bioenergy Research, R&D Centre, IOCL, Faridabad, India. She obtained M.Sc. in Biotechnology from GG University, Bilaspur, India, and Ph.D. in Biotechnology from the Cochin University of Science and Technology, Cochin, India. She has worked at CSIR-NIIST, Trivandrum, India; EPFL, Lausanne, Switzerland; and Blaise Pascal University, Clermont-Ferrand, France. Her major research interest is in the areas of Microbial and Enzyme Technologies, with current focus on Biofuels from Lignocellulosic Biomass. She has authored 34 publications in peer-reviewed national and international journals and 21 book chapters along with several conference communications to her credit. She is the recipient of Elsevier Best Paper Award in 2007, AU-CBT Excellence Award of BRSI in 2008 and IFIBiop Young Scientist Award in 2013. She has been conferred Best Reviewer Certificate from *Bioresource Technology* journal for 2 years. She has served as Guest Editor of Renewable Energy and Bioresource Technology journals for a special issue. She is Editorial Board Member of Biofuel Research Journal. She is the *Editor* of Bioresource Technology Reports, an *Elsevier Journal*. She has received Most Cited Paper Award from Bioresource Technology, Elsevier, in 2015.

She has received the prestigious "DBT—Energy Bioscience Overseas Fellowship" from the Department of Biotechnology, Ministry of Science and Technology, Government of India, for working on Bioenergy in India.

Dr. Rashmi Avinash Agarwal is Young Scientist Fellow at the Indian Institute of Technology Kanpur, Kanpur-208016, India. She completed her doctoral degree in Inorganic Chemistry from July 2008 to May 2014 at the Indian Institute of Technology Kanpur under the supervision of Prof. P. K. Bhardwaj. The Ph.D. thesis's title was "Coordination Polymers of Transition Metal Ions with Benzimidazole Based Ligands: Single-Crystal to Single-Crystal (SC-SC) Transformations, Gas Storage and Magnetic Studies." Prior to her doctoral degree, she completed her M.Sc. in Organic Chemistry from Rajasthan University, Jaipur, in 2002 and B.Sc. in Chemistry from Kanoria College, Rajasthan University, Jaipur, in 2000. She has expertise in Supramolecular Chemistry, Coordination Chemistry, Coordination Polymers, Organic Synthesis, Inorganic Synthesis, Hydrothermal Synthesis of Coordination Polymers, Crystallization, Crystal Structure Determination, Topology, Hydrogen Bonding, Structural Analysis, Florescence, SC (Single-Crystal)-to-SC transformation, FTIR, Thermogravimetric, NMR, Mass and PXRD Analysis, Column Chromatography, Synthesis of Nanoparticles Utilizing Porous Materials, TEM Analysis. She has published over 20 research papers in leading international journals.

Prof. R. Praveen Kumar is graduate in **Chemical Engineering** and postgraduation in **Industrial Biotechnology from Annamalai University**. He is working as **Head of the Department** in **Department of Biotechnology, Arunai Engineering College, Tiruvannamalai, Tamil Nadu, India**. His **area of research** includes **Renewable Energy from Biomass and Municipal Waste**. He was the **Chair for first, second and third international conferences on Bioenergy, Environment and Sustainable Technologies (BEST2013, BEST2015 and BEST2017)** and has organized various national-level symposiums and conferences. He has served as **scientific advisory committee member in several national and international events.** He has visited various countries including the USA, Australia, China, Italy, Hongkong and Bangladesh to attend various conferences. He has **chaired sessions and delivered invited talks** in various national and international conferences. He has done provisional registration for **four patents** and published more than 52 papers in peer-reviewed journals. He is a life member in various professional societies which includes Biotech Research Society, India (BRSI), Indian Institute of Chemical Engineering (IICHE), International Forum on Industrial Bioprocesses (IFIBiop), Biogas Forum of India (BigFin), Indian Society for Technical Education (ISTE), Engineers Without Borders, India (EWB-India). He has **served as Management Council Member in Biotech Research Society, India (BRSI)**, and is currently serving as **Vice-President** in **Engineers Without Borders—India (EWB-India) Chennai Chapter.** He was awarded with **Outstanding Young Investigator Award for excellence in research and teaching in a rural setting, awarded by Raise.Rural.**

Dr. Rajeev K. Sukumaran is Senior Scientist at CSIR-National Institute for Interdisciplinary Science and Technology (CSIR-NIIST), Trivandrum, India. Currently, he heads the Biofuels and Biorefineries Section, Microbial Processes and Technology Division, CSIR-NIIST. He did his master's degree and Ph.D. in Biotechnology at the Cochin University of Science and Technology, Cochin, India (1993–2000). He was Technical Member of Bigtec Pvt Ltd, Bangalore, where he additionally served as Consultant at the Vittal Mallya Scientific Research Foundation (United Breweries Group's R&D Centre). He had postdoctoral trainings in Molecular Immunology and Stem Cell Biology at the Icahn School of Medicine, Mount Sinai, New York and National University Hospital, Singapore, respectively, before joining CSIR in November 2004. Since then, he has been working on biomass conversion to biofuels, especially on enzymes for biomass hydrolysis.

He played a leading role in setting up the Centre for Biofuel's lignocellulosic ethanol pilot plant and the solid-state enzyme production pilot plant at NIIST. His work on cellulases has led to a technology transfer on the production of this enzyme and development of highly efficient beta glucosidase enzyme for biomass hydrolysis. His current research interests include—Developing Enzymes for Biomass Conversion, Glucose-tolerant β-glucosidases, Heterologous Protein Expression in Fungi and the Molecular Biology of Cellulase Gene Regulation. He has published more than 80 international peer-reviewed journal publications and about 120 conference papers, besides several book chapters and reports.

Introduction

With the increasing population, dwindling resources and rapidly deteriorating environment, it is becoming increasingly evident that sustainable development is the only way to move forward and for the very survival of the human species on this planet. Stress on the environment due to human activities has increased to an alarming level, and the increase in the atmospheric carbon dioxide levels has already started showing its effects on global temperatures. If such relentless exploitation and degradation of nature and its resources continue, it would not be long before we start experiencing the serious consequences of global warming and environmental deterioration. About 2 billion tonnes of municipal solid waste (MSW) is generated annually besides the 7–10 billion tonnes of urban solid waste from households, commerce, industry and construction as the 2015 United Nations Environment Programme (UNEP) report summarizes. Worldwide, the generation of waste is increasing and the MSW generation is projected to double in another 15–20 years. While there has been significant progress in waste collection and disposal in the developed nations, a large fraction of the population in developing nations do not have access to solid waste collection; accumulation of the same is a major public health concern. Waste management is considered as a basic human necessity and often can interfere with provision of other essential services including potable water, health care, energy and other basic needs and has to be seriously looked upon. There has been umpteen number of case studies where unmanaged waste has created problems concerning public health, choking of water bodies and streams, environmental and atmospheric pollution and even navigation. There are even instances of hazardous wastes impacting public utilities like water supply and agricultural produce creating serious implications. Hence, it is the need of the hour to take quick action for waste mitigation and organized management of waste following the principles of the three R's—Reduce, Reuse and Recycle. While the earlier attempts on waste management were primarily focused on treating waste after it was generated, the modern approach is to address this at the source itself. Preventing the generation of wastes, reducing the usage of substances wherever feasible, use of environmentally benign materials for consumer products that will reduce the management efforts after their designed lifetime, separation of waste at source for its effective reuse and recycle and even designing products with the aim of recycling them are now being proposed/implemented. The target has now been shifted to integrated waste management, which incorporates a radical rethinking of wastes as resources against waste disposal and treatment. This necessitates concerted efforts to identify the resource potential of wastes, be it MSW, agricultural, industrial or even domestic wastes, and find avenues for their value addition as part of a "circular economy." The monograph is hence an attempt to catalogue various approaches to sustainable development, creating "wealth" from waste, primarily looked upon from the angle of organic wastes, be it MSW, agricultural wastes, domestic wastes or wastes from industries that handle organic feedstocks like food and feed processing or biomass-based resources.

While a comprehensive coverage of all waste valorization prospects is not the intention, the monograph describes organic waste valorization to generate chemicals, biopolymers, bioactives, fertilizers, etc., and even bioelectricity, management of wastewaters and wastes through microalgae with co-production of high-value metabolites and biomass, and conversion of lignocellulosic wastes/residues into fuels and chemicals. Organic wastes are generated by food and feed industries, besides agriculture and aquaculture. Highly perishable wastes include those generated from hotels and restaurants, vegetable and kitchen wastes and sewer wastes, fish and meat processing rejects, animal excreta. Often the categorizations overlap and so do the methods of their valorization. While in several cases, treatment of the wastes becomes a primary concern as for the sewer wastewaters, there are also options to generate value from these before, during or post-treatment of such wastes. These include algal cultivation in wastewaters which yields algal biomass from which oils and biomass itself can be recovered and used. Similarly, food and feed wastes can serve as raw materials for the generation of products through microbial fermentation. These can also be used to generate biogas and energy through various means including but not limited to methods like combustion, pyrolysis or through microbial action. Crude glycerol generated as a by-product of biodiesel industry is now increasingly finding applications in production of various high-value compounds through microbial fermentation. Microbial conversion of glycerol has been used successfully for the production of single-cell proteins, organic acids like citric acid, platform chemicals like succinic acid, DHA, 1, 3-PDO, etc. Organic waste streams/residues from a range of different industries have also been explored as feedstock for the production of such compounds. Apart from several such relatively smaller molecules, waste streams have also been used for microbial production of biopolymers and complex class of compounds like biosurfactants and bioactives.

Biopolymers are one of the main waste components in agricultural and industrial processes. There is a huge variety of biopolymers that can be recycled from these wastes, primarily polysaccharides, proteins, bioplastics, polyamides, polyesters and others. The functional advantages of biopolymers over synthetic polymers include biofunctionality (stereoselectivity and enantioselectivity), bioinertia, biocompatibility, bioactivity, biostability, biodegradability and restricted molecular weight. Also, biopolymers have a wide spectrum of mechanical and chemical properties with the advantage of being biodegradable, very cheap and biocompatible in many cases. Finding novel applications for these materials involves reconverting biopolymer wastes into value-added products. Different types of biopolymers, their properties and their potential uses in biomedicine such as drug delivery, tissue engineering and regeneration and other health-based applications are being explored. One of the highlights of the monograph includes the discussions on recovery of biopolymers from organic wastes like meat and fish processing residues, food and feed industry wastes and rejects, agrowastes, etc., and their potential applications in medicine. Biomedical applications of biopolymers are discussed in detail in the chapter by Bayon et al.

Biopolymers from food and feed resources like fish processing have a direct value in many applications. This includes compounds like chitin and chitosan recovered from shellfish processing industry, collagen and gelatin from meat processing, pectin from fruit processing, etc. The residues from such food processing, which are largely underutilized, are now being considered as high-value resources. Residues left after the processing for main product(s) are processed to recover such biopolymers. Depending on the level of purity, such compounds command very high value and are used even in biomedical and therapeutic applications. Food and feed processing waste is an important resource, which is rich in carbon (often as carbohydrates) and sometimes protein, and could serve as an excellent feedstock, not just for recovery of biopolymers, but also for fermentative production of chemicals and metabolites, and also for anaerobic digestion to generate methane (biogas). There are also some unique applications of such wastes—e.g. the use of fish bone waste for generation of hydroxyapatite used in biomedical applications. This includes generation of coatings for transplants and prosthetics which are biocompatible, and by incorporation of silver nanoparticles, applications also exist in treatment of wounds and burns where the silver nanoparticles can act as antimicrobial and the matrix can serve as a biocompatible scaffold.

Similar to biopolymers, another important product that can be recovered and used from wastes is biosurfactants. Biosurfactants find applications in various fields ranging from medicine to environmental pollution management (e.g. clean-up of oil spills), and production of biosurfactants is a rapidly growing industrial proposition. A wide spectrum of feedstock has been tried for the production of glycolipid biosurfactants which include molasses, cassava flour wastewater, wheat bran, waste glycerol, cashew apple processing waste, waste oils and oil cakes, etc. Industrial wastes like whey from dairy industry, distillery wastes, etc., have also been employed for the production of biosurfactants through fermentation. Production of surfactants from such wastes or residues would require different fermentation conditions than those normally used for classical media. An understanding of the process parameters and biology of the organism can be important in the production of biosurfactants using waste streams. A description of the various waste feedstocks used for the production of such biosurfactants and their method of production are described in the monograph.

The conversion of food waste to high-value products is now practiced on a larger scale, and there are numerous examples of generation of unique products. Enzymes are one of these product possibilities, and worldwide, there are several examples of the use of food waste being used as substrate for the production of enzymes. The monograph includes very promising examples of such possibilities where the potential to use coffee husk waste for the production of tannase, cellulase, phytase, polygalacturonase, etc., is described. Citrus fruit waste has been converted into antioxidents through solid-state fermentation techniques. A concept for biorefinery mode of operation is presented where hemicellulose from avacado residues is used for production of high-value compounds. One of the finest examples of a near-complete utilization of food process waste is presented in one of the monographs in the form of seafood waste utilization. Fish processing wastes have enormous potential for utilization, which is highlighted here. Possibilities of products from such wastes include—fish meals, silages, etc., which can serve as additives to aquaculture and animal feeds; oils, biopolymers like chitin and chitosan, lipids and fatty acids that have applications in both food and feed industry; enzymes from the gastric content of fishes and shellfishes; pigments, flavouring agents and even calcium and other minerals. Needless to say, there are also the possibilities to do anaerobic fermentation to generate biogas, volatile fatty acids and biohydrogen. Emerging possibilities include generation of bioelectricity through microbial fuel cells, especially using wastewaters containing high organic load.

Wastewater treatment using microalgae is a mature technology practiced worldwide, but the recent attention has shifted towards its valorization through cultivation of algae. One of the most commonly advocated uses of microalgae is in the generation of oil that can be converted into biodiesel. The low cell densities achieved in large-scale freshwater cultivation of algae, low photosynthetic efficiencies and the need for added nutrients often become limitations in algal biodiesel. Heterotropic or mixotrophic cultivation of microalgae in wastewaters rich in minerals and organic content at least partially addresses this issue by achieving high cell densities and cheaper cost of production. Since this also serves to treat the wastewater, it possesses dual advantage. Algal biomass generated in wastewater phyco-treatment facilities can serve as a source of oil for biodiesel, and the residues generated are a rich source of cheap protein and biomass that may be converted by other means into fuels and chemicals. The section on waste valorization through algae discusses these aspects, especially in the context of waste utilization. There are also elaborate discussions on the range of products possible through algal cultivation in waste streams, which include oil, polyunsaturated fatty acids, algal polysaccharides, chlorophyll, pigments, bioactives and a lot more.

Biorefinery mode of operation for lignocellulosic (LC) waste conversion to fuels and chemicals is the accepted way forward for second-generation fuels. Conversion of lignocellulose to bioethanol typically involves several unit operations like pretreatment, hydrolysis, fermentation and distillation. Each of these steps generates by-products and waste streams, especially wastewater with high organic load which can create an effluent treatment nightmare. However, lignocellulose process wastes/residues have great potential to be converted into high-value chemicals and fuels. There is now a huge amount of literature on such possibilities which include separation and use of lignin for multitude of applications, generation of fuels and platform chemicals like HMF, furfural, 1, 3-PDO, etc., through catalytic biological route. Key to the effective utilization of such resources is a proper understanding of the processes involved in biomass conversion and the technologies for by-product conversion. The monograph describes the various technologies or pretreatment including some novel approaches using auxin, and the enzymes for hydrolysis of biomass. The latter is especially important, considering the fact that enzymes contribute majorly to the cost of bioethanol. The section on LC residue conversion also describes consolidated bioprocessing as a potent method for bio-conversion of biomass to fuels. In addition to the biochemical route for LC biomass conversion, the section also describes pyrolysis technology for biomass conversion to fuels and chemicals, along with an economic analysis.

While there is specific literature on handling MSW and solid wastes in general, some of these manuscripts do describe the conversion technologies available, including vermicomposting and thermochemical methods. Biochar, which is one of the products of such conversion, and its value as a soil-quality modulator are discussed. Finally, there is also a section on the economics of solid waste management which gives valuable insights into the economics of such processes which will help to make learned decisions on the route to be taken for waste valorization or whether to do it at all.

The research monograph covers a range of topics in waste valorization or organic wastes. This describes the state of art in conversion of organic wastes by various technologies, range of products and the applications. It even discusses the economic balances in waste to value-added products so as to allow learned decisions before embarking on a particular route to convert waste streams into value-added products. The specific topics covered include the following:

- Types of organic wastes, their features and potential for generating value-added by-products
- Value-addition propositions for organic wastes and the range of products possible from given wastes/by-products
- Potential for utilization of specific streams and the technologies available to do so
- Food waste conversion—waste parameters and how to handle them in the context of value addition
- Organic wastes as fertilizers and soil-quality enhancers
- Non-conventional waste conversion and value-addition strategies—bioelectricity and biochar
- Agricultural wastes and agricultural applications of conventional waste streams and novel waste streams
- Algae as effective source for treatment of wastewater and its value addition
- Range of products from algal cultivation
- Lignocellulosic biomass conversion technologies—pretreatment and hydrolysis
- Enzymes for lignocellulosic biomass hydrolysis
- Consolidated bioprocessing
- Thermochemical conversion of biomass to fuels and chemicals—pyrolysis

Chapter 1
Biopolymers from Wastes to High-Value Products in Biomedicine

Bernardo Bayón, Ignacio Rivero Berti, Ana M. Gagneten and Guillermo R. Castro

God forgives, men sometimes, Nature never
Los Piojos (2007)

Abstract Biopolymers comprise a large variety of molecules with diverse chemical structures involving from polysaccharides to proteins, bioplastics, polyamides, polyesters, among others. Also, biopolymers are very abundant in nature and one major wastes of industrial and agricultural activities contributing to the environmental pollution. On the other side, the recent advances in micro- and nano-technologies in the biomedical field open a new window for their use in drug delivery, tissue engineering and many health-associated technologies. The production and physicochemical properties of the main biopolymers used in biomedical applications reported in the literature are reviewed in this chapter.

Keywords Biopolymers · Biomass wastes · Green chemistry · Highly added value products · Biomedical applications

1.1 Introduction

Since ancient times, biomass wastes have been reused as structural materials and fertilizer or have been burned to produce heat in households and some industries. However, based on the intensive and extensive use of natural resources and on the rapid expansion of man-made materials in the last century, e.g., plastics and synthetic rubbers, a huge number of biomaterials are now being discarded and converted into pollutants. Biopolymer is one of the main waste components in agricultural and

B. Bayón · I. R. Berti · G. R. Castro (✉)
Laboratorio de Nanobiomateriales, CINDEFI—Departamento de Química,
Facultad de Ciencias Exactas, Universidad Nacional de La Plata—CONICET
(CCT La Plata), Calles 47 Y 15, B1900AJI La Plata, Buenos Aires, Argentina
e-mail: grcastro@gmail.com

A. M. Gagneten
Laboratorio de Ecotoxicología, Facultad de Humanidades Y Ciencias, Ciudad Universitaria,
Universidad Nacional Del Litoral, Paraje El Pozo, C.P. 3000 Santa Fe, Argentina

R. R. Singhania et al. (eds.), *Waste to Wealth*, Energy, Environment, and Sustainability, https://doi.org/10.1007/978-981-10-7431-8_1

industrial processes. However, there is a huge variety of biopolymers that can be recycled from these wastes, principally polysaccharides, proteins, bioplastics, polyamides, polyesters, and others. The functional advantages of biopolymers over synthetic polymers include various "bios": biofunctionality (stereoselectivity and enantioselectivity), bioinertia, biocompatibility (i.e., most of them are non-toxic), bioactivity (specific tailored residues), biostability, biodegradability (non-toxic degradation products), and restricted molecular weight. Also, biopolymers have a wide spectrum of mechanical and chemical properties with the advantage of being biodegradable, very cheap, and biocompatible in many cases. Thereby, billions of tons of potentially useful materials are discarded and contribute to the world contamination. Besides, green chemistry has already demonstrated examples of the so-called life cycle innovation, that is, improvements at all stages of the product or process life cycle. The best way to solve this issue is finding novel applications for these materials and reconverting biopolymer wastes into highly added value products. Biomedicine including medical and pharmaceutical applications brings plenty of opportunities to be fulfilled. Different types of biopolymers, their properties, and their potential uses in biomedicine, such as drug delivery, tissue engineering and regeneration, and other health-based applications, will be discussed in the present chapter.

1.2 Biopolymers: Environmental Aspects

From the beginning of life in Earth, life works by itself; now, human beings are realizing that instead of wondering how to manage huge quantities of wastes, organic matter from biological origin (dead bodies of animals and plants) can be reconverted into useful materials to improve human being quality both by reducing the amount of discarded materials and by taking advantages of them. In the middle, there is a need of knowledge to be developed on biotechnological processes to achieve this goal at a large scale. They must be promoted, explored, and tested. The present chapter will deal with environmental aspects of biopolymers and some proposals to fill these gaps.

The continuing growth population and goods consumption imply the increase in global food demand for at least another 40 years (Godfray et al. 2010). Food waste production raises with the increment of population, as does new demands for materials. Parfitt et al. investigated the potential of food waste by considering the demand to feed nine billion people in 2050 (Parfitt et al. 2010). They observed that post-harvest wastes are higher in developing countries, while post-consumption wastes are higher in developed countries.

One of the main objectives in the society is to achieve sustainable development. It involves the rational use of resources and the maintenance of ecosystem services (Millennium Ecosystem Assessment 2005). Along these lines, it is necessary to improve the efficiency of production methods to make industrial development compatible with social welfare and environmental protection, to apply environmentally friendly and low-impact methodologies in manufacturing processes, and to reduce the maximum wastes produced together with the design of new methods to convert waste into valuable products (Catalina et al. 2012).

As was pointed out by Dhillon et al. (2011a), a serious problem at the global level is the management and disposal of food and agro-industry wastes. Most of these wastes contain vegetal or animal biopolymers. Methods currently in use include landfill, incineration, cattle feed, or land farming. All these methods can generate environmental impact by the production of greenhouse gases or unpleasant odors, aquifer contamination by infiltration or runoff, and health effects on human health by the dissemination of vectors of diseases. On the other hand, these methods generate high costs related to their treatment and transportation.

The type of waste produced by the society reflects its culture: What and how much each human being consumes daily in developed and in developing countries are indicative of cultural differences in the quantity and quality of the garbage: For example, annual plastics consumption in the USA, Europe, and India is approximately 150, 20, and 5 kg, respectively (Nayak and Swain 2002). The EU produces more than 67 million tons of packaging waste per year—a third of urban waste—and in the UK, it is equivalent to more than 12% of total household waste (Klingbeil 2000; EC Packaging 2004; Davis and Song 2006). By contrast, developing countries generate more organic wastes, which is also related to the prevailing climate of the different regions worldwide (e.g., countries in could organic garbage decomposes slowly than in humid and warmer areas).

There is strong scientific evidence that emissions of greenhouse gasses (GHGs), such as carbon dioxide (CO_2), methane (CH_4), and nitrous oxide (N_2O), arising from fossil fuel combustion and land-use change because of human activities, contribute to the upset of the Earth's climate (Piemonte 2011). In this scenario, the overgrowing environmental pressure caused by the widespread consumption of petroleum-based polymers and plastics has hastened the development of biodegradable and environmentally acceptable materials. In contrast, biopolymers derived from various natural resources such as proteins, cellulose, starch, and other polysaccharides are regarded as the alternative and useful materials. Biodegradable polymeric materials derived from renewable sources are the most promising materials because of their easy availability and cost-effectiveness (Kumari et al. 2010).

The term "biopolymer" generally refers to polymers that have been fully synthesized by cells, and there are three main subsets: polysaccharides, polynucleotides, and polypeptides. Indeed, ASTM D6866 defines biopolymers as polymers produced by living organisms, that is, polymers synthesized in nature by enzyme-catalyzed chain growth polymerization reactions by cell metabolic processes during their growth cycles (Chandra and Rustgi 1998).

Biopolymers can disintegrate over time through biological degradation and mineralizing completely into carbon dioxide or methane and water. The biodegradable containers are made up of polymers and biopolymers depending on whether a synthetic polymer or biologically derived polymer predominates. Among commercially available biodegradable packaging materials based on natural raw materials, those based on polysaccharides (e.g., starch and its derivatives) are currently the front runners. This is mainly attributable to the facts that starch is annually renewable and is abundant around 15 million tons/year in Europe and about 50% is used for non-food applications (Löckes 1998) and inexpensive—around 0.5–1.0 kg^{-1} (Petersen et al. 2001).

Developed countries look with interest the high production of biomass generated by the less developed ones, especially biopolymers such as cellulose, starch, pectin, sugarcane, wood, chitin/chitosan (i.e., exoskeletons of crustaceans), and derivatives of industries linked with the processing of raw material of vegetal and animal origins. The increasing interest aroused for two reasons:(1) for being sources of energy almost inexhaustible, e.g., products with C finally motorized by solar energy; (2) to give added value to the enormous amount of waste annually produced that is contaminating soils, oceans, and inland water bodies with negative consequences for the aquatic and terrestrial biota included human beings.

Some biopolymers especially selected by worldwide abundance and relevance as wastes will be discussed below. Gautam et al. (2010) affirmed that polysaccharides are the "material of the future." What is this strong statement based on? Since they are the most abundant and diverse molecules on planet Earth. Basically, it is renewable and biodegradable biomass. Few molecules—if any—meet these conditions. They constitute the cell walls of plant cells where they also reserve energy in the form of starch or glycogen in the liver cells of animals. Due to its abundance, availability, and permanent renewal, it is logical to assume that given the depletion of non-renewable resources, attention should be focused on reusing and/or modifying polysaccharides for specific applications

A notable feature of polysaccharides is the tendency to be very bioactive. In general, they are derived from agro-industrial wastes, such as cellulose, starch, and pectin, or from crustacean shells, such as chitin and chitosan. Chitin, after cellulose, is the most abundant polysaccharide, available in more than 10 gigatons per year (Sahoo and Nayak 2011).

1.3 Biopolymers from Animal Wastes

The article 3 of Regulation (EC) 1069/2009 of the European Parliament and the Council of the European Union restricts the use and controls the disposal of animal by-products. Such wastes include catering waste, used cooking oil, former foodstuffs and butcher waste, blood, feathers, wool, hides, shells and skins, and fallen stock, among others. These wastes are given three categories for disposal of which Category 3, low-risk materials, includes the remains of animals, which are approved for human consumption, after passing through slaughterhouses. These parts are not eaten, and some inedible by-products are treated for animal feed, fertilizer, or protein-based adhesives. Some tissues are used to produce composite bone-cum-protein meals or individual products such as bone meal, meat meal, and blood meal. The total amount of waste is 10–15% of the live weight killed (LWK) animal (Verheijen et al. 2013).

In 2011, the worldwide meat production reached 244 Tg, where 42% corresponds to pork, 35% to poultry, and 23% to beef (USDA 2012). Approximately 25 Tg of Category 3 meat waste is produced annually in the world. Proteins from animal sources include collagen which is composed of different peptides, mainly glycine, proline, hydroxyproline, and lysine. They are enzymatically degradable and can be obtained from animal waste such as bones and skin (Vroman and Tighzert 2009).

Associated with meat production is the leather industry, one of the most polluting using chromium salts in the first process of elimination of fat, collagen, and cow hair, thereby producing a lot of waste that is dumped to the river with little or no control. Effluents from tanneries can generate strong decrease in biodiversity, as previously reported (Gagneten and Paggi 2009). Leather processing includes the tannery process, which generates a large amount of waste that can be revalued as collagen biopolymers to make fibers, gels, films, etc., by techniques such as cross-linking with various compounds. In this way, biopolymers have been obtained with improved application properties in cosmetics, medicine, or veterinary applications (Catalina et al. 2012). Chitosan can be recovered in large quantities from wastewater from meat, dairy, and seafood-processing plants. These proteins, followed by drying and sterilization processes, are used as dietary supplements on farms (Sahoo and Nayak 2011). On the other hand, it was found that chitin and chitosan can decrease aquatic contamination by their high affinity for dyes, which are very resistant to degradation (Crini 2006). Another use of chitosan is the stabilization of municipal and industrial sludge which results in great economic advantages (No and Meyers 2000).

1.3.1 Chitin and Chitosans

Regarding seafood, the shellfish catch consists approximately of 30% of crustaceans and 70% of mollusks. Crustacean processing waste compromises 40% of exoskeletons (shell), while the mollusk processing waste consists of 65% shells. Several industries of seafood discarded crab and shrimp shells in Chile, Japan, Norway, Mexico, and USA. In 2010, 20.8 Tg of mollusk and 11.8 Tg of crustaceans were caught, producing 18.2 Tg of shell waste (FAOSTAT 2013). This type of wastes contains approximately 10% of chitin on dry weight (Thirunavukkarasu et al. 2011). The α-chitosan is the most commonly obtained form of chitosans that present in approximately 70% of the organic compounds obtained from the residues of shrimp and crab shells. Finally, the peptide-modified chitosan derivatives obtained by tyrosinase should be environmentally friendly and biodegradable (Aberg et al. 2002; Wu et al. 2002). Chitin and chitosan are natural polymers composed by amino-polysaccharides that present mostly in crustacean's shells. Chitin is widely distributed in marine invertebrates, insects, fungi, and yeast (Aberg et al. 2004; Dhillon et al. 2012a, b).

In particular, chitin is a linear cationic heteropolymer of (1–4)-linked N-acetyl-β-D-glucosamine. Because chitin has an intractable character and very poor solubility, its direct use as a macromolecular material is very limited. However, it can be chemically modified by partial alkaline N-deacetylation at high temperatures which generates the corresponding primary amine groups. The percentage conversion of acetylglucosamine into glucosamine is described as the degree of deacetylation. This influences the physical, chemical, and biological properties and consequently their applications (Hirano et al. 1989).

From structural point of view, chitin is associated with proteins and, therefore, is high in protein content. The industrial way of producing chitin is an acidic extraction of the polymer from crustacean's shells, and the next step consists in an alkaline extraction to make proteins soluble. Moreover, the result is decolorized to obtain a colorless product. These treatments are dependent of the chitin source. The resulting chitin needs to be meticulously separate from residual protein because they can produce an allergic response in biomedical applications. Under alkaline conditions, the chitin is deacetylated, thereby producing chitosan that is one of the most important and useful chitin derivates in biomedical applications (Peniston and Johnson 1980).

Other authors (Muzzarelli 1997; Younes and Rinaudo 2015) report that the shell of crustaceans is composed of 30–40% protein, 30–50% calcium carbonate, and calcium phosphate, and 20–30% chitin.

Among the major chitin producers of the hydrosphere—marine, freshwater, and brackish water—are arthropods. Therefore, a correct valuation of its contribution as a source of chitin is of utmost importance for the rational exploitation of chitin for industrial and commercial purposes. Chitin is between 3 and 16% of total crustacean body dry weight. It is estimated that the annual production of chitin is approximately 10^{11} tons (Elieh-Ali-Komi and Hamblin 2016).

Interestingly, Abeg et al. (2004) inquires: "Why incur waste management costs if investments in manufacturing could yield salable by-products?" The authors examine how by-products from food-processing wastes can be converted into value-added functional polymers. The main source of chitin is from seafood-processing waste which produces around 18 Tg of shell waste.

Chitosan, the *N*-deacetylated product of chitin, is only accepted when the degree of deacetylation permits its solubility in acidic media. At 40% level of acetylation, the polysaccharide chains become moderately soluble, forming stable aggregates in which the *N*-acetyl groups are unevenly distributed. When it is higher than 60%, it becomes insoluble and acquires structural flexibility (Franca et al. 2011). Chitosan can make high-quality films on glass, quartz, metals, and other hydrophilic surfaces and was recently used with novel applications. An easy procedure for the preparation of high-quality, large area, single or few-layer N-doped grapheme using chitosan as precursor a valueless biomass waste that acts as simultaneous source of C and N was reported (Primo et al. 2012). The new procedure is based on the unprecedented finding that chitosan forms films of excellent quality of nanometric dimensions and sub-nanometric roughness on hydrophilic surfaces by spin coating of aqueous solutions of this biopolymer. The method is derived from the ability of chitosan to form high-quality, conformal thin films and the ease of carbohydrates to undergo carbonization. Chitin and chitosan have singular structures and a wide range of applications in biomedical and other industrial fields (Sashiwa and Aiba 2004; Elnashar and Kahil 2014). These polymers are considered of great interest not only as an underutilized resource, but also as a functional material with a huge potential in various fields, which is the reason why research on chitin and chitosan has increased exponentially in past years (Pillai et al. 2009; Cheung et al. 2015; Usman et al. 2016).

Chitin, a natural polymer, is a nitrogenous polysaccharide (poly(b-(1–4)-N-acetyl-D-glucosamine)). It is composed by 2-acetamido-2-deoxy-b-D-glucose bounds by a b (1–4) linkage. This biopolymer can be found mainly in the shells of two marine crustaceans, shrimps and crabs; and is the second polymer, in order of abundance, in the world after cellulose and is really like its structure, with hydroxyl at position C-2 replaced by an acetamide group. Chitin also is present in nature as a structural component in arthropods exoskeleton or in fungus cell walls and produced by many other organisms, and always with structural functionality, it may be compared to the protein keratin. Chitin can be degraded by chitinase, a hydrolytic enzyme that breaks down glycosidic bonds (Fig. 1.1).

Chitosan is produced by the N-deacetylation of chitin, although this deacetylation is hardly ever complete. Chitosan is a natural linear polymer composed by β-(1–4)-D-glucosamine with deacetylation degree between 60 and 100% and average MW of 3.8–20 kDa (Rinaudo 2006).

Amino groups are found in chitin and chitosan which are protonated in acid media and converts them in the only cationic polymer found nature (Hench 1998). As a cationic polymer and water soluble, chitosan has an extremely good adhesion to mucosal membranes that are negatively charged (Islam et al. 2015).

There are many procedures to obtain chitin, and the most common way to obtain it is by processing crabs or shrimps shells. There are two alternative ways: enzymatic or chemical treatments for the step of deproteinization. The most common procedure involves the dissolution of calcium carbonate, present in crustacean shells, after deproteinization. One of the most common ways for chitosan synthesis is the deacetylation of chitin by sodium hydroxide aqueous solutions. The alkali removes simultaneously proteins and deacetylates chitin (Ravi Kumar 2000). Generally, chemical treatments have many disadvantages: It changes chemical and physical properties of chitin and also produces variations in molecular weight and degree of deacetylation. Another drawback is the presence of several chemical wastes produced by the chemical treatment of chitin which seriously contribute to contamination. On the other hand, enzymatic synthesis is a simple, safer, and greener option, all with an excellent reproducibility. Besides, it is more expensive and difficult to scale-up (Younes and Rinaudo 2015). The main enzymes used in enzymatic processing are chitinases, deacetylases, and proteases. Besides, there are many applications of chitin and chitosan that have been reported in the last years.

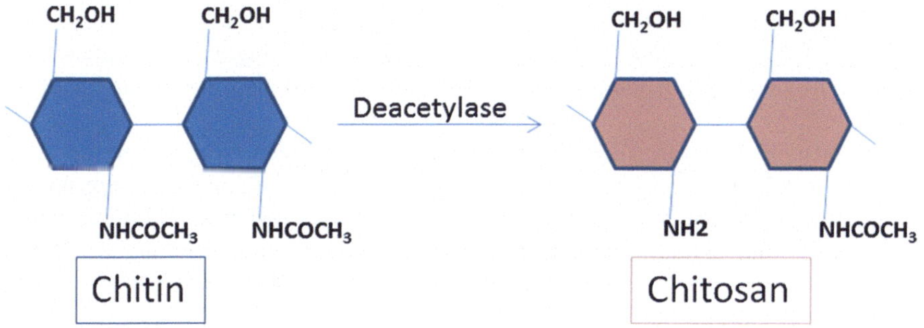

Fig. 1.1 Chitin deacetylation to chitosan through an enzyme deacetylase

These biopolymers have unique structural properties and could be modified chemically and/or mechanically to generate different polymers with novel properties. These properties are biocompatibility, absence of toxicity, low allergic response, bioadhesivity, antimicrobial activity, and good biodegradability. One of the major disadvantage of using chitin is the poor solubility, even though many types of chitin and chitosan have a great number of biomedical uses (Ravi Kumar 2000; Sashiwa and Aiba 2004; Younes and Rinaudo 2015).

1.3.2 Collagen

Collagen is the most abundant protein in the animal organisms, comprises approximately 30% of the whole body, and has principally a structural role in connective tissues. On the other hand, collagen is absent in plants and unicellular organisms where it is functionally replaced by cellulose or other polysaccharides. Many tissues, with clearly different properties, are made of collagen as bone, tendon, or cartilage. The polymer is usually in the form of elongated fibrils, but different dispositions and complex organization of these fibrils make it possible to be adapted to many different types of tissues as muscles, bones, blood vessels, cartilage, tendons, teeth, etc. The fibroblast is the principal type of cell that produces collagen (Ramshaw et al. 2009; Yamada et al. 2014; Silvipriya et al. 2015).

Gelatin is the result of collagen hydrolysis and used in a myriad of industrial products. Collagen and gelatin are extensively used in the food, pharmaceutical, and cosmetic fields. Collagen is biodegradable, has no toxicity, rarely can produce immune response (depending of the polymer source), and has a good adhesion and interaction with cells (Naito et al. 2013). Collagen may also be processed into a variety of formats, including porous sponges, gels, and sheets, and can be cross-linked with chemical reagents to make it stronger or to alter its degradation rate. Applications of collagen are numerous; it has been explored for use not only in various types of surgery, cosmetics, and drug delivery, but also in implants and tissue engineering (Ramshaw et al. 2009; Silva et al. 2014). One of the most important characteristics of collagen is that cells can be grown on collagen surface and sometimes with a similar behavior they have in the organism, which is the reason why collagen is so important in tissue engineering field.

Collagen could be extracted from several sources, including wastes from animals for human consumption mainly bovine and porcine, but also fish (Catalina et al. 2013).

The primary structure of collagen protein is recognized by three amino acid, collagen protein is recognized by the characteristic pattern of amino acid Gly-Pro-X or Gly-X-Hyp, where X could be any other amino acid residues. The basic structural unit of collagen is tropocollagen, and many of these units form larger collagen aggregates, called fibrils. Tropocollagen is composed of three polypeptide strands called alpha peptides, and these strains are ordered in a triple helix with two identical chains and the third which has minimal chemical differences. In case of the tertiary structure, the alpha chains are linked to each other, building the characteristic left-handed triple helix of types I, II, and III collagens. In the triple helix, the glycine residues are positioned around a central axis, while larger amino acids belonging to the X and Y residues (usually proline and hydroxyproline) occupy outer positions (Brinckmann 2005; Ramshaw et al. 2009; Sherman et al. 2015). These three left-handed helices are twisted together into a right-handed triple helix, a cooperative quaternary structure stabilized by many hydrogen bonds. So, the quaternary structure of collagen finally forms the collagen fibrils (Fig. 1.2) (Ferreira et al. 2012).

Today, approximately 28 different types of collagen have been reported. All the different types have the triple helix but differ in the length, size, and nature of the residues portion (Miller 1984; Brinckmann 2005; Ricard-Blum 2011). The main types are listed below with their respective tissues of the organism where they are found:

- Collagen I: It is the principal type and found in the human body such as tendons, muscle fibers, bones, teeth, skin, and cartilage.
- Collagen II: It is found in hyaline cartilage and vitreous humor of the eye.
- Collagen III: It is found in skin, muscle, blood vessels, and mainly in reticulate fibers.
- Collagen IV: It is found principally in basal lamina, eye lens, and with filtration function in capillaries and renal system.
- Collagen V: It is found in hair, cell surfaces, and placenta.

The function of collagen is the main supporting structure of the extracellular matrix in almost every tissue. It is present characteristically in connective tissue, as it has considerable tensile strength; thus, it is the main compound in skin, tendons,

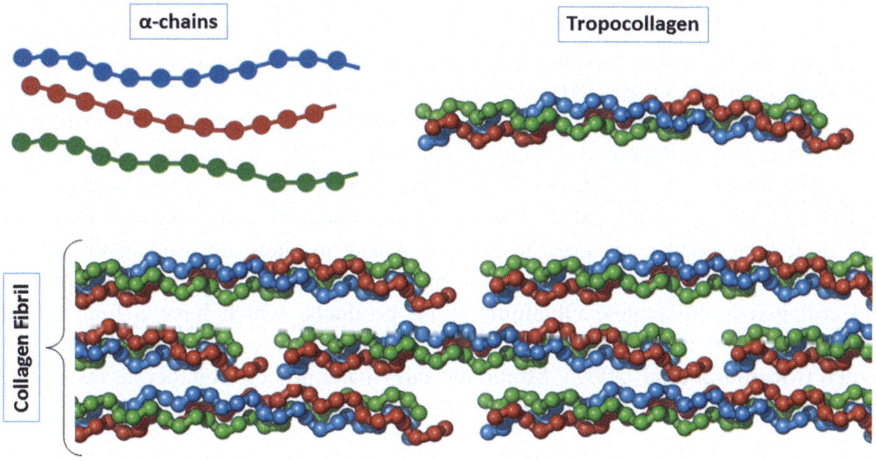

Fig. 1.2 Collagen structure

bones, cartilage, fascia, etc. It has an important role in tissue wound healing or tissue regeneration (Buehler 2006).

The major sources of commercial collagen are:

- Bovine are one of the most important industrial sources of collagen from leather and bones of cow. Besides, a lot of other cow tissues is generally used, such as tendons, placenta, and nasal or articular cartilage. Sometimes depending on cow diseases, like "mad cow" epidemic, the use of collagen from cow is discontinued. Nevertheless, it helps to control contamination as it used rests of animal after meat processing (Paschalis et al. 2001; Silvipriya et al. 2015).
- Porcine tissues are like human, collagen is not the exception, and porcine source is widely used to obtain collagen for industrial purpose. Pigs skin and bones are utilized to obtain collagen, and one advantage is that it does produce almost non-allergic response (Browne et al. 2013). It is also compromised with pig diseases (Zheng et al. 2004; Yang and Shu 2014).
- Collagen obtained from marine sources is considered as an interesting alternative. Presently, it is one of the safest ways for obtaining collagen, considering religious restrictions as pork consumption in Muslim and Jews societies, and animal diseases as mad cow disease. Based on this collagen extraction from marine sources, it is an interesting alternative to obtain the polymer. As a marine source of collagen, there could be used the wastes of many sea living products, from fishes skin or skeleton to prawn, in order to reduce environmental pollution (Dong et al. 2010; Krishnan et al. 2012; Mahboob 2014; Kittiphattanabawon et al. 2015). Notwithstanding the collagen extraction from marine sources, it is easy, cheap, and safe, but it has some disadvantages such as low structural denaturation (Subhan et al. 2015).

The chemical treatment to obtain collagen from animal sources is based on a purification of the polymer using acid or alkaline solutions to eliminate no useful molecules. The next step consists in an acid hydrolysis using organic acids such acetic, citric, or lactic acids, but also hydrochloric acid could be used. After this, the solution is filtered and the collagen is precipitate with NaCl, ending with a dialysis procedure (Schmidt et al. 2016).

In case of enzymatic extraction, the crude material that can have some residue of acidic extraction is added to a solution containing selected enzymes such as protease. The mixture is continuously stirred for about a couple of days and followed by filtration (Wang et al. 2014). The filtrate is subjected to precipitation and dialysis, as for chemical extraction. In general, waste products are used in the manufacture of fertilizers, feed, and fuel, but are currently being used to obtain protein and collagen hydrolysates. Obtaining those products, which have a high added value, is a better alternative to use these wastes, which would otherwise be discarded (Morimura et al. 2002). Moreover, enzymatic processes generate less waste and may reduce the processing time, but they are more expensive.

1.4 Biopolymers from Food Industry

New biocompatible and biodegradable biopolymers are produced from plants, microbes, animals, renewable agricultural wastes, and feedstocks (Niranjan Raj et al. 2010). The food waste generated annually is approximately 1.3 Pg. It contains raw material that could be diverted to materials production as the mineral oil price rises, incurring no conflict over land use and providing new wealth-creating opportunities for food-producing countries.

Among food wastes, polysaccharides are of the utmost importance, the most abundant natural biopolymers (Annarita et al. 2011) compared to about 140 M tons of synthetic polymers. Huge amounts of debris such as shells, seeds, and pulps are generated as products from the manufacture of juices and canned goods. Between 5 and 30% of the raw material of fruits and vegetables is discarded for not meeting the standards established by an increasingly demanding market. The management of such waste generates costs—in strictly economic terms as well as in environmental ones—by the high content in organic matter. Most by-products from the food industry could be reused by applying modern, eco-compatible technologies to extract high-value-added chemicals such as polysaccharides, pigments, flavors, and antioxidants with multiple industrial applications (Annarita et al. 2011).

1.4.1 Potato

Potato (*Solanum tuberosum L.)* is particularly popular in Europe, South America, and China. Approximately 25% of the input to a potato processing plant emerges as waste, consisting of a portion of the peel and whole or cut potatoes discarded due to size, blemishes, or failure to meet quality standards. China is now the biggest potato producer at 72 Tg per year, and almost one-third potatoes are harvested in China and Russia. Starch is the main resource at 68% based on dry weight and used in polymers and as a precursor to furan derivatives.

Large amounts of starch as a residue from potato chips industry are discarded daily. Moreover, potato wastes are left in open fields to decompose and some are used to feed cattle which can consume up to 12% of their body weight of fresh potatoes daily: Potato delivers four times the energy value of cereal grain for beef cattle. Enzymatic hydrolysis of potato processing wastes was proposed as a possible source of a fermentable substrate to produce the plastic biopolymer polyhydroxybutyrate (PHB). After this procedure, potato starch waste could be converted with high yield to a concentrated glucose solution (Rusendi and Sheppard 1995).

Several novel projects have addressed the opportunities provided by the extensive availability of potato waste. Digested potato waste may enable a biogas plant to provide electricity to the public grid and to preheat industrial dryers. Another application is the use of potato waste as a medium to produce xanthan gums (Bilanovic et al. 2010), a thickening agent used in food products and oil recovery which is currently obtained expensively from sugars. Polylactic acid (PLA) has also been produced from potato waste which can be used as a non-petroleum-based polymer (Sanchez-Vazquez et al. 2013).

1.4.2 Corn

Corn has been a dietary staple in the Americas since prehistoric times. In 2009, 819 Tg of corn were harvested and the world's corn production (USDA 2013)—per country production/Tg—was 333.0 (40%) in USA; 164.1 (20%) in China; 51.2 (6.3%) in Brazil; 20.1 (2.5%) in Mexico; 17.6 (2.1%) in Indonesia; and 16.7 (2.0%) in India, followed by 15.3 (1.9%) in France and 13.1 (1.6%) in Argentina (Sanchez-Vazquez et al. 2013; Food and Agriculture Organization, the United Nations 2017).

Corn waste (or corn stover) compromises five parts: nodes, leaves, shell, core, and sheath. At harvest, the grain represents only 15% of the weight; the rest is treated as food waste. This means that around 696 Tg of total corn waste is produced each year. Corn stover is the most abundant lignocellulose renewable resource in the world due to its chemical composition and the enormous quantity that is produced annually worldwide: 70% of the total corn stover is composed of lignocellulose corresponding to 487 Tg.

The main application of corn stovers is as fertilizer, but new opportunities for its use are emerging, among them research on enzymatic hydrolysis and solid-state fermentation (Hongzhang et al. 2011) and as a renewable source for ethanol production using *Pichia stipitis*, a cellulolytic extremophile (Zambare et al. 2011) or with biocatalysts (Ryu and Karim 2011). Another potential application is in the production of biopolymers. Polyhydroxyalkanoates (PHA) and polylactic acid (PLA) can be produced from corn grains, but there are attempts to produce it in combination with corn stover (Dale and Bruce 2005).

1.4.3 Pectin

Citrus waste provides interesting potential precursors for materials manufacture such as pectins. The world citrus production is divided into four categories: orange (*Citrus sinensis*), mandarins/tangerines (*Citrus nobilis*), grapefruit (*Citrus paradise*), and lemon (*Citrus limon*)/lime (*Citrus aurantifolia*). The total production of citrus in 2009–2010 according to the United States Department of Agriculture (USDA) was 82 Tg (FAOSTAT 2013).

The world orange production is mostly attributed to five countries which contribute 71% of all world production, Brazil being the largest producer (33%). For mandarin/tangerine production, 64% is produced by China. Grapefruit and lemon/

lime have a lower worldwide production. Even though Brazil only produces orange, the quantity of 16.2 Tg makes it the second biggest producer of citrus (20%).

The citrus peel represents about 15% of the total fruit weight, meaning that the worldwide production of total waste from citrus crops is potentially of the order 12.3 Tg per year. According to USDA (2013), China produces per year 3.4 Tg of citrus waste, while Brazil produces 2.4 Tg followed by EU-27 with 1.6 Tg, Mexico with 0.9 Tg, and Turkey with 0.5 Tg. The main source of commercial pectin is the residue of the citrus and apple industry for the extraction of juices. This is one of the main food industries, producing annually between $3.0–4.2 \times 10^6$ and 15.6×10^6 M tons of waste (Thibault and Ralet 2003).

Pectin is a complex mixture of polysaccharides that make up about one-third of the cell wall of dry matter of higher plants and is commercially used as a thickener in processed foods (May 1990). Commercial pectins can be extracted from citrus peel or apple pomace. Both sources are common wastes from juice and cider manufacturing, which in many cases represents the mayor destination of those fruits production. However, a significant part of these wastes ends up in a landfill anyway, regardless of the possibility of taking advantage of processing them (van Heerden et al. 2002). Other sources of pectin drew interest over the past years and were searched for alternative industrial exploitation. However, at present time, the provided results so far do not show any significant commercial use. Some examples include pectins obtained from sugar beet pomace, raw papaya peel, sunflower head residues, and olive pomace (Fissore et al. 2014)

In recent years, pectin has been used and studied for high-value medical and pharmaceutical applications due to the biopolymer characteristics and properties (see below). This could increase their demands and the market value of pectin and reduce to a minimum biomass waste of citrus and apples (Srivastava and Malviya 2011).

From a molecular point of view, pectin has large linear segments of poly($1 \rightarrow 4$) α-D-galactopyranosyluronic acid with some of the carboxyl groups esterified with methanol and side branches or "hairy regions" with either 1,4-linked β-D-galactose or 1,5-α-linked L-arabinose. The molecular weight and proportion of each residue will differ from molecule to molecule (polymolecular) and from sample to sample depending on the source and the conditions of isolation and purification (Rolin 1993).

All pectin molecules contain linear segments of ($1 \rightarrow 4$)-linked α-D-galactopyranosyluronic acid units with some of the carboxyl groups esterified with methanol and 1,2-linked rhamnose with side branches of either 1,4-linked β-D-galactose or 1,5-α-linked L-arabinose. However, the structure of pectin is not completely understood and one possible structure is depicted in Fig. 1.3.

A typical average molecule weight for a commercial pectin sample is in the order of 100 kDa, but large differences may exist between samples and between molecules within a sample, and estimates may differ between methods of measurement (Rolin 1993). Also, some companies are providing pectin for specific uses and defined molecular weight.

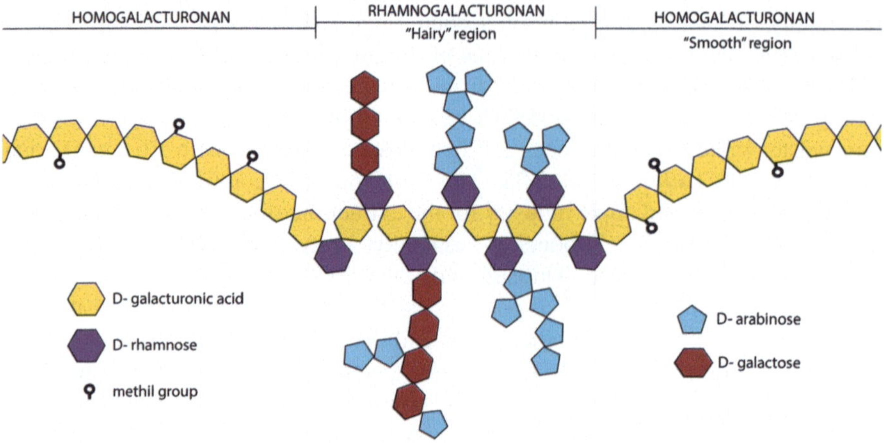

Fig. 1.3 Typical model of pectin structure. In addition, pectins from some sources have some of the hydroxyl groups of the galacturonosyl units esterified with acetic acid

In nature, about 80% of the carboxyl groups of galacturonic acid are esterified, although this proportion can change in the process of extraction of pectins. The ratio between esterified/non-esterified galacturonic acids gives pectin different behaviors, so pectins are usually classified as HM (high methoxylated), MM (medium methoxylated), or LM (low methoxylated) depending on whether they possess more than half of their esterified carboxyl groups.

MM and LM pectins undergo ionotropic gelation in the presence of polyvalent cations such as Ca(II) or Zn(II), but other factors such as pH, sugar co-solutes, and temperature determine whether gelation can occur and influence the characteristics of the gel. HM pectin gelation involves several kinds of intermolecular interactions, which are promoted in high concentrations of sucrose and low pH; unlike MM and LM pectins, HM pectins do not undergo ionotropic gelation in the presence of polyvalent cations because of the lack of free carboxyl groups.

Pectins are soluble in pure water. Solubility in aqueous systems is affected by the same molecular and environmental factors that affect gelation, but in the inverse manner. Pectins (like other gel-forming polymers) are not soluble in an aqueous system in which it would have formed a gel under the same conditions. Dilute pectin solutions possess Newtonian behavior, but at moderate concentration, they exhibit the non-Newtonian, pseudo-plastic behavior characteristics. Similarly, the viscosity of a pectin solution is related to the molecular weight, degree of esterification, concentration in solution, and the pH and ionic strength (including the presence of counterions in the solution). Viscosity, solubility, and gelation are generally related. For example, factors that increase gel strength will increase the tendency to gel, decrease solubility, and increase viscosity, and vice versa (Rolin 1993)

Production of pectin involves extraction and purification processes, which differs significantly between laboratory and industry. The processes of extraction and purification of pectin differ significantly in the laboratory from the industry. Pectin extraction in the laboratory and for research purposes usually implies soft experimental conditions and more specific steps than those required at the industrial level. The harsh conditions of extraction at industrial scale generally involve an erosion of the "hairy regions" of the molecules and the consequent loss of neutral sugar side chains and molecular mass.

Purification at laboratory scale involves two main methods: direct boiling and microwave heating. Direct boiling is the conventional method of pectin extraction, which involves heating the plant tissues under acid aqueous solutions (pH up to 2) incubated for two to four hours, followed by precipitation with cold isopropyl alcohol. Many conditions and additives, such as EDTA and CDTA, have been tested to improve the yield of this simple method, but due to a relatively long period of direct heating, the extracted pectin undergoes thermal degradation anyway. The other method, microwave heating, takes normally fifteen minutes, and because of that, it normally gives a better yield and quality. The chelating agents can be used in both procedures (Srivastava and Malviya 2011).

Large-scale pectin production begins with the dried citrus peel that is mainly obtained from the juice industry. Before the pectin extraction, the peels are dried below 110 °C. The world largest producers of this raw material are in Argentina and Mexico, but producers are located also in Peru, Spain, and Bolivia. In Germany, on the other hand, pectin is usually extracted from apple pomace (Pagliaro et al. 2016). Regardless of the type of raw material, the current manufacturing procedures are based on extraction via acid hydrolysis in hot water; following the separation from the spent peel, the pectin extract is filtered and precipitated with isopropyl alcohol that can be recovered by distillation later. Later, the pectin is washed, dried, and sold in powder form.

This conventional method of extraction has a great disadvantage, which is the cost of eliminating the waste produced by it. This cost proves to be so high that, when in the USA the Environmental Protection Agency passed new regulation in the early 1990s, the last US-based pectin production plant was relocated from Florida to Mexico (Hui 2006). New methods have been developed recently, and at lower economic and environmental costs (Fidalgo et al. 2016), this will prove to be of great importance in the future, since it is estimated that the production of pectin will be a business of U$ 2.4 billion by 2020.

1.4.4 Grape Waste

The total world production of grapes (*Vitis vinifera*) is around 15 Tg per year, and the main producers are listed in Table 13. According to the *Organisation Internationale de la Vigne et du Vin* (OIV, France 2007), 65% is used for wine and juice (9.8 Tg), 23% as fresh fruit (3.5 Tg), and 12% as dried fruit (1.8 Tg). The main wine producers (2010 data) are France, Italy, and Spain, sometimes known as the "Big Three" of wine (Sanchez-Vazquez et al. 2013). Approximately 30 kg of waste is produced in the production of 100 L of wine, and each liter of wine needs on average 1.3 kg of grapes so that 23% of the grapes used become available waste. From the available information, during 2009, 2.6 Tg of wine residue was produced for which a range of uses has been proposed (Arvanitoyannis et al. 2006).

1.4.5 Sugar Bagasse

Sugarcane (*Saccharum L.*) is mainly cultivated for sugar and ethanol production, and world production in 2010–2011 was 130 Tg, with Brazil and India being the main producers with 50% of total production. The sugarcane process is divided into two steps: milling and refinery. The industrial waste from milling is the sugarcane remaining, known as sugar bagasse, and represents 28% of the dry weight of the original. The main constituents of the sugar bagasse are cellulose (40–45%), lignin (20–30%), and xylan (30–35%) (Cardona et al. 2010). Annual production of bagasse is thus approximately 36.5 Tg and much is currently used in distillery plants as a source of energy for pulp and the paper industry to produce particle board, fiber board, cardboard, furfural, microcrystalline cellulose, hydrolyzed bagasse, predigested pith, molasses-urea-pith, furfural cement, and compost (Almazan et al. 1998).

Sugarcane is a lignocellulosic material and hence an attractive feedstock for ethanol production. The world production of sugarcane during 2010/2011 (Country Production/Tg) was 39.40 in Brazil, 11.72 in China, and 25.70 in India (FAOSTATS 2013).

1.4.6 Banana Waste

Banana (*Musa sapientum*) is one of the most popular tropical fruit in the world available throughout the year. The peel represents 40% of the banana fruit generating annually around 22 Tg of peels, much of which is domestic waste (Agri-Food Business Development Center via Nation Master 2000). The four main producers are India, Brazil, Ecuador, and China who generate 27 Tg of bananas per year.

1.4.7 Avocado Seed

Avocado (*Persea Americana*) is the fruit of a native Mexican tree and is mainly cultivated in tropical climates. According to FAO, the world production is around 3.9 Tg per year where 76% of the total production is controlled by 10 countries. As it originated in Mexico, this country is the biggest producer with 32% of world output.

1.4.8 *Apple and Apple Products*

The increase in the human population generates a greater demand of fruits. World production of apple and its products exceeded 69,603,640 tons (Food and Agriculture Organization of the United Nations 2017). The processing industries to produce juices and jams generate enormous amounts of residues in the form of shell, pulp, and seeds (Dhillon et al. 2013). They are biodegradable (Shalini and Gupta 2010; Dhillon et al. 2011b), so their elimination represents an environmental problem, since only 20% is recovered as animal feed. The rest is used in composting or incinerated, generating greenhouse gases. The technological advance offers alternatives for the direct extraction of bioactive compounds and the bio-production of products of highly added value, such as enzymes, organic acids, and biofuels, among other products (Singh et al. 2014).

1.4.9 *Carrot Waste*

The carrot (*Daucus carota*) after the potato is globally the second most popular vegetable. The world carrot production is 33.6 Tg, China being the main producer at 36%. Approximately 40–30% of carrot pulp is produced after the extraction of juice, leaving a high potential total world production of by-product. The leaf has the major quantity of crude protein and carbohydrates, while the pomace contains more than 60% of fiber. Lipids and ash are present in similar amounts in each of the carrot by-products, and the by-products are being studied to produce antioxidants (Bardiya et al. 1996).

1.4.10 *Peanut Husk*

Peanut (*Arachis hypogaea L.)* is widely harvested and used in most cultures. The worldwide production of peanut in 2011 reached 34 Tg. The grain constitutes about 30%, indicating that 10 Tg of total residues is produced in the form of husks. The annual production totals approximately 29 million metric tons (http://www.soyatech.com/peanut_facts.htm), being (14.6 Tg), India (6.0 Tg), USA (1.8 Tg), and Nigeria (1.6 Tg) the main producing countries. Nepote et al. (2002) investigated the extraction of antioxidant components from peanut skins.

1.5 Biopolymers from Agricultural Wastes

Nowadays, agricultural production generates wastes produced annually in such amount that it turned to be named as "agricultural waste industry." However, the agricultural wastes depend on regions, countries, and cultures which manage different crops and provide different types of rich polymers sources. Typical examples are USA, India, Brazil, China, and Argentina.

1.5.1 Cereals Straw

According to FAO in 2011, the worldwide production of cereals was of 2.6 Pg. This category includes the production of wheat, rice paddy, barley, maize, popcorn, rye, oats, millets, sorghum, buckwheat, quinoa, canary seed, mixed grains (mixture of cereal species that are sown and harvested together), and minority cereals (cereal crops that are not identified separately because of their minor relevance at the international level). The top producer countries are as follows: China is the main producer with 20% of the world production, followed by the USA and India with 15 and 11%, respectively. These three countries produce 42% of worldwide production.

The cereal crop residues after harvesting comprise 50–75% of the total production (Bauder 2013). This indicates that at least 2.6 Pg of residues is produced after harvesting the crops. As an example, wheat straw mainly contains fiber, cellulose, and hemicelluloses that could be used for biotechnological applications. The country production/Tg in 2011 was 520.9 in China; 386.8 in USA; 285.5 in India; 91.8 in Russia, and 83.4 in Indonesia. Some of this residue is left on the soil to reduce its erosion. I also fertilize the soil through the incorporation of organic matter. Post-harvest losses include crop losses that could not be harvested (due to technical constraints, a proportion of the crop left in the field) as well as losses associated with drying, storage, and transport (FAO 2017).

Stubble is composed by the stems and leaves that remain on the ground after cutting a crop. Stubble is often confused with remains of little value. However, the stubble is a very good resource to protect the soil from the impact of erosive precipitation and the consequent runoff. The presence of stubble on the ground is like a water trap, which facilitates infiltration and reduces evaporation losses by keeping the surface of the soil cooler and protected. This conception is the core of a correct system of cultivation with zero tillage. However, this system may or may not involve the retention of high amounts of stubble, or only enough to protect the soil.

The amount of stubble remaining on the crop soil varies by crop. Cereals in most cases contribute considerably more stubble than oilseeds. Crops such as maize, wheat, sorghum, and barley provide more stubble than soy, rape, or sunflower. As an example, Menéndez and Hilbert (2013) estimated 12–43 mill Ton/ha corn stubble after 2.5 mill hectares sown in 2008–2011 in Argentina. The huge amounts of stubble could be used for biotechnological purposes with the incorporation of highly added value.

Another usual practice of farmers was burning stubble to acidify the soil. In the last years, diverse studies recommend the maintenance of the stubble to avoid erosion and loss of much organic matter. After the cereal harvest, the volume of the residual dry material is huge: 5.50–11.0 Tn/hectare (Moreno et al. 2014).

Crop residues are an important source of fodder and energy (Fischer and Schrattenholzer 2001; Montico 2010). However, the use of these materials as inputs of biorefineries to obtain energy is under permanent questioning by global and regional decision-makers because of the concern for the conservation of soils (Lal et al. 2007; Montico 2010). The work of Sheehan et al. (2004) estimates that in Iowa (USA), approximately 40% of the maize residue could be used for energy production, while maintaining the risks of erosion below tolerable limits. However, novel biotechnological applications could be more profitable and less aggressive for the environment. An important aspect to be considered in the final balance for the different alternatives of biomass utilization is the return of the necessary amount of the wastes to the productive system/soil.

1.5.2 Starch

Starch is one of the most abundant of plant products and also a major food reserve which provides bulk nutrients and energy source since the beginning of human civilization. Starch is produced from agricultural plants, mainly from potatoes, rice, maize, and wheat in the form of hydrophilic crystallites with dimensions ranging from 1 to 100 μm (BeMiller 1993). It appears as a food waste, however, mainly in potato and to a lesser extent mango seed. It is a hydrocolloid biopolymer comprised of two types of α-glucan: amylose (poly-α-1,4-D-glucopyranoside), a linear polymer and amylopectine (poly-α-1,4-D-glucopyranoside and α-1,6-D-glucopyranoside) as shown in Fig. 1.8. Depending on the botanical source, the percentage of each polymer varies, as well as the morphology and molecular structure.

A polysaccharide-based biopolymer has been produced recently from the starch extracted from mango seed waste. Mango (*Mangifera indica*) is a native tropical fruit from southern Asia which is now cultivated in most frost-free tropical and warmer subtropical climates. Seven countries are responsible for almost 74% of the entire world production of mango. India is the largest producer with 13.6 Tg equivalent to 39% of world production. During the processing of mango, peel and almond are generated as waste materials and they represent around 40–50% of the total fruit weight so that the total world production yields approximately 15.7 Tg of mango waste per annum (Nilani et al. 2010).

Starches from whatever source are composed of one or more glucans and are water-soluble polymers that produce viscous dispersions, solutions, or gels at low concentrations and are used extensively in practical applications. They are normally extracted in the form of granules that can be used to identify the source of the starch because of its characteristics, i.e., size and morphology. Starch is a hydrocolloid biopolymer comprised of two types of α-glucan: amylose (poly-α-1,4-D-glucopyranoside), a linear polymer, and amylopectine (poly-α-1,4-D-glucopyranoside and α-1,6-D-glucopyranoside) as shown in Fig. 1.4. The molecular weight of amylose is reported to be 1.6×10^5 to 2.6×10^6 Da with degree of polymerization (DP) in the range of 1.0×10^3 to 1.6×10^4], depending on the source and method of preparation and to some extent on the maturity and growing conditions of the plant (Whistler and BeMiller 1993). In the solid state, it probably exists most often as a left-handed, sixfold helix. In solution, it seems to be a loosely wound and extended helix that behaves as an almost spherical random coil (Hsein-Chih and Sarko 1978). The amylose helix has a hydrophobic interior that enables the molecule to complex with nonpolar molecules, including fatty acids and their salts, mono- and diglycerides, and various detergents (Bulpin et al. 1982).

Amylopectin molecules contain clusters of branches (Fig. 1.5). The molecular weight of amylopectin is reported to be 5.0×10^7 to 4.0×10^8 Daltons (DP 3.0×10^5 to 2.5×10^6), depending on the source and method of preparation and to some extent on maturity and growing conditions. It is claimed that amylopectin molecules in aqueous solution are two-dimensional, which is consistent with the coating and other properties of starch (BeMiller 1993).

Native starch occurs mostly in the form of semicrystalline granules, which have a very complex hierarchical structure. Starch granules are generally composed of an amorphous bulk core area surrounded by concentric semicrystalline growth rings alternating with amorphous growth rings. When starch is heated in the presence of water and subsequently cooled, the disrupted amylose and amylopectin chains can gradually reassociate into a differently ordered structure in a process termed as retrogradation (Wang et al. 2015). Retrogradation is only one of the many structural and chemical modifications that have been studied in starch to improve its characteristics. For use in its natural form, it should be modified to overcome the poor thermal, shear, and acid stability as well as high rates and extents of retrogradation. It has two available functional groups for modification: the nucleophilic hydroxyl groups and ether bonds. Starch can be chemically modified; for example, partial

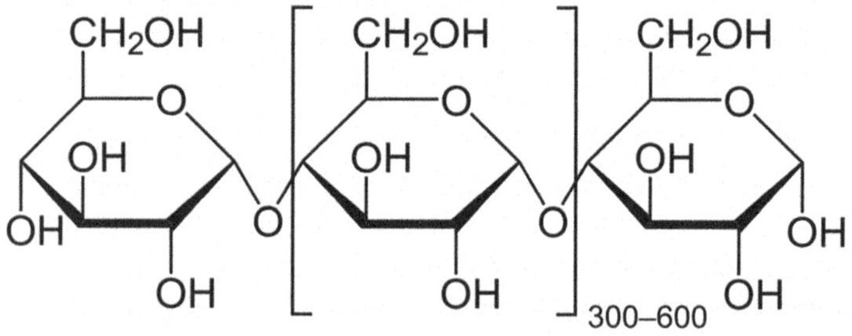

Fig. 1.4 Haworth projection of amylose composed by linear α-1,4 glucose units

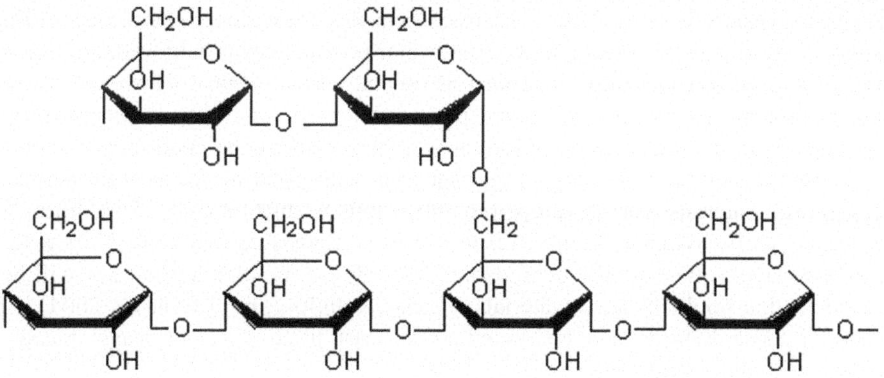

Fig. 1.5 Haworth projection of amylopectin composed by linear α-1,4 glucose units and α-1,6 glucose unit in the branch points

acid hydrolysis on the amorphous regions of the starch granules generates starch nanocrystals (Angellier-Coussy et al. 2009), or it can be physically modified by hydrothermal treatment, where the starch structure and properties are changed without destroying its granular structure (Hoover 2010).

Starch in its native form should be modified for its applications as a material, unless used as a filler. One of these modifications is the restructured starch. To produce this material normally, starch is hydrated and heated. Because of this process, the organization of the granule is destroyed, most of the hydrogen bonds are disrupted, and the starch forms a gel. With this procedure, both melting temperature (Tm) and glass transition temperature (Tg) decrease, and by decreasing the moisture content (20 wt%), the melting temperature tends to be close to the degradation temperature. Thermoplastic starches were synthesized by the addition of exogenous molecules such as PEG or sorbitol; or nitrogen compounds like urea o amines. In this way, Tm is decreased without the inconvenience of being too proximal to the degradation temperature (Avrous 2004).

TPS has enormous advantages such as cheap, abundant, and biodegradable, but it has two major disadvantages such as poor mechanical strength properties and high moisture sensitivity. There are some solutions to this: TPS can be mixed with appropriate fillers such as nanoparticles, the surface can be chemically modified, or it can be blended with a hydrophobic polymer, such as polyvinyl alcohol, poly (ethylene-co-vinyl alcohol), PLA, polycaprolactone (PCL), poly(butylene succinates) (PBS), polyhydroxybutyrate (PHB), and poly(3-hydroxybutyrate-co-3-hydroxyvalerate) (PHBV) (Sanchez-Vazquez et al. 2013).

As stated above, pure native starch has multiple applications. But for biomedical applications, starch is seldom used alone or without any previous modification. PLS materials have some promising properties: They are very cheap compared to fossil-derived plastics, are fully compostable, and can be processed using traditional plastic process such as injection molding and extrusion. However, PLS materials have several disadvantages such as poor mechanical properties and moisture sensitivity (Kaseem et al. 2012). To overcome this difficulties, chemical modification of starch has been made in the past, water resistance was achieved, and some industrial applications were proposed (Fringant et al. 1996). However, these modification forms involve additional purification steps and alter the mechanical properties due to a decrease in molecular weight caused by the chemical reactions.

As compostable and biocompatible materials with good mechanical properties, PLS hybrid materials were developed to obtain low production costs. Two different materials can be associated with PLS to obtain compostable materials: biodegradable polyesters (other non-biodegradable materials can be blended but not will be treated in this book) or agro-materials (lignins, cellulose, etc.) (Avrous 2004). In addition, these hybrids can be mixed in at least three different ways: blends, composites, and multilayers.

Blending consists in mixing different polymers in an intimate way without any organization. It is the simplest way to associate polymers and a powerful tool to improve materials in both performance and cost. Besides, blends can also be used to test the compatibility and interactions between different polymeric phases.

1.5.3 Cellulose

Cellulose is an organic compound, which minimum formula is $(C_6H_{10}O_5)_n$, and is a polysaccharide consisting of a chain with many thousands of β(1 → 4)-glycosidic bonds. Cellulose is characterized by its poor solubility due to the strong intra- and intermolecular hydrogen bonds within and among the individual chains (Fig. 1.6).

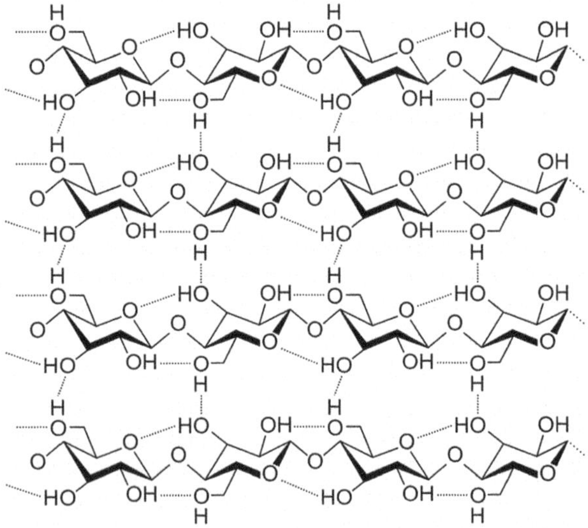

Fig. 1.6 Cellulose polysaccharide chains and its bonds. *Source* image from Wikimedia commons GNU FDL

It has several hydrogen bonds between chains form microfibrils that plenty contribute to tensile strength in cell walls (McNamara et al. 2015).

Cellulose is a polysaccharide composed by glucose molecules in a linear chain, and the glucose molecules are connected using acetal linkages between the C1 and C4 carbons of the glucopyranose rings (Fig. 1.6), thereby forming a high molecular weight polymer. This complex structure gives cellulose some properties as hydrophilicity, biodegradability, and chirality, among others. It is a relatively stable polymer because it has many intra- and interhydrogen bonds between polysaccharide chains; that is the reason why cellulose is the main structural component of many plants. The polymer has a highly ordered polymer region (crystalline) but also an amorphous region.

Cellulose is the principal natural polymer that presents on the Earth: Not only plants synthesize this polymer, but also many bacteria produce it as an extracellular component with structural functions. The cell walls of plant generally have cellulose as the main component. Cellulose is present, apart from plants, in several other organisms such as algae, fungi, bacteria, and in some marine animals called tunicates. The main cellulose sources are listed and briefly summarized here: Plants are the principal cellulose source, and due to its biomass, it is mostly composed of cellulose and its costs of production are relatively low, using mainly wood and cotton. The polymer could be also obtained from sugarcane bagasse and cereal plant wastes.

Cellulose is an important component of algae cell walls. Many types of algae are commonly used in cellulose production. The different species of algae produced different types of cellulose. Bacterial cellulose is produced by certain species of bacteria, and one of the most popular is *Komagataeibacter xylinus*, commonly found as contaminant during vinegar fermentation. This bacterium produces a cellulose flat membrane on the interface between air and growth medium (Rajwade et al. 2015; Jozala et al. 2016). The cellulose membranes are composed by pure cellulose along with a large amount of water. Bacterial cellulose has unique properties over plant/algae cellulose:

- It is unique and more intricate nanostructure.
- There is no hemicellulose or lignin to be removed, and it has high purity.
- It has stronger and longer fibers, and higher stability.
- It could be shaped in any form, depending on the container where the bacteria have been grown.
- It has higher water swelling.
- The bacteria could use several substrates to grow and produce cellulose.
- Membranes properties and yields could be varied depending on which substrate and which bacteria strain are used

Microbial cellulose and plants cellulose are identical in molecular formula and polymeric structure, but differs in the way the glycosyl units are positioned, modifying the crystallinity. In comparison with plant cellulose, bacterial cellulose has a high degree of polymerization and better mechanical properties. Besides, there are some issues that have stopped the increase of bacterial cellulose production: high price of production (about 50 times higher than vegetal cellulose), low yields of production, low scale production, and longer times (Stephen 1995).

It is a water-insoluble polymer and the most relevant in maintaining the structure of cell walls in plants, because of its toughness. Other interesting properties about cellulose are its biodegradability and biocompatibility (George and Sabapathi 2015). The reactivity of cellulose is affected by the morphology and degree of crystallinity, which varies according to the origin and pretreatment of the material (Krassig 1990).

Various industries such as wine, textile, paper, and fruit have cellulose as waste. The whole world is imploring biodegradable materials with low toxicity. Natural cellulose is mainly used to produce paperboard and paper. Cellulose-based materials have been exploited by human for thousands of years as construction materials, textile fibers, combustible sources, etc. In biofuels field, the ethanol produced from cellulose is an important alternative energy source. Cellulose filter is a modern application of this polymer (McKinnon and Avis 1993), e.g., additives in paints and coatings, biomedical applications (Lavoine et al. 2012; Mondal 2017), food additives (Stephen 1995), and renewable energy (McNamara et al. 2015).

The strong mechanical properties, its flexibility, and biocompatibility are intrinsic properties of cellulose materials; however, they have some complex bioengineering applications that need different forms of cellulose as nanocelullose and bacterial cellulose that are acquiring much importance (Moon et al. 2011).

It is a relatively stable polymer because it has many intra- and interhydrogen bonds between polysaccharide chains; that is the reason why cellulose is the main structural component of many plants. The polymer has a highly ordered polymer region (crystalline) but also an amorphous region (George and Sabapathi 2015; McNamara et al. 2015).

1.5.4 Soy Hull

Agro-food production, which characterizes developing countries such as Argentina and Brazil, has undergone extraordinary changes in productive modes in the last two decades, through the unleashing of unprecedented technological advances, and brought mainly by bioengineering and new agricultural techniques, besides mechanical, computer, and chemical tools. Consequently, the agricultural frontier continues to expand, especially for soybean monoculture with an increase of 185.3% in the area sawn (INTA 2015). Then, the high availability of soy residues could be an abundant and cheap source of pectin (approximately 25%), in amounts like that recovered with citrus (Kalapathy and Proctor 2001).

1.5.5 Lignin

Lignin is a three-dimensional network formed by three types of monomers: *p*-coumaryl, coniferyl, and sinapyl alcohols. Lignin is a complex highly branched structure and irregular macromolecule in which basic blocks can be defined as "C9" units; however, the structure varies according to the vegetable source. In lignocellulosic materials, lignin is the matrix that surrounds cellulosic fibers. Lignin is viewed as a waste material, available in large quantities from peanut husks, citrus peels, sugar bagasse, and corn stover, and is also derived from wood pulp. Commercially, lignins are available as co-products whose main derivatives are lignosulfates and kraft lignins (Lebo et al. 2001). Their main structure is based on lamellar macromolecular complexes which link through non-covalent interactions (Rouilly and Rigal 2002).

1.6 Biopolymers Produced from Fermented Wastes

Polyhydroxyalkanoates (PHAs) are polyesters produced by microbes in nature or by fermentation of sugar and/or lipids. In general, PHAs are commercially produced using expensive carbon sources: sucrose (Biomatera in Canadá), sugarcane (Biocylce in Brazil), and corn (Telles in the USA and Tanian in China) (Shen et al. 2009). Besides, some reports demonstrated that PHAs can be produced from several sources of waste-related, such as olive mill wastes (Morganti et al. 2016), residual banana (Naranjo et al. 2014), residual vegetable oil (Song et al. 2007), and other agro-industry wastes (Oliveira et al. 2004).

Furthermore, wastewater from municipal, pulp paper, starch, and dairy wastewater treatment plants was used as carbon source for PHA's production (Yan et al. 2006). The utilization of alternative carbon sources not only responds to be environmental-friendly but reduces the costs of production that today halts the widespread utilization of this plastic materials in oppose to those derived from petroleum (Koller et al. 2010).

Originally, the main purpose of PHA's industrial developments was the replacement of polypropylene or polyethylene in plastic bags and containers, because of its thermoplasticity. All types of containers were made of PHAs. But the great biocompatibility, plus the biodegradability of PHA's plastics and their composites, has been used in the development of materials for biomedical applications, for tissue regeneration and hard tissue replacements.

PHAs are lineal polyesters that comprise a complex class of polyoxoesters. The majority of prokaryotes synthesize polyhydroxybutyrate (PHB), a biopolymer first isolated and characterized in 1925 by French microbiologist Maurice Lemoigne (Lemoigne 1926).

PHA is large family of hydroxyalkanoic acids comprising more than 150 different polymers because of the substrate promiscuity of one of the key enzymes, the

PHA synthases. Different PHAs exhibit a wide variety of mechanical properties from hard crystalline to elastic, depending on the composition of monomer units (Rudnik 2008). The generic formula for PHAs is displayed in Fig. 1.7, where R can be hydrogen or hydrocarbon chains of up to C15 in length.

The value "n" can vary from 100 to 30,000 depending on the organism used to produce the bioplastic (Lee 1996). The table exemplifies some of the variety around PHAs, their names, and common abbreviations and also indicates the PHA type referred to the side chain (Table 1.1).

Short-chain-length polyhydroxyalkanoates (scl-PHA), such as poly (3-ydroxybutyrate) (P(3HB)), are highly crystalline materials. The melting point of P(3HB) is around 177 °C, close to that of polypropylene, showing other similar properties such as thermoplasticity and UV resistance, although the biopolymer is stiffer and more brittle. P(3HB) is water insoluble and relatively resistant to hydrolytic degradation. This differentiates P(3HB) from most other currently available biobased plastics which are either moisture or water soluble. Because of the crystallinity of this material, one of the largest uses of biodegradable polymers, the application to a flexible film, cannot be achieved. To improve the characteristics of this bioplastic, attempts have been made to copolymerize PHB with several different copolymers. When the copolymerization occurs with other hydroxy acid units, the mechanical properties can be improved. For example, the addition of 3-hydroxyvalerate co-monomer (3HV) enhances the impact strength and the polymer becomes more flexible.

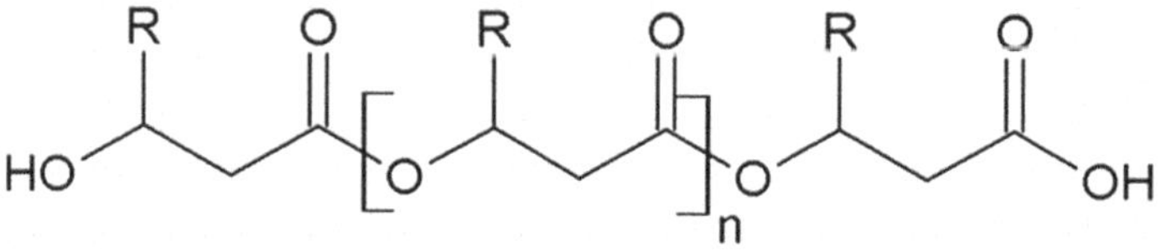

Fig. 1.7 Cartoon of chemical structure of PHAs. *Source* http://www.polyfermcanada.com/pha.html

Table 1.1 Classification of polyhydroxyalkanoates based on their R-groups

PHA type	R group	PHA name	Abbreviation
PHA-scl	R = H	Poly(3-hydroxypropionate)	P3HP
	R = CH_3	Poly(3-hydroxybutyrate)	P3HB
	R = C_2H_5	Poly(3-hydroxyvalerate)	P3HV
PHA-mcl	R = C_3H_7	Poly(3-hydroxyhexanoate)	P(3HHx)
	R = C_5H_{11}	Poly(3-hydroxyoctanoate)	P(3HO)
PHA-lcl	R = $C_{15}H_{31}$	Poly(3-hydroxyoctadecanoate)	P3HOD

scl short chain length; *mcl* medium chain length; *lcl* long chain length. *Source* (http://www.polyfermcanada.com/pha.html)

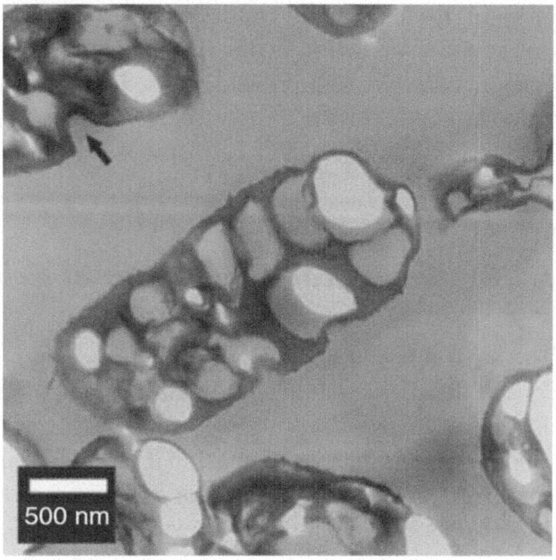

Fig. 1.8 Micrograph of wild-type *Bacillus megaterium strain uyuni S29* isolated from Uyuni salt field (Bolivia) containing PHB's granules after 21-h fermentation. *Source* Rodríguez-Contreras et al. (2013)

As the size of the chain length of the monomer increases, the material becomes softer and rubbery. So, medium-chain-length PHAs (mcl-PHA) are normally elastomers and sticky materials. This way, the addition of monomers of medium or long side chain length have a range of properties depending on composition, being hard, with some elasticity when the percentage of 3-hydroxyhexanoate (3HHx) co-monomer is low, too soft, and rubbery when the 3HHx percentage is higher. Consequently, a copolymer PHB-P(3HHx) combines properties normally associated with synthetic plastics, such as the strength and flexibility of polyethylene and chemical compatibility of polyesters (printability, dyeability) (Rudnik 2008; Bugnicourt et al. 2014).

From microbial physiological point of view, PHAs are storage compounds produced in large quantities in the presence of culture media with carbon excess (Fig. 1.8), and it is natural for microorganisms to have an enzyme to degrade it and to recover this stored carbon source when in the surrounding environment it is scarce, so PHAs work as functional carbon depot that can serve as a source of carbon and/or energy for microorganisms during periods of fasting or nutritional stress.

1.7 Biomedical Application of Biopolymers

Bioconversion of industrial and agricultural wastes in useful compounds like biopolymers not only gives some additional value to products but also reduces pollution creating, virtuous circles. However, sometimes, the value of the recycled products is not enough to economically support biotransformation and downstream processes at large scale. In these cases, specific applications of such molecules can bring high-value output for the commercial processes and social benefits. Typically, this is the case of biomedical uses of biopolymers in biomedicine such as regenerative, curative, and preventive medicines, drug delivery, and tissue engineering, among others. In the present section, some relevant biomedical applications of selected biopolymers will be displayed.

1.7.1 Biomedical Applications of Cellulose

The most relevant applications of cellulose in biomedical uses are described below:

Artificial cornea: Several cellulose biocomposites have been described to fully adapt the material's properties for eye therapeutics. Distinct functionalization is reported to produce tridimensional structures more adequate to cell growth to be employed as artificial cornea or relieve glaucoma. Thus, many related cellulose composites can support the growth of corneal stromal cells while preserving a full vision to the patient; these materials present a great potential to be used as eye scaffolds and replace the less biocompatible systems currently utilized in the clinic. Moreover, cellulose lenses could be loaded with drugs to maintain adequate concentrations during the treatment of eye infections and allergies (Jia et al. 2009; Fadel Picheth et al. 2017).

Cartilage or meniscus implants: The bacterial cellulose has mechanical properties like those of the cartilage, even more than the collagen which is the material that is usually used for the manufacture of scaffolds. In the case of damage to the meniscus of the knee, which is a common injury, the menisci are removed and do not regenerate again. For this reason it is necessary the development of implants to replace them (Bodin et al. 2007; Rajwade et al. 2015).

Dental implants: The development of new materials for bone and soft tissue regeneration has been changed by research in the field of biomaterials. Cellulose has a structure that permits cell growing and is also biocompatible with the damage tissue. Usually, the cellulose matrix is functionalized and combined with other materials as ceramics or other biopolymers, in order to change mechanical properties for this application (Grumezescu 2016).

Drug delivery applications: Cellulose-based drug delivery systems attracted attention as a promising material in the biomedical field because of their outstanding properties such as hydrophilicity, biocompatibility, biodegradability, and high surface area. Cellulose has an internal structure that allows the loading of some molecules such as pharmaceutical drugs. Drug delivery nowadays has been accepted as a promising strategy to prolong the residence time and to improve specific localization of drug delivery systems on various biological membranes (Yordanov et al. 2013; Almeida et al. 2014; Cacicedo et al. 2016). Another advantage of these drug delivery systems is their potential to extend the residence time at the site of drug absorption, and thus, they can reduce the dosing frequency in controlled release drug formulations. These dosage forms can also intensify the contact of their drug contents with underlying mucosal barrier and improve the epithelial transport of drugs across mucus membranes, especially in the case of poorly absorbed drugs. Moreover, cellulose is a green material that is innocuous for the human body (Edgar 2006; Dash et al. 2012; Shokri and Adibki 2013; George and Sabapathi 2015).

Filtration: One of the oldest and most common uses of cellulose is to use thin layers of the biopolymer as filters. It takes advantage of the intricate fiber structure; in biomedicine, cellulose filters are used to remove microorganisms from solutions (McKinnon and Avis 1993).

Wound dressing: Following the same reasoning as in filters, cellulose membranes (plants or bacterial) maintain a physical barrier that not allow the passage of bacteria due to its microfibrillar structure that provides flexibility, high water retention capability, and gas exchange. These properties are extremely important in skin burn/wound treatment. The cellulose membranes also could be loaded with a drug to treat the injured tissue. Currently, the main commercial utilization of BC membranes is as wound dressing devices, commercialized under several trademarks (Portal et al. 2009).

1.7.2 Biomedical Applications of Chitin and Chitosan

Chitin and chitosan have excellent properties for biomedical applications because of high biocompatibility and biodegradability, non-toxicity, molecules adsorption properties, antimicrobial activity, water absorption, ecological safety, etc. The most relevant application of chitin and chitosan is summarized below:

Antimicrobial agent: As it was previously discussed, chitosan has interesting antimicrobial properties; it was proved that it inhibited the growth of *E. coli, S. aureus* and also some fungi organisms (Martins et al. 2014).

Contact lens: The antimicrobial properties of chitosan along with the possibility of making clear thin films makes chitosan an adequate material for the development of ocular dressings or contact lens (Silva et al. 2016; Anirudhan et al. 2016).

Cosmetics: Lots of cosmetics must be dissolved in acid medium, and chitosan has the unique property of being a cationic polymer that could be dissolved under acid conditions. That is the reason why chitosan is used as thickener in creams and lotions, and also as an antimicrobial agent (Jimtaisong and Saewan 2014).

Drug delivery systems: Many polymers are used as a material for drug delivery and controlled release devices. Chitosan as a cationic polymer allows the encapsulation of molecules with negative charge. The drug delivery concept is the predictable and controlled release of molecules (antibiotics drugs, antineoplastic drugs, enzymes, etc.) in specific environmental conditions, during a determined period at established concentrations. This procedure can minimize secondary effects of several drugs and the dose numbers and maximize the treatment efficacy. Something that is usually pursued in this field of research is a biodegradability of the material and a low toxicity. All these interesting properties of chitosan make this natural polymer an ideal candidate for controlled drug release formulations (Janes and Alonso 2003; Duttagupta et al. 2015).

Enzymes, bacteria, and cells immobilization: As alginate, collagen, and others, chitosan could also immobilize biological systems, such as enzymes, bacteria, and cells. This is really useful in bioreactors or bioprocess where the biological compounds must be collected and reutilized after production (Trabelsi et al. 2014; Rana et al. 2014; Rangel-Rodríguez et al. 2014).

Sutures: Chitin is a convincing material for wound sutures because of its elasticity, biocompatibility, biodegradability, and low toxicity in combination with its antimicrobial activity and low immunogenicity, points to immense potential for future development of absorbable sutures (Medovent 2016).

Wound dressing: Chitin and chitosan materials for wound healing are in continuous research. Several wound dressing polymeric membranes were developed, with combination of diverse polymers including gelatin and alginate. Chitosan shows an excellent adhesion to skin. It was found that skin regeneration is promoted in the contact area of the dressing. Other advantage of the chitosan in a wound dressing is the antimicrobial activity (Upadhyaya et al. 2013; Patrulea et al. 2015).

1.7.3 Biomedical Applications of Collagen

Cosmetic surgery: Collagen has been widely used in cosmetic, as a healing aid for burn patients, for reconstruction of damage tissues, and for surgical purposes. Both human and bovine collagen are widely used for the treatment of skin aging. When it is used as a cosmetic, there could be a redness caused by allergic reactions; however, this can be tested before using the cosmetic (Gomez-Guillen et al. 2011; Liu et al. 2015).

Tissue regeneration: Collagen scaffolds are used in tissue regeneration, whether it adopts several forms of sponges, thin sheets, or gels. Collagen has really good properties that look for in tissue engineering materials, such as porosity, permeability, hydrophilicity, and biocompatibility. Collagen scaffolds are also ideal for the seeding of cells, such as osteoblasts and fibroblasts, allowing normal growth once inserted in the tissue. Biomaterials based on collagen are widely used in tissue engineering such as injectable matrices and scaffolds intended for bone regeneration (Hoyer et al. 2012).

Collagen scaffolds play an important role in helping tissue regeneration, mainly skin, bone, and cartilage (Bomhard et al. 2013). The scaffolds are implanted when the tissue damage is severe, and the organism is unable to heal it.

Reconstructive surgical uses: Collagens are widely employed in the construction of the artificial skin substitutes used in the management of severe burns and wounds and also in the reconstruction of the skin tissue after a surgery (Chvapil et al. 1973; Lazovic et al. 2005). These collagens may be derived from bovine, equine, porcine, or even human sources and are sometimes used in combination with silicones, polysaccharides, fibroblasts, growth factors, and other substances.

Wound treatment: Collagen is one of the skin tissue components so it can plenty benefit the wound healing and cicatrization process. Collagen is a natural product; therefore, it is used as a natural wound dressing and has properties that artificial wound dressings do not have. There are some works that show that collagen avoids bacterial infections which is relevant in a wound dressing material. When collagen is used as a burn dressing, collagen fibers serve to guide fibroblasts causing formation of fibrillar structures and helping it to heal rapidly (Chattopadhyay and Raines 2015).

Pharmaceutical industries: Collagen is used in pharmaceutical industries as microparticles, injectable dispersions, membranes, and materials for drug delivery systems. Its application in the pharmaceutical as well as biomedical field is due to its characteristics such as low toxicity, cell adhesion ability, biodegradability, and biocompatibility (Leitinger and Hohenester 2007; Pamfil et al. 2015).

Drug delivery: Collagen has been used in drug delivery systems: for example, collagen shields in ophthalmology, mini-pellets and tablets for protein delivery, gel formulation in combination with liposomes as controlling material for transdermal delivery, and nanoparticles for gene delivery (Kittiphattanabawon et al. 2015).

1.7.4 Biomedical Applications of Pectin

Since its discovery, pectin has been extensively used as a food additive. The main use of pectin remains as a gelling agent, thickening agent, and stabilizer in foods. However, pectin applications in other fields such biomedicine and health care could provide the necessary market value to make its recovery from plant wastes profitable.

Food additive: It is recognized that consumption of pectin as a dietary supplement provides many health benefits. There is clear evidence that pectin can lower cholesterol levels (Brown et al. 1999) and glucose serum levels (Behall and Reiser 1986). Intake of pectin also reduces food intake because of its properties to extend the residence time in the upper digestive tract, so it has been used to treat overeating. Pectin and its degradation products have also showed apoptosis induction activity in colonic adenocarcinoma cells, and this could explain the protective effect of fruits in colorectal cancer (Olano-Martin et al. 2003).

Drug delivery: In the pharmaceutical area, pectins have been used as a binding and blending agent for controlled release matrix formulations (Sriamornsak et al. 2007; Costas et al. 2012; Dini et al. 2014; Islan et al. 2015). In nasal drug delivery, the conventional formulations have a residence time around 15 min; hence, gel-like formulations have been tested and pectin fentanyl spray was proven to improve the analgesic onset, treatment efficacy, and acceptability to treat breakthrough cancer pain (Fisher et al. 2010). Ocular drug delivery fronts similar obstacles; therefore, an alternative approach was the application of pectin systems for an *in situ* gel formation where pectin is applied in a liquid form and then gels in the eye. The gelation is mainly triggered by the pH of the tears and by the presence of electrolytes in the tear film. Pectin piroxicam microspheres have been tested in this sense (Giunchedi et al. 1999).

However, it is in the development of specific site-directed formulations in which pectin has truly promising pharmaceutical applications, for example, in colon-specific drug delivery, for systemic action, or topical treatments of diseases such as ulcerative colitis, Crohn's disease, and colon carcinomas (Costas et al. 2012). The development of smart hybrid pectin alginate microspheres to release antibiotic and destroy microbial biofilms in the intestine was reported successfully (Islan et al. 2015). The rationale for the use of pectins is based in the degradation by the pectinolytic enzymes produced by colon flora, which will delay the cargo release and protect the active component in the upper digestive tract due to its insolubility in acid medium and the absence of enzymes capable of degrading it in the stomach and the small intestine in humans. For this use of pectins, several articles were published, for a variety of active compounds, i.e., antioxidants (Bermúdez-Oria et al. 2017), water-soluble and water-insoluble drugs (Glinsky and Raz 2009; Liu et al. 2003), enzymes (Bourgeois et al. 2006; Islan et al. 2015), probiotics (Anal and Singh 2007), and SPIONs (Dutta and Sahu 2012).

Wound dressing: The medical research on pectin does not end here, because its biocompatibility and gel-forming ability that functionalized biomaterials based in pectin have been tested in wound dressing (Munarin et al. 2012) and tissue engineering (Coimbra et al. 2011).

Also, pectin- and pectinate-based systems were developed in other areas of research. Pectin gel beads were used to remove cadmium and copper from aqueous solutions (Cataldo et al. 2013). Pectin-synthetic polymers were developed to generate biodegradable films and packages (Cavallaro et al. 2013; Nesic et al. 2011).

1.7.5 Biomedical Applications of PHBs

PHAs are generally well tolerated by the mammalian immune system, PHB fibers in particular were tested to suture wounds, and the physiological reaction was equal to silk, a completely biocompatible polymer and used in the development of tissue scaffolds (Volova et al. 2003; Zhao et al. 2003).

The immense variety of mechanical properties, and the biodegradability and biocompatibility of the PHA have made that in the last 20 years various applications have been developed in the field of medicine. Changing the PHA composition can provide for materials with different degradation times within desirable time frames under specific physiological conditions (Chen and Wu 2005). For example, in surgical devices such as sutures, meniscus repair devices, screws, bone plates, repair patches, and orthopedic pins (Dai et al. 2009), PHAs have been used also for the replacement or scaffold for damaged tissue in tendon repair devices, nerve guides (Bian et al. 2009), vein valves, bone marrow scaffolds, skin substitutes, and wound dressing (Wang et al. 2008).

Another homopolymers and copolymers such as poly(lactic-co-glycolic acid) (PLGA) or poly(glycolic acid) (PGA) are widely used in commercially available sustained release products for drug delivery. However, lactate and glycolate copolymers are degraded by bulk hydrolysis; hence, drug release cannot be fully controlled. Instead, PHA is hydrophobic and can be molded into a variety of forms such as films, microcapsules, nanoparticles, and porous matrices. This way, several drugs have been trapped in PHA matrices, antibiotics, antitumoral agents, vaccines, and hormones (Nobes et al. 1998; Orts et al. 2008).

In 2005, Peters and Rehm demonstrated that formation of PHA granules was not affected by GFP fusion with the N-terminal PHA synthase (Peters and Rehm 2005). This opened the door to PHA modification to allow functionalized PHA particles to acquire new properties, including target delivery. PHA particles were surface modified to provide ligand–receptor specificity, for example, mannosylated human 1-acid glycoprotein (hAGP) and human epidermal growth factor (hEGF) for targeting cancer cells or macrophages, respectively (Yao et al. 2008).

1.7.6 Biomedical Applications of Starches

Application of starches in biomedicine is based on blends within other polymers to provide specific properties for designed purposes. Blends with agro-materials like pectin for drug release in colon were reported (Fishman et al. 2000; Dimantov et al. 2004). Nevertheless, the patents and literature on starch blending with biodegradable plastics are more diverse and abundant. Poly(vinyl alcohol) (PVA) blends have demonstrated good biodegradability (Ishigaki et al. 1999) and excellent mechanical properties in the formation of films encapsulating drugs or proteins for biomedical applications (Jayasekara et al. 2004; Kenawy et al. 2014). Hydrogels can be produced with starch and PVA for its use in wound treatment (Kunal et al. 2006). Results on blends using different polyhydroxyalkanoates (PHA) and starch were compiled by Shogren 2009, but the applications developed for these materials target disposable containers, textiles, and adhesives or coatings.

Starch composites and nanocomposites containing cellulosic or lignin and cotton fibers were included in PLS in order to improve the mechanical properties, such as tensile strength and water intake, but their biocompatibility was not evaluated (Wan et al. 2009; Prachayawarakorn et al. 2010; Grande et al. 2009).

1.8 Conclusion and Final Remarks

In synthesis, although there are many agricultural species which yield food production and industrial wastes, the raw materials that issue from them are few, principally fatty acids and terpenes, sugars, celluloses, starches, lignin, proteins, polyphenols, and fibers. The limited range is, however, compensated by the high tonnage of food wastes available annually worldwide from the food-processing industry.

Finally, although biopolymers from renewable sources are friendlier to the environment than synthetic ones, it should be kept in mind that they can also leave an important environmental footprint. For example, the growing of agricultural crops may involve the use of huge quantities of pesticides. Unless proper management practices are used, depletion of soil nutrients and microbes may occur. In addition, the preparation and purification of biopolymers requires chemical and biochemical processes, with the use of water, energy, and chemical or biological additives. Green chemistry is full of examples of the so-called life cycle innovation, which implies to achieve improvements at all stages of the product or process life cycle (Anastas and Lankey 2000). The challenge, then, is to convert natural biopolymers of discard into useful materials that can be used in the pharmaceutical industry and in biomedicine, keeping in mind the environmental integrity. However, designing a molecule that avoids one type of toxic mechanism of action without increasing the likelihood of initiating any of the others is a very big achievement (Anastas and Zimmerman 2016).

In each case, it should be considered the desirability of applying a biotechnological process through a life cycle analysis (LCA). The LCA allows to evaluate the interactions that a product or service has with the environment, considering its whole life cycle that includes the preproduction points (extraction and production of raw materials), production, distribution, use (including reuse and maintenance), recycling, and final disposal. In brief, the activity throughout the entire life cycle involving all the stages of the product chains is considered. The concept was summarized in the phrase "from cradle to grave" (Gentil et al. 2010; Piemonte 2011; Guo et al. 2013).

If we want our economy to rest on a sustainable basis, an urgent replacement of fossil carbon as chemical industry feedstocks must be done (Andrady and Neal 2009; Pei et al. 2011). Humans are concerned about the availability and real cost of oil, the accumulation of waste in natural environments, including the sea with negative effects on marine fauna through plastic ingestion, the leaching of plastic products with the potential to transfer chemicals to human beings and wildlife (Thompson et al. 2009).

Valorization of waste to valuable products (e.g., materials, chemicals, and energy) holds a significant sustainability potential for future generations. Apart from being an interesting alternative for waste disposal or landfilling, it will also increase its value through the production of highly added value products, resulting in a great progress in both environmental and economic terms (Catalina et al. 2012). The use of food and agro-industrial wastes for bioproduction of valuable bioproducts is economically important and can minimize various environmental hazards. In this scenario, environmental biotechnology is emerging as a good option to tackle adverse impacts.

However, the transition from a fossil fuel-based economy to a biobased economy requires the exploitation of synergies, scientific innovations and breakthroughs, and step changes in the infrastructure of chemical industry. Sustainable production of chemicals and biopolymers should be dependent entirely on renewable carbon (Koutinas et al. 2014). Particularly, the novel trends in biomedicine are potential applications for biopolymers such as scaffolds, controlled drug release, surgical sutures, gauzes, bandages, bone plates, and wound care and structural materials in personal hygiene products and packaging. Also, by combining emerging micro- and nanotechnologies with biopolymers, novel platforms to develop hybrid biopolymeric systems usable in many industries will be created for specific applications and highly added values.

Finally, human beings have the key. The reuse of biopolymers will depend on developing technologies and the responsibility of human being toward himself and the environment. That in turn will partly depend on how strongly society is committed to the concepts of resource conservation and reuse, environmental preservation, restoration, and the use of sustainable technologies. There are growing signs that people indeed want to live in greater harmony with nature and leave future generations a healthy planet.

In this line, we can agree with Trevors and Saier (2010a, b), who pointed out: "Social rules can be broken, but the laws of nature can't." All agree that human activities are the cause of most of our environmental problems. Educate, manage humans and their damaging actions -not just manage the environment- and we have a chance for sustainability and species survival.

References

Aberg CM, Chen T, Payne GF (2002) Renewable resources and enzymatic processes to create functional polymers. Adapting materials and reactions from food processing. J Polym Environ 10:77–84

Aberg CM, Chen T, Olumide A, Raghavan SR, Payne GF (2004) Enzymatic grafting of peptides from casein hydrolysate to chitosan. Potential for value-added byproducts from food-processing wastes. J Agric Food Chem 52:788–793

Agri-Food Business Development Center via Nation Master. Agriculture statics. Banana production by country (2000). http://www.nationmaster.com/graph/agr ban pro-agriculture banana- production

Almazan O, Gonzalez L, Galvez L (1998) The sugar cane, its by-products and co-products; *Asociación de Técnicos Azucareros de Cuba*. Food and Agricultural Research Council, Mauritius, pp 13–25

Almeida IF, Pereira T, Silva NHCS et al (2014) Bacterial cellulose membranes as drug delivery systems: An in vivo skin compatibility study. Eur J Pharm Biopharm 86:332–336

Anal AK, Singh H (2007) Recent advances in microencapsulation of probiotics for industrial applications and targeted delivery. Trends Food Sci Technol 18:240–251

Anastas PT, Lankey RL (2000) Life cycle assessment and green chemistry: the yin and yang of industrial ecology. Green Chem 2:289–295. https://doi.org/10.1039/b005650m

Anastas PT, Zimmerman JB (2016) Safer by design. Green Chem 18:4324

Andrady AL, Neal MA (2009) Applications and societal benefits of plastics. Phil Trans R Soc B 364:1977–1984

Angellier-Coussy H, Putaux J-L, Molina-Boisseau S et al (2009) The molecular structure of waxy maize starch nanocrystals. Carbohydr Res 344:1558–1566

Anirudhan TS, Nair AS, Parvathy J (2016) Extended wear therapeutic contact lens fabricated from timolol imprinted carboxymethyl chitosan-g-hydroxy ethyl methacrylate-g-poly acrylamide as a onetime medication for glaucoma. Eur J Pharm Biopharm 109:61–71

Annarita P, Anzelmo G, Fiorentino G, Nicolaus B, Tommonaro G, Di Donato P (2011) Polysaccharides from wastes of vegetable industrial processing: New opportunities for their eco-friendly re-use. In: Elnashar M (ed) Biotechnology of Biopolymers. InTech. https://doi.org/10.5772/16387

Arvanitoyannis IS, Ladas D, Mavromatis A (2006) Potential uses and applications of treated wine waste: a review. Int J Food Sci Technol 41:475–487

Avrous L (2004) Biodegradable Multiphase Systems Based on Plasticized Starch: a Review. J Macromol Sci Part C Polym Rev 44:231–274

Bardiya N, Somayaji D, Khanna S (1996) Biomethanation of banana peel and pineapple waste. Biores Technol 58:73–76

Bauder J (2013) Cereal crop residues and plant nutrients. Montana State University Communications Services. http://www.montana.edu/cpa/news/wwwpb-archives/ag/baudr230.html

Behall K, Reiser S (1986) Effects of pectin on human metabolism. In: Chemistry and function of pectins. Am Chem Soc Symp 310:248–265

BeMiller JN (1993) Starch Based Gums. In: BeMiller JN, Whistler RL (ed) Industrial Gums, 3rd ed. Academic Press, pp 579–600

Bermúdez-Oria A, Rodríguez-Gutiérrez G, Rubio-Senent F, Lama-Muñoz A, Fernández-Bolaños J (2017) Complexation of hydroxytyrosol and 3,4-dihydroxyphenylglycol with pectin and their potential use for colon targeting. Carbohydr Polym 163:292–300

Bian Y-Z, Wang Y, Aibaidoula G, et al (2009) Evaluation of poly(3-hydroxybutyrate-co-3-hydroxyhexanoate) conduits for peripheral nerve regeneration. Biomaterials 30:217–225. https://doi.org/10.1016/j.biomaterials.2008.09.036

Bilanovic DD, Malloy SH, Remeta P (2010) Solid or semi-solid state fermentation of xanthan on potato or potato waste. Assgn: Bemidji State University Foundation.US Patent Number: 07727747

Bodin A, Concaro S, Brittberg M, Gatenholm P (2007) Bacterial cellulose as a potential meniscus implant. J Tissue Eng Regen Med 1:406–408

Bourgeois S, Gernet M, Pradeau D Andremont, A Fattal E (2006) Evaluation of critical formulation parameters influencing the bioactivity of β-lactamases entrapped in pectin beads. Sel Pap 15th Int Microencapsul Symp, vol 324, pp. 2–9

Brinckmann J (2005) Collagens at a glance. Top Curr Chem 247:1–6

Brown L, Rosner B, Willett WW, Sacks FM (1999) Cholesterol-lowering effects of dietary fiber: a meta-analysis. Am J ClinNutr 69:30–42

Browne S, Zeugolis DI, Pandit A (2013) Collagen: finding a solution for the source. Tissue Eng Part A 19:1491–1494

Buehler MJ (2006) Nature designs tough collagen: explaining the nanostructure of collagen fibrils. Proc Natl Acad Sci USA 103:12285–12290

Bugnicourt E, Cinelli P, Lazzeri A, Alvarez V (2014) Polyhydroxyalkanoate (PHA): Review of synthesis, characteristics, processing and potential applications in packaging. Express Polym Lett 8:791–808

Bulpin PV, Welsh EJ, Morris ER (1982) Physical characterization of amylose-fatty acid complexes in starch granules and in solution. Starch 34:335–339

Cacicedo ML, León IE, Gonzalez JS, Porto LM, Alvarez VA, Castro GR (2016) Modified bacterial cellulose scaffolds for localized doxorubicin release in human colorectal HT-29 cells. Colloid Surface B 140:421–429

Cardona CA, Quintero JA, Paz IC (2010) Production of bioethanol from sugarcane bagasse: Status and perspectives. Biores Technol 101:4754–4766

Cataldo S, Cavallaro G, Gianguzza A, Lazzara G, Pettignano A, Piazzese D, Villaescusa I (2013) Kinetic and equilibrium study for cadmium and copper removal from aqueous solutions by sorption onto mixed alginate/pectin gel beads. J Environ Chem Eng 1:1252–1260

Catalina M, Cot J, Balu AM, Serrano-Ruiz JC, Luque R (2012) Tailor-made biopolymers from leather waste valorization. Green Chem 14:308–312

Catalina M, Cot J, Borras M et al (2013) From waste to healing biopolymers: Biomedical applications of bio-collagenic materials extracted from industrial leather residues in wound healing. Mater (Basel) 6:1599–1607

Cavallaro G, Lazzara G, Milioto S (2013) Sustainable nanocomposites based on halloysite nanotubes and pectin/polyethylene glycol blend. Polym Degrad Stab 98:2529–2536

Chandra R, Rustgi R (1998) Biodegradable polymers. Prog Polym Sci 23:273–1335

Chattopadhyay S, Raines RT (2015) Collagen-Based Biomaterials for Wound Healing. Biopolymers 101:821–833

Chen GQ, Wu Q (2005) The application of polyhydroxyalkanoates as tissue engineering materials. Biomaterials 26:6565–6578

Cheung RCF, Ng TB, Wong JH, Chan WY (2015) Chitosan: An update on potential biomedical and pharmaceutical applications. Mar Drugs 13:5156–5186

Chvapil M, Kronenthal L, Van Winkle W (1973) Medical and surgical applications of collagen. Int Rev Connect Tissue Res 6:1–61

Coimbra P, Ferreira P, de Sousa HC, Batista P, Rodrigues MA, Correia IJ, Gil MH (2011) Preparation and chemical and biological characterization of a pectin/chitosan polyelectrolyte complex scaffold for possible bone tissue engineering applications. Int J Biol Macromol 48:112–118

Costas L, Pera LM, Gómez López A, Mechetti M, Castro GR (2012) Controlled release of sulfasalazine release from smart pectin gel microspheres under physiological simulated fluids. Appl Biochem Biotechnol 167:1396–1407

Crini G (2006) Non-conventional low-cost adsorbents for dye removal: A review. Biores Technol 97:1061–1085

Dai Z-W, Zou X-H, Chen G-Q (2009) Poly(3-hydroxybutyrate-co-3-hydroxyhexanoate) as an injectable implant system for prevention of post-surgical tissue adhesion. Biomaterials 30:3075–3083. https://doi.org/10.1016/j.biomaterials.2009.02.015

Dale SK, Bruce E (2005) Life cycle assessment study of biopolymers (polyhydroxyalkanoates) derived from no-tilled corn. Int J Life Cycle Assess 10:200–210

Dash R, Ragauskas AJ, Banker GS et al (2012) Synthesis of a novel cellulose nanowhisker-based drug delivery system. RSC Adv 2:3403–3409

Davis G, Song JH (2006) Biodegradable packaging based on raw materials from crops and their impact on waste management. Ind Crop Prod 23:147–161

Dhillon GS, Oberoi HS, Kaur S, Bansal S, Brar SK (2011a) Value-addition of agricultural wastes for augmented cellulase and xylanase production through solid state tray fermentation employing mixed-culture of fungi. Ind Crop Prod 34:1160–1167

Dhillon GS, Brar SK, Varma M, Tyagi RD (2011b) Apple pomace ultrafiltration sludge, a novel substrate for fungal bioproduction of citric acid: optimization studies. Food Chem 128:864–871

Dhillon GS, Brar SK, Kaur S, Sabrine M, M'hamdi N (2012a) Lactoserum as a moistening medium and crude inducer for fungal cellulases and hemicellulose induction through solid-state fermentation of apple pomace. Biomass Bioenerg 41:165–174

Dhillon GS, Kaur S, Sarma SJ, Brar SK, Surampalli RY (2012b) Recent development in applications of important biopolymer chitosan in biomedicine, pharmaceuticals and personal care products. Curr Tissue Eng 2:20–40

Dhillon GS, Kaur S, Kaur Brar S (2013) Perspective of apple processing wastes as low-cost substrates for bioproduction of high value products: A review. Renew Sust Energ Rev 27:789–805

Dimantov A, Kesselman E, Shimoni E (2004) Surface characterization and dissolution properties of high amylose corn starch–pectin coatings. Food Hydrocoll 18:29–37

Dini C, Islan GA, Castro GR (2014) Stability analysis of natural biopolymeric matrices for phage microencapsulation. Appl Biochem Biotechnol 174:2031–2047

Dong Z, Liu D, Keesing JK (2010) Jellyfish blooms in China: Dominant species, causes and consequences. Mar Pollut Bull 60:954–963

Dutta RK, Sahu S (2012) Development of oxaliplatin encapsulated in magnetic nanocarriers of pectin as a potential targeted drug delivery for cancer therapy. Results. Pharma Sci 2:38–45

Duttagupta DS, Jadhav VM, Kadam VJ (2015) Chitosan: a propitious biopolymer for drug delivery. Curr Drug Deliv 12:369–381

EC Packaging 2004: http://eur-lex.europa.eu/legal-content/EN/ALL/?uri=CELEX%3A32004L0012. Last access June 13, 2017

Edgar KJ (2006) Cellulose esters in drug delivery. Cellulose 14:49–64

Elieh-Ali-Komi D, Hamblin MR (2016) Chitin and chitosan: production and application of versatile biomedical nanomaterials. Int J Adv Res (Indore) 4:411–427

Elnashar MMM, Kahil T (2014) Biopolymeric formulations for biocatalysis and biomedical applications. Biomed Res Int 2014:1–10

Fadel Picheth G, Luiz Pirich C, Rita Sierakowski M et al (2017) Bacterial cellulose in biomedical applications: a review. Int J Biol Macromol 104:97–106

FAOSTAT (2013) Production. Food and Agriculture Organization of the United Nations (FAOSTAT) and United State Department of Agriculture (USDA) in the "Foreign Agricultural Service http://faostat.fao.org

Ferreira AM, Gentile P, Chiono V, Ciardelli G (2012) Collagen for bone tissue regeneration. Acta Biomater 8:3191–3200

Fidalgo A, Ciriminna R, Carnaroglio D, Tamburino A, Cravotto G, Grillo G, Ilharco LM, Pagliaro M (2016) Eco-friendly extraction of pectin and essential oils from orange and lemon peels. ACS Sustain Chem Eng 4:2243–2251

Fischer G, Schrattenholzer L (2001) Global bioenergy potentials through 2050. Biomass Bioenergy 20:151–159

Fisher A, Watling M, Smith A, Knight A (2010) Pharmacokinetics and relative bioavailability of fentanyl pectin nasal spray 100-800 μg in healthy volunteers. Int J Clin Pharmacol Ther 48:860–867

Fishman M, Coffin D, Konstance R, Onwulata C (2000) Extrusion of pectin/starch blends plasticized with glycerol. Carbohydr Polym 41:317–325

Fissore EN, Basanta M, Nieto Calvache JE (2014) Chemical composition and rheological behaviour of pectins from unconventional sources. In: Hauppauge (ed) Pectin: Chemical Properties, Uses and health benefits. Nova Science Publishers Inc., NY, pp 187–202

Food and Agriculture Organization (2011) http://faostat.fao.org. last access June 12, 2017

Food and Agriculture Organization (2017) FAOSTAT. http://www.fao.org/faostat/en/#data/QC. Last access Nov 8, 2017

Franca EF, Freitas LCG, Lins RD (2011) Chitosan molecular structure as a function of N-acetylation. Biopolymers 95:448–460

Fringant C, Desbrieres J, Rinaudo M (1996) Physical properties of acetylated starch-based materials: Relation with their molecular characteristics. Polymer (Guildf) 37:2663–2673

Gagneten AM, Paggi JC (2009) Effects of heavy metal contamination (Cr, Cu, Pb, Cd) and eutrophication on zooplankton in the lower basin of the Salado River (Argentina). Wat Air Soil Pollut 198:317–334

Gautam S, Sharon A, Sagar P (2010) Grafted polysaccharides: Smart materials of the future, their synthesis and applications. In: Kalia S, Avérous L (eds) Biopolymers: Biomedical and environmental applications. Wiley, Hoboken, NJ, pp 35–57

Gentil EC, Damgaard A, Hauschild M, Finnveden G, Eriksson O, Thorneloe S, Ozge Kaplan P, Barlaz M, Muller O, Matsui Y, Li R, Christensen TH (2010) Models for waste life cycle assessment: Review of technical assumptions. Waste Manage 12:2636–2648

George J, Sabapathi SN (2015) Cellulose nanocrystals: Synthesis, functional properties, and applications. Nanotechnol Sci Appl 8:45–54

Giunchedi P, Conte U, Chetoni P, Saettone M (1999) Pectin microspheres as ophthalmic carriers for piroxicam: evaluation in vitro and in vivo in albino rabbits. Eur J Pharm Sci 9:1–7

Glinsky VV, Raz A (2009) Modified citrus pectin anti-metastatic properties: one bullet, multiple targets. Pectin Struct Funct 344:1788–1791

Godfray HCJ, Beddington JR, Crute IR, Haddad L, Lawrence D, Muir JF, Pretty J, Robinson S, Thomas SM, Toulmin C (2010) Food security: The challenge of feeding 9 billion people. Science 327:812–818
Gomez-Guillen MC, Gimenez B, Lopez-Caballero ME, Montero MP (2011) Functional and bioactive properties of collagen and gelatin from alternative sources: A review. Food Hydrocoll 25:1813–1827
Grande CJ, Torres FG, Gomez CM et al (2009) Development of self-assembled bacterial cellulose-starch nanocomposites. Mater Sci Eng, C 29:1098–1104
Grumezescu AM (2016) Nanobiomaterials in soft tissue engineering: applications of nanobiomaterials. Elsevier, Amsterdam, p 528
Guo M, Stuckey DC, Murphy RJ (2013) End-of-life of starch-polyvinyl alcohol biopolymers. Biores Technol 127:256–266
Hench LL (1998) Biomaterials: a forecast for the future. Biomaterials 19:1419–1423
Hirano S, Tsuchida H, Nagao N (1989) N-acetylation in chitosan and the rate of its enzymatic hydrolysis. Biomaterials 10:574–576
Hongzhang C, Hongqiang L, Liying L (2011) The inhomogeneity of corn stover and its effects on bioconversion. Biomass Bioenerg 35:1940–1945
Hoover R (2010) The impact of heat-moisture treatment on molecular structures and properties of starches isolated from different botanical sources. Crit Rev Food Sci Nutr 50:835–847
Hoyer B, Bernhardt A, Heinemann S et al (2012) Biomimetically mineralized salmon collagen scaffolds for application in bone tissue engineering. Biomacromol 13:1059–1066
Hsein-Chih HW, Sarko A (1978) The double-helical molecular structure of crystalline b-amylose. Carbohydr Res 61:7–25
Hui YH (2006) Handbook of food science, technology, and engineering (Taylor & Francis)
INTA (2015) National Institute of Agricultural Technology (INTA). Pesticides added to the soil and their destiny in the environment 1st ed. Balcarce, Buenos Aires, p 73
Ishigaki T, Kawagoshi Y, Ike M, Fujita M (1999) Biodegradation of a polyvinyl alcohol-starch blend plastic film. World J Microbiol Biotechnol 15:321–327
Islam MA, Park T-E, Reesor E et al (2015) Mucoadhesive chitosan derivatives as novel drug carriers. Curr Pharm Des 21:4285–4309
Islan GA, Dini C, Bartel LC, Bolzán AD, Castro GR (2015) Characterization of smart auto-degradative hydrogel matrix containing alginate lyase to enhance levofloxacin delivery against bacterial biofilms. Int J Pharm 496:953–964
Janes KA, Alonso MJ (2003) Depolymerized chitosan nanoparticles for protein delivery: Preparation and characterization. J Appl Polym Sci 88:2769–2776
Jayasekara R, Harding I, Bowater I et al (2004) Preparation, surface modification and characterisation of solution cast starch PVA blended films. Polym Test 23:17–27
Jia H, Jia Y, Wang J, et al (2009) Potentiality of bacterial cellulose as the scaffold of tissue engineering of cornea. In: 2009 2nd Int Conf Biomed Eng Infor IEEE, pp 1–5
Jimtaisong A, Saewan N (2014) Utilization of carboxymethyl chitosan in cosmetics. Int J Cosmet Sci 36:12–21
Jozala AF, de Lencastre-Novaes LC, Lopes AM et al (2016) Bacterial nanocellulose production and application: a 10-year overview. Appl Microbiol Biotechnol 100:2063–2072
Kalapathy U, Proctor A (2001) Effect of acid extraction and alcohol precipitation conditions on the yield and purity of soy hull pectin. Food Chem 73:393–396
Kaseem M, Hamad K, Deri F (2012) Thermoplastic starch blends: a review of recent works. Polym Sci Ser A 54:165–176
Kenawy E-R, Kamoun EA, Mohy Eldin MS, El-Meligy MA (2014) Physically crosslinked poly (vinyl alcohol)-hydroxyethyl starch blend hydrogel membranes: synthesis and characterization for biomedical applications. Arab J Chem 7:372–380
Kittiphattanabawon P, Nalinanon S, Benjakul S, Kishimura H (2015) Characteristics of pepsin-solubilised collagen from the skin of splendid squid (*Loligo formosana*). J Chem 2015: Article ID 482354, 8 pages

Klingbeil M (2000) Working document of biodegradable waste management. European Commission, Brussels
Koutinas AA, Vlysidis A, Pleissner D, Kopsahelis N, Lopez Garcia I, Kookos K, Papanikolaou S, Kwanb TH, Ki Lin CS (2014) Valorization of industrial waste and by-product streams via fermentation for the production of chemicals and biopolymers. Chem Soc Rev 43:2587–2627
Krishnan S, Sekar S, Katheem MF et al (2012) Fish scale collagen-a novel material for corneal tissue engineering. Artif Organs 36:829–835
Kumari A, Kaith BS, Singha AS, Kalia S (2010) Polysaccharide graft copolymers—synthesis, properties and applications. In: Kalia S, Avérous L (eds) Biopolymers: Biomedical and Environmental Applications. Wiley, Hoboken, NJ, pp 35–57
Kunal P, Banthia AK, Majumdar DK (2006) Starch based hydrogel with potential biomedical application as artificial skin. Afr J Biomed Res 9:23–29
Lal R, Follett RF, Stewart BA, Kimble JM (2007) Soil carbon sequestration to mitigate climate change and advance food security. Soil Sci 172:943–956 https://doi.org/10.1097/ss.0b013e31815cc498
Lavoine N, Desloges I, Dufresne A, Bras J (2012) Microfibrillated cellulose—its barrier properties and applications in cellulosic materials: A review. Carbohydr Polym 90:735–764
Lazovic G, Colic M, Grubor M, Jovanovic M (2005) The application of collagen sheet in open wound healing. Ann Burns Fire Disasters 18:151–156
Lebo Jr. SE, Gargulak JD, McNally TJ (2001) Lignin. In: Kirk-Othmer (ed) Encyclopedia of Chemical Technology. Wiley, Malden, MA. vol 15, 1–32
Leitinger B, Hohenester E (2007) Mammalian collagen receptors. Matrix Biol 26:146–155
Lemoigne M (1926) Produits de dehydration et de polymerisation de l'acide ß-oxobutyrique. Bull Soc Chim Biol 8:82
Liu L, Fishman ML, Kost J, Hicks KB (2003) Pectin-based systems for colon-specific drug delivery via oral route. Biomaterials 24:3333–3343
Liu D, Nikoo M, Boran G et al (2015) Collagen and gelatin. Annu Rev Food Sci Technol 6:527–557
Löckes J (1998) Properties and applications of compostable starch-based plastic materials. Polym Degrad Stab 59:245–249
Mahboob S (2014) Isolation and characterization of collagen from fish waste material- skin, scales and fins of *Catla catla* and *Cirrhinus mrigala*. J Food Sci Technol 52:4296–4305
Martins A, Facchi S, Follmann H et al (2014) Antimicrobial activity of chitosan derivatives containing N-quaternized moieties in its backbone: a review. Int J Mol Sci 15:20800–20832
May CD (1990) Industrial pectins: sources, production and applications. Carbohydr Polym 12:79–99
McKinnon BT, Avis KE (1993) Membrane filtration of pharmaceutical solutions. Am J Hosp Pharm 50:1921–1936
McNamara JT, Morgan JLW, Zimmer J (2015) A molecular description of cellulose biosynthesis. Annu Rev Biochem 84:895–921
Medovent RM (2016) Chitosan based suture focusing on the real advantages of an outstanding. Adv Chitin Sci 14:211–216
Menéndez JE, Hilbert JA (2013) Cuantificación y uso de Biomasa de residuos de cultivos en Argentina para bioenergía. In: Hilbert JA (ed). INTA. Buenos Aires p 48
Millennium Ecosystem Assessment (2005) http://www.millenniumassessment.org/es/ last access: June 12, 2017
Miller A (1984) Collagen: the organic matrix of bone. Philos Trans R Soc Lond B Biol Sci 304:455–477
Mondal S (2017) Preparation, properties and applications of nanocellulosic materials. Carbohydr Polym 163:301–316
Montico S (2010) Valoración económica de la utilización de los residuos de cosecha como fuente bioenergética en el sur de Santa Fe, Argentina. Jornada de Biocombustibles. Aportes para el análisis de la sustentabilidad de los biocombustibles
Moon RJ, Martini A, Nairn J et al (2011) Cellulose nanomaterials review: structure, properties and nanocomposites. Chem Soc Rev 40:3941–3994

Moreno J, Moral R, García Morales JL, Pascual JA, Bernal MB (2014) De residuo a recurso, el camino de la sostenibilidad. Red española de compostaje. Ed. Mundi Prensa
Morganti P, Cinelli P, Lazzer A (2016) La bioplastica dagli scarti di lavorazione dell'olio di oliva. Ultim Front Nat 67–71
Morimura S, Nagata H, Uemura Y et al (2002) Development of an effective process for utilization of collagen from livestock and fish waste. Process Biochem 37:1403–1412
Munarin F, Tanzi MC, Petrini P (2012) Advances in biomedical applications of pectin gels. Int J Biol Macromol 51:681–689
Muzzarelli RAA (1997) Chitin. Pergamom, Oxford
Naito H, Yoshimura M, Mizuno T et al (2013) The advantages of three-dimensional culture in a collagen hydrogel for stem cell differentiation. J Biomed Mater Res, Part A 101:2838–2845
Naranjo JM, Cardona CA, Higuita JC (2014) Use of residual banana for polyhydroxybutyrate (PHB) production: case of study in an integrated biorefinery. Waste Manag 34:2634–2640. https://doi.org/10.1016/j.wasman.2014.09.007
Nayak P, Swain S (2002) Plastics and pollution: biodegradable polymers. Pop Plast Pkgg 47:66–78
Nepote V, Grosso NR, Guzman CA (2002) Extraction of antioxidant components from peanut skins. Grasas Aceites 53:391–395
Nesic AR, Trifunovic SS, Grujic AS, Velickovic SJ, Antonovic DG (2011) Complexation of amidated pectin with poly(itaconic acid) as a polycarboxylic polymer model compound. Carbohydr Res 346:2463–2468
Nilani P, Raveesha P, Kasthuribai NRN, Duraisamy B et al (2010) Formulation and evaluation of polysaccharide based biopolymer: An ecofriendly alternative for synthetic polymer. J Pharm Sci Res 2:178–184
Niranjan Raj S, Lavanya SN, Sudisha J, Shekar Shetty H (2010) Applications of biopolymers in agriculture with special reference to role of plant derived biopolymers in crop protection. In: Kalia S, Avérous L (eds) Biopolymers: Biomedical and Environmental Applications. Wiley, Hoboken, NJ, pp 461–481
No HK, Meyers SP (2000) Application of chitosan for treatment of waste waters. Rev Environ Contam Toxicol 163:1–28
Nobes GAR, Maysinger D, Marchessault RH (1998) Polyhydroxyalkanoates: materials for delivery systems. Drug Deliv 5:167–177. https://doi.org/10.3109/10717549809052032
Olano-Martin E, Rimbach GH, Gibson GR, Rastall RA (2003) Pectin and pectic-oligosaccharides induce apoptosis in vitro human colonic adenocarcinoma cells. Anticancer Res 23:341–346
Oliveira FC, Freire DMG, Castilho LR (2004) Production of poly(3-hydroxybutyrate) by solid-state fermentation with Ralstonia eutropha. Biotechnol Lett 26:1851–1855. https://doi.org/10.1007/s10529-004-5315-0
Organisation Internationale de la Vigne et du Vin (OIV) (2007) World Vitivinicultural Statistics 2007, Structure of the World Vitinicultural Industry in 2007
Orts WJ, Nobes GAR, Kawada J, et al (2008) Poly(hydroxyalkanoates): biorefinery polymers with awhole range of applications. The work of Robert H. Marchessault. Can J Chem 86:626–640
Pagliaro M, Ciriminna R, Abrantes Fidalgo AM, Ilharco LM (2016) Pectin production and global market. Agro Food Ind Hi Tech 27
Pamfil D, Nistor MT, Vasile C (2015) Collagen-Based Materials for Pharmaceutical Applications. Handbook of Polymers for Pharmaceutical Technologies. Wiley, Hoboken, NJ, USA, pp 439–481
Parfitt J, Barthel M, Macnaughton S (2010) Food waste within food supply chains: quantification and potential for change to 2050. Phil Trans R Soc B 365:3065–3081
Paschalis EP, Verdelis K, Doty SB et al (2001) Spectroscopic Characterization of collagen cross-links in bone. J Bone Miner Res 16:1821–1828
Patrulea V, Ostafe V, Borchard G, Jordan O (2015) Chitosan as a starting material for wound healing applications. Eur J Pharm Biopharm 97:417–426
Pei L, Schmidt M, Wei W (2011) Conversion of biomass into bioplastics and their potential environmental impacts. In: Elnashar M (ed) Biotechnology of Biopolymers. InTech, pp 57–74
Peniston QP, Johnson E (1980) US patent 4,195–175

Peters V, Rehm BHA (2005) In vivo monitoring of PHA granule formation using GFP-labeled PHA synthases. FEMS Microbiol Lett 248:93–100. https://doi.org/10.1016/j.femsle.2005.05.027

Petersen K, Nielsen PV, Olsen MB (2001) Physical and mechanical properties of bio-based materials-starch polylactate and polyhydroxybutyrate. Starch 53:356–361

Piemonte V (2011) Bioplastic wastes: the best final disposition for energy saving. J Polym Environ 19:988–994

Pillai CKS, Paul W, Sharma CP (2009) Chitin and chitosan polymers: chemistry, solubility and fiber formation. Prog Polym Sci 34:641–678

Portal O, Clark WA, Levinson DJ (2009) Microbial cellulose wound dressing in the treatment of nonhealing lower extremity ulcers. Wounds a Compend. Clin Res Pract 21:1–3

Prachayawarakorn J, Sangnitidej P, Boonpasith P (2010) Properties of thermoplastic rice starch composites reinforced by cotton fiber or low-density polyethylene. Carbohydr Polym 81:425–433

Primo A, Atienzar P, Sanchez E, Delgado JM, García H (2012) From biomass wastes to large-area, high-quality, N-doped graphene: catalyst-free carbonization of chitosan coatings on arbitrary substrates. Chem Commun 48:9254–9256

Rajwade JM, Paknikar KM, Kumbhar JV (2015) Applications of bacterial cellulose and its composites in biomedicine. Appl Microbiol Biotechnol 99:2491–2511

Ramshaw JAM, Peng YY, Glattauer V, Werkmeister JA (2009) Collagens as biomaterials. J Mater Sci Mater Med 20:3–8

Rana M, Kumari A, Chauhan GS, Chauhan K (2014) Modified chitosan microspheres in non-aggregated amylase immobilization. Int J Biol Macromol 66:46–51

Rangel-Rodríguez AM, Conxita S, Susana V et al (2014) Immobilization of pectinesterase in genipin-crosslinked chitosan membrane for low methoxyl pectin production. Appl Biochem Biotechnol 174:2941–2950

Ravi Kumar MN (2000) A review of chitin and chitosan applications. React Funct Polym 46:1–27

Ricard-Blum S (2011) The collagen family. Cold Spring Harb Perspect Bio

Rinaudo M (2006) Chitin and chitosan: properties and applications. Prog Polym Sci 31:603–632

Rolin C (1993) Pectin. In: Whistler RL (ed) Industrial gums polysaccharides and their derivatives. Whistler Center for Carbohydrate Research Purdue University West Lafayette. Academic Press, Indiana, pp 258–288

Rouilly A, Rigal L (2002) Agro-materials: A bibliographic review, Polymer Reviews. J Macromol Sci Part C—Polymer Rev C42:441–479

Rudnik E (2008) Compostable polymer materials. Elsevier

Rusendi D, Sheppard JD (1995) Hydrolysis of potato processing waste for the production of Poly-fl-Hydroxybutyrate. Biores Technol 54:191–196

Ryu S, Karim MN (2011) A whole cell biocatalyst for cellulosic ethanol production from dilute acid-pretreated corn stover hydrolyzates. Appl Microbiol Biotechnol 91:529–542

Sahoo D, Nayak PL (2011) Chitosan: The most valuable derivative of chitin. In: Kalia S, Avérous L (eds) Biopolymers: Biomedical and Environmental Applications. Wiley, Hoboken, NJ, pp 129–166

Sanchez-Vazquez SA, Hailes HC, Evans JRG (2013) Hydrophobic polymers from food waste: Resources and synthesis. Polymer Rev 53:627–694

Sashiwa H, Aiba SI (2004) Chemically modified chitin and chitosan as biomaterials. Prog Polym Sci 29:887–908

Schmidt MM, Dornelles RCP, Mello RO et al (2016) Collagen extraction process. Int Food Res J 23:913–922

Shalini R, Gupta DK (2010) Utilization of pomace from apple processing industries: a review. J Food Sci Technol 47:365–371

Sheehan J, Aden A, Paustian K, Killian K, Brenner J, Walsh M, and Nelson R (2004) Energy and environmental aspects of using corn stover for fuel ethanol. J Ind Ecol 7(3–4). Massachusetts Institute of Technology and Yale University

Shen L, Haufe J, Patel MK (2009) Product overview and market projection of emerging bio-based plastics

Sherman VR, Yang W, Meyers MA (2015) The materials science of collagen. J Mech Behav Biomed Mater 52:22–50
Shogren R (2009) Starch-Poly(hydroxyalkanoate) composites and blends. In: Yu L (ed) Biodegradable polymer blends and composites from renewable resources. Wiley, Hoboken, NJ, USA, pp 209–226
Shokri J, Adibki K (2013) Application of Cellulose and Cellulose Derivatives in Pharmaceutical Industries. In: Cellulose—Medical, Pharmaceutical and Electronic Applications. InTech,
Silva TH, Moreira-Silva J, Marques ALP et al (2014) Marine origin collagens and its potential applications. Mar Drugs 12:5881–5901
Silva D, Pinto LFV, Bozukova D et al (2016) Chitosan/alginate based multilayers to control drug release from ophthalmic lens. Colloid Surface B 147:81–89
Silvipriya KS, Krishna Kumar K, Bhat AR et al (2015) Collagen: Animal sources and biomedical application. J Appl Pharm Sci 5:123–127
Singh B, Guldhe A, Rawat I, Bux F (2014) Towards a sustainable approach for development of biodiesel from plant and microalgae. Renew Sust Energ Rev 29:216–245
Song JH, Jeon CO, Choi MH, et al (2007) Polyhydroxyalkanoate (PHA) production using waste vegetable oil by Pseudomonas sp. strain DR2. J Microbiol Biotechnol 18:1408–1415
Sriamornsak P, Thirawong N, Weerapol Y, Nunthanid J, Sungthongjeen S (2007) Swelling and erosion of pectin matrix tablets and their impact on drug release behavior. Eur J Pharm Biopharm 67:211–219
Srivastava P, Malviya R (2011) Sources of pectin, extraction and its applications in pharmaceutical industry—An overview. Indian J Nat Prod Resour 2:10–18
Stephen AM (1995) Food polysaccharides and their applications. Marcel Dekker, Inc
Subhan F, Ikram M, Shehzad A, Ghafoor A (2015) Marine Collagen: An Emerging Player in Biomedical applications. J Food Sci Technol 52:4703–4707
Thibault J-F, Ralet M-C (2003) Physico-chemical properties of pectins in the cell walls and after extraction. In: Advances in pectin and pectinase research. Springer Netherlands, Dordrecht, pp 91–105
Thirunavukkarasu N, Dhinamala K, Inbaraj RM (2011) Production of chitin from two marine *Stomatopods Oratosquilla* spp. (*Crustacea*). J Chem Pharm Res 3:353–359
Thompson RC, Moore CJ, vom Saal FS, Swan SH (2009), Plastics, the environment and human health: current consensus and future trends. Phil Trans R Soc B 364:2153–2166
Trabelsi I, Ayadi D, Bejar W et al (2014) Effects of *Lactobacillus plantarum* immobilization in alginate coated with chitosan and gelatin on antibacterial activity. Int J Biol Macromol 64:84–89
Trevors JT, Saier MH Jr (2010) Three Laws of Biology. Wat Air Soil Pollut 205(Suppl 1):S87–S89
Trevors JT, Saier Jr MH (2010a) Manage Humans, not the Environment. Wat Air Soil Pollut (2010a) 205(Suppl 1):S93–S95
United Nations. Management of waste from animal product processing. http://www.fao.org/WAIRDOCS/LEAD/X6114E/x6114e00.htm#Contents
United States Department of Agriculture (2012) Office of Global Analysis. Livestock and Poultry: World Markets and Trade. Foreign Agricultural Service, pp 1–25
Upadhyaya L, Singh J, Agarwal V, Tewari RP (2013) Biomedical applications of carboxymethyl chitosans. Carbohydr Polym 91:452–466
Usman A, Zia KM, Zuber M et al (2016) Chitin and chitosan based polyurethanes: A review of recent advances and prospective biomedical applications. Int J Biol Macromol 86:630–645
van Heerden I, Cronjé C, Swart SH, Kotzé JM (2002) Microbial, chemical and physical aspects of citrus waste composting. Bioresour Technol 81:71–76
Verheijen LAHM, Wiersema D, Pol LWH (2013) Food and Agriculture Organization
Volova T, Shishatskaya E, Sevastianov V, et al (2003) Results of biomedical investigations of PHB and PHB/PHV fibers. Biochem Eng J 16:125–133. https://doi.org/10.1016/S1369-703X(03)00038-X
von Bomhard A, Veit J, Bermueller C et al (2013) Prefabrication of 3D cartilage contructs: towards a tissue engineered auricle? a model tested in rabbits. PLoS ONE 8:e71667
Vroman I, Tighzert L (2009) Biodegradable polymers. Materials 2:307–344

Wan YZ, Luo H, He F et al (2009) Mechanical, moisture absorption, and biodegradation behaviours of bacterial cellulose fibre-reinforced starch biocomposites. Compos Sci Technol 69:1212–1217

Wang Y, Bian Y-Z, Wu Q, Chen G-Q (2008) Evaluation of three-dimensional scaffolds prepared from poly(3-hydroxybutyrate-co-3-hydroxyhexanoate) for growth of allogeneic chondrocytes for cartilage repair in rabbits. Biomaterials 29:2858–2868. https://doi.org/10.1016/j.biomaterials.2008.03.021

Wang Z, Wang L, Lin S et al (2014) Isolation and characterization of collagen from the muscle of Amur sturgeon (*Acipenser schrenckii*). Biotechnol Bioprocess Eng 19:935–941

Wang S, Li C, Copeland L et al (2015) Starch Retrogradation: a comprehensive review. Compr Rev Food Sci Food Saf 14:568–585

Wu LQ, Embree HD, Balgley BM, Smith PJ, Payne GF (2002) Utilizing renewable resources to create functional polymers: chitosan-based associative thickener. Environ Sci Technol 36:3446–3454

Yan S, Tyagi RD, Surampalli RY (2006) Polyhydroxyalkanoates (PHA) production using wastewater as carbon source and activated sludge as microorganisms. Water Sci Technol 53:175–180

Yamada S, Yamamoto K, Ikeda T, et al (2014) Potency of fish collagen as a scaffold for regenerative medicine. Biomed Res Int. 2014: Article ID 302932, 8 p

Yang H, Shu Z (2014) The extraction of collagen protein from pigskin. J Chem Pharm Res 6:683–687

Yao Y-C, Zhan X-Y, Zhang J, et al (2008) A specific drug targeting system based on polyhydroxyalkanoate granule binding protein PhaP fused with targeted cell ligands. Biomaterials 29:4823–4830. https://doi.org/10.1016/j.biomaterials.2008.09.008

Yordanov G, Evangelatov A, Skrobanska R (2013) Epirubicin loaded to pre-polymerized poly (butyl cyanoacrylate) nanoparticles: preparation and in vitro evaluation in human lung adenocarcinoma cells. Colloid Surface B 107:115–123

Younes I, Rinaudo M (2015) Chitin and chitosan preparation from marine sources. Structure, properties and applications. Mar Drugs 13:1133–1174

Zambare VP, Bhalla A, Muthukumarappan K, Sani RK, Christopher LP (2011) Bioprocessing of agricultural residues to ethanol utilizing a cellulolytic extremophile. Extremophiles 15:611–618

Zhao K, Deng Y, Chun Chen J, Chen G-Q (2003) Polyhydroxyalkanoate (PHA) scaffolds with good mechanical properties and biocompatibility. Biomaterials 24:1041–1045. https://doi.org/10.1016/S0142-9612(02)00426-X

Zheng F, Lin Y, Verbeken E et al (2004) Host response after reconstruction of abdominal wall defects with porcine dermal collagen in a rat model. Am J Obstet Gynecol 191:1961–1970

Chapter 2
Biosurfactants from Processed Wastes

Seba George and K. Jayachandran

Abstract Better waste management and economic biosurfactant production have been achieved when scientific studies were piloted concentrating on the three 'R' principles: Reduce, Reuse, and Recycle. Waste products from various industries and processes when selected as low-cost substrates having proper nutrient balance for biosurfactant production, biotechnological research contributed a substantial share to its environmental preservation strategies. Utilization of a variety of natural waste materials as alternative cost-effective carbon sources for the economic production of biosurfactants generates a high-value biotechnological product with the potential industrial application and, moreover, a process that can contribute to decreasing the disposal of wastes into the environment. Even though it is a fact that these biosurfactants derived from renewable raw materials are coming progressively on to the market, their growth and development need extensive cooperation across disciplines in order to fully characterize them and identify their potential uses in various sectors and industries.

Keywords Biosurfactants · Rhamnolipids · Agrowastes · Bioprocess

2.1 Introduction

Even as nature is ultimately the provider of all his resources, modern man tends to treat it as the ultimate bin for all his waste. In this scenario as well, science does its magic by taking advantage of both these facts and shows that waste can be applied as a resource, through the application of biotechnology for the production of one of the most versatile process chemicals used in industry: biosurfactants, an environmental-friendly green chemical. They are microbiological compounds that exhibit high surface activity, emulsifying activity, low toxicity, biodegradability,

S. George · K. Jayachandran (✉)
School of Biosciences, Mahatma Gandhi University,
P.D. Hills, Athirampuzha, Kottayam, Kerala 686560, India
e-mail: jayansbs@gmail.com

R. R. Singhania et al. (eds.), *Waste to Wealth*, Energy, Environment, and Sustainability, https://doi.org/10.1007/978-981-10-7431-8_2

ecological acceptance and stability at extremes of temperature, pH, and salinity. To spread the application of microbial surfactants by making them competitive with their chemical counterparts, many efforts have been done which mainly concentrated on the use of alternative low-cost substrates. This research is mainly centered on tropical agro-industrial crops and residues and the waste generated by the oil and fat industries. Processed agro-industrial wastes reported to be found as substrates for biosurfactants production are: starch-rich waste from potato processing industry and potato peels, cassava flour wastewater, wheat bran, grape marc, distillery and whey waste, orange peelings, carrot peel waste, lime peelings, coconut oil cake, and banana waste, paneer/cottage cheese whey waste, and rice straw. Oil-derived waste materials reported as similar substrates are olive oil mill effluent, soybean oil refinery wastes, used olive oil, waste lubricating oil, vegetable oil refinery waste of canola, soybean and corn oil, waste frying coconut oil, and groundnut oil refinery residue. Industrial waste has attracted considerable interest from researchers as low-cost substrates for biosurfactant production, as the substrate—accounts for up to 50% of the final production cost. Accounts of the studies of the waste material usage for biosurfactant production indicate that the two categories (i) the agro-industrial wastes with high contents of carbohydrates or lipids and (ii) the hydrophobic wastes are being extensively concentrated for this purpose.

2.1.1 Biosurfactant Production on Agro-Industrial Wastes

Potential substrates for biosurfactant production are made available in the potato processing industry. A study by Fox and Bala (2000) highlighted the potential environment threat and economic liability of starch-rich wastes from potato processing industries. An established potato medium, simulated liquid and solid potato waste media, and a commercially prepared potato starch in mineral salts medium were evaluated in shake flask experiments to establish growth, surface tension and carbohydrate reduction capacities. A drop in surface tension to 28.3 from 71.3 mN/m in simulated solid potato medium from a methylene chloride extract of the biosurfactant was observed. Thompson et al. (2000) used high solids (HS) and low solids (LS) potato effluents as substrates for surfactin production. They used effluents diluted 1:10, unamended and amended with trace minerals or corn steep liquor. *Bacillus subtilis* 21332 grew on all three potato substrates regardless of the addition of exogenous nutrients. Das and Mukherjee (2007) produced lipopeptide biosurfactants using two thermophilic strains of *B. subtilis* grown on a substrate derived from waste potato peels.

Another available agro-industrial waste used for biosurfactant production was molasses. It is one by-product of the sugar industry, rich in various nutrients and a major raw material for the production of baker's yeast, citric acid, feed yeasts, acetone, organic acids, and amino acids. Makkar and Cameotra (1997) used minimal medium supplemented with molasses as a carbon source for biosurfactant production by two strains of *B. subtilis* and observed maximum biosurfactant production in late stationary phase and exhibited good emulsification index. Production of rhamnolipid was studied by growing *Pseudomonas aeruginosa* EBN-8 mutant on varying concentrations of clarified blackstrap molasses as a sole carbon and energy source and biosurfactant yield were observed as 1.45 g/l (Raza et al. 2007a). Wide spectrum of comparative studies on different sources of molasses like sugarcane, beet and soy molasses using for a range of biosurfactants was reviewed by Banat et al. (2014).

Nitschke et al. (2004) have evaluated and compared cassava flour wastewater as a potential alternative culture medium for biosurfactant production and Veenanadig et al. (2000), reported the surfactant production by *Bacillus subtilis* cultivated on wheat bran. 46 g/l rhamnolipid yield was reported by Neto et al. (2008), by *P. aeruginosa* UFPEDA 614 in a solid-state culture using a 50:50 mixture of sugarcane bagasse and sunflower seed meal, supplemented with impregnating solution containing 10% (v/v) glycerol as substrate. They reported the production of rhamnolipid in solid-state culture (SSC) for the first time in the journal literature. In another study, *B. subtilis* LAMI008 strain isolated from the tank of chlorination at the wastewater treatment plant in Brazil was screened for surfactin production in mineral medium containing clarified cashew apple juice (MM-CAJC). Surfactin concentration of 3.5 mg/l was obtained when (MM-CAJC), supplemented with yeast extract was used, indicating that it is feasible to produce surfactin from clarified cashew apple juice. Same authors reported the use of natural cashew apple juice, a by-product of the cashew nut industry, as fermentation medium for biosurfactant production by *Acinetobacter-calcoaceticus* (Rocha et al. 2006). In our work, we carried out the fermentative production of rhamnolipid biosurfactant from *P. aeruginosa* MTCC 2297 using various cost-effective waste materials such as orange peelings, carrot peel waste, lime peelings, coconut oil cake, and banana waste. The orange peel was found to be the best substrate and was utilized by *P. aeruginosa* MTCC 2297 as a very effective carbon source generating 9.18 g/l of rhamnolipid biosurfactant (George and Jayachandran 2009).Carrot peel waste, lime peelings, coconut oil cake, and banana waste were also tested as low-cost carbon sources (George and Jayachandran 2009; Rane et al. 2017). Products such as straw of wheat, straw of rice-(Zhu et al. 2013), hull of soy, -corn, -rice, -rice water (by-product-from-domestic-cooking-and rice-processing-industry) are representative candidates of agro industrial wastes (Banat et al. 2014).

Industrial wastes such as distillery and whey wastes were—utilized successful for batch kinetic studies on rhamnolipid biosurfactant production by Sudhakar-Babu et al. (1996) and Dubey and Juwarkar (2001). Another good substrate for biosurfactant production is lactic whey which is composed of high levels of lactose (75% of dry matter), 12–14% protein, organic acids, and vitamins. Koch et al. (1988) used the lactose-utilizing capability of *Escherichia coli* by cloning the lactose gene Lac ZY from *E. coli* in *P. aeruginosa*, which then grew on whey for biosurfactant production. Daniel et al. (1998) achieved production of high concentrations of sophorolipids using a two-stage cultivation process: first, deproteinized whey concentrate (DWC) was used for the cultivation of the yeast *Cryptococcus curvatus* ATCC 20509; cells were then disrupted, autoclaved and the resulting crude cell extract containing the single cell oil served as a substrate for growth of *Candida bombicola* ATCC 22214 and for sophorolipids production in a second stage. A recent report shows that 4.8 g/l rhamnolipid biosurfactant, which was detected to be nontoxic against mouse fibroblastic cell line L292, was produced by *P. aeruginosa* SR17 using paneer (cottage cheese) whey waste as substrate (Patowary et al. 2016).

Grape marc, a useless agricultural residue from wineries, is produced after pressing the crushing grapes in white wine-making technology or after fermentation and maceration on red wine-making technology. Grape marc is usually distilled in wineries to recover ethanol, further used to produce spirituous liquors, giving huge amounts of distilled grape marc without an efficient use after the wine-making process. This distilled grape marc was used to obtain sugar solutions by diluted acid hydrolysis, which after nutrient supplementation was used to efficiently convert monomeric hemicellulosic sugars into lactic acids and biosurfactants by *Lactobacillus pentosus* (Rivera et al. 2007). In this study, they observed the production of 4.8 mg/L of intracellular biosurfactant, measured as biosurfactin, by *L. pentosus*.

2.1.2 Biosurfactant Production on Oils and Oil-Containing Wastes

Vollbrecht et al. (1999) investigated the production of biosurfactant using domestic vegetable oils in order to convert renewable resources into higher value products. In their study with *Tsukamurella* species DSM 44370, they obtained the best growth and glycolipid production using natural vegetable oils rather than complex media and hydrophobic carbon sources. Of the vegetable oils tested, oleic acid-rich oils and rapeseed oil gave the best results. On sunflower oil, a yield of around 5 g/l glycolipid was obtained. Sarubbo et al. (1997) evaluated the production of bioemulsifiers by two strains of *Candida lipolytica* using media supplemented with 5% BabaCu oil and 1% glucose as carbon source. They observed the production of bioemulsifiers as secondary metabolites at the end of the exponential growth phase and beginning of the stationary growth phase. Mercade et al. (1993) reported the use of olive oil mill effluent (OOME) as a new substrate for rhamnolipid production by *Pseudomonas* sp. JAMM. OOME is black liquor containing the water-soluble fraction of ripe olives and water that is used in the process of olive oil extraction, and it is a major pollutant of the agricultural industry in Mediterranean countries. Of the 22 strains tested, only 15 showed growth. In another study using *Pseudomonas* strain 42A2 and a subproduct from the distillation of non-specific mixtures of vegetable oils, the same authors observed the production of a new biosurfactant 7, 10—dihydroxy—8E—octadecanoic acid. Abalos et al. (2001) have reported the use of soybean oil refinery wastes for the production of new rhamnolipid by *P. aeruginosa* AT10. They reported a final production level of 9.5 g/l of rhamnolipids, which was reached in two main stages. Rahman et al. (2002) reported rhamnolipid biosurfactant production by strains of *P. aeruginosa* using low-cost raw materials. An oil-degrading strain, *P. aeruginosa* DS10-129, produced maxima of 4.31, 2.98, and 1.77 g/l rhamnolipid using soybean oil, safflower oil, and glycerol, respectively. Marsudi et al. (2008) reported the use of palm oil as the sole carbon source for the simultaneous production of PHAs and rhamnolipids using *P. aeruginosa* IFO3924. The highest yields of PHA were 0.086 g/g and rhamnolipid was 0.063 g/g at 6.9 g/l palm oil. A marine bacterium, *P. aeruginosa* BYK-2 (KCTC 18012P), was immobilized by entrapment in 10% polyvinyl alcohol beads and optimized for the continuous production of rhamnolipid. A modified basal salt medium containing 1% fish oil (Pollack–liver) was utilized as a carbon source in a bioreactor, and about 6 g/l rhamnolipid was obtained after 8 days (Jeong et al. 2004). Meat processing industries such as food and leather produce significant quantities of animal fat. Demand for animal fats is considerably less than vegetable oils, and hence the huge unspent amount of animal fat becomes a problem-for-utilization-as-well-as-for-their disposal. Using animal fat and corn steep liquor, Santos et al. (2013) achieved maximum glycolipid production by the yeast *C. lipolytica* UCP 0988. The authors also report that the product has uses in bioremediation as well as oil mobilization and recovery. *P. aeruginosa* J4 isolated from wastewater of a petrochemical factory was used to produce rhamnolipid from a variety of carbon substrates, including hydrophilic substrates, vegetable oils, and mineral oils. The *P. aeruginosa* J4 strain was able to assimilate in the seven carbon substrates examined (namely glucose, glycerol, olive oil, sunflower oil, grape seed oil, diesel, and kerosene), but it grew less efficiently in mineral oils (especially kerosene). Among the seven carbon substrates examined, olive oil was the most efficient one for rhamnolipid production at a concentration of 10% (Wei et al. 2005). Rapeseed oil (Trummler et al. 2003), palm oil (Pornsunthorntawee et al. 2008), and soybean oil (Abdel-Mawgoud et al. 2009) were utilized as substrates for the production of rhamnolipids by different strains of *P. aeruginosa*. The yeast strain CLOA 72 isolated from the effluent of an aeration tank at an activated sludge wastewater treatment plant of a dairy industry, in Brazil and identified as *Trichosporon*-montevideense, was able to grow and produce a glycolipid biosurfactant when cultured on a mineral medium (MM) with sunflower oil as the carbon source (Monteiro et al. 2009). Sobrinho et al. (2008) described a low-cost medium for the production of a surfactant by the yeast *C. sphaerica*. The medium was formulated only with distilled water supplemented with 5.0% groundnut oil refinery residue plus 2.5% corn steep liquor as substrates. The isolated biosurfactant was formed with a yield of 4.5 gl.

Frying oil is produced in large quantities for use both in the food industry and at the domestic scale. After being used, cooking oil changes its composition and contains more than 30% of polar compounds (Kock et al. 1996) depending on the variety of food, the type of frying and the number of times it has been used. Haba et al. (2000) compared the composition of used olive and sunflower oils with the standard unused oils and screened 36 microorganisms for the production of biosurfactants in submerged culture with waste olive or sunflower oil as carbon sources. Same authors reported the simultaneous production of rhamnolipids and polyhydroxy alkanoates when grown with residual or low-value oils as carbon substrate (Haba et al. 2007). In another study, Raza et al. (2007a, b) produced rhamnolipid by using distant carbon sources and four varieties of substrates, viz hydrocarbons (paraffin, kerosene, n-hexadecane); waste frying oils (canola, soybean, corn), vegetable oil refinery wastes (canola, soybean, corn), and molasses under single- and fed-batch cultivation setups. Hydrocarbons and waste frying oils gave good rhamnolipid production, whereas vegetable oil refinery wastes and molasses promoted poor yields. Soybean and corn waste frying oils were observed to be preferred carbon sources followed by kerosene and paraffin oils. The waste frying oils gave the highest production of rhamnolipids ranging from 2.1 to 4.1 g/l and the best results were obtained with soybean waste frying oil as carbon source producing 4.1 g/l rhamnolipids at seven days of incubation. The possibility of waste frying coconut oil to be used as a very effective alternate substrate for the economic production of rhamnolipid by a newly isolated *P. aeruginosa* D was reported by George and Jayachandran (2012). Used olive oil was a better substrate for cell growth as well as for rhamnolipid production. Growth and biosurfactant production by various *Bacillus* strains were not as good as with *Pseudomonas* strain. A new lipopeptide biosurfactant produced by *Klebsiella* sp. Y6-1, isolated from waste soybean oil, utilized crude oil as the sole carbon source, and inoculated to the medium with waste soybean oil suspension. The final concentration of crude biosurfactant produced was 100 mg/m (Lee et al. 2008). Spent oils are usually abundantly available oils that are quite difficult to dispose of due to environmental concerns including persistence and resistance to biodegradation. They include waste vegetable oil, used motor oil, lubricating oils, jet fuels all of which can act as a cheaper source for microbial processes such as biosurfactant production. Usage of such kind substrates is usually encouraged as a pollution control strategy (Banat et al. 2014). Recently, Saravanan and Subramaniyan (2014) isolated P. *aeruginosa* PB3A strain from oil-contaminated soil and examined biosurfactant production on various substrates namely, castor oil, coconut oil, rapeseed oil, corn oil, motor oil, sunflower oil, olive oil, olein, barley bran, rice bran peanut cake, potato waste, and wheat bran instead of routine carbon sources. Corn oil and cassava waste flour were found to be highly effective.

Waste or used lubricating oils have become a serious environmental problem. Mercade et al. (1996) reported the screening and selection of microorganisms capable of utilizing waste lube oil for producing biosurfactants. Only 10% of the strains isolated, produced biosurfactants. Further characterization of these strains showed production of Trehalose glycolipids from *Rhodococcus* species and lipopeptide from *Bacillus* species. Antarctic marine bacterial isolates of genus *Rhodococcus* isolated from the Ross Sea were able to utilize diesel fuel as the sole carbon and energy source for the biosurfactant production (Pini et al. 2007). Qiao and Shao (2010) reported the isolation and characterization of a novel linear lipoamino biosurfactant produced by a marine oil-degrading bacterium *Alcanivoraxdieselolei* B-5 growing with diesel oil as the sole carbon and energy source. Thermophilic bacterial culture isolated from a hot spring environment (sulphataric hot springs in Viterbo, Italy) was identified as *P. aeruginosa* AP02-1 and was tested for the ability to utilize a range of hydrocarbons both n-alkanes and polycyclic aromatic hydrocarbons as sole carbon source for the production of rhamnolipid. Strain AP02-1 had an optimum growth temperature of 45 °C and degraded 99% of crude oil 1% (v/v) and diesel oil 2% (v/v) when added to a basal mineral medium within seven days of incubation (Perfumo et al. 2006). Bento and Gaylarde (1996) observed the production of surfactants by *Pseudomonas* in the presence of mineral salts and glucose medium with an increase in emulsifying activities of the surfactant by the addition of sterile diesel oil to the medium. In a similar study, Muriel et al. (1996) observed the production of extracellular biosurfactants by *Cladosporium-resinae* when growing on jet fuel JP8. Soap stock is the residue from oil refinery that is generated in large quantities by the vegetable oil processing industry. Sunflower oil soap stock was assayed by Benincasa et al. (2002) as the carbon source for rhamnolipid production by *P. aeruginosa* LBI strain giving a final surfactant concentration of 12 g/l in a shaker and 16 g/l in bioreactor experiments. Thus, producing biosurfactant from vegetable oils, used vegetable oil, and used motor oil is a sound strategy of waste management for the food and auto industries to reduce the generation of waste. Although these biosurfactants derived from renewable raw materials are coming progressively on to the market, their development needs extensive cooperation across disciplines in order to fully characterize them and identify their potential uses.

The cost-effective substrates, a microbial source, biosurfactant type with their yield are listed in Table 2.1.

Table 2.1 Cost-effective substrates, microbial sources, and biosurfactant types with their yield

Low-cost substrates	Microorganism	Biosurfactant	Yield (g/l)	Reference
Potato effluents	*Bacillus subtilis*	Surfactin	*nd*[a]	Thompson et al. (2000), Makkar and Cameotra (1997)
Waste potato peels	*B. subtilis*	Lipopeptide	*nd*	Das and Mukherjee (2007)
Wheat bran	*B. subtilis*	*Nd*	*nd*	Veenanadig et al. (2000)
Rice straw	*B. amyloliquefaciens XZ-173*	Surfactin	15.03 mg/gds	Zhu et al. (2013)
Molasses	*Pseudomonas putida* 300-B	Rhamnolipid	0.7–1.2	Raza et al. (2007a)
Molasses and Corn steep liquor	*P. aeruginosa* GS3	Rhamnolipid	0.24	
Molasses	*B. subtilis*	Surfactin	*nd*	Makkar and Cameotra (1997)
Sugar cane, Beet and Soy molasses	*B. subtilis*	Surfactin	*nd*	Banat et al. (2014)
Clarified Blackstrap Molasses	*P. aeruginosa* EBN-8	Rhamnolipid	1.45	Raza et al. (2007b)
Sugarcane bagasse + Sunflower seed meal + Glycerol	*P. aeruginosa* UFPEDA 614	Rhamnolipid	46	Neto et al. (2008)
Orange peelings	*P. aeruginosa* MTCC 2297	Rhamnolipid	9.18	George and Jayachandran (2009)
Clarified cashew apple juice and yeast extract	*B. subtilis* LAMI008	Surfactin	0.0035	
Distilled grape marc	*Lactobacillus pentosus*	Biosurlactin	0.048	Rivera et al. (2007)
Natural cashew apple juice	*Acinetobacter calcoaceticus*	*Nd*	*nd*	Rocha et al. (2006)

(continued)

Table 2.1 (continued)

Low-cost substrates	Microorganism	Biosurfactant	Yield (g/l)	Reference
Distillery and whey wastes	*P. aeruginosa* BS2	Rhamnolipid	0.92	Dubey and Juwarkar (2001)
Lactic whey	*Cryptococcus curvatus, Candida bombicola*	Sophorolipid	*nd*	Daniel et al. (1998)
Paneer (cottage cheese) whey waste	*P. aeruginosa SR17*	Rhamnolipid	4.8	Patowary et al. (2016)
Carrot peel waste, lime peelings, coconut oil cake, banana waste	*P. aeruginosa*	Rhamnolipid	2.1–5.7	George and Jayachandran (2009), Rane et al. (2017)
Olive oil mill effluent (OOME)	*P. aeruginosa* JAMM	Rhamnolipid	6.4	Mercade et al. (1993)
Groundnut oil refinery residue + corn steep liquor	*C. sphaerica*	Surfactant	4.5	Sobrinho et al. (2008)
Sunflower oil	*Tsukamurella* sp.	Glycolipid	5	Vollbrecht et al. (1999)
Soybean oil refinery waste	*P. aeruginosa* AT10	Rhamnolipid	9.5	Abalos et al. (2001)
Waste frying coconut oil	*P. aeruginosa* D	Rhamnolipid	3.550	George and Jayachandran (2012)
Waste frying or used olive oil	*P. aeruginosa*47T2 *Bacillus* sp.	Rhamnolipid Lipopeptides	2.7 *nd*	Haba et al. (2000)
Used lubricating oil	*Rhodococcus* sp., *Bacillus* sp.	Trehalose glycolipids Lipopeptide	*nd*	Mercade et al. (1996)
Sterile diesel oil and glucose	*Pseudomonas* sp.	*nd*	*nd*	Bento and Geylarde (1996)

(continued)

Table 2.1 (continued)

Low-cost substrates	Microorganism	Biosurfactant	Yield (g/l)	Reference
Jet fuel JP8	*Cladosporium resinae*	Extracellular biosurfactant	*nd*	Muriel et al. (1996)
Soapstock (sunflower oil)	*P. aeruginosa* LBI	Rhamnolipid	12 (Shaker) 16 (Bioreactor)	Benincasa et al. (2002)
Soybean oil	*P. aeruginosa*	Rhamnolipid	4.31 *nd*	Rahman et al. (2002)
Safflower oil	*P. aeruginosa* DS10-129	Rhamnolipid	2.98	Rahman et al. (2002)
Diesel fuel	*Rhodococcus* sp.	*Nd*	*nd*	Pini et al. (2007)
Crude oil and waste soybean oil	*Klebsiella* sp.Y6-1	Lipopeptide	100 mg/m	Lee et al. (2008)
Palm oil	*P. aeruginosa* IFO3924	Rhamnolipid	0.063 g/g	Marsudi et al. (2008)
Fish oil (pollack liver)	*P. aeruginosa* BYK-2	Rhamnolipid	6	Jeong et al. (2004)
Animal fat + corn steep liquor	*C. lipolytica*UCP0988	Anionic Glycolipid	*nd*	Santos et al. (2013)
Soybean oil	*P. aeruginosa* BS20	Rhamnolipid	*nd*	Abdel-Mawgoud et al. (2009)
Rapeseed oil	*Pseudomonas* sp. DSM 2874	Rhamnolipid	45 g/l	Trummler et al. (2003)
Palm oil	*B. subtilis* PT2 *P. aeruginosa* SP4	SurfactinRhamnolipid	*nd*	Pornsunthorntawee et al. (2008)
Diesel oil	*Alcanivoraxdieselolei* B-5	Proline Lipid	*nd*	Qiao and Shao (2010)

(continued)

Table 2.1 (continued)

Low-cost substrates	Microorganism	Biosurfactant	Yield (g/l)	Reference
Sunflower oil	*Trichosporonmontevideense*CLOA 72	Glycolipid	*nd*	Monteiro et al. (2009)
Waste frying oils (canola, soybean, corn)	*P. putida* 300-B	Rhamnolipid	2.1–4.1	Raza et al. (2007a)
Vegetable oil refinery wastes (canola, soybean, corn)	*P. putida* 300-B	Rhamnolipid	1.8–2.9	Raza et al. (2007a)
Olive oil, sunflower oil, grape seed oil, diesel, kerosene	*P.s aeruginosa* J4	Rhamnolipid	*nd*	Wei et al. (2005)
Waste frying soybean oil	*P. aeruginosa* EBN-8	Rhamnolipid	9.3	
Olive oil, soybean oil	*P. aeruginosa* EMI	Rhamnolipid	3.70 2.63	
Castor oil, coconut oil, rapeseed oil, corn oil, motor oil, sunflower oil, olive oil, olein, barley bran, rice bran peanut cake, potato waste, wheat bran, corn oil and cassava waste flour	*P. aeruginosa* PB3A	Rhamnolipid	0.28–0.62	Saravanan and Subramaniyan (2014).

[a]*nd*—Not detected

References

Abalos A, Pinazo A, Infante MR, Casals M, Garcia F, Manresa A (2001) Physicochemical and antimicrobial properties of new rhamnolipids produced by *Pseudomonas aeruginosa* AT10 from soybean oil refinery wastes. Langmuir 17:1367–1371

Abdel-Mawgoud AM, Aboulwafa MM, Hassouna NA (2009) Characterization of rhamnolipid produced by *Pseudomonas aeruginosa* isolate Bs20. Appl Biochem Biotechnol 157(2):329–345

Banat IM, Satpute SK, Cameotra SS, Patil R, Nyayanit NV (2014) Cost effective technologies and renewable substrates for biosurfactants' production. Front Microbiol 5:697

Benincasa M, Contiero J, Manresa MA, Moraes IO (2002) Rhamnolipid production by *Pseudomonas aeruginosa* LBI growing on soap stock as the sole carbon source. J Food Eng 54:283–288

Bento FM, Gaylarde CC (1996) The production of interfacial emulsions by bacterial isolates from diesel. Int Biodeterior Biodegradation 38:31–33

Daniel HJ, Reuss M, Syldatk C (1998) Production of sophorolipids in high concentration from deproteinized whey and rapeseed oil in a two stage fed batch process using *Candida bombicola* ATCC 22214 and *Cryptococcus curvatus* ATCC 20509. Biotech Lett 20:1153–1156

Das K, Mukherjee AK (2007) Comparison of lipopeptide biosurfactants production by *Bacillus subtilis* strains in submerged and solid state fermentation systems using a cheap carbon source: some industrial applications of biosurfactants. Process Biochem 42:1191–1199

Dubey K, Juwakar A (2001) Distillery and curd whey wastes as viable alternative sources for biosurfactant production. World J Microbiol Biotechnol 17:61–69

Fox SL, Bala GA (2000) Production of surfactant from *Bacillus subtilis* ATCC 21332 using potato substrate. Biores Technol 75:235–240

George S, Jayachandran K (2009) Analysis of rhamnolipid biosurfactants produced through submerged fermentation using orange fruit peelings as sole carbon source. Appl Biochem Biotechnol 158(3):694–705

George S, Jayachandran K (2012) Production and characterization of rhamnolipid biosurfactant from waste frying coconut oil using a novel *Pseudomonas aeruginosa* D. J Appl Microbiol 114:373–383

Haba E, Espuny MJ, Busquets M, Manresa A (2000) Screening and production of rhamnolipids by *Pseudomonas aeruginosa* 47T2 NCIB 40044 from frying oils. J Appl Microbiol 88:379–387

Haba E, Vidal-Mas J, Bassas M, Espuny MJ, Llorens J, Manresa A (2007) Poly 3-(hydroxy alkanoates) produced from oily substrates by *Pseudomonas aeruginosa* 47T2 (NCBIM 40044): effect of nutrients and incubation temperature on polymer composition. Biochem Eng J 35:99–106

Jeong H-S, Lim D-J, Hwang S-H, Ha S-D, Kong J-Y (2004) Rhamnolipid production by *Pseudomonas aeruginosa* immobilized in polyvinyl alcohol beads. Biotech Lett 26:35–39

Koch A, Resier KJ, Kapelli O, Fiechter A (1988) Genetic construction of lactose-utilizing strains of *P. aeruginosa* and their applications in biosurfactants production. Biotechnology 6:1335–1339

Kock JLF, Botha A, Bloch J, Nigam S (1996) Used cooking oil: science tackles a potential health hazard. S Afr J Sci 92:513–514

Lee SC, Lee SJ, Kim SH, Park IH, Lee YS, Chung SY, Choi YL (2008) Characterization of new biosurfactant produced by *Klebsiella* sp. Y6-1 isolated from waste soybean oil. Biores Technol 99:2288–2292

Makkar RS, Cameotra SS (1997) Utilization of molasses for biosurfactant production by two *Bacillus* strains at thermophilic conditions. J Am Oil Chemist's Soc (JAOCS) 74:887–889

Marsudi S, Unno H, Hori K (2008) Palm oil utilization for the simultaneous production of polyhydroxyalkanoates and rhamnolipids by *Pseudomonas aeruginosa*. Appl Microbiol Biotechnol 78:955–961

Mercade ME, Manresa A, Robert M, Espuny MJ, deAndres C, Guinea J (1993) Olive oil mill effluent (OOME), a new substrate for biosurfactant production. Biores Technol 43:1–6

Mercade ME, Monleon L, Andres C, Rodon I, Martinez E, Espuny MJ (1996) Screening and selection of surfactant-producing bacteria from waste lubricating oil. J Appl Microbiol 81:161–166

Monteiro AS, Coutinho JOPA, Junior AC, Rosa CA, Siqueira EP, Santos VL (2009) Characterization of new biosurfactant produced by *Trichosporon montevideense* CLOA 72 isolated from dairy industry effluents. J Basic Microbiol 49:553–563

Muriel JM, Bruqu JM, Olfas JM, Jimenez- Sanchez A (1996) Production of biosurfactant by *Cladosporiumresinae*. Biotech Lett 18:235–240

Neto DC, Meira JA, de Araújo JM, Mitchell DA, Krieger N (2008) Optimization of the production of rhamnolipids by *Pseudomonas aeruginosa* UFPEDA 614 in solid-state culture. Appl Microbiol Biotechnol 81:441–448

Nitschke M, Ferraz C, Pastore GM (2004) Selection of microorganisms for biosurfactant production using agro industrial wastes. Braz J Microbiol 35:81–85

Patowary R, Patowary K, Kalita MC, Deka S (2016) Utilization of paneer whey waste for cost-effective production of rhamnolipid biosurfactant. Appl Biochem Biotechnol 180(3):383–399

Perfumo A, Banat IM, Canganella F, Marchant R (2006) Rhamnolipid production by a novel thermophilic hydrocarbon-degrading *Pseudomonas aeruginosa* AP02-1. Appl Microbiol Biotechnol 72:132–138

Pini F, Grossi C, Nereo S, Michaud L, Giudice AL, Bruni V, Baldi F, Fani R (2007) Molecular and physiological characterisation of psychrotrophic hydrocarbon-degrading bacteria isolated from Terra Nova Bay (Antarctica). Eur J Soil Biol 43:368–379

Pornsunthorntawee O, Wongpanit P, Chavadej S, Abe M, Rujiravanit R (2008) Structural and physicochemical characterization of crude biosurfactant produced by *Pseudomonas aeruginosa* SP4 isolated from petroleum-contaminated soil. Biores Technol 99:1589–1595

Qiao N, Shao Z (2010) Isolation and characterization of a novel biosurfactant produced by hydrocarbon-degrading bacterium *Alcanivorax dieselolei* B-5. J Appl Microbiol 108(4):1207–1216

Rahman KSM, Rahman TJ, McClean S, Merchant R, Banat IM (2002) Rhamnolipid biosurfactant production by strains of *Pseudomonas aeruginosa* using low-cost raw materials. Biotechnol Prog 18:1277–1281

Rane AN, Baikar VV, Ravi Kumar DV, Deopurkar RL (2017) Agro-industrial wastes for production of biosurfactant by Bacillus subtilis ANR 88 and its application in synthesis of silver and gold nanoparticles. Front Microbiol 8:492

Raza ZA, Khan MS, Khalid ZM (2007a) Evaluation of distant carbon sources in biosurfactant production by a gamma ray-induced *Pseudomonas putida* mutant. Process Biochem 42:686–692

Raza ZA, Khan MS, Khalid ZM (2007b) Physicochemical and surface-active properties of biosurfactant produced using molasses by a *Pseudomonas aeruginosa* mutant. J Environ Sci Health Part A 42:73–80

Rivera OMP, Moldes AB, Torrado AM, Dominguez JM (2007) Lactic acid and biosurfactants production from hydrolyzed distilled grape marc. Process Biochem 42:1010–1020

Rocha MVP, Oliveira AHS, Souza MCM, Goncalves LRB (2006) Natural cashew apple juice as fermentation medium for biosurfactant production by *Acinetobacter calcoaceticus*. World J Microbiol Biotechnol 22:1295–1299

Santos DKF, Rufino RD, Luna JM, Santos VA, Salgueiro AA, Sarubbo LA (2013) Synthesis and evaluation of biosurfactant produced by *Candida lipolytica* using animal fat and corn steep liquor. J Petrol Sci Eng 105:43–50

Saravanan V, Subramaniyan V (2014) Production of bio surfactant by *Pseudomonas aeruginosa* PB3A using agro-industrial wastes as a carbon source. Malays J Microbiol 10:57–62

Sarubbo LA, Marcel MCR, Campose-Takai GM (1997) Comparative study on bioemulsifiers produced by *Candida lipolytica* strains. Arquivos de Biologia e Tecnologia 40:707–720

Sobrinho HBS, Rufino RD, Luna JM, Salgueiro AA, Campos-Takaki GM, Leite LFC (2008) Utilization of two agroindustrial by-products for the production of a surfactant by *Candida sphaerica* UCP0995. Process Biochem 43:912–917
Sudhakar-Babu P, Vaidya AN, Bal AS, Kapur R, Juwarkar A, Khanna P (1996) Kinetics of biosurfactant production by *Pseudomonas aeruginosa* strain BS2 from industrial wastes. Biotech Lett 18:263–268
Thompson DN, Fox SL, Bala GA (2000) Biosurfactants from potato process effluents. Appl Biochem Biotechnol 84–86:917–930
Trummler K, Effenberger F, Syldatk C (2003) An integrated microbial/enzymatic process for production of rhamnolipids and L-(+)-rhamnose from rapeseed oil with *Pseudomonas* sp. DSM 2874. Eur J Lipid Sci Technol 105:563–571
Veenanadig NK, Gowthaman MK, Karanth NGK (2000) Scale up studies for the production of biosurfactants in packed column bioreactor. Bioprocess Biosyst Eng 22:95–99
Vollbrecht E, Rau U, Lang S (1999) Microbial conversion of vegetable oils to surface-active di-, tri-, and tetrasaccharide lipids (biosurfactants) by the bacterial strain *Tsukamurella* spec. Fett/ Lipid 101:389–394
Wei Y-H, Chou C-L, Chang J-S (2005) Rhamnolipid production by indigenous *Pseudomonas aeruginosa* J4 originating from petrochemical wastewater. Biochem Eng J 27:146–154
Zhu Z, Zhang F, Wei Z, Ran W, Shen Q (2013) The usage of rice straw as a major substrate for the production of surfactin by *Bacillus amyloliquefaciens*XZ-173 in solid-state fermentation. J Environ Manage 127:96–102

Chapter 3
Synthesis of Value Added Biomimetic Material of Hydroxyapatite Using Aqueous Calcareous Fish Wastes

M. Sutha, K. Sowndarya, M. Chandran, D. Yuvaraj, B. Bharathiraja and R. Praveen Kumar

Abstract Hydroxyapatite is the natural bone mineral found in abundance in human bones and teeth. Fish and bovine bone also have the same property as human hydroxyapatite. Thus, it can be replaced by bovine bones and fish bones. The gel of *Aloe vera* naturally has wound healing property. The main objective is to produce a cost-effective biomimetic material using hydroxyapatite with a natural wound healing agent for biomedical purposes. The objective is to be accomplished by producing a thin film of HAp scaffolds using fish bone as a source of hydroxyapatite and Aloe vera as a natural wound healing agent. Combining these two properties is expected to improve wound healing and faster bone growth. In this work, hydroxyapatite was obtained from fish bone waste. It is then added with natural *Aloe vera* gel of different concentration. The thin scaffold was prepared using sodium alginate and calcium chloride. The obtained scaffold is to be tested using SEM and XRD. A further test is to be done to check wound healing and bone ingrowth.

Keywords Hydroxyapatite · *Aloe vera* · Biomimetic · Calcareous waste

3.1 Introduction

Hydroxyapatite powder (HAp) a natural bone mineral of human beings can be synthetically prepared and derived from natural sources of bones of bovine, pigs, fish, etc. Hydroxyapatite is a preferred biomimetic and biomaterial for orthopedic treatment due to its biocompatibility, bioactivity, osteoconductivity, non-toxicity, and non-inflammatory nature. *Aloe vera* naturally has wound healing property, and

M. Sutha · K. Sowndarya · M. Chandran · D. Yuvaraj · B. Bharathiraja (✉)
Vel Tech High Tech Dr. Rangarajan Dr. Sakunthala Engineering College Avadi, Chennai 600062, India
e-mail: btrbio@gmail.com

R. Praveen Kumar
Arunai Engineering College, Tiruvannamalai, India

R. R. Singhania et al. (eds.), *Waste to Wealth*, Energy, Environment, and Sustainability, https://doi.org/10.1007/978-981-10-7431-8_3

it is known to promote the renewal of healthy and noncancerous cells; it is being expected that those properties of *Aloe vera* gel may increase the efficiency of hydroxyapatite in acting as a perfect and effective bone transplant. Synthesis of HAp from chemical sources is a complicated or biologically unsafe process (Guo et al. 2003).

Therefore, people turn to the other option from natural sources like fish bones (Ho et al. 2013), bovine bones (Bahrololoom et al. 2009), and teeth and bones of pig (Russias et al. 2006). Chemical analysis has shown that these products are bio-wastes with abundant resources of calcium in the form of carbonate and oxide (Guo et al. 2003; Russias et al. 2006). Extraction of HAp from these wastes not only considered as economically beneficial but also environmentally-friendly. About 50% of the total weight generated by the fish processing industry is a waste (Mohamadi et al. 2007). Fish processing industry is one of the world's largest industries; several tons of fish were processed for food every year. As India is having large coastal area, it is one of the largest producers of fish with a contribution of 5.43% in global production. In 2009–2010, fish production was around 78.51 lakh tones. Every year a billion tons of fish are utilized for edible purpose; waste non-edible parts include head, viscera, dorsal fins, tail, skin, and liver. These waste non-edible parts are considered to be worthless garbage and discarded without any recovery of valuable products by dumping on land or hauling into the ocean.[1] India alone generates 2 million metric tonnes of fish waste while processing. These wastes may pose adverse environmental effects such as generation of toxic hydrogen sulfide gas, increased gathering of scavengers in discharge locations, and noxious conditions caused by odors, bacteria and waste decomposition (FAO and WHO 2013), when not managed properly. These wastes should be managed properly, as it has valuable organic compounds; they may be used for the production of value-added products. The most important environment-friendly and profitable option for utilization of fish waste includes animal feed supplements, aquaculture feed, fish meal, and fish silage, renewable energy in the form of biodiesel and biogas (Tanimu et al. 2014), composting for production of organic fertilizers, extraction of natural pigments, extraction of novel and industrial enzymes like proteases (Kumaran et al. 2013), cosmetics, pharmaceutical industries such as collagen, fish protein hydrolysate, fish bone extracts, and polyunsaturated fatty acids (Bimbo 2011; Tanwar et al. 2013).

There are several types of calcium phosphate salts formed in fish bones and scales as a biological response to their extreme physiological environment and surprisingly, these salts can be used as starting material in producing HAp (Hsu et al. 2005). The preparation of HAp from fish scale was first reported by Sankar et al. (2008). According to Huang (2007) and Sadat-Shojai et al. (2013), HAp can be prepared at low temperature (90–100 °C) as well as after heat treatment at higher temperatures of 500 °C and above (Zandi et al. 2010; Shiny et al. 2000). Unfortunately, most methods involve very elaborate and cost-effective process

[1]http://lipidlibrary.aocs.org/processing/marine/index.htm (Accessed on January 30th, 2014)

since usage of many varieties of organic chelating agents (Pradeesh et al. 2005). In 2004, Santos was able to synthesize the HAp just at room temperature but the obtained HAp reported was not pure (Zandi et al. 2010; Pradeesh et al. 2005). Calcination at high temperature improves the properties of HAp which promoted a better packing of the ceramic grains and promoted more fluidity capabilities (Zandi et al. 2010; Gopi et al. 2012). The temperature must be high enough to overcome the activation energy and to improve CaO or $Ca(OH)_2$ dissolution which is considered as impurities (Gopi et al. 2012; Gleeson et al. 2010). These impurities will affect the HAp crystal structure, crystallinity, and solubility, thus limiting the application of HAp, especially in biomedical applications (Simon et al. 2007; Tanaka et al. 2017). Considering these wastes as a potential source, the present research aims to extract HAp from unexplored bio-wastes of fish which are abundant in nature. In this paper, the effects of various calcination temperatures were studied on the features of HAp.

Recent researches have shown that the hydroxyapatite has immense potential in the field of bone tissue engineering, as it has a natural tendency to develop tight binding with bone tissue.

Moreover, this biomimetic material has no adverse effects (Guo et al. 2003) thereby preventing immune rejection for successful transplantation. In our current work, *Aloe vera* gel is employed to produce a combination of hydroxyapatite and *Aloe vera* gel. Since *Aloe vera* has properties of wound healing (Ho et al. 2013) and it is known to promote the renewal of healthy and noncancerous cells, it is being expected that those properties of *Aloe vera* gel may increase the efficiency of hydroxyapatite in acting as a perfect and effective bone transplant. The main application of HAp is to be used as a coating material for bone implants, used as scaffolds for inducing osteoblast proliferation and differentiation, and quick healing of bone tissue injuries can be achieved and also provides a biological environment for cells thereby promoting the success of the transplant.

3.2 Preparation

Fish bone wastes, *Aloe vera*, HCl, NaOH, sodium alginate, calcium chloride ($CaCl_2$). Composites of *Aloe vera* and hydroxyapatite can be prepared in four steps, after collection of calcareous waste (fish bones) deproteinization, precipitation, and scaffold preparation.

3.2.1 Calcareous Waste

Fish bones (catfish) were collected from a local market. Collected bones were washed with water and ethanol and dried in hot air oven at 100 °C.

The sample is deproteinized by external washing with 1 N HCl solution for 24 h at room temperature (35 ± 2 °C). Then it is rinsed with water. Then the samples were crushed into a fine powder using a grinder. The precipitate was filtered and treatment with 5%(w/v) NaOH for 5 h at 70°. Then these samples were calcined at a temperature of 800–1000 °C using muffle furnace. The resulting hydroxyapatite is mixed with different concentrations of *Aloe vera* gel, and its mixed ratio is optimized for the best result.

3.2.2 Preparation of Scaffold

About 3% of sodium alginate was added with 20 ml distilled water. Then 2.5% HAp was added with 1% *Aloe vera* gel. This mixture is added to the above-prepared sodium alginate solution by continuous mixing using a magnetic stirrer for 45 min. Then the mixture is poured in the casting plate and dispersed with 2% calcium chloride solution to produce a thin scaffold of Hap.

3.3 Conclusion

The scaffold obtained should be further characterized using SEM, XRD for physical characterization. This could enable to gain more scientific knowledge on the scaffold which might give better result in further studies. A test for cell proliferative activity and wound healing activity may be done to know the efficiency. If the biomaterial is successfully produced and characterized, the product will be made useful for medical purposes after proper testing. This is expected to improve wound healing and faster bone deformation.

References

Bahrololoom ME, Javidi M, Javadpour S, Ma J (2009) Characterisation of natural hydroxyapatite extracted from bovine cortical bone ash. J Ceram Process Res 10:129–138

Bimbo AP (2011) The production and processing of marine oils. The AOCS lipid library. http://lipidlibrary.aocs.org

FAO and WHO (2013) Codex Alimentarius Commission joint FAO/WHO food standards programme codex committee on fats and oils Twenty-third Session, Langkawi, Malaysia, 25 February–1 March

Gleeson JP, Plunkett NA, Brien FJO (2010) Addition of hydroxyapatite improves stiffness, interconnectivity and osteogenic potential of a highly porous collagen-based scaffold for bone tissue regeneration. European Cells and Materials 20:218–230

Gopi D, Indira J, Kavitha L, Sekar M, Mudali UK (2012) Synthesis of hap nanoparticles by a novel ultrasonic assisted with mixed hollow sphere template method. Spectrochimacta part A 93:131–134

Guo H, Khor KA, Boey YC, Miao X (2003) Laminated and functionally graded hydroxyapatite/yttria stabilized tetragonal zirconia composites fabricated by spark plasma sintering. Biomaterials 24(4):667–675

Ho WF, Hsu HC, Hsu SK, Hung CW, Wu SC (2013) Calcium phosphate bioceramics synthesized from eggshell powders through a solid-state reaction. Ceram Int 39(6):6467–6473

Hsu FY, Tsai SW, Lan CW, Wang YJ (2005) An in vivo study of a bone grafting material consisting of hydroxyapatite and reconstituted collagen. J Mater Sci Mater Med 16(4):341–345

Kumaran E, Mahalakshmipriya A, Rajan S (2013) Effect of fish waste based bacillus protease in silver recovery from waste X-ray films. Int J Curr Microbiol App Sci 2(3):49–56

Mohamadi Y, Mirzadeh H, Moztarzadeh F, Soleimani M, Jabbari E (2007) Design and fabrication of biodegadable porous chitosan/gelatin/tricalcium phosphate hybrid scaffolds for tissue engineering. Iran J Polym Sci Technol (In Persion) 20:297–308

Pradeesh TS, Sunny MC, Varma HK, Ramesh P (2005) Preparation of microstructured hydroxyapatite microspheres using oil in water emulsions. Bull Mater Sci 28(5):383–390

Russias J, Saiz E, Nalla RK, Gryn K, Ritchie RO, Tomsia AP (2006) Fabrication and mechanical properties of PLA/HA composites: a study of in vitro degradation. Mate Sci Eng C 26 (8):1289–1295

Shiny V, Ramesh P, Sunny MC, Varma HK (2000) Extrusion of hydroxyapatite to clinically significant shapes. Mater Lett 46:142–146

Simon JL, Michna S, Lewis JA, Rekow ED, Thompson VP, Smay JE, T (2007) In vivo bone response to 3D periodic hydroxyapatite scaffolds assembled by direct ink writing, J Biomed Mater Res A 83(3):747–58

Tanaka M, Haniu H, Kamanaka T, Takizawa T (2017) Physico-chemical, in vitro, and in vivo evaluation of a 3D unidirectional porous hydroxyapatite scaffold for bone regeneration, Materials (Basel) 10(1):33

Tanimu MI, Ghazi TI, Harun RM, Idris A (2014) Effect of carbon to nitrogen ratio of food waste on biogas methane production in a batch mesophilic anaerobic digester. Int J Innov Manage Technol 5(2):116

Tanwar D, Tanwar A, Sharma D et al (2013) Production and characterization of fish oil methyl ester. Int J Innov Technol Res 1(3):209–217

Zandi M, Mirzadeh H, Mayer C, Urch H, Eslaminejad MB, Bagheri F, Mivehchi H (2010) Biocompatibility evaluation of nano rod hydroxyapatite/gelatin coated with nano HAp as a novel scaffold using mesenchymal stem cells. J Biomed Mater Res A 92(4):1244–55

Chapter 4
Utilization of Crude Glycerol from Biodiesel Industry for the Production of Value-Added Bioproducts

S. Chozhavendhan, R. Praveen Kumar, S. Elavazhagan, B. Barathiraja, M. Jayakumar and Sunita J. Varjani

Abstract The rapid economic development and industrialization mainly depends on fossil fuels for their energy sources. Diminishing oil reserves, high cost and other environmental problems lead the researchers to ripen carbon-unbiased and sustainable alternative to fossil fuels. When compared with other alternate fuels, biodiesel gained its popularity. The dynamic growth of biodiesel production results in colossal generation of glycerol as a waste by-product of about 10% of the total biodiesel production. Disposal of crude glycerol tainted with salt, methanol, free fatty acids, etc., causes severe economic and environmental challenge. However, high cost of crude glycerol purification and market saturation urges the biodiesel producers to look for new, cheaper, and alternate solution for utilizing it. Presence of impurities and its composition in crude glycerol can have a negative influence in the fermentation process by inhibiting microbial growth and product formation. Microbial fermentation of crude glycerol represents a notable alternate and competitive to add value for the biodiesel producers. Hence, pre-treatment of crude glycerol becomes requisite before using it as carbon source. The higher degree reduction of carbon atom in glycerol confers that it could be used as a sole organic carbon and energy source for the production of higher value chemicals such as citric acid, ethanol, DHA, glyceric acid.

S. Chozhavendhan (✉) · R. Praveen Kumar
Arunai Engineering College, Tiruvannamalai, Tamilnadu, India
e-mail: scv.ibt@gmail.com

S. Elavazhagan
Shantha Biotechnics Private Limited, Hyderabad, Telangana, India

B. Barathiraja
Vel Tech High Tech Dr.RR Dr.SR Engineering College, Chennai, Tamilnadu, India

M. Jayakumar
Institute of Technology, Haramaya University, Dire Dawa, Ethiopia

S. J. Varjani
Gujarat Pollution Control Board, Gandhinagar, Gujarat, India

R. R. Singhania et al. (eds.), *Waste to Wealth*, Energy, Environment, and Sustainability, https://doi.org/10.1007/978-981-10-7431-8_4

Keywords Crude glycerol · Biodiesel · Ethanol · DHA · Citric acid
Microbial fermentation

4.1 Introduction

World's anticipated energy prerequisite in the year 2030 shall be 50% more than it is today. On one hand, the rapid economic development and industrialization mainly be contingent on fossil fuels for transportation of goods and services (Bhaskar et al. 2014). In fact, 96% of the shipping sector are reliant on fossil fuels (Yahaya et al. 2014). Increasing human population has also utilized the superfluous amount of world's diminishing oil resources for transportation and other purposes which has led to the confronting energy crisis for the last two decades. The global energy demand in the transportation sector is anticipated to grow by 2% per year (Wilson and Lee 2012). Significant increase in energy demand and emission of unburned hydrocarbons from fossil fuels forced the researchers to work on renewable energy (Noshadi et al. 2012). The renewable sources of energy, including wind, ocean, solar, hydropower, geothermal, and biomass, have the competency to afford unconventional sources of clean energy (Schiermeier et al. 2008). These renewable energy sources which are place-specific and hence can be utilized only at locations are suitable for their harnessing. The limitations and drawbacks in other forms of renewable energy paves the way for higher quantity of biofuel production, in particular biodiesel gained attention worldwide owing to its reliability and sustainability which can contribute to reduce pollution (Quispe et al. 2013).

4.1.1 Production of Biodiesel and a By-product Glycerol

Biodiesel is produced from a wide variety of raw materials (comprising of triglycerides as major component) by either esterification or transesterification or both, depending on the acid value of the feedstock and the type of catalyst (acid or alkaline) used for its synthesis (Sun et al. 2010) Three different types of catalyst play an important role, namely alkalis, acids, and enzymes (Fjerbaek et al. 2009; Laque et al. 2008; Sharma et al. 2008). During biodiesel production process by transesterification reaction, approximately 10% wt of glycerol was collected as by-product (Teng et al. 2014). The transesterification reaction is reversible and takes place in step-wise conversion of triglycerides into diglycerides and diglycerides into monoglycerides and finally as glycerol. To favor the forward transesterification reaction, sufficient quantity of alcohol is usually added in the process. Figure 4.1 shows the chemical equation for glycerol production by transesterification of

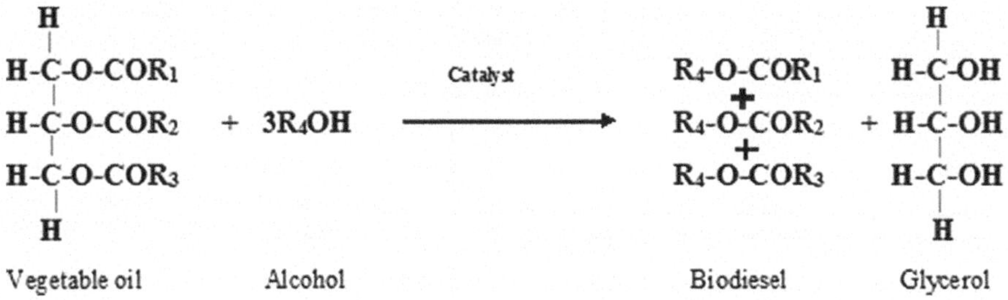

Fig. 4.1 Equation shows transesterification of large branched triglyceride molecule to biodiesel and glycerol

triglycerides (Ngo et al. 2011). As per stoichiometry, the reaction necessitates 3:1 alcohol-to-oil molar ratio, but in practical up to six molecules of alcohols are added (Sivasamy et al. 2009; Vincente et al. 2004). However, an excess amount of alcohol is generally added to drive the equilibrium in the direction of the product side (Mythili et al. 2014). It is assessed that biodiesel market will reach 40 billion gallons by 2020 with the annual growth of 42%, which indirectly producing 4 billion gallons of waste glycerol stream as a by-product. Thus, waste glycerol stream can also be called as raw glycerol, waste glycerol, and crude glycerol (Chozhavendhan et al. 2016). Figure 4.2 shows the diagrammatic representation production of biodiesel and crude glycerol from vegetable oils and animal fats.

4.1.2 General Properties of Glycerol

Glycerol is a simple, cheap, and abundant alcohol with the molecular formula of $C_3H_8O_3$. It is also known as propane 1, 2, 3, triol (Andre et al. 1991). When glycerol concentration is above 95%, it is commercially known as glycerin and many commercial grades are also available. The chemical composition and physiochemical properties of glycerol vary from other fuels (Agarwal 1990; Chozhavendhan et al. 2014). The hydrophilic hydroxyl groups of glycerol make it miscible in water, and it is a colorless, odorless, highly viscous liquid. Glycerol is very viscous at room temperature and remains as viscous liquid even at absolute concentration without crystallizing. Its melting point and boiling point are 17.9 and 290 °C, respectively. Its high calorific value finds varied application and used as supplementary fuel in boilers.

Glycerol can be produced either by microbial production or chemical synthesis from petrochemical feedstocks or as a primary co-product obtained during transesterification of fat and vegetable oils. It is used as a major carbon source in the culture medium for the cultivation of microorganism in the industrial fermentation

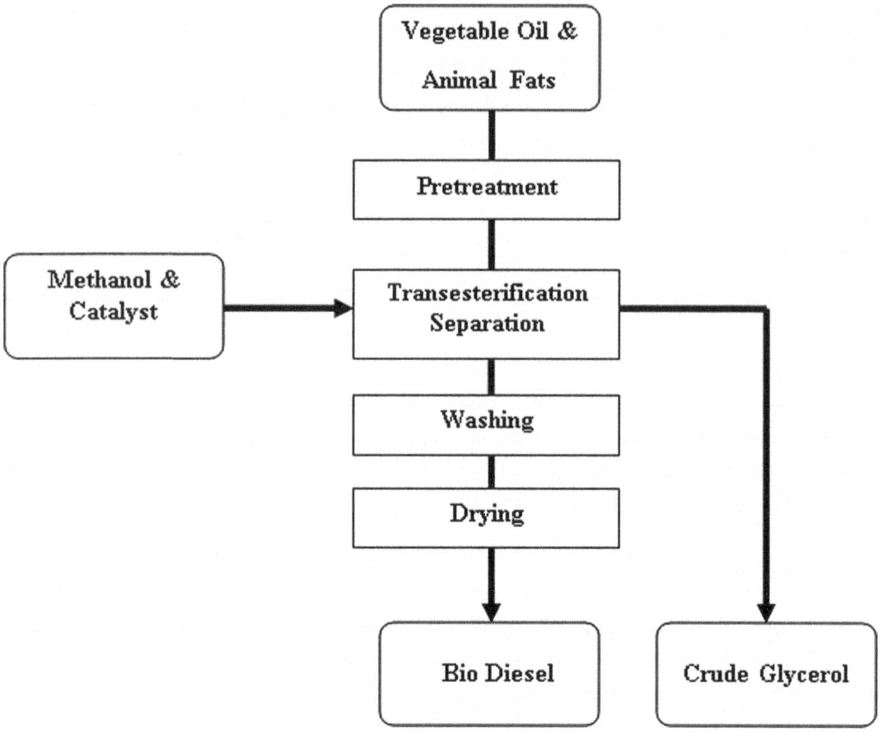

Fig. 4.2 Production of biodiesel and crude glycerol from vegetable oils and animal fats

process. Many organisms are able to convert glycerol into a series of metabolic intermediates to the one obtained by using sugar as substrate. In particular, platform chemicals such as glyceric acid, tartonic acid, and dihydroxyacetone are obtained by simple oxidation or fermentation process (Solomon et al. 1995; Barbirato and Bories 1997; Colin et al. 2001).

4.1.3 Glycerol Market Scenario

The glycerol comes from conversion actions based on vegetable oil, fat, animal tallow, and biorefineries. Worldwide consumption of glycerol in 2008 was about 750 thousand tons and forecasted as 2 million tons in 2018. Till 2004, the glycerol production and prices were stable and in 2005, the firm market has been radically transformed by the biodiesel industries, and the prices of glycerol were shrunken. In 2006, the price of refined glycerol in Asian market is around 1200–1800 US $ per ton. It has faced strong falling trend and sold at 440–660 US $. Meanwhile, crude glycerol prices around 200–220 US$ (Erin et al. 2016). Volatile commodity glycerol's market in worldwide is unpredictable, and it can be stabilized by the subsidy policy and regulation posed.

The saturated growth of biodiesel industry faced a great demand of glycerol. Europe and the USA are the largest importers, and now China also surged into the market of the glycerol. Brazil, Malaysia, and Indonesia are the largest exporters of glycerol. Glycerol market is expected to recover and will ensure the sustained growth in upcoming years. In addition, utilization of glycerol in various new fields is expected to improve the glycerol ultimatum in future.

4.1.4 Limitation of Crude Glycerol

The surplus production of crude glycerol in biorefineries leads to deterioration of pure glycerol value (Yazdani and Gonzalez 2007). Crude glycerol obtained from biodiesel or oleochemical plant is generally occurring in dark brown color liquid, faulty smell with higher pH (Ayoub and Abdullah 2012). Crude glycerol consists of copious amount of impurities such as methanol, soap, oils, and other solid organic materials depending upon technology imposed by various biodiesel plants as well as the nature of the oil utilized as the starting material (Wen et al. 2009). NaOH commonly used as a catalyst during the biodiesel manufacturing process causes harmful effect to the biotic civics. High alkaline condition of crude glycerol creates an unpleasant odor, which leads to air pollution (Vasudevan and Briggs 2008).

The glycerol can be easily transformed into glucose for the energy production in the animal liver because of its high absorption rate. Hence, crude glycerol was an excellent energy source and can be used as animal feed. The surplus glycerol in the animal diet may disturb the normal physiological metabolism. Different chemical composition and the varying impurity level have distracted the attention of animal producers to utilize crude glycerol as feed for animal food (Kijora and Kupsch 1996; Schieck et al. 2010)

Crude glycerol, when directly assorted with the microbial culture as substrate for fermentation process, soap precipitates to form bubbles in the liquid media, which have been found as destructive to cell growth by reducing the oxygen transfer rate (Stelmachowski 2011). Free fatty acids or phospholipids provide a deleterious effect on the transesterification procedure and make downstream processing work dreary on recovering the final product (Leung et al. 2010). Methanol, which is considered as hazardous waste, non-biodegradable, it can also exert an inhibitory action on microbial growth (Sneha et al. 2009; Taconi et al. 2009).

High viscous, high self-ignition temperature and formation of acrolein on burning of crude glycerol restrict its usage as alternate to fossil fuel. The high salt content in crude glycerol causes corrosion problem in metal contacts which sorts to avoid its usage as co-firing in thermal energy processes (Bohon et al. 2011).

Presence of water and methanol in crude glycerol makes it incompetent of maintaining firm flame in the conservative burners. Pure glycerol is sold as an important commodity because it is the vivacious raw material for food, pharmaceuticals, and cosmetic manufacturing industries (Guerrero-Perez et al. 2009). Crude glycerol cannot be disposed without treatment process as it creates several environmental concerns. The low-level purity and presence of ample amount of impurities in crude glycerol limit its usage as feedstock in various industries and reduce its market value (Pagliaro et al. 2007; Waala et al. 2016). Therefore, it is necessary to establish various purification processes to use it as an important feedstock in food and cosmetic industries.

4.2 Step-Wise Purification Process

At present, numerous techniques are adapted for purification of crude glycerol to commercial grade which is energy intensive and economically challenging (Isahak 2010). A simple purification process such as distillation, acidification, adsorption, centrifugation, and other physiochemical treatments helps to remove major impurities or combination of these techniques to achieve a high purity level of about 50–60%, which can be used as a sole carbon source for the production of high-value chemical products. Yield and recovery of products are the two important aspects which directly affect the economic viability of the product.

4.2.1 Distillation

Distillation is a process of using thermal energy for separating the component or substances from a liquid mixture by selective evaporation and condensation. Distillation may result in essentially complete or partial separation of selected components of the mixture. Usually, excess methanol is used as alcohol in the transesterification process for the production of biodiesel. The methanol reacts with triacylglycerol to give three molecules of biodiesel and one molecule of glycerol. Excess unreacted methanol settles in the glycerol phase which can be distant by controlled simple distillation process and reused in the same in biodiesel production (Canakci and Sanli 2008). The boiling point of methanol is 65 °C which is far less than boiling point of water and glycerol. Hence, biodiesel manufactures habitually make an effort to convalesce excess non-reactive methanol because rescuing methanol is economic than using new one.

4.2.2 *Acidification*

In the chemical reaction, the term acidification/neutralization is used in a reaction between an acid and a base or alkali. During the course of reaction, salt is formed when a cation (positive ion) of an acid forms a compound with the anion (negative ion) of a base. The neutralization (neutral pH) with a pH of 7 was achieved with a strong acid and a strong base. Crude glycerol was acidified with different mineral acids, such as hydrochloric acid, sulfuric acid, and phosphoric acid. When any of these acids reacts with soluble soap molecules present in the crude glycerol, it forms insoluble free fatty acids and sodium/potassium salts according to their reaction. In this process, phase separation time, precipitation time, and amount of acid consumed play a vital role. At pH in the range of 4–5, glycerol liberation is more when compared with other pH. Three distinct layers were formed during phase separation: top layer as free floating fatty acids, middle layer as glycerol-rich phase, and bottom layer as precipitated inorganic salts (Nanda et al. 2014).

4.2.3 *Adsorption*

It is the adhesion of ions, atoms, or molecules from a liquid, gas, or suspended solid to a surface. Adsorption creates a film of the adsorbate on the exterior of the adsorbent. The process of adsorption is quite different from absorption, in which a fluid (the absorbate) is dissolved by or pervades a liquid or solid (the absorbent), respectively. This process of adsorption was usually carried out with activated charcoal in order to reduce the color and absorb some unwanted amino acids which causes smell. Then the samples were filtered using Whatmann filter paper 1 to remove oversize solids in the fluids (Christy Mathelin et al. 2015).

4.2.4 *Centrifugation*

This process comprises the application of the centrifugal force to accelerate the sedimentation process of varied mixed compounds. This process is used to separate two miscible substances, which results in the formation of two distinct phases like sediment and centrifugate. After series of purification steps, the glycerol phase was centrifuged at 5000 rpm for 5 min to remove some traces of precipitated fatty acids and suspended solids based on different densities (Zhanyou et al. 2007).

In most commercial applications, the quality of glycerol needs to be enhanced until it has an acceptable purity that is entirely different from those attained from biodiesel industries.

4.2.5 Microbes Used in Bioconversion of Crude Glycerol

As discussed earlier, disposal of crude glycerol causes many environmental problems because of impurities present in it. The plummeted crude glycerol can be reduced by utilizing in the production of high-value compounds through biotechnological processes (Shen et al. 2009). A variety of unicellular and multicellular microorganisms have the promising ability to utilize crude glycerol as a only or appendage carbon source for the production of superfluity metabolic products such as single-cell protein (SCP), organic acid, solvents, and alcohols. Initially, the fermentation of glycerol was witnessed in pathogens such as Klebsiella and Citrobacter (Homann et al. 1990). Later, many organisms such as *Clostridium butyricum* (Wilkens et al. 2012), *Citrobacter freundii* from the Enterobacteriaceae family (Metsoviti et al. 2012a), *Ustilago maydis* (Liu et al. 2011), *Kluyvera cryocrescens* (Choi et al. 2011), *Zobellella denitrificans* (Ibrahim and Steinbüchel 2010), *Actinobacillus succinogenes* (Vlysidis et al. 2011), *Yarrowia lipolytica* (Auta et al. 2014), *Saccharomyces cerevisiae, E. coli* (Mattam et al. 2013), *Klebsiella pneumonia* were found capable of utilizing glycerol in the production of other highly valuable metabolic products.

4.3 Production of Value-Added Products

Bioconversion of crude glycerol offers safer, more viable, and more economical than any other process involved in treating it (Leoneti et al. 2012; Schultz et al. 2014). Biological metabolism of glycerol is shown in Fig. 4.3. The properties of glycerol have fashioned to a handy range of products. Reduced nature of glycerol can yield series of metabolites with yields which are comparable to the ones attained from using glucose as substrate. Till the year 2000, there were more than fifteen hundred end uses of glycerol in the chemical industry (Soap and Detergent Association, 2000). On the contemporary, now it is assessed that there are more than two thousand end uses of glycerol in industries. However, it is used in small quantities in majority of the products and only in few needs used as large amount.

4.3.1 Ethanol

Ethanol (EtOH) is a straight-chain alcohol with the molecular formula of C_2H_5OH. It is mainly used as a substitute fuel for conveyance and can be used as a chemical intermediate. Industrial production of ethanol was mainly focused on fermentation process by utilizing baker's yeast as a sole microbes and sugarcane sucrose or corn starch or cassava as a substrate (Gray et al. 2006). An extensive study was made on

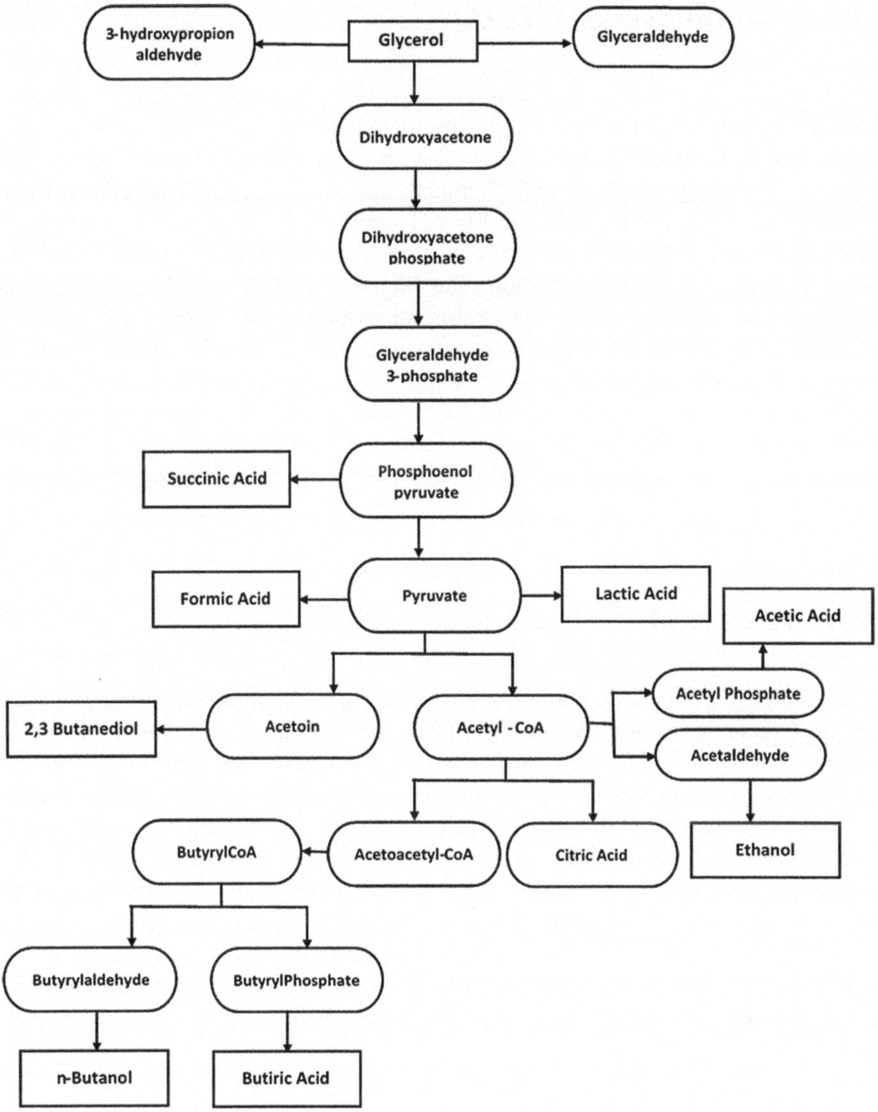

Fig. 4.3 Value-added products obtained from bioconversion metabolism of glycerol

the lignocellulosic feedstock to reach maximum yield of ethanol (Trinch et al. 2008; Yamano et al. 2008). Despite of the reduced nature of glycerol as a sole or a complementary carbon source, purification cost, commercial price, and bioconversion cost permit the production of ethanol by various microbes *such as S. cerevisiae, E. coli, K. oxytoca, Kluyvera cryocrescens, Zymomonas mobilis* (Hong et al. 2010).

4.3.2 *Dihydroxyacetone (DHA)*

Dihydroxyacetone is a non-chiral and non-toxic three carbon sugar ($C_3H_6O_3$) with a colossal quantity of commercial applications (Mishra et al. 2008). DHA is commonly used in cosmetics as an artificial browning agent (Green et al. 1961) and also serves as a building block for fine chemicals such as 1, 2 propylene glycerol and methotrexate. DHA can be produced from glycerol via either microbial or chemical route. The microbial bioconversion was more competent when compared to the chemical method of DHA production. The chemical method of DHA production is expensive and produces more side chain and requires more safety measurements (Pagliaro et al. 2007). Various microorganisms such as *Gluconobacter oxydans, Acetobacter aceti, Gluconacetobacter xylinus, Schizochytrium limacinum, Pichia membranifaciens* are capable of utilizing the crude glycerol for the production of DHA through fermentation process. Due to its common use, the global demand for DHA is successively increasing and its application is still emerging (Lidia et al. 2014).

4.3.3 *Citric Acid*

Citric acid ($C_6H_8O_7$) is also known as 2-hydroxy-1, 2, 3-propane tricarboxylic acid. It is most versatile, natural chemical produced and consumed throughout the World (Soccol et al. 2006). Citric acid is extensively used as organic acid in the field of food (70%), pharmaceuticals (12%), and other applications (18%) (Vandenberghe et al. 1999; Shah et al. 1993). In 1874, citric acid was first isolated from the lemon juice by Karls Scheels in England. It is a weak organic acid that is commercially produced from fungal fermentation with sucrose or molasses (Anastassiadis et al. 2008). A wide variety of microorganisms, including bacteria, yeasts, and fungi, have been employed for citric acid production. Glycerol is appraised as an impending substrate for the production of citric acid when cultured with *Y. lipolytica* yeast strain (Rymowicz et al. 2006; Imandi et al. 2007; Kamzolova et al. 2011).

4.3.3.1 1, 3-Propanediol

1, 3-Propanediol or 1, 3 PDO ($C_3H_8O_2$) is colorless, water miscible liquid with boiling point 214 °C, an emerging commodity chemical and renewed interest in the expansion of new polypropylene terephthalate, polyester (Reynaud et al. 2003, Gonzalez-pajuelo et al. 2006). Products obtained by 1, 3-PDO polymerization are considered as good biodegradability, better specificity, and higher industrial safety.

The present scenario of 1, 3-PDO production is expensive and hazardous as derived from acrolein, a detrimental agent obtained from petroleum fuels. Hence, 1, 3-PDO production was carried out by the fermentation process using microbes such as *Klebsiella, Clostridium, Enterobacter,* and *Citrobacter* which were also studied (Reimann et al. 1998; Hirschmann et al. 2005). Glycerol and many other carbon sources were used in the production of 1, 3-propanediol.

4.3.4 Glyceric Acid

Glyceric acid with molecular formula $C_3H_6O_4$ (GA) (2R, 3 hydroxy propanoic acid) is naturally occurring organic acid (Handa et al. 1986). Glyceric acid serves as a building block for numerous chemical compounds used in pharmaceuticals and cosmetic industries (Rosseto et al. 2008; Elina et al. 2013). The usage of glyceric acid is restricted due to the high cost of chemical synthesis. Glyceric acid is obtained as the outcome of metallic oxidation of the primary hydroxyl group of glycerol or by microbial fermentation process (Bianchi et al. 2005). It is produced as a by-product of dihydroxyacetone. Glyceric acid has shown to confer propitious effects on liver stimulation and cholesterolytic actions in dog.

4.3.5 Succinic Acid

Succinic acid or butanedioic acid is a dicarboxylic acid with the molecular formula of C_4H_6O (Millard et al. 1996). Purified succinic acid is produced from amber by Georgius Aaricola in 1546; hence, it is also known as amber acid. Succinic acid is widely used in food and pharmaceutical industry. The linear saturated dicarboxylic acid of succinic acid is widely used as a transitional chemical for the conversion of n-methyl pyrrolidone, tetrahydrofuran, 1, 4-butenediol, and linear aliphatic esters (Zeikus et al. 1999). At present, succinic acid is produced from butane via maleic anhydride; the natural succinic acids produced by fermentation process are vended in the food market. The most documented succinic acid producers are wild strains of *Actinobacillus Succinogenes, Mannheimia succiniciproducens, Anaerobiospirillum succiniciproducens, Corynebacterium glutamicum* recombinant *E. coli, Lactobacillus, S. cerevisiae,* which utilize crude glycerol and other carbon source (Min et al. 2014; Beauprez et al. 2010; Song and Lee 2006). The production of succinic acid is expensive through chemical route, and hence, bioconversion route is a better choice.

Apart from the above-mentioned products, lot of other high-value bioproducts such as butanol, propionic acid, lactic acid can be obtained from crude glycerol fermentation process by a variety of microorganisms and few shown in Table 4.1.

Table 4.1 Microbial conversion of glycerol into value-added products

Product	Organisms	Substrate	Productivity (g/L/h)	Yield (substrate/product) (g/g)	References
Ethanol	*Escherichia coli*	Glycerol	0.051	1	Yazdani and Gonzalez (2007)
	Klebsiella oxytoca FMCC-197	Crude glycerol	0.29	0.2	Metsoviti et al. 2012a, b
	Citrobacter freundii FMCC-197	Crude glycerol	0.66	0.45	Metsoviti et al. 2012a, b
	Enterobacter aerogenes HU-101	Crude glycerol	0.83	0.4	Ito et al. 2005
1,3 Propanediol	*Klebsiella pneumoniae* DSM2026	Crude glycerol	1.57	0.52	Chen et al. 2003
	Clostridium butyricum VPI 3266	Pure glycerol	2.98	0.62	Gonzalez-Pajuelo et al. 2004
		Crude glycerol	3.02	0.6	Gonzalez-Pajuelo et al. 2004
	Clostridium acetobutyricum DG1	Pure glycerol	1.77	0.54	Gonzalez-Pajuelo et al. 2005
Citric acid	*Yarrowia lipolytica* Wratislavia AWG7	Crude glycerol	1.16	0.33	Rywinska et al. 2009
	Yarrowia lipolytica Wratislavia K1	Pure glycerol	–	0.45	Rywinska et al. 2009
		Crude glycerol	–	0.43	Rywinska et al. 2009

(continued)

Table 4.1 (continued)

Product	Organisms	Substrate	Productivity (g/L/h)	Yield (substrate/product) (g/g)	References
Succinic acid	*Anaerobiospirillum succiniciproducens*	Pure glycerol	0.16	1.6	Lee et al. 2001
	Escherichia coli	Crude glycerol	0	0.69	Blankschein et al. 2010
	Yarrowia lipolytica	Crude glycerol	0.45	–	Yuzbashev et al. 2010
	Basfia succiniciproducens DD1	Crude glycerol	0.09	1.02	Yuzbashev et al. 2010
Dihydroxyacetone	*Gluconobacter* sp. NBRC1034565	Crude glycerol		0.16	Habe et al. 2009
Lactic acid	*E. coli AC-521*	Crude glycerol	0.49	0.9	Hong et al. 2009
Propionic acid	*Propionibacterium acidipropionici*	Pure glycerol	0.1	0.66	Zhang and Yang 2009
		Crude glycerol	0.085	0.88	Zhang and Yang 2009

4.4 Conclusion

Increasing awareness of environmental pollution, diminishing oil resources, and high-cost petroleum products leads to production of biodiesel. Glycerol is an ingenious carbon as well as energy source employed in the production of high-value chemicals. For the same low cost, high degree of reduction and availability makes glycerol as an attractive carbon sources for the production of fuels and high-value chemicals such as ethanol, butanol, citric acid, glyceric acid. Wild-type natural producers and genetic engineered strains are proficient and sustainable on utilizing crude glycerol as a carbon source for the bioconversion process. The major drawback with the bioconversion process is the inability of microorganism to use crude glycerol directly due to the various major and minor impurities and low product specificity. Utilization of crude glycerol will greatly benefit not only the economy of biodiesel industry, but also the bio-based industries completely. If biodiesel industries itself proceed for purification process, the amount invested on purification process was very less and economical when compared with the market value of product produced by bioconversion process. Also, those industries enjoy numerous benefits such as self-disposal of crude glycerol, zero liquid discharge, and elimination of the risk contamination followed by legal sanctions. This technology will have great substantial impact on economic and environmental sectors.

References

Agarwal GP (1990) Glycerol. Adv Biochem Eng Biotechnol 41:95–128

Anastassiadis S, Morgunov IG, Kamzolova SV, Finogenova TV (2008) Citric acid production patent review. Recent Pat Biotechnol 2:107–123

Andre L, Hemming A, Adler L (1991) Osmoregulation in *Saccharomyces cervisiae*. Studies on the osmotic induction of glycerol production and glycerol-3-phosphate dehydrogenase (NADC). FEBS Lett 286(1–2):13–17

Auta HS, Abidoye KT, Tahir H (2014) Sesan Abiodun Aransiola citric acid production by *Aspergillusniger* cultivated on Parkiabiglobosa fruit pulp. Int Sch Res Not 762021:8

Ayoub M, Abdullah AZ (2012) Critical review on the current scenario and significance of crude glycerol resulting from biodiesel industry towards more sustainable renewable energy industry. Renew Sustain Energy Rev 16:2671–2686

Barbirato F, Bories A (1997) Relationship between physiology of Enterobacteragglomerans CNCM 1210 grown anaerobically on glycerol and the culture conditions. Res Microbiol 148:475–484

Beauprez JJ, DeMey M, Soetaert WK (2010) Microbial succinic acid production, natural versus metabolic engineered producers, process. Biochem 45:1103–1114

Bhaskar S, Abhishek G, Ismail R, Faizal B (2014) Towards a sustainable approach for development of biodiesel from plant and microalgae. Renew Sustain Energy Rev 29:216–245

Bianchi CL, Canton P, Dimittratos N, Porta F, Parti L (2005) Selective oxidation of glycerol with oxygen using mono and bimetallic catalyst based on Au, Pd, and Pt metals. Catal Today 102:203–212

Blankschein MD, Clomburg JM, Gonzalez R (2010) *Escherichia coli* strains engineerined for homofermentative production of D-Latic acid from glycerol. Metabolic Eng 12:409–419

Bohon MD, Metzger BA, Linak WP, King CJ, Roberts WL (2011) Glycerol combustion and emissions. Proc Combust Inst 33:2717–2724

Canakci M, Sanli H (2008) Biodiesel production from various feedstocks and their effects on the fuel properties. J Ind Microbiol Biotechnol 35:431–441

Chen X, Zhang DJ, Qi WT, Goa SJ, Xiu ZL, Xu P (2003) Microbial fed batch production of 1,3 propanediol by Klebsiella pneumoniae under micro aerobic conditions. Appl Microbiol Bioethanol 63:143–146

Choi WJ, Hartono MR, Chan WH, Yeo SS (2011) Ethanol production from biodiesel-derived crude glycerol by newly isolated Kluyvera cryocrescens. Appl Microbiol Biotechnol 89: 1255–1264

Chozhavendhan S, Praveenkumar R, Sivarathanakumar S (2014) Comparative study on Candida sp for the production glycerol. Int J Chemtech Res 6:5058–5063

Chozhavendhan S, Praveen Kumar R, Bharathiraja B, Jayakumar M (2016) Recent progress on transforming crude glycerol into high value chemicals: a critical review. Biofuels. https://doi.org/10.1080/17597269.2016.1174018

Christy Mathelin R, Sumitha V, Sivanandham M (2015) Semi purification and characterization of biodiesel waste derived crude glycerol. Int J Innovative Res Technol Sci Eng 1(4):141–146

Colin T, Bories A, Lavigne C (2001) Effect of acetate and butyrate during glycerol fermentation by Clostridium butyricim. Curr Microbiol 43:238–243

Elina S, Svetlana C, Svetlana DK, Kampars ZV, Vera Z, Lidija K, Eriks P (2013) Selective liquid phase oxidation of glycerol to glyceric acid over novel supported Pt catalyst. J Serb Chem Soc 78(9):1359–1372

Erin J, Tahereh S, Sascha K, Garret M, Lars R (2016) Consolidating biofuel platforms through the fermentative bioconversion of crude glycerol to butanol. World J Microbiol Biotechnol 32:1–14

Fjerbaek L, Christensen KV, Norddahl B (2009) A review of the current state of biodiesel production using enzymatic transesterification. Biotechnol Bio eng 102:1298–1315

Gonzalez-Pajuelo M, Andrade JC, Vasconcelos I (2004) Production of 1, 3-propanediol by *Clostridium butyricum* VPI 3266 using a synthetic medium and raw glycerol. J Ind Microbiol Biotechnol 31:442–446

Gonzalez-Pajuelo M, Meynial-Salles I, Mendes F, Soucaille P, Vasconcelos I (2005) Metabolic engineering of *Clostridium acetobutrylicum* in the industrial production of 1, 3 propanediol from glycerol. Metab. Engg. 7:3329–3336

Gonzalez-pajuelo M, Meynial-Salles I, Mendis F, Soucallie P, Vasconclo I (2006) Microbio; conversion of glycerol to 1, 3 propanediol: physiological comparison of a natural producer *Clostridium butyricum* VPI 3266 and a engineered strain *Clostridiumacetobutylicum*DG1 (pSPD5). Appl Env Microbio 72:96–101

Gray KA, Zhao L, Emptage M (2006) Bioethanol Curr Opin Chem Biol 10:141–146

Green SR, Whalen EA, Molokie E (1961) Dihydroxyacetone: Production and uses. J Biochem Microb Technol Eng 3:351–355

Guerrero-Perez MO, Juana MR, Jorge B, Jose RM, Tomas C (2009) Recent inventions in glycerol transformations and processing. Recent Patents Chem Eng 2:11–21

Habe H, Shimada Y, Fukuoka T, Kitamoto D, Itagaki M, Watanabe K, Yanagishita H, Sakaki K (2009) Production of glyceric acid by Gluconobacter sp. NBRC3259 using raw glycerol. Biosci Biotechnol Biochem 73:1799–1805

Handa SS, Sharma A, Chakraborti KK (1986) Natural products and plants as a liver protecting drugs. Fitoterapia 57:307–351

Hirschmann S, Baganz K, Koschik I (2005) Development of an integrated bioconversion process for the production of 1, 3 propanediol from raw glycerol waters. Landbauforsch Volk 55: 261–267

Homann T, Tag C, Biebl H (1990) Fermentation of glycerol to 1, 3-propanediol by Klebsiella and Citrobacter strains. Appl Microbiol Biotechnol 53:435–440

Hong AA, Kk Cheng, Peng F, Zhou S, Son Y, Liu CM, Liu DH (2009) Strain isolation and optimization of process parameter for bioconversion of glycerol to lactic acid. J Chem Technol Biot 84:1576–1581

Hong WK, Kim CH, Heo SY, Luo L, Oh BR, Seo JW (2010) Enhanced production of ethanol from glycerol by engineered Hansenulapolymorpha expressing pyruvate decarboxylase and aldehyde dehydrogenase genes from Zymomonas mobilis. Biotechnol Lett 32:1077–1082

Ibrahim MHA, Steinbüchel A (2010) Zobellella denitrificans strain MW1, a newly isolated bacterium suitable for poly (3-hydroxybutyrate) production from glycerol. J ApplMicrobiol 108:214–225

Imandi SB, Bandaru VVR, Somalanka SR (2007) Optimization of medium constituents for the production of citric acid from byproduct glycerol using Doehlert experimental design. Enzyme Microb Technol 40:1367–1372

Isahak WNRWIM (2010) Purification of crude glycerol from transesterification RBD palm oil over homogenous catalysts for the bio lubricant preparation. Appl Sci 10:2590–2595

Ito T, Nakashimada Y, Senba K, Matsui T, Nishio N (2005) Hydrogen and ethanol production glycerol containing waste discharged fter biodiesel manufacturing process. J Biosi Bioeng 100:260–265

Kamzolova SV, Fatykhova AR, Dedyukhina EG (2011) Citric acid production by yeast grown on glycerol-containing waste from biodiesel industry. Food Technol Biotechnol 49:65–74

Kijora C, Kupsch RD (1996) Evaluation of technical glycerols from Biodiesel production as a feed component in fattening of pigs. Fett/Lipid 98:240–245

Laque R, Herrero-Davila L, Campelo JM, Clark JH, Hidalgo JM, Luna D, Marinas JM, Romero AA (2008) Biofuels: a technological perspective. Energy Environ Sci 1:542–564

Lee PC, Lee WG, Lee SY, Chang HN (2001) Succinic acid production with reduced byproduct formation in the fermentation of *Anerobiospirullum succinoproducens* using glycerol as a carbon source. Biotechnol Bioeng 72:41–48

Leoneti AB, Aragoa- Leoneti V, Valle S, Oliveria WB (2012) Alternatives for the use of unrefined glycerol. Renew Energy 45:138–145

Leung DYC, Wu X, Leung MKH (2010) A review on biodiesel production using catalyzed transesterification. Appl Energy 87:1083–1095

Lidia S, Rozanska, Stanisław B, Iwona G (2014) Effect of glycerol and dihydroxyacetone concentrations in the culture medium on the growth of acetic acid bacteria Gluconobacter *oxydans*. ATCC 621 Eur Food Res Technol 239:453–461

Liu T, Koh CMJ, Ji L (2011) Bioconversion of crude glycerol to glycolipids in Ustilagomaydis. Bioresour Technol 102:3927–3933

Mattam AJ, Clomburg JM, Gonzalez R (2013) Fermentation of glycerol and production of valuable chemical and biofuel molecules. Biotechnol Lett 35:831–842

Metsoviti M, Paramithiotis S, Drosinos EH, Galiotou-Panayotou M, Nychas GYE, Zeng AP, Papanikolaou S (2012a) Screening of bacterial strains capable of converting biodiesel-derived raw glycerol into 1, 3-propanediol, 2, 3-butanediol and ethanol. Eng Life Sci 12:57–68

Metsoviti M, Paraskevaidi K, Koutinas A, Zeng A, Papanikolaou S (2012b) Production of 1, 3 propanediol, 2, 3 butanediol and ethanol by newly isolated Klebsiella oxytoca strain growing on biodiesel derived glycerol based media. Process Biochem 47(12):1872–1882

Millard CS, Chao YP, Liao JC, Donnelly MI (1996) Enhanced production of succinic acid by overexpression of phophoenolpyruvate carboxylase in *E.coli*. Appl Environ Microbiol 62(5):1808

Min J, Wenyu D, Xi YL, Wu M, Xiangping K, Ma JF(2014). Succinic acid production from sucrose by *Actinobacillus succinogenes*. Bioresour Technol 153:327–332

Mishra R, Jain S, Kumar R (2008) Microbial production of dihydroxyacetone. Biotechnol Adv 26:293–303

Mythili R, Venkatachalam P, Subramanian P, Uma D (2014) Recovery of side streams in biodiesel production process. Fuel 117:103–108

Nanda MR, Yuan Z, Qin W, Poirier MA, Chunbao X (2014) Purification of crude glycerol using acidification: effects of acid types and product characterization. Austin J Chem Eng 1:1–7

Ngo TA, Kim MS, Sim SJ (2011) High-yield biohydrogen production from biodiesel manufacturing waste by Thermotoganeapolitana. Int J Hydrogen Energy 36:5836–5842

Noshadi I, Amin NAS, Parnas Richard S (2012) Continuous production of biodiesel from waste cooking oil in a reactive distillation column catalyzed by solid heteropolyacid: optimization using response surface methodology (RSM). Fuel 94:156–164

Pagliaro M, Ciriminna R, Kimura H, Rossi M, Pina CD (2007) From glycerol to value-added products. Glycerol Chem Angewandte Chemie Int Ed 46:4434–4440

Quispe CAG, Coronado CJR, Carvalho JA Jr (2013) Glycerol: production, consumption, prices, characterization and new trends in combustion. Renew Sustain Energy Rev 27:475–493

Reimann A, Biebl H, Deckwer WD (1998) Production of 1, 3-propanediol by Clostridium butyricum in continuous culture with cell recycling. Appl Microbiol Biotechnol 49:359–363

Reynaud C, Sarcabal P, Meynial-Salles I, Croux Soucallie P (2003) Molecular characterization of the 1, 3 Propanediol operon of *Clostridium butyricum*. Proc Natl Acd Sci 100:5010–5015

Rosseto R, Tcacenco CM, Ranganathan R, Hajdu (2008) Synthesis of phosphatidylcholine analogues derived from glyceric acid: a new class of biologically active phospholipid compounds. Thtrahedron let 49:3500–503

Rymowicz W, Rywinska A, Zarowska B (2006) Citric acid production by acetate mutants of Yarrowia lipolytica. Chem Pap 60(5):391–394

Rywinska A, Rymowicz W, Żarowska B, Wojtatowicz M (2009) Biosynthesis of citric acid from glycerol by acetate mutants of *Yarrowia lipolytica* in fed-batch fermentation. Food Technol Biotechnol 47:1–6

Schieck SJ, Shurson GC, Kerr BJ, Johnston LJ (2010) Evaluation of glycerol, a biodiesel coproduct, in grow-finish pig diets to support growth and pork quality. J Anim Sci 88:3927–3935

Schiermeier Q, Tollefson J, Scully T, Witze A, Morton O (2008) Energy alternatives: electricity without carbon. Nature 454:816–823

Schultz EL, Souza DT, Damso MCT (2014) The glycerol refinery: a purpose for Brazillian biodiesel production. Chem Biol Technol Agric 1:1–9

Shah ND, Chattoo BB, Baroda RM, Patiala VM (1993) Starch hydrolysate, an optimal and economical source of carbon for the secretion of citric acid by Yarrow lipolytica. Starch 45:104–109

Sharma YC, Singh B, Upadhyay SN (2008) Advancements in development and characterization of biodiesel: a review. Fuel 87:2355–2373

Shen Y, Pei Z, Yuan W, Mao E (2009) Effect of nitrogen and extraction method on algae lipid yield. Int J Agric Biol Eng 2:51–57

Sivasamy A, Cheah KY, Fornasiero P, Francis K, Zinoviev S, Stanislav M (2009) Catalytic applications in the production of biodiesel from vegetable oils. Chem Sus Chem 2:278–300

Sneha KA, Rafael AG, Zhiyou W (2009) Use of biodiesel-derived crude glycerol for producing Eicosapentaenoic acid (EPA) by the fungus Pythiumir regular. J Agric Food Chem 57(7):2739

Soccol CR, Vandenberghe LPS, Rodrigues C, Pandey A (2006) New Perspectives for citric acid production and application. Food Technol Biotechnol 44(2):141–149

Solomon BO, Zeng AP, Biebl H (1995) Comparison of the energetic efficiency of hydrogen and oxychemicals formation in Klebsiella pneumonia and Clostridium butyricum during anaerobic growth on glycerol. J Biotechnol 39:107–117

Song H, Lee SY (2006) Production of succinic acid by bacterial fermentation. Enzyme Microb Technol 39:352–361

Stelmachowski M (2011) Utilization of glycerol, a by-product of the transesterification process of vegetable oils: are view. Ecol Chem Eng 18:9–30

Sun J, Caton JA, Jacobs TJ (2010) Oxides of nitrogen emissions from biodiesel-fuelled diesel engines. Prog Energy Combust 36:677–695

Taconi KA, Venkataramanan KP, Johnson DT (2009) Growth and solvent production by *Clostridium pasteurianum* ATCC 6013 utilizing biodiesel-derived crude glycerol as the sole carbon source. Environ Prog Sustain Energy 28:100–110

Teng WK, Ngoh GC, Yusoff R, Arouna MK (2014) A review on the performance of glycerol carbonate production via catalytic transesterification: effect of influencing parameter. Energy Convers Manage 88:484–497

Trinch CT, Carlson R, Wlaschin A, Srienc F (2008) Minimal *Escherichia coli* cells for the most efficient production of ethanol from hexoses and pentoses. Appl Enviorn Microbiol 74: 3634–3643

Vandenberghe LPS, Soccol CR, Pandey A, Lebeault JM (1999) Review: microbial production of citric acid. Braz Arch Biol Technol 42:263–276

Vasudevan PT, Briggs M (2008) Biodiesel production-current state if the art and challenges. J Ind Microbiol Biotechnol 35:421–430

Vincente G, Martinez M, Aracil J (2004) Integrated biodiesel production: a comparison of different homogeneous catalysts systems. Bioresour Technol 92(3):297–305

Vlysidis A, Binns M, Webb C, Theodoropoulos C (2011) Glycerol utilization for the production of chemicals: conversion to succinicacid, a combined experimental and computational study. Biochem Eng J 58–59:1–11

Waala AE, Aymen SY, Amal EA, Magdy AM (2016) Utilization of crude glycerol as a substrate for the production of Rhamnoliid by *Pseudomonas aeruginosa*. Biotech Research Int 1–9

Wen Z, Pyle DJ, Athalye SK (2009) Glycerol waste from biodiesel manufacturing. In: Aggelis G (ed) Microbial conversions of raw glycerol. Nova Science, New York, pp 1–7

Wilkens E, Ringel AK, Hortig D, Willke T, Vorlop KD (2012) High level production of 1, 3-propanediol form crude glycerol by Clostridium butyricum AKR102a. Appl Microbiol Biotechnol 93:1057–1063

Wilson K, Lee AF (2012) Rational design of heterogeneous catalysts for biodiesel synthesis. Catal Sci Technol 2:884–897

Yahaya MS, Wan Mohd Ashri WD, Abdul Aziz AR (2014) Activity of solid acid catalysts for biodiesel production: a critical review. Appl Catal A 470:140–161

Yamano LP, York SW, Zhou S, Shanmugam KT, Ingram LO (2008) Re-engineering *Escherichia coli* for ethanol production. Biotechnol Letter 30:2097–2103

Yazdani SS, Gonzalez R (2007) Anaerobic fermentation of glycerol: a path to economic viability for the biofuels industry. Curr Opin Biotechnol 18:213–219

Yuzbashev TV, Yuzbasheva EY, Sobolevskaya TI, Laptev IA, Vybornaya TV, Larnia AS, Matsui K, Fukui K, Sp Sineoky (2010) Production of succinic acid at low pH by a recombinant strain of yeast *Yarrowia lipolytica*. Biotechnol Bioeng 107:673–682

Zeikus JG, Jain MK, Elankovan P (1999) Biotechnology of succinic acid production and markets for the dervied industrial products. Appl Microbial Biotechnol 51:545–552

Zhang A, Yang, ST (2009) Propionic acid production from glycerol by metabolically engineered *Propionibacterium acidipropionici*. Process Biochem 44:1346–1351

Zhanyou C, Denver P, Zhiyou W, Craig F, Shulin C (2007) A laboratory study of producing docosahexaenoic acid from biodiesel-waste glycerol by microalgal fermentation. Process Biochem 42:1537–1545

Chapter 5
Utilization of Citrus Waste Biomass for Antioxidant Production by Solid-State Fermentation

Ramón Larios-Cruz, Rosa M. Rodríguez-Jasso, Hector A. Ruiz, Arely Prado-Barragán, Jorge E. Wong-Paz, Raúl Rodríguez-Herrera, Julio C. Montañez and Cristóbal N. Aguilar

Abstract Citrus fruits such as lemon, orange, grapefruit, and tangerine are consumed for their flavor, low cost, and human health benefits. However, citrus juice extraction generates by-products that are mostly unused and is discharged in landfills. In this study, the by-products of lemon, orange, grapefruit, and tangerine were subjected to solid-state fermentation (SSF) using *Fusarium oxysporum*, *Penicillium purpurogenum* GH2, *Trichoderma harzianum* T1-04, and *Aspergillus niger* GH1 to enhance their antioxidant activity. After fermentation, ethanol extracts were obtained and tested for their antioxidative activity by employing three techniques, 2,2-diphenyl-1-picrylhydrazyl (DPPH$^{\cdot}$), ferric reducing antioxidant power (FRAP), and lipid oxidation inhibition (LOI). An increase in antioxidant activity from 33.13 to 41.62 mg/g$_{msi}$ of antioxidants after fermentation of tangerine by-products by *A. niger* GH1 was observed. Major compounds present in ethanol extracts obtained after fermentation by *A. niger* GH1were identified by HPLC-MS, and their m/z corresponded to chlorogenic acid, didymin, naringin, and hesperidin. These results indicated that SSF is a suitable method to enhance antioxidant activity of citrus by-products.

Keywords Lemon · Orange · Grapefruit · Tangerine · *Aspergillus niger* GH1 Ethanol extracts

---

R. Larios-Cruz · R. M. Rodríguez-Jasso · H. A. Ruiz · R. Rodríguez-Herrera · J. C. Montañez · C. N. Aguilar (✉)
Group of Bioprocesses, Food Research Department, School of Chemistry, Universidad Autónoma de Coahuila, 25280 Saltillo, Coahuila, Mexico
e-mail: cristobal.aguilar@uadec.edu.mx

A. Prado Barragán
Biotechnology Department, Universidad Autónoma Metropolitana, Unidad Iztapalapa, CDMX, Mexico

J. E. Wong-Paz
Departamento de Ingenierías, Instituto Tecnológico de Ciudad Valles. Tecnológico Nacional de México, 79010 Ciudad Valles, San Luis Potosi, Mexico

R. R. Singhania et al. (eds.), *Waste to Wealth*, Energy, Environment, and Sustainability, https://doi.org/10.1007/978-981-10-7431-8_5

5.1 Introduction

Worldwide, the most important citrus fruits are lemon, orange, grapefruit, and tangerine. These citrus fruits are consumed in different ways such as fresh fruit, juices or beverages derived from concentrate juice (Li et al. 2006b; Liu et al. 2012). Currently, Mexico is the fifth largest citrus producer of the world (Liu et al. 2012), and the national production of orange is more than 50% of total citrus production, followed by lemon, grapefruit, and tangerine (Table 5.1). Among the states of Mexico, Veracruz is the major citrus producer, and in 2015 recorded 50% of the National production in orange, grapefruit, and tangerine (SIAP 2016).

The fresh fruit consumption in the word is over 75% of total production; in Mexico, almost 80% of the total citrus production is for internal market consumption (SIAP 2016). These citrus fruits are characterized by high carbohydrates content and presence of important nutritional elements such as fiber, organic acids, protein, lipids, micronutrients, vitamins, and compounds with antioxidant activity (Liu et al. 2012). The citrus juice has phenolic compounds which offer benefits due to its antioxidant capacity. However, juice extraction process generates by-products, which are underutilized or unused and there is potential presence of bioactive compounds such as antioxidants, which can add value to these by-products (Li et al. 2006a; Wang et al. 2011). Citrus by-products have several applications including production of ethanol, methane, limonene, pectin (Pourbafrani et al. 2010), pectinases (Biz et al. 2016; Tao et al. 2011), endoglucanase, β-glucosidase, and xylanase (Tao et al. 2011). Extraction of bioactive compounds from citrus by-products is important as it enhances value and further generates employment. Several methods have been developed for extraction, and extraction from citrus by-products mostly involves liquid–solid extraction, in which diverse solvents at different concentrations are used (Barros et al. 2012; Lagha-Benamrouche and Madani 2013; Tounsi et al. 2011), and in some cases only water is used (Xu et al. 2008). Ultrasound (Dahmoune et al. 2013; Sun et al. 2013), microwave (Dahmoune et al. 2013; Sun et al. 2013), enzymatic processing (Li et al. 2006b), and SSF (Correia et al. 2004; Yang et al. 2013) are some other extraction methods employed.

The SSF is considered as an alternative for by-products utilization and to add value by recovering its biological compounds and/or activity (Ng et al. 2014) or

Table 5.1 Production of the principal citrus fruits in Mexico

Types	Production (t)[a]				
	2011	2012	2013	2014	2015
Lemon	2,132,921.78	2,055,208.89	2,187,257.20	2,120,612.50	2,326,068.34
Orange	4,079,677.74	3,666,789.65	4,533,427.86	4,409,967.62	4,515,520.33
Grapefruit	397,266.70	415,470.85	424,678.08	425,432.97	424,315.36
Tangerine	231,167.16	272,426.07	297,326.45	323,617.37	291,078.27

[a]Servicio de Información Agroalimentaria y Pesquera (SIAP 2016)

enzyme production (Mohseni et al. 2012). Bhanja Dey and Kuhad (2014a) employed SSF using *Rhizopus oryzae* RCK2012 to enhance the antioxidant activity of wheat grains and reported that the recovery of antioxidants with different solvents were improved after the fermentation process. Later, Bhanja Dey and Kuhad (2014b) reported enhancement of the antioxidant content in cereals by SSF with filamentous fungi. Total phenolic content was found to be enhanced after fermentation of cereals and also reported an increase of antioxidants, which indicated that SSF is a suitable technique to enhance the content of antioxidants. Analysis of ultra-performance liquid chromatography of unfermented and fermented extracts showed that the profile of phenolic compounds was different in both samples, some of the compounds are unknown, and most of the unknown were present in the fermented extracts. Hence, the suitability of SSF to increase the value of citrus by-products by way of enhancing the presence of antioxidant compounds in the extracts obtained after fermentation was evaluated, along with identification of polyphenols present in fermented citrus extracts.

5.2 Materials and Methods

5.2.1 Chemicals

2,2-diphenyl-1-picrylhydrazyl (DPPH˙), 2,4,6-tris(2-pyridyl)-s-triazine (TPTZ), linoleic acid, and (±)-6-hydroxi-2,5,7,8-tetramethylchromane-2-carboxylic acid (Trolox) were purchased from Sigma (St Louis, USA). All others chemicals were analytical grade.

5.2.2 Citrus by-Products and Microorganisms

Citrus by-products from lemon, orange, grapefruit, and tangerine were used as the substrates for SSF in order to extract biological compounds with antioxidant capacity. Citrus by-products were donated by a juice producer at the Fruit Market in Saltillo, Coahuila, Mexico. By-products (peel and bagasse) were collected immediately after juice extraction. The material was dried in an oven at 60 °C for 48 h. Then, samples were pulverized in a mill to a mesh particle size between of 40 and 100 (420–150 μm) and stored at room temperature (27 ± 3 °C) in dark until used.

Four fungal (*Fusarium oxysporum*, *Penicillium purpurogenum* GH2, *Aspergillus niger* GH1, and *Trichoderma harzianum* T104) strains were used in this study. The first three strains were obtained from the DIA–UAdeC microbial collection (Saltillo, México), while *T. harzianum* T104 was obtained from the UAAAN microbial collection (Saltillo, México). Each fungal strain was stored at −20 °C in a medium-containing glycerol-skimmed milk. Potato dextrose agar (PDA) was used as a medium for activation of the fungal strains.

5.2.3 Solid-State Fermentation

5.2.3.1 Radial Growth

To determine fungal growth on citrus by-products, 3 g of sample was placed in a petri dish and distributed uniformly along the plate. To achieve 70% (W/V) of humidity, the substrates were supplemented with distilled water in one treatment and in another with minimal medium (Czapek-Dox). The minimal medium was prepared as follows: K_2HPO_4 (1 g L^{-1}), $MgSO_4$ (0.5 g L^{-1}), KCl (0.5 g L^{-1}), $FeSO_4$ (0.1 g L^{-1}), and 1 mL L^{-1} of oligo-elements solution ($Na_2B_4O_7 \cdot 10H_2O$, 0.1 g L^{-1}; $MnCl_2 \cdot 3H_2O$, 0.05 g L^{-1}; $Na_2MO_4 \cdot 2H_2O$, 0.5 g L^{-1}; $CuSO_4 \cdot 5H_2O$, 0.25 g L^{-1}). Inoculation of citrus by-product was done at the center of the petri dish (1×10^5 spores g^{-1}), and the fungal growth was measured at four points at every 8 h during a fermentation time of 120 h. Citrus by-products were not sterilized prior to inoculation. All experiments were done in triplicate.

5.2.3.2 Antioxidant Capacity of Citrus by-Products by Solid-State Fermentation

To determine antioxidant activity of the extracts, fermentation was performed again with distilled water supplement and sterilized before inoculation (2×10^6 spores g^{-1}). *A. niger* GH1 was the fungus selected for fermentation of all citrus by-products. The fermentation time was also selected, based on the results of a preliminary experiment where *A. niger* GH1 elicited the highest antioxidant capacity on each different citrus by-products, 96 h for lemon and grapefruit and 120 h for orange and tangerine. Samples were taken at every 24 h during a total fermentation period of 120 h. For extraction of biological compounds, 15 mL of 70% ethanol was added to the fermented by-products and agitated at 320 rpm for 15 min. Afterward, the samples were pressed to obtain the extracts and the extracts were then filtered through cotton and filter paper (Whatman 41). The filtered extracts were stored in amber bottles against light at −20 °C until its analysis. All experiments were done in duplicate.

5.2.4 Antioxidant Capacity Assays

5.2.4.1 DPPH˙-Scavenging Capacity Assay

Ethanol extracts were analyzed for their antioxidant capacity according to the method mentioned by Meléndez et al. (2014). About 193 μL of 60 μM DPPH˙ radical was added to 7 μL of ethanol extract, and the mixture was left in dark for

30 min, and then absorbance was read at 517 nm in a spectrophotometer (Tecan Sunrise, Grödig, Austria). Ethanol was used as blank. Trolox was used as standard compound. Trolox equivalents of extracts were calculated with the standard curve and expressed in μM of Trolox g^{-1} of initial citrus by-product in dry weight.

5.2.4.2 Ferric Reducing Antioxidant Power (FRAP) Assay

The FRAP assay was done according to the method described by Benzie and Strain (1996). Briefly, 10 μL of sample were placed in a 96-well microplate with 290 μL of freshly prepared FRAP reactive. FRAP reactive was prepared by mixing 10 mM 2,4,6-tris(1-pyridyl)-5-triazine diluted in 40 mM HCl, 20 mM $FeCl_3$, and 0.3 M acetate buffer (pH 3.6) in a proportion of 1:1:10 (TPTZ:$FeCl_3$:Acetate buffer). Reaction was incubated for 15 min at 37 °C. Then, absorbance was read at 593 nm in a spectrophotometer (Tecan Sunrise, Grödig, Austria). Distilled water was used as blank. $FeSO_4 \cdot 7H_2O$ was the standard compound. The ferrous sulfate equivalents of extracts were calculated with the standard curve and expressed in μM of Fe (II) g^{-1} of initial citrus by-product in dry weight.

5.2.4.3 Lipid Oxidation Inhibition (LOI) Assay

The LOI assay was done according to the method reported by Martínez-Ávila et al. (2012). The linoleic acid solution was prepared with 560 mg of linoleic acid in 8 mL of 96% ethanol and 1.5 g of Tween 20. For the reaction, 50 μL of extract was mixed with 100 μL of linoleic acid solution and 1.5 mL of 0.02 M acetate buffer (pH 4.0). Distilled water was used as control. After homogenization and incubation (37 °C, 1 min), 750 μL of 0.5 mM $FeCl_2$-EDTA 1:1 (Rhee 1978) was added to the reaction mixture in order to induce oxidation of linoleic acid and incubated at 37 °C for 1 and 24 h. After that 250 μL of the reaction mixture was added with 1 mL of 0.1 M NaOH in 10% ethanol and 2.5 mL of 10% ethanol. This step was performed to stop the oxidation process. Samples were read at 232 nm against 10% ethanol as blank. The percentage of antioxidant activity was calculated according to the equation described by Toivonen and Sweeney (1998), which is as follows:

$$\%\text{LOI} = \left(\frac{\Delta D_{\text{control}} - \Delta D_{\text{extract}}}{\Delta D_{\text{control}}}\right) \times 100 \tag{1}$$

where $\Delta D_{\text{control}}$ is the difference in absorbance between 24 and 1 h of reaction time in the controls and $\Delta D_{\text{extract}}$ is the difference in absorbance between 24 and 1 h of reaction time in the samples.

5.2.4.4 Total Antioxidant Activity Calculations

Results of DPPH˙, FRAP, and LOI assays were expressed in mg g_{idw}^{-1} (idw corresponds to initial dry weight) to display three graphs in one. For DPPH˙ and FRAP assays, molar concentration was calculated, the following Eq. (2) was used to determine antioxidant activity of ethanol extracts. For LOI assay, the percentage was expressed in (g) (100 $g_{idw})^{-1}$. After mg g_{idw}^{-1} was calculated, all assays were added up to a total antioxidant activity of extracts (mg) based on initial dry weight of citrus by-products (g_{idw}).

$$\mathrm{Aa} = \frac{M * v * \mathrm{Mw}}{w} \quad (2)$$

Where Aa is antioxidant activity (mg g_{idw}^{-1}), M is molar concentration (mmol L^{-1}), v is volume of compounds (L) extracted, Mw is molar weight of standard compound (mg $mmol^{-1}$), and w is the initial dry weight of citrus by-products (g).

5.2.5 *High-performance Liquid Chromatography–Mass Spectrometry (HPLC–MS) Analysis*

Ethanol extracts for HPLC–MS analysis were selected according to their total antioxidant capacity.

5.2.5.1 HPLC–MS Procedure

Analyses were performed using an HPLC–MS system, which included an autosampler (Varian ProStar 410, USA), a ternary pump (Varian ProStar 230I, USA), and a PDA detector (Varian ProStar 330, USA) set at 280 nm and coupled to Workstation Multi-Instrument (V. 6.2) for data acquisition. For mass spectrometry analysis, a liquid chromatography ion trap mass spectrometer (Varian 500-MS IT Mass Spectrometer, USA) equipped with an electrospray ion source was used.

The column employed was a C18 5 μm (150 mm × 2.0 mm) and was maintained at 30 °C. The mobile phase consisted of 3% acetic acid (A) acetonitrile (B). The flow rate was 0.3 mL min^{-1} with an injection sample volume of 5 μL. The elution gradients for A were as follows: initial 97%, 5 min 91%, 15 min 84%, 45 min 50%, 48 min 10%, 55 min 97%, and 60 min 97%. The whole effluent (0.3 mL/min) was injected into the source of mass spectrometer, without splitting. All mass experiments were carried out in the negative mode $[M\text{-}H]^{-1}$. Nitrogen was used as nebulizing gas and helium as damping gas. The ion source parameters were spray voltage (3.5 kV), capillary voltage (90.0 V), and temperature (350 °C). Data were collected and processed using MS Workstation Software (V 6.9). Full scan spectra were acquired in the m/z range 50–2000.

5.2.6 *Statistical Analysis*

The ANOVA was done with a significance level of $p < 0.05$. When needed, the Tukey's multiple range test was used for treatment means comparison. All statistical analyses were performed using Statistica 7.0 (Stat Soft, Tulsa, OK, USA) software.

5.3 Results and Discussion

5.3.1 *Radial Fungal Growth*

Of the four fungal strains tested, *A. niger* GH1 presented the highest growth on citrus by-products, followed by *T. harzianum* T1-04, *P. purpurogenum* GH2, and *F. oxysporum* (Fig. 5.1). All fungi presented the same growth behavior on the different citrus by-products, which reflected their ability for growth on all tested substrates. On the other hand, a long lag phase especially for those two fungi with less growth could be due to the low quantity of spores used for inoculation. Similar to *A. niger* GH1, *T. harzianum* T1-04 was the fungus with a high invasion capacity on the petri dish completely with its growth in 120 h.

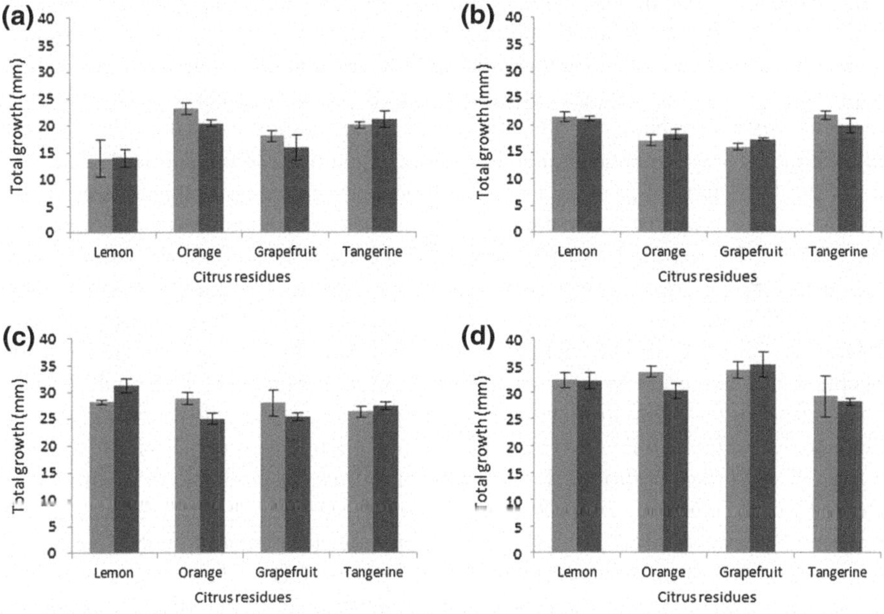

Fig. 5.1 Fungal growth on citrus by-products. Results are after a fermentation time of 120 h. ■ Not supplemented medium ■ Supplemented medium. **a** *F. oxysporum*, **b** *P. purpurogenum* GH2, **c** *T. harzianum* T1-04, and **d** *A. niger* GH1 in combination with all citrus by-products. Lines over the bars represent the standard deviation

Results showed that citrus by-products contained enough quantity of macro- and micronutrients for suitable fungal growth, since there was no significant growth difference between the treatments supplemented with water and minimal medium. Hence, supplementation with minimal medium was discarded for fermentation and could also aid in bringing down the cost of fermentation. Liu et al. (2012) reported that citrus contain sugars in abundance and various macro- and microelements important to improve the fruit nutritional characteristics (Li et al. 2006a). Elements such as potassium, the most abundant, calcium, magnesium, and others microelements such as copper, iron, manganese, and zinc are found in citrus. Those elements could be present in peel (Xu et al. 2008) and pulp. They are reported in more quantities in peel than in pulp (Cano and Bermejo 2011) and in more quantity in flavedo than in albedo (Barros et al. 2012). Citrus seeds also have minerals (El-Adawy et al. 1999) and most of them remain in enough quantity in citrus by-products and could support fungal growth.

5.3.2 Increase in Antioxidant Capacity of Citrus by-Products After SSF

Growth of all tested fungal strains on each citrus by-product was observed during SSF. Liberation of compounds with antioxidant capacity from the citrus by-products after SSF was also observed (Fig. 5.2). Extracts from tangerine by-products were found to possess the highest antioxidant activity. and this was observed with all tested fungi. However, it should be noted that tangerine extracts always presented higher antioxidant activity (before and after fermentation) than the other citrus by-products, but it was with grapefruit the recovery of antioxidants was higher than tangerine after fermentation. On the other hand, lemon presented the lowest antioxidant activity, whereas orange by-products lost most of the antioxidants at the end of the fermentation process. Li et al. (2006a) determined antioxidant activity of lemon, orange, grapefruit, and tangerine peels after extraction with different solvents. They also reported that they obtained the highest antioxidant activity with grapefruit citrus extracts, followed by tangerine, lemon, and orange. Further, they observed that the most suitable solvent for extraction of these compounds was ethanol at 72%. A similar behavior was observed in the present study: with exception that tangerine by-product recorded the highest values. This could be due to tangerine cultivar, environmental conditions of tangerine cultivation, degree of fruit ripeness, which could have affected the final chemical composition of the fruit.

Fungal enzymes hydrolyzed cell wall of citrus by-products over time and released more compounds with antioxidant activity. Then, recovery of these compounds was enhanced with ethanol solution after fermentation with three out of the four tested fungal strains. With *A. niger* GH1, *F. oxysporum*, and *P. purpurogenum* GH2, the increment in antioxidant activity of extracts was almost similar to that of during the fermentation (Fig. 5.2). Among these three, *A. niger* GH1 recorded the highest enhancement of antioxidant activity. Whereas,

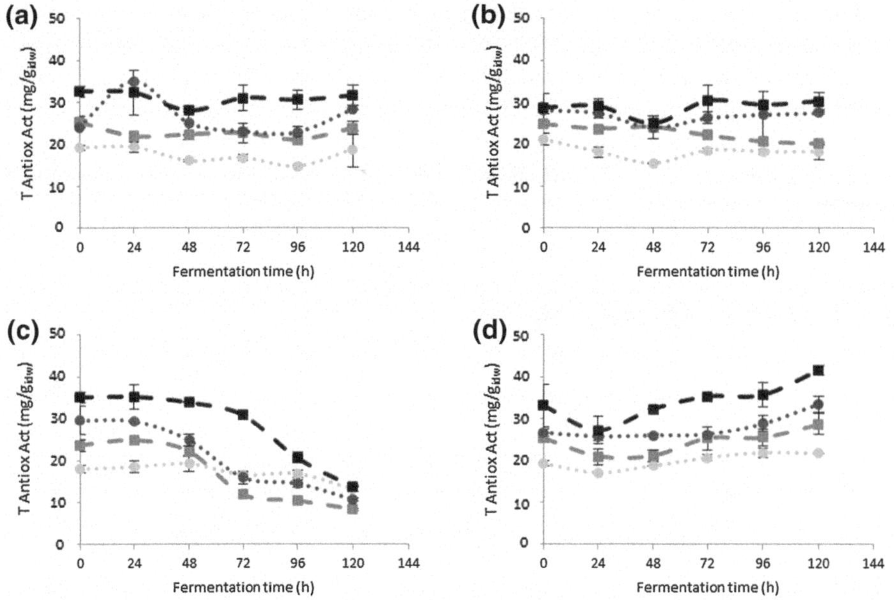

Fig. 5.2 Antioxidant activity of four citrus by-products after SSF. ● Lemon, ▪ orange, ● grapefruit, ■ tangerine. **a** *F. oxysporum*, **b** *P. purpurogenum* GH2, **c** *T. harzianum* T1-04, and **d** *A. niger* GH1 in combination with all citrus by-products. Lines over the bars represent the standard deviation. T Antiox Act: total antioxidant activity

T. harzianum T1-04 showed a decrease in antioxidant activity at the end of the fermentation, and it could be due to consumption or biotransformation of antioxidant compounds during fermentation. Distribution of compounds with antioxidant activity is higher in citrus peel than in pulp (Barros et al. 2012; Xu et al. 2008). Similarly, it was reported that flavedo contains more antioxidant compounds than albedo (Cano and Bermejo 2011). These results suggest that the increase in antioxidant activity of citrus peel extracts is due to fungal enzymatic activity during SSF.

It can be inferred that enzyme production by the tested fungi hydrolyzed and degraded citrus by-product, which has been demonstrated earlier with different agro-industrial by-products (Pandey et al. 1999). Li et al. (2006b) applied commercial enzymes to increase antioxidant activity in citrus peel extracts. A similar behavior was also expected in the present work, and it was observed that nearly half of the initial substrate (on weight basis) was degraded by *A. niger* GH1 at the end of fermentation. Thus, enzymatic activity could be applied as pre treatment in the recovery of antioxidants. The SSF used in the present work was a pre-treatment due to the enzymes produced by the microorganisms. However, different factors could affect the fermentation process and not all the fungi enhanced the production of antioxidants. The metabolism of the microorganism with initial presence of antioxidants may play an important role in the differences with *T. harzianum* T1-04.

The effect of fungal strains tested in increasing the antioxidant activity of the obtained ethanol extracts after SSF was determined based on calculations mentioned earlier in methods section. Results showed that SSF with *A. niger* GH1 yielded the highest increase in antioxidant activity of all citrus by-products, and hence this strain was selected for further studies. In contrast, the extracts obtained after fermentation with *T. harzianum* T1-04 recorded the lowest antioxidant activity.

5.3.3 HPLC–MS Analysis

Extracts of unfermented and fermented citrus by-products were analyzed using HPLC-MS. Profile of antioxidant compounds presented in both unfermented and fermented citrus by-product extracts is presented in Fig. 5.3. It was observed that some compounds decreased in their absorbance units during fermentation, while others appear only at the end of fermentation process. These changes could be due to consumption or biotransformation of phenolic compounds by fungal metabolism (Barz and Hösel 1975; Cao et al. 2014). Hence, it is possible that some of the compounds observed at the end of fermentation could be metabolites formed from others compounds (Li et al. 2006b).

Polyphenols are widely present in most of the citrus fruits, and principal polyphenols in lemon are didymin, hesperidin (Peterson et al. 2006a), narirutin (Nogata et al. 2006), pinoresinol, and medioresinol (Peñalvo et al. 2005; Peterson et al. 2010). In the case of orange, didymin, hesperidin (Peterson et al. 2006b), narirutin, eriocitrin (Nogata et al. 2006), pinoresinol, medioresinol (Peñalvo et al. 2005; Peterson et al. 2010), ferulic acid and chlorogenic acid (Kelebek et al. 2009) are the major compounds. While in grapefruit, naringin, hesperidin, narirutin (Peterson et al. 2006b; Sameer et al. 2013; Xi et al. 2014a), didymin (Xi et al. 2014a; Sameer et al. 2013) have been reported. Whereas, tangerine has been reported to contain hesperidin, narirutin (Nogata et al. 2006; Peterson et al. 2006b), didymin (Peterson et al. 2006b), pinoresinol (Milder et al. 2005), and chlorogenic acid (Xi et al. 2014b; Zhang et al. 2014). In the present study, most of these polyphenols were identified in both unfermented and fermented ethanol extracts.

Polyphenols identified in ethanol extracts were chlorogenic acid with m/z of 353 $[M-H]^-$; didymin, m/z of 593 $[M-H]^-$; apigenin 7-O-apiosyl-glucoside (apiin), m/z of 563 $[M-H]^-$; pinoresinol, m/z of 357 $[M-H]^-$; medioresinol, m/z of 387 $[M-H]^-$; hesperidin, m/z of 609 $[M-H]^-$; naringin, m/z of 579 $[M-H]^-$; and an unknown compound with m/z of 289 $[M-H]^-$ (Fig. 5.3). It is important to mention that a compound with m/z of 289 $[M-H]^-$, which is similar to epicatechin standard was observed in all citrus by-products after SSF. However, the UV-vis spectrum and the retention time were completely different (data not shown). Further studies are needed to obtain a clear identification of this compound with m/z 289 $[M-H]^-$.

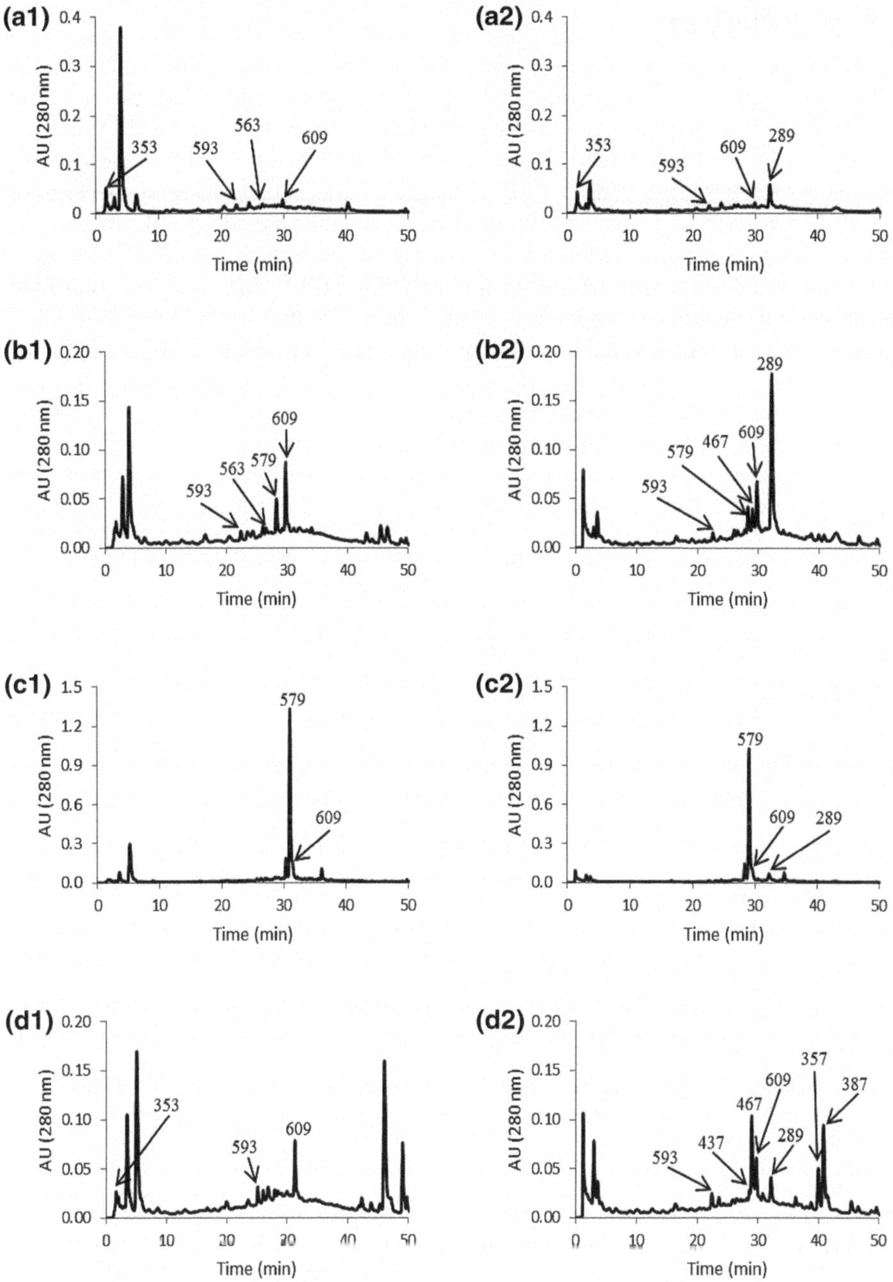

Fig. 5.3 HPLC–MS analysis of ethanol extracts of citrus residues after SSF. Figures with number 1 (left) are from zero time and figures with number 2 (right) are at the end of fermentation time. **a** Lemon (0 and 96 h), **b** orange (0 and 120 h), **c** grapefruit (0 and 96 h), and **d** tangerine (0 and 120 h). The m/z is presented in negative mode $[M\text{-}H]^-$ in top of each HPLC peak. All citrus by-products were inoculated with *A. niger* GH1

5.4 Conclusions

All citrus by-product extracts obtained after SSF showed antioxidant activity, and the increase in antioxidant activity was maximum using *A. niger* GH1. Whereas, *T. harzianum* T1-04 recorded consumption or biotransformation of these compounds. Extracts obtained after SSF of tangerine by-product showed the highest antioxidant activity. It was also observed that SSF changed the profile of antioxidant compounds without reduction of antioxidant properties and could be due to release or metabolites formed during fermentation. HPLC–MS analysis confirmed that some of phenolic compounds were conserved during the SSF, but other compounds with low molecular weight appeared after fermentation. Further studies are required in order to identify some of the compounds formed during the fermentation, which can the explain changes in the content of polyphenols of the obtained extracts after SSF of citrus by-products.

Acknowledgements RLC thanks the Mexican Council for Science and Technology (CONACYT), Mexico, for the financial support offered for his M.Sc. studies. Authors thank Dr. Nagamany Balagurusamy (School of Biological Science, Universidad Autonoma de Coahuila, Mexico) for the technical support to improve the quality of the manuscript.

References

Barz W, Hösel W (1975) Metabolism of flavonoids. In: Harborne JB, Mabry TJ, Mabry H (eds) The flavonoids. Springer, USA, pp 916–969. https://doi.org/10.1007/978-1-4899-2909-9_17

Benzie IFF, Strain JJ (1996) The ferric reducing ability of plasma (FRAP) as a measure of "antioxidant power": the FRAP assay. Anal Biochem 239(1):70–76. https://doi.org/10.1006/abio.1996.0292

Bhanja Dey T, Kuhad RC (2014a) Enhanced production and extraction of phenolic compounds from wheat by solid-state fermentation with *Rhizopus oryzae* RCK2012. Biotechnol Rep 4:120–127. https://doi.org/10.1016/j.btre.2014.09.006

Bhanja Dey T, Kuhad RC (2014b) Upgrading the antioxidant potential of cereals by their fungal fermentation under solid-state cultivation conditions. Lett Appl Microbiol 59(5):493–499. https://doi.org/10.1111/lam.12300

Biz A, Finkler ATJ, Pitol LO, Medina BS, Krieger N, Mitchell DA (2016) Production of pectinases by solid-state fermentation of a mixture of citrus waste and sugarcane bagasse in a pilot-scale packed-bed bioreactor. Biochem Eng J 111:54–62. https://doi.org/10.1016/j.bej.2016.03.007

Cano A, Bermejo A (2011) Influence of rootstock and cultivar on bioactive compounds in citrus peels. J Sci Food Agric 91(9):1702–1711. https://doi.org/10.1002/jsfa.4375

Cao H, Chen X, Jassbi AR, Xiao J (2014) Microbial biotransformation of bioactive flavonoids. Biotechnology advances (in press). http://dx.doi.org/10.1016/j.biotechadv.2014.10.012

Correia RTP, McCue P, Magalhaes MMA, Macedo GR, Shetty K (2004) Phenolic antioxidant enrichement of soy flour-supplemented guava waste by *Rhizopus oligosporus*-mediated solid-state bioprocessing. J Food Biochem 28(5):404–418. https://doi.org/10.1111/j.1745-4514.2004.05703.x

Dahmoune F, Boulekbache L, Moussi K, Aoun O, Spigno G, Madani K (2013) Valorization of *Citrus limon* residues for the recovery of antioxidants: evaluation and optimization of microwave and ultrasound application to solvent extraction. Ind Crops Prod 50:77–87. https://doi.org/10.1016/j.indcrop.2013.07.013

de Moraes Barros HR, de Castro Ferreira TAP, Genovese MI (2012) Antioxidant capacity and mineral content of pulp and peel from commercial cultivars of citrus from Brazil. Food Chem 134(4):1892–1898. https://doi.org/10.1016/j.foodchem.2012.03.090

El-Adawy TA, Rahma EH, El-Bedawy AA, Gafar AM (1999) Properties of some citrus seeds. Part 3. Evaluation as a new source of protein and oil. Food/Nahrung 43(6):385–391. https://doi.org/10.1002/(sici)1521-3803(19991201)43:6<385:aid-food385>3.0.co;2-v

Kelebek H, Selli S, Canbas A, Cabaroglu T (2009) HPLC determination of organic acids, sugars, phenolic compositions and antioxidant capacity of orange juice and orange wine made from a Turkish cv Kozan. Microchem J 91(2):187–192. https://doi.org/10.1016/j.microc.2008.10.008

Lagha-Benamrouche S, Madani K (2013) Phenolic contents and antioxidant activity of orange varieties (*Citrus sinensis* L. and *Citrus aurantium* L.) cultivated in Algeria: peels and leaves. Ind Crops Prod 50:723–730. https://doi.org/10.1016/j.indcrop.2013.07.048

Li BB, Smith B, Hossain MM (2006a) Extraction of phenolics from citrus peels I. Solvent extraction method. Sep Purif Technol 48(2):182–188. https://doi.org/10.1016/j.seppur.2005.07.005

Li BB, Smith B, Hossain MM (2006b) Extraction of phenolics from citrus peels II. Enzyme-assisted extraction method. Sep Purif Technol 48(2):189–196. https://doi.org/10.1016/j.seppur.2005.07.019

Liu Y, Heying E, Tanumihardjo SA (2012) History, global distribution, and nutritional importance of citrus fruits. Compr Rev Food Sci Food Saf 11(6):530–545. https://doi.org/10.1111/j.1541-4337.2012.00201.x

Martínez-Ávila GC, Aguilera-Carbó AF, Rodríguez-Herrera R, Aguilar CN (2012) Fungal enhancement of the antioxidant properties of grape waste. Ann Microbiol 62(3):923–930. https://doi.org/10.1007/s13213-011-0329-z

Meléndez NP, Nevárez-Moorillón V, Rodríguez-Herrera R, Espinoza JC, Aguilar CN (2014) A microassay for quantification of 2,2-diphenyl-1-picrylhydracyl (DPPH) free radical scavenging. Afr J Biochem Res 8(1):14–18. https://doi.org/10.5897/AJBR2013.0669

Milder IEJ, Arts ICW, Bvd Putte, Venema DP, Hollman PCH (2005) Lignan contents of Dutch plant foods: a database including lariciresinol, pinoresinol, secoisolariciresinol and matairesinol. Br J Nutr 93(03):393–402. https://doi.org/10.1079/BJN20051371

Mohseni S, Najafpour GD, Vaseghi Z, Mahjoub S (2012) Solid state fermentation of agricultural residues for lipase production in a tray-bioreactor. World Appl Sci J 16(7):1034–1039

Ng IS, Wu X, Lu Y, Yao C (2014) *Trichoderma reesei* cellulase complex in hydrolysis of agricultural waste of grapefruit peel and orange peel. BioResources 9(4):6420–6431

Nogata Y, Sakamoto K, Shiratsuchi H, Ishii T, Yano M, Ohta H (2006) Flavonoid composition of fruit tissues of citrus species. Biosci Biotechnol Biochem 70(1):178–192. https://doi.org/10.1271/bbb.70.178

Pandey A, Selvakumar P, Soccol CR, Nigam P (1999) Solid state fermentation for the production of industrial enzymes. Curr Sci 77(1):149–162

Peñalvo JL, Haajanen KM, Botting N, Adlercreutz H (2005) Quantification of lignans in food using isotope dilution gas chromatography/mass spectrometry. J Agric Food Chem 53 (24):9342–9347. https://doi.org/10.1021/jf051488w

Peterson J, Dwyer J, Adlercreutz H, Scalbert A, Jacques P, McCullough ML (2010) Dietary lignans: physiology and potential for cardiovascular disease risk reduction. Nutr Rev 68 (10):571–603. https://doi.org/10.1111/j.1753-4887.2010.00319.x

Peterson JJ, Beecher GR, Bhagwat SA, Dwyer JT, Gebhardt SE, Haytowitz DB, Holden JM (2006a) Flavanones in grapefruit, lemons, and limes: a compilation and review of the data from the analytical literature. J Food Compos Anal 19, Supplement (0):S74–S80. http://doi.org/10.1016/j.jfca.2005.12.009

Peterson JJ, Dwyer JT, Beecher GR, Bhagwat SA, Gebhardt SE, Haytowitz DB, Holden JM (2006b) Flavanones in oranges, tangerines (mandarins), tangors, and tangelos: a compilation and review of the data from the analytical literature. J Food Compos Anal 19, Supplement (0): S66–S73. http://doi.org/10.1016/j.jfca.2005.12.006

Pourbafrani M, Forgács G, Horváth IS, Niklasson C, Taherzadeh MJ (2010) Production of biofuels, limonene and pectin from citrus wastes. Biores Technol 101(11):4246–4250. https://doi.org/10.1016/j.biortech.2010.01.077

Rhee KS (1978) Factors affecting oxygen uptake in model systems used for investigating lipid peroxidation in meat. J Food Sci 43(1):6–9. https://doi.org/10.1111/j.1365-2621.1978.tb09723.x

Sameer M, Shruti S, Archana S, Aruna J, Vilasrao K (2013) *Citrus paradisi*: an overview. Indo Am J Pharm Res 3(10):8070–8077

SIAP (2016) Servicio de Información Agroalimentaria y Pesquera, México. http://infosiap.siap.gob.mx/aagricola_siap_gb/ientidad/index.jsp. Accessed Diciembre 2016

Sun Y, Qiao L, Shen Y, Jiang P, Chen J, Ye X (2013) Phytochemical profile and antioxidant activity of physiological drop of citrus fruits. J Food Sci 78(1):C37–C42. https://doi.org/10.1111/j.1750-3841.2012.03002.x

Tao N-g, Shi W-q, Liu Y-j, Huang S-r (2011) Production of feed enzymes from citrus processing waste by solid-state fermentation with *Eupenicillium javanicum*. Int J Food Sci Technol 46 (5):1073–1079. https://doi.org/10.1111/j.1365-2621.2011.02587.x

Toivonen PMA, Sweeney M (1998) Differences in chlorophyll loss at 13 °C for two broccoli (*Brassica oleracea* L.) cultivars associated with antioxidant enzyme activities. J Agric Food Chem 46(1):20–24. https://doi.org/10.1021/jf970490n

Tounsi MS, Wannes WA, Ouerghemmi I, Jegham S, Njima YB, Hamdaoui G, Zemni H, Marzouk B (2011) Juice components and antioxidant capacity of four Tunisian *Citrus* varieties. J Sci Food Agric 91(1):142–151. https://doi.org/10.1002/jsfa.4164

Wang Z, Shang Q, Wang W, Feng X (2011) Microwave-assisted extraction and liquid chromatography/mass spectrometry analysis of flavonoids from grapefruit peel. J Food Process Eng 34(3):844–859. https://doi.org/10.1111/j.1745-4530.2009.00513.x

Xi W, Fang B, Zhao Q, Jiao B, Zhou Z (2014a) Flavonoid composition and antioxidant activities of Chinese local pummelo (*Citrus grandis* Osbeck.) varieties. Food Chem 161:230–238. https://doi.org/10.1016/j.foodchem.2014.04.001

Xi W, Zhang Y, Sun Y, Shen Y, Ye X, Zhou Z (2014b) Phenolic compositions in the pulps of Chinese wild mandarin pulps (*Citrus reticulata* Blanco.) and their antioxidant properties. Ind Crops Prod 52:466–474. https://doi.org/10.1016/j.indcrop.2013.11.016

Xu GH, Chen JC, Liu DH, Zhang YH, Jiang P, Ye XQ (2008) Minerals, phenolic compounds, and antioxidant capacity of citrus peel extract by hot water. J Food Sci 73(1):C11–C18. https://doi.org/10.1111/j.1750-3841.2007.00546.x

Yang F-C, Ma T-W, Lee Y-H (2013) Reuse of citrus peel to enhance the formation of bioactive metabolite-triterpenoid in solid-state fermentation of *A. cinnamomea*. Biochem Eng J 78:59–66. https://doi.org/10.1016/j.bej.2013.03.013

Zhang Y, Sun Y, Xi W, Shen Y, Qiao L, Zhong L, Ye X, Zhou Z (2014) Phenolic compositions and antioxidant capacities of Chinese wild mandarin (*Citrus reticulata* Blanco) fruits. Food Chem 145:674–680. https://doi.org/10.1016/j.foodchem.2013.08.012

Chapter 6
Coffee Husk: A Potential Agro-Industrial Residue for Bioprocess

Swaroop S. Kumar, T. S. Swapna and Abdulhameed Sabu

Abstract Agro-industrial processing always generates waste materials. Coffee is one of the largest commodities in the world, and its processing yields coffee husk as waste by-product. Its disposal without proper treatment can cause serious environmental problems. Coffee husk contains carbohydrates including fermentable sugars, polyphenols such as tannins, lipids. Various microorganisms including filamentous fungi are reported to grow on it despite the presence of antimicrobial compounds. Chemical nature of coffee husk makes it a suitable and inexpensive source for solid-state fermentation. Several studies have been reported on the application of coffee husk in bioprocess. It acts as a substrate as well as carbon source during fermentation. Production of various enzymes such as xylanases, cellulases, polygalacturonases, polyphenol oxidases, tannases in high titers can be achieved by fermentation using coffee husk as substrate. Apart from production of enzymes, bioconversion of coffee husk is also achieved during bioprocess which in turn favors sustainable utilization of waste products. Production of citric acid, gibberellic acid, gallic acid, polyhydroxyalkanoates (PHA), and bacterial cellulose is reported by fermentation using coffee husk as substrate. Mass production of microorganisms is another advantage of using coffee husk in bioprocess. It is excellent for the growth of various biocontrol agents such as *Trichoderma* sp. Besides economic production, prolonged shelf life of biocontrol agents multiplied on coffee husk makes it more attractive. Biopesticides such as *Bacillus sphaericus* and *B. thuringiensis* can also be produced by solid-state fermentation on coffee husk. The sustainable management of agro-industrial waste like coffee husk through bioprocess makes it as an attractive source of wealth.

Keywords Coffee husk · Coffee pulp · Fermentation · Value addition
Sustainable production · Detoxification

S. S. Kumar · A. Sabu (✉)
Department of Biotechnology & Microbiology, School of Life Sciences,
Kannur University, Thalassery Campus, Kannur, Kerala 670661, India
e-mail: drsabu@gmail.com

T. S. Swapna
Department of Botany, University of Kerala Kariavattom,
Thiruvananthapuram, Kerala 695581, India

R. R. Singhania et al. (eds.), *Waste to Wealth*, Energy, Environment,
and Sustainability, https://doi.org/10.1007/978-981-10-7431-8_6

6.1 Introduction

Processing of agricultural products in agro-industries generates huge amount of waste materials. These waste materials include husk, straw, leaves, peel, pulp, bagasse. Either burning or usage as animal feed is the common practice for elimination of these agro-industrial wastes which is not adequate for proper eradication. These wastes are generated throughout the year and contain sugars, proteins, pectin, tannins etc. This makes agro-industrial residues suitable as raw materials for production of value-added compounds while reducing the production cost. But most of the agro-industrial waste also contains materials with toxic potential such as phenolic compounds which when released to environment is hazardous, and some of them are not suitable to be used as animal feed. Thus, recycling these wastes has not only acquired environmental benefit but also has got great economic potential.

Coffee is one of the highest selling agro-industrial commodities, and major producers include Brazil, India, Mexico, Indonesia, Ethiopia, Colombia. Waste generated while processing coffee is much higher as about 50% of the processed coffee (Matos 2008). Coffee husks and coffee pulp are agro-industrial residues obtained during its processing. Elimination of these residues is a major environmental issue to be addressed. They are known to contain carbohydrates, lipids, proteins, and polyphenols. Hence, besides the environmental perspective of successful waste management, the economical point of view for development of value-added products also plays a crucial role. Extraction of bioactive compounds such as pigments, production of absorbents (Baquero et al. 2003), ethanol production, biomethanation, and enzyme production are some of the successful use of coffee husk as raw material for value addition through sustainable development. Their reuse in fermentation is not only an excellent alternative for the bioremediation but also reduces the production cost during the process. It becomes economically feasible when coffee husk is used in solid-state fermentation (SSF), where it serves as solid support as well as source of carbon and nitrogen for microbial growth. In this chapter, we are focusing on recycling coffee agro-industrial residues through fermentation for value addition in compliance with sustainable development. First of all, generation of coffee agro-industrial wastes and their chemical composition are explained followed by its potential usage fermentation process (Fig. 6.1).

6.2 Chemical Composition of Coffee Agro-Industrial Wastes

First of all, we will begin with a brief understanding of coffee processing. The coffee cherry contains two cotyledons (beans) covered by testa (silver skin) which is again covered by parchment. This is surrounded by mucilaginous pulp called mesocarp which in turn covered by outer skin called exocarp (Fig. 6.2). The cherry

Fig. 6.1 Ripened coffee berries

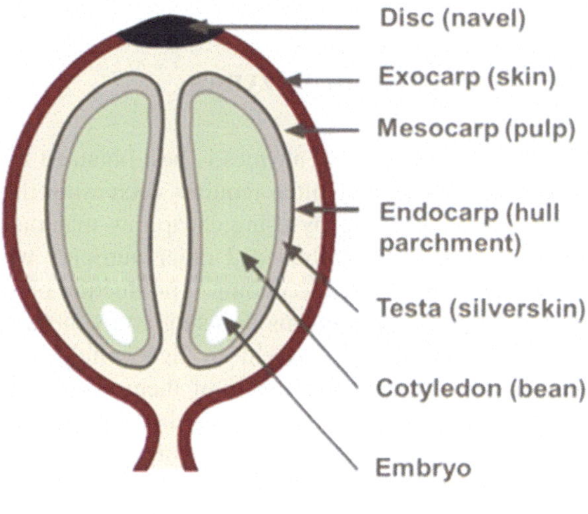

Fig. 6.2 Cross sectional view of coffee cherry. (Image is taken from http://www.yara.in)

is processed to obtain fine coffee. Generally, two types of processing exist, i.e., dry and wet processing. In dry processing, the cherries are dried either by natural method (sun drying) or artificial method. After drying, the de-hulling of the cherries is carried out in a de-hulling machine where beans get separated from the covering material and the residues generated are called coffee husk (CH), whereas in wet processing, exocarp and mucilaginous pulp are detached mechanically leaving the residue as coffee pulp (CP).

CHs and CPs are rich source of nutrients and organic matter. They contain carbohydrates including fermentable sugars, proteins, caffeine, tannins, and polyphenols. Tannins are usually considered as antinutritional factor and thus making coffee by-products less suitable for its reuse as animal feed. They contain almost 50–85% total carbohydrates of which 14–24% is reducing sugars. Besides, it contains 4–12% proteins, 0.5–3% lipids, 3–10% minerals, 1–9% tannins, and approximately 1% caffeine (g/100 g dry weight). Composition varies for CHs and CPs method of processing, storage conditions. Also, they vary according to the species and geographical distribution of coffee plants (Gouvea et al. 2009; Adams and Dougan 1987; Clifford and Ramirez-Martinez 1991; Navya and Pushpa 2013).

6.3 Reuse of Coffee Husk in Fermentation

Coffee pulp and husk are known to harbor many microorganisms including bacteria, yeasts, and fungi; filamentous fungi especially basidiomycetes are more suitable for cultivation upon them. Among the microflora, coffee husk showed predominant fungal population, whereas coffee pulp shown mixed population of fungi, bacteria, and yeast in similar proportions (Roussos et al. 1995; Pandey et al. 2000). One of the most efficient ways to recycle coffee agro-industrial by-products is by utilizing them in fermentation. It can be used for bioremediation of waste materials as well as development of value-added compounds.

6.4 Production of Enzymes

Microbial production of enzymes is best obtained through fermentation. Industrial production of enzymes often requires decreasing the cost of fermentation process which can be achieved by using cheap raw materials. Coffee husks and pulps are excellent sources of carbon and other nutrients which makes them suitable for fermentation process and economical production of enzymes. Some of the enzymes produced by microorganisms by using coffee husk for fermentation are discussed here. β-1,4-xylans are present in lignocellulosic materials, and xylanases are enzymes required for hydrolysis of them. These enzymes are used in food processing along with cellulases. They find applications as clarifying agent in fruit juices and wines, biopulping, degumming of plant fibers etc. (Christov et al. 1999;

Lemos and Pereira Junior 2002). Coffee husk, pulp, spent coffee, and silver skin known for lignocellulosic materials were used as sole source of carbon for production of xylanase from *Penicillium* sp. CFR 303 by SSF. All coffee agro-industrial residues were found to be respectable as carbon source and coffee pulp pre-treated with steam produced maximum xylanase enzyme (23,494 U/g) from the fungus (Murthy and Naidu 2012). *P. glabrum* using brewer's spent grain as substrate yielded xylanase under solid-state fermentation (Knob et al. 2013).

Gluconacetobacter hansenii UAC09 has known to produce different enzymes, viz. exopolygalacturonase, tannase, and polyphenol oxidase by submerged fermentation while using coffee husk extract as sole carbon source. These enzyme causes transformation of polyphenols and pectin from coffee husk and thus its effective removal (Rani and Appaiah 2012). *Rhizopus stolonifer* is another fungal species which was able to utilize coffee husk as sole carbon source during solid-state fermentation. The strain was effective in producing β-glucosidase enzyme (Navya et al. 2012). *P. verrucosum* was able to grow on coffee pulp while used in SSF and produces tannase enzyme. This is also another application of fermentation for biotransformation agro-industrial waste and value addition (Bhoite et al. 2013).

Aspergillus niger was known for its potential to grow on coffee industrial waste. They are one of the predominant fungal strains isolated from coffee waste. Various strains of *A. niger* produce pectinase enzyme. *A. niger* C28B25 when cultivated on coffee pulp yielded high pectinase production (138 Units/g dry pulp), and its mutant shown hyper production of about (228 Units/g dry pulp) (Antier et al. 1993). *Penicillium* sp. isolated from coffee by-products also shown comparable pectinase production to *Aspergillus* while SSF was performed using coffee pulp as substrate (Boccas et al. 1993).

Coffee by-products were found excellent for production of α-amylase from *Neurospora crassa* CFR 308 by SSF. Fermentation yielded α-amylase with an activity of 7084 Units/g dry substrate (Murthy et al. 2009). *A. oryzae* CFR305 isolated from coffee residues yielded protease by SSF while using coffee cherry husk as substrate. A total production of 12,236 Units/g dry substrate was achieved from the strain (Murthy and Naidu 2010). β-fructofuranosidase is an enzyme responsible for production of fructooligosaccharides (FOS) from sucrose. Under SSF, *A. japonicas* produced *β*-fructofuranosidase enzyme with an activity (71.3 U/ml) using coffee by-products as substrate (Mussatto and Teixeira 2010). As coffee husk can harbor many organisms, its potential for production of other enzymes is yet to be exploited.

6.5 Production of Other Value-Added Products

During fermentation using coffee by-products, apart from enzyme, microorganisms also produce other substances making sustainable development more practical. Gallic acid, a phenolic antioxidant produced from tannins, finds application as adjuvants, food supplements etc. Acid hydrolysis of tannic acid yields gallic acid. Microbial tannase enzyme also produces gallic acid. The fungus *P. verrucosum* produces gallic acid while performing biotransformation of coffee pulp tannins (Bhoite et al. 2013).

Another important finding is the production of bacterial cellulose (BC) by *Gluconacetobacter* sp. while using coffee husk for fermentation. Bacterial cellulose is superior to plant cellulose as the latter is often associated with hemicellulose and lignin. BC is highly crystalline in nature and shows stability toward temperature and chemical treatments. It also shows high mechanical strength. Production of BC is of great demand because of its use as artificial skin (Krystynowicz et al. 2002), scaffold for tissue engineering of cartilages (Svensson et al. 2005), etc. But bacterial cellulose production is not economically feasible because of lower yield by bacterium and high cost of the production medium (Moon et al. 2006). Instead of using synthetic media, coffee husk extract was used for fermenting *Gluconacetobacter hansenii* UAC09 for bacterial cellulose production and conditions were optimized for maximal production of BC. It was found that while using coffee husk extract for production, a threefold increase in the yield of BC was observed (Rani and Appaiah 2013). Thus, by substituting the synthetic media with agro-industrial waste residues, the cost of production for BC can be reduced.

Use of biopolymers as an alternative to petroleum-based plastics attracted environmental perspective due to its biodegradable nature. Polyesters of hydroxyalkanoates called polyhydroxyalkanoates (PHAs) are produced by microorganisms as an intracellular carbon and energy source. PHAs are used as biopolymers and used in making disposable razors, cups, surgical stitches, etc. Biopol, Nodax, Degra pol are some of the PHAs available commercially (Anjum et al. 2016). However, the cost of production is the limiting factor for industrial application of these polymers. Bacterium *Enterobacter aerogenes* is able to produce PHAs by using coffee husk extract for fermentation. Thus, coffee husk serves as a cheap substrate for PHA production and, thereby reducing the cost of production of biopolymers as well paving the way for bioremediation of coffee by-products (Chandrika et al. 2015).

Citric acid is an organic acid with high commercial value due to its large market potential. It finds application in cosmetic, pharmaceutical, and food industries. Production of citric acid can be achieved by solid-state fermentation. Coffee husk was used as substrate for the first time to produce citric acid from *A. niger* CFTRI 30. Higher citric acid production as high as about 1.5 g/10 g dry coffee husk was obtained from the strain (Shankaranand and Lonsane 1994). 187.54 g citric acid/kg dry coffee husk was obtained from a mutant of *A. niger* (*A. niger* RCNM 17), when basal coffee husk medium was used for SSF (Ramachandra et al. 2013).

Production of flavoring and fragrance substances by bioprocess is quite demanding nowadays and finds its application in food and cosmetic industries. Microbial production of aromatic compounds has been reported from various sources. A fungal strain *Ceratocystis fimbriata* was used for aroma production using coffee husk as substrate for SSF. During the fermentation ethyl acetate, ethanol and acetaldehyde were produced (Medeiros et al. 2003). Few of the other microbial producers includes *Kluyveromyces* sp. (Jiang 1995; Fabre et al. 1995), *Neurospora* sp. (Pastore et al. 1994; Yamauchi et al. 1989), *Zygosaccharomyces rouxii* (Sugawara et al. 1994), *Aspergillus* sp. (Ito et al. 1990), *Trichoderma viride* (Gervais and Sarrette 1990), and *R. oryzae* (Bramorski et al. 1998). Most of these strains are known to grow on coffee agro-industrial residues. Gibberellic acid (GA) often called gibberellins is a pentacyclic diterpenoid acid which is growth regulators of plants. They are produced by both plants and fungi. A large variety of gibberellic acid is known of which GA3 is more popular. It is used for regulation of plant growth in agriculture field but yet expensive. Industrial production of GAs is achieved by SSF. Production of GA3 was achieved from fungus *Gibberella fujikuroi* LPB-06 under SSF using coffee husk as substrate. The final yield was about 492.5 mg of GA3/kg of dry coffee husk (Machado et al. 2002).

Development of prebiotic products is a promising area, and fructooligosaccharides (FOS) are one of the prebiotic products characterized by its low-colorific value and bifidobacterial growth promotion in colon. *A.japonicus* under SSF conditions produces FOS with a yield of 128.7 g/l (Mussatto and Teixeira 2010).

6.6 Production of Biocontrol Agents

Biocontrol agents are used for pest and disease management. *Trichoderma* sp. is biocontrol agents used for management of fungal diseases and growth promotion in plants (Inbar et al. 1994). A wide variety of *Trichoderma* sp. are identified and known to show antagonistic activity against various pathogens (Paulitz and Bélanger 2001). *Rhizoctonia, Fusarium, Alternaria, Ustilago, Venturia* and *Colletotrichum, Pythium* and *Phytophthora* are some among them. Most widely used species are *T. harzianum* and *T.viride. Trichoderma* formulation based on coffee husk was developed to manage *phytophthora* foot rot of black pepper (Sawant et al. 1995). Most of these fungal strains have ability to grow on acidic pH. Coffee husk normally shows acidic pH, and hence, they can mass multiply in coffee husk. Another advantage of coffee husk-based *Trichoderma* formulation is its increased shelf life up to 18 months compared to other formulation thus making this formulation more suitable for economical production and distribution.

Mosquito vector control programs have always been relied on chemical insecticides but with the development of biocontrol agents scenario changed. *Bacillus sphaericus* (*Bs*) and *B. thuringiensis* subspecies *israelensis* (*Bti*) are two well-known mosquitocidal biocontrol agents. Coffee husk waste filtrate was used to culture *Bti* H-14 (IPS-82) and *Bs* (IAB-59) under submerged fermentation. The bacteria produced in coffee husk filtrate medium was analyzed for its activity against larvae of major mosquito vectors *Culex quinquefasciatus, Aedes aegypti,* and *Anopheles stephensi* and found to show insecticidal activity (Poopathi and Abidha 2011). It was found that toxins produced by them in this medium are comparable with that of synthetic medium. Thus, it shows another effective utilization of coffee agro-industrial waste.

6.7 Production of Biogas (Biomethanation)

Biomethanation is the conversion of organic matter into biogas (mainly methane) by anaerobic digestion. It is an excellent method for sludge and other waste treatment. Coffee husk rich in polyphenols and acid in nature (pH 4.3), and this makes coffee husk unsuitable for biogas production. Coffee husk was pre-treated with thermophilic fungus *Mycotypha*. Coffee husk was fermented for 15 days and later inoculated with methanogenic inoculum. The anaerobic digestion caused production of biogas from coffee husk treated with *Mycotypha* (Jayachandra et al. 2011).

6.8 Production of Ethanol

Bioethanol is ethanol produced by fermenting sugar from plant and agro-industrial residues and considered as renewable energy source. Decreasing amount of fossil fuels demands alternate energy sources. Bioethanol is considered as a substitute for these non-renewable fossil fuels. But currently, the production cost and yield are two major limiting factors for industrial production of these biofuels. Sugarcane, cereal grains, sugarcane bagasse, etc., are used for fermentation to produce ethanol. A large amount of carbohydrates in coffee industrial waste makes it a suitable substrate for fermentation. Apart from producing energy source, it is also a good practice to remove the waste materials. Dry sticky coffee husk was used for production of ethanol from baker's yeast *Saccharomyces cerevisiae,* and about 13.6 g ethanol/L was obtained which was reasonable (Gouvea et al. 2009). According to Kefale et al., baker yeast yielded 7.4 g ethanol/L when coffee pulp was used for fermentation under optimal conditions (Kefale et al. 2012). Bioethanol production based on coffee husk as substrate has to be explored more for developing new techniques for biofuel synthesis.

6.9 Detoxification by Fermentation

The presence of antinutritional factors and toxic substances in coffee agro-industrial residues demands the need of detoxification. Detoxification method involves physical, chemical, and microbial methods (Shiono et al. 2017). Here, detoxifications by fermentation using microorganisms are discussed. Previously discussed fermentation strategies often involve the bioconversion of coffee husk apart from production value-added compounds and thus aiding detoxification. Also, detoxified coffee wastes are more suitable as substrate for bioprocess. One of the toxic compounds in these wastes is caffeine. Decaffeination is one of the challenges in these detoxification methods. Many of the fungal and bacterial species are known to use caffeine as energy source and degrade them. Few of the microorganisms for caffeine degradation are *Penicillium* sp., *Aspergillus* sp., *Pseudomonas* sp. etc. (Asano et al. 1993; Hakil et al. 1999). *P. roqueforti* and *Stemphylum* sp. degrade caffeine in liquid medium (Kurtzman and Schwimmer 1971). *Aspergillus* sp. LPBx was able to degrade 90% of caffeine and by SSF (Brand et al. 2002). Fungal strain *Rhizopus delemar* was able to degrade caffeine in packed bed column bioreactor using coffee husk as substrate (Tagliari et al. 2003). Under optimum conditions, caffeine was degraded to theobromine by *Pseudomonas putida* with a total yield of 92% (Asano et al. 1993). These detoxifications reduce the toxicity of these waste materials which can be used as animal feed after processing. Reducing the pollution caused by these waste materials is the primary concern for development of these methods (Table 6.1).

Table 6.1 Some of the cultivated microorganisms on coffee agro-industrial residues with products obtained or its functional application by fermentation

Microorganism	Product obtained/Function	References
Aspergillus niger	Pectinase	Murthy and Naidu (2011)
Penicillium sp.	Pectinase	Boccas et al. (1993)
Neurospora crassa	Amylase	Murthy et al. (2009)
Paecilomyces variotii	Tannase	Battestin and Macedo (2007)
Gluconacetobacter hansenii	Tannase	Rani and Appaiah (2012)
G. hansenii	Exopolygalacturonase	Rani and Appaiah (2012)
Penicillium verrucosum	Tannase	Bhoite et al. (2013)
G. hansenii	Polyphenol oxidase	Rani and Appaiah (2012)
Rhizopus stolonifer	β-glucosidase	Navya et al. (2012)
A. oryzae	Protease	Murthy and Naidu (2010)
Penicillium sp.	Xylanase	Murthy and Naidu (2012)
P. glabrum	Xylanase	Knob et al. (2013)
A. japonicus	β-furctofuranosidase	Mussatto and Teixeira (2010)
Ceratocystis fimbriata	Flavoring agents (Aroma compounds)	Medeiros et al. (2003)
Kluyveromyces sp.	Flavoring agents (Aroma compounds)	Jiang (1995), Fabre et al. (1995)
Zygosaccharomyces rouxii	Flavoring agents (Aroma compounds)	Sugawara et al. (1994)
Fusarium moniliforme	Gibberellic acid	Machado et al. (2002)
Gibberella fujikuroi	Gibberellic acid	Machado et al. (2002)
Aspergillus niger	Citric acid	Shankaranand and Lonsane (1994)
A. japonicus	Fructooligosaccharides	Mussatto and Teixeira (2010)

(continued)

Table 6.1 (continued)

Microorganism	Product obtained/Function	References
G. hansenii	Bacterial cellulose	Rani and Appaiah (2013)
Enterobacter aerogenes	Polyhydroxyalkanoates	Chandrika et al. (2015)
Saccharomyces cerevisiae	Bioethanol	Gouvea et al. (2009)
Methanogenic consortium	Biogas	Jayachandra et al. (2011)
P. roqueforti	Detoxification	Kurtzman and Schwimmer (1971)
Stemphylum sp.	Detoxification	Kurtzman and Schwimmer (1971)
R.delemar	Detoxification	Tagliari et al. (2003)
Pseudomonas putida	Detoxification	Asano et al. (1993)
Trichoderma harzianum	Biocontrol agents	Sawant et al. (1995)
T. viride	Biocontrol agents	Sawant et al. (1995)
Bacillus sphaericus (*Bs*)	Biocontrol agents	Poopathi and Abidha (2011)
Bacillus thuringiensis subspecies *israelensis* (*Bti*)	Biocontrol agents	Poopathi and Abidha (2011)

6.10 Conclusions

Coffee by-products disposal is major problem faced by coffee producing countries, and though many of the disposal mechanisms are proposed, lack of adequate disposal and economic usage remained as a challenge. Effective utilization of these by-products was done for isolation of bioactive compounds, production of mushrooms, development of adsorbents, as raw material for production of enzymes, value-added products, biofuels, aroma, etc. In this chapter, only fermentation aspect of coffee waste was focused. Coffee husks and coffee pulp are rich in organic matter, and this makes it suitable for fermentation. Significant cost reduction was achieved by the use of these agro-industrial wastes for production of enzymes and other products such as flavoring agents, citric acid, gibberellic acid. Also, it is very much helpful in reducing the waste removal by bioconversion of the toxins in these wastes. Solid-state fermentation turns to be one of the best methods for effective utilization of these agro-industrial residues. Antinutritional and toxic substances such as caffeine and tannins restrict use of coffee husk as animal feed. However, the detoxification by fermentation achieves bioconversion of this organic matter making it suitable for use as feed. However, the detoxification strategies sum up the cost of feed produced. Still, it is considered significant as the direct release of these toxic substances into environment increases the risk of pollution. Coffee husk as alternate renewable source of energy by production of bioethanol is quite promising. Further, extensive research is needed for the economical production and commercialization of biofuels. Biomethanation of coffee husk was performed for gas production. But it still does not seem to be economical. Cost-effective production of biogas from these wastes has yet to be technically developed. Biocontrol agent production is another greatest advantage of using coffee husk in fermentation with increased shelf life and economic feasibility of mass production. Though many of the aspects are discussed here for reuse of coffee by-products in fermentation, they are yet to be exploited for production of newer compounds and development of more economically feasible techniques.

References

Adams MR, Dougan J (1987) Waste products. In: Clarke RJ, Macrae R (eds) Coffee. Springer, Netherlands, pp 257–291

Anjum A, Zuber M, Zia KM, Noreen A, Anjum MN, Tabasum S (2016) Microbial production of polyhydroxyalkanoates (PHAs) and its copolymers: a review of recent advancements. Int J Biol Macromol 89:161–174

Antier P, Minjares A, Roussos S, Raimbault M, Viniegra-Gonzalez G (1993) Pectinase-hyperproducing mutants of *Aspergillus niger* C28B25 for solid-state fermentation of coffee pulp. Enzyme Microb Technol 15:254–260

Asano Y, Komeda T, Yamada H (1993) Microbial production of theobromine from caffeine. Biosci Biotechnol Biochem 57:1286–1289

Baquero MC, Giraldo L, Moreno JC, Suárez-García F, Martínez-Alonso A, Tascón JMD (2003) Activated carbons by pyrolysis of coffee bean husks in presence of phosphoric acid. J Anal Appl Pyrol 70:779–784

Battestin V, Macedo GA (2007) Tannase production by *Paecilomyces variotii*. Biores Technol 98:1832–1837

Bhoite RN, Navya PN, Murthy PS (2013) Statistical optimization of bioprocess parameters for enhanced gallic acid production from coffee pulp tannins by *Penicillium verrucosum*. Prep Biochem Biotechnol 43:350–363

Boccas F, Roussos S, Gutierrez M, Serrano L, Viniegra GG (1993) Production of pectinase from coffee pulp in solid state fermentation system: selection of wild fungal isolate of high potency by a simple three-step screening technique. J Food Sci Technol 31:22–26

Bramorski A, Christen P, Ramirez M, Soccol CR, Revah S (1998) Production of volatile compounds by the edible fungus *Rhizopus oryzae* during solid state cultivation on tropical agro-industrial substrates. Biotech Lett 20:359–362

Brand D, Pandey A, Rodriguez-Leon JA, Roussos S, Brand I, Soccol CR (2002) Relationship between coffee husk caffeine degradation and respiration of *Aspergillus* sp. LPBx in solid-state fermentation. Appl Biochem Biotechnol 102–103:169–177

Chandrika SP, Sabarinathan D, Preethi K (2015) Bioremediation of coffee husk through polyhydroxyalkanoates (PHA) production for a greener environment. Int J Recent Sci Res 6:4857–4860

Christov LP, Szakacs G, Balakrishnan H (1999) Production, partial characterization and use of fungal cellulase-free xylanases in pulp bleaching. Process Biochem 34:511–517

Clifford MN, Ramirez-Martinez JR (1991) Tannins in wet-processed coffee beans and coffee pulp. Food Chem 40:191–200

Fabre CE, Duviau VJ, Blanc PJ, Goma G (1995) Identification of volatile flavour compounds obtained in culture of *Kluyveromyces marxianus*. Biotech Lett 17:1207–1212

Gervais P, Sarrette M (1990) Influence of age of mycelium and water activity of the medium on aroma production by *Trichoderma viride* grown on solid substrate. J Ferment Bioeng 69:46–50

Gouvea BM, Torres C, Franca AS, Oliveira LS, Oliveira ES (2009) Feasibility of ethanol production from coffee husks. Biotech Lett 31:1315–1319

Hakil M, Voisinet F, Viniegra-González G, Augur C (1999) Caffeine degradation in solid state fermentation by *Aspergillus tamarii*: effects of additional nitrogen sources. Process Biochem 35:103–109

Inbar J, Abramsky M, Cohen D, Chet I (1994) Plant growth enhancement and disease control by *Trichoderma harzianum* in vegetable seedlings grown under commercial conditions. Eur J Plant Pathol 100:337–346

Ito K, Yoshida K, Ishikawa T, Kobayashi S (1990) Volatile compounds produced by the fungus *Aspergillus oryzae* in rice koji and their changes during cultivation. J Ferment Bioeng 70:169–172

Jayachandra T, Venugopal C, Appaiah KAA (2011) Utilization of phytotoxic agro waste-coffee cherry husk through pretreatment by the ascomycetes fungi *Mycotypha* for biomethanation. Energy Sustain Dev 15:104–108

Jiang J (1995) Changes in volatile composition of *Kluyveromyces lactis* broth during fermentation. Dev Food Sci 37:1073–1086

Kefale A, Redib M, Asfaw A (2012) Bioethanol production and optimization test from agricultural waste: The case of wet coffee processing waste (Pulp). In J Renew Energy Res 2:446–450

Knob A, Beitel SM, Fortkamp D, Terrasan CRF, de Almeida AF (2013) Production, purification, and characterization of a major *Penicillium glabrum* xylanase using Brewer's spent grain as substrate. BioMed Res Int 728735

Krystynowicz A, Czaja W, Wiktorowska-Jezierska A, Gonçalves-Miśkiewicz M, Turkiewicz M, Bielecki S (2002) Factors affecting the yield and properties of bacterial cellulose. J Ind Microbiol Biotechnol 29:189–195

Kurtzman RH, Schwimmer S (1971) Caffeine removal from growth media by microorganisms. Experientia 27:481–482

Lemos JLS, Pereira Junior N (2002) Influence of some sugars on xylanase production by *Aspergillus awamori* in solid state fermentation. Brazilian Arch Biol Technol 45:431–437

Machado CMM, Soccol CR, de Oliveira BH, Pandey A (2002) Gibberellic acid production by solid-state fermentation in coffee husk. Appl Biochem Biotechnol 102–103:179–191

Matos AT (2008) Tratamento de resíduos na pós-colheita do café (residues disposal in coffee post-processing). In: Borém FM (ed) Pós-colheita do Café (coffee post processing). Lavras (Brazil), Editora UFLA, pp 161–201

Medeiros ABP, Christen P, Roussos S, Gern JC, Soccol CR (2003) Coffee residues as substrates for aroma production by *Ceratocystis fimbriata* in solid state fermentation. Brazilian J Microbiol 34:245–248

Moon SH, Park JM, Chun HY, Kim SJ (2006) Comparisons of physical properties of bacterial celluloses produced in different culture conditions using saccharified food wastes. Biotechnol Bioprocess Eng 11:26

Murthy PS, Naidu MM (2010) Protease production by *Aspergillus oryzae* in solid-state fermentation utilizing coffee by-products. World Appl Sci J 8:199–205

Murthy PS, Naidu MM (2011) Improvement of Robusta coffee fermentation with microbial enzymes. European J Appl Sci 3:130–139

Murthy PS, Naidu MM (2012) Production and application of xylanase from *Penicillium* sp. utilizing coffee by-products. Food Bioprocess Technol 5:657–664

Murthy PS, Naidu MM, Pullabhatla S (2009) Production of α-amylase under solid-state fermentation utilizing coffee waste. J Chem Technol Biotechnol 84:1246–1249

Mussatto SI, Teixeira JA (2010) Increase in the fructooligosaccharides yield and productivity by solid-state fermentation with *Aspergillus japonicus* using agro-industrial residues as support and nutrient source. Biochem Eng J 53:154–157

Navya PN, Pushpa SM (2013) Production, statistical optimization and application of *endoglucanase* from *Rhizopus stolonifer* utilizing coffee husk. Bioprocess Biosyst Eng 36:1115–1123

Navya PN, Bhoite RN, Murthy PS (2012) Improved β-glucosidase production from *Rhizopus stolonifer* utilizing coffee husk. Int J Current Res 4:123–129

Pandey A, Soccol CR, Nigam P, Brand D, Mohan R, Roussos S (2000) Biotechnological potential of coffee pulp and coffee husk for bioprocesses. Biochem Eng J 6:153–162

Pastore GM, Park YK, Min DB (1994) Production of fruity aroma by *Neurospora* from beiju. Mycol Res 98:1300–1302

Paulitz TC, Bélanger RR (2001) Biological control in greenhouse systems. Annu Rev Phytopathol 39:103–133

Poopathi S, Abidha S (2011) Coffee husk waste for fermentation production of mosquitocidal bacteria. J Econ Entomol 104:1816–1823

Ramachandra YL, Narayanamurthy G, Jois S, Chavan A, Satwadi PR (2013) Production of citric acid in basal coffee husk medium by *Aspergillus niger* under solid state fermentation. Adv Biol Res 7:234–240

Rani MU, Appaiah KAA (2012) *Gluconacetobacter hansenii* UAC09-mediated transformation of polyphenols and pectin of coffee cherry husk extract. Food Chem 130:243–247

Rani MU, Appaiah KAA (2013) Production of bacterial cellulose by *Gluconacetobacter hansenii* UAC09 using coffee cherry husk. J Food Sci Technol 50:755–762

Roussos S, de los Angeles Aquiáhuatl M, del Refugio Trejo-Hernández M, Gaime Perraud I, Favela E, Ramakrishna M, Raimbault M, Viniegra-González G (1995) Biotechnological management of coffee pulp-isolation, screening, characterization, selection of caffeine-degrading fungi and natural microflora present in coffee pulp and husk. Appl Microbiol Biotechnol 42:756–762

Sawant IS, Sawant SD, Nanaya KA (1995) Biological control of *Phytophthora* root-rot of coorg mandarin (*Citrus reticulata*) by *Trichoderma* species grown on coffee waste. Indian J Agr Sci 65:842–846

Shankaranand VS, Lonsane BK (1994) Coffee husk: an inexpensive substrate for production of citric acid by *Aspergillus niger* in a solid-state fermentation system. World J Microbiol Biotechnol 10:165–168

Shiono T, Yamamoto K, Yotsumoto Y, Yoshida A (2017) Caffeine adsorption of montmorillonite in coffee extracts. Biosci Biotechnol Biochem 1–7

Sugawara E, Hashimoto S, Sakurai Y, Kobayashi A (1994) Formation by yeast of the HEMF (4-hydrpxy-2 (or 5)-ethyl-5 (or 2)-methyl-3 (2H)-furanone) aroma components in Miso with aging. Biosci Biotechnol Biochem 58:1134–1135

Svensson A, Nicklasson E, Harrah T, Panilaitis B, Kaplan DL, Brittberg M, Gatenholm P (2005) Bacterial cellulose as a potential scaffold for tissue engineering of cartilage. Biomaterials 26:419–431

Tagliari CV, Sanson RK, Zanette A, Franco TT, Soccol CR (2003) Caffeine degradation by *Rhizopus delemar* in packed bed column bioreactor using coffee husk as substrate. Brazilian J Microbiol 34:102–104

Yamauchi H, Akita O, Obata T, Amachi T, Hara S, Yoshizawa K (1989) Production and application of a fruity odor in a solid-state culture of *Neurospora* sp. using pregelatinized polished rice. Agric Biol Chem 53:2881–2886

Chapter 7
Sustainable Valorization of Seafood Processing By-Product/Discard

P. V. Suresh, Tanaji G. Kudre and Lidiya C. Johny

Abstract Seafood is one of the major food products for human consumption and is an important part of the diet in many countries. However, huge amount of non-edible by-products (viscera, head, skin, scales, bones, etc.) are generated from the seafood processing operations annually. Generally, some of the seafood by-products are utilized as low-price ingredients through the mass transformation into a single product such as fish meal, fish oil, fertilizer/manure, fish silage, fish sauce, and protein hydrolysates, but the main bulk is dumped at sea/river or landfill, creating both disposal and pollution problems. It has been noted that these by-products contain valuable components including bioactive peptides, collagen and gelatin, oligosaccharides, fatty acids, enzymes, calcium, water-soluble minerals, vitamin, carotenoids, chitin, chitosan, biopolymers. These seafood-derived components have potential application in food, cosmetic, pharmaceutical, environment, biomedical, and other industries. Other than that, they are promising source for biofuel production. Thus, present chapter summarizes and highlights the need for an efficient by-product/discard reduction strategies and by-product/discard valorization that can provide viable and profitable options for seafood by-product/discard processing.

Keywords Seafood · Seafood processing · Seafood by-products
By-product valorization

7.1 Introduction

Seafood is one of the major food products for human consumption and is an important part of the diet in many countries. In 2013, approximately 17% of animal protein and 6.7% of all the protein consumed by humans came from seafood was

---

P. V. Suresh (✉) · T. G. Kudre · L. C. Johny
Meat and Marine Sciences Department, CSIR-Central Food Technological Research Institute, Academy of Scientific and Innovative Research, Mysuru 570 020, India
e-mail: drsureshpv@cftri.res.in; drsureshpv@hotmail.com

R. R. Singhania et al. (eds.), *Waste to Wealth*, Energy, Environment, and Sustainability, https://doi.org/10.1007/978-981-10-7431-8_7

reported (FAO 2016). In addition, seafood provided more than 3.1 billion people with almost 20% of their average per capita consumption of animal protein (FAO 2016). A significant portion of the seafood meant for human intake is disposed of as by-product/discard throughout the supply chain of seafood, beginning with harvesting and finishing with consumption. Further, industrial processing of seafood generates an enormous amount of non-edible portion (50–80%, w/w, of the total weight) as by-products depending on the type of seafood and level processing, which is wasted or many parts of the world. In addition, huge amounts of total catches are discarding every year from the fish farm as well as capture fishing as bycatch or unintentional catches fish species. Several approaches (physical, mechanical, chemical, and biological) have been attempted by seafood industries to treat or minimize the disposal of seafood by-product/discard (Guerard 2007; Bhaskar et al. 2011; Suresh and Prabhu 2013; Sachindra and Mahendrakar 2015). In general, these by-product/discard are rich sources of various valuable components with significant commercial applications in the different area. In this chapter, the scope for the sustainable valorization of the enormous quantity of seafood processing by-product/discard is exploring.

7.2 Seafood and Seafood Processing By-Products

The term 'seafood' generally defines a group of biologically divergent edible animals (excluding mammals) consisting both finfish (species of bony and cartilaginous fish as well as hagfishes and lampreys) and shellfish (species of Arthropoda, Mollusca, Cephalopods, and Echinodermata) from aquatic sources whether of freshwater, estuarine, or marine (Venugopal 2006; Suresh and Prabhu 2013; Sachindra and Mahendrakar 2015; FAO 2016). This term is applied to both edible animals (excluding mammals) collected from aquatic sources by fishing and produced by aquaculture. In addition, aquatic animals such as frogs and turtles that are served as food and eaten by humans are also considered as seafood (FAO 2016). Edible seaweeds and other aquatic plants are also considered as seafood, and they are widely consumed as sea vegetables around the world, especially in Asian countries (Suresh and Prabhu 2013; FAO 2016). Throughout this chapter, the term seafood is considered with a broad sense that comprises both finfish and shellfishes.

In recent decades, the fisheries and aquaculture have been evolving rapidly as food production industries. Seafood is now the most internationally traded food product, with some 37% by volume traded across national borders (FAO 2012a, b). Seafood is one of the essential food products for human consumption and is a significant part of the diet in many nations. Apart from rich in high-value protein, seafood is usually low in saturated fats, carbohydrates, and cholesterol and also provides a board range of essential micronutrients (FAO 2012a, b; Suresh and Prabhu 2013; FAO 2016). Thus, even in small quantities, provision of seafood can be useful in addressing food and nutritional security among the poor and vulnerable populations around the world (FAO 2012a, b; World Bank 2013). Many millions of people all over the world find a source of income and livelihood in the fisheries and aquaculture field. The very recent evaluations indicate that 56.6 million people were involved in the primary sector of capture fisheries and aquaculture in 2014. Of this total, 36% involved full time, 23% part-time, and the remainder was either occasional fishers or unspecific status (FAO 2016).

According to Food and Agriculture Organization of the United Nations (FAO 2016), world seafood production (by capture fisheries and aquaculture) was 167.2 million tonnes (MT) in 2014 (excluding aquatic plant), of which 146.3 MT (87.5%) is used for direct human consumption. The remaining (20.9 MT, 12.5%) were intended for non-food goods such as fish meal and fish oil. In 2014, 46% (67 MT) of the seafood for direct human consumption was in the form of live, fresh, or chilled fish. The remaining of the production for edible purposes was in different processed forms, with about 12% (17 MT) in dried, salted, smoked, or other cured forms, 13% (19 MT) in prepared and preserved forms, and 30% (about 44 MT) in frozen form. Freezing is the primary method of processing seafood for human consumption, and it accounted for 55% of total processed seafood for human consumption and 26% of total seafood production in 2014 (FAO 2016). Large portions of seafood landed for human consumption are processing due to some reasons in some ways or the other, all over the world in the food supply chain. This processing of seafood creates a huge volume of non-edible portion (50–80%) as by-product/discard/visceral waste, which is thrown away or underutilized in many parts of the world (Guerard 2007; Bhaskar et al. 2011; Suresh and Prabhu 2013; Ghosh et al. 2016). Subsequent capture or harvesting and processing is the primary phase in the supply chain of seafood where most by-product/discard take place. In general, seafood is processing immediately after harvest either in onboard processing facilities of fishing fleets or in offshore processing factories. Since the seafood includes various kinds of aquatic animals, the non-edible part produced varies significantly in their volume and composition (Table 7.1). Finfish processing generates 10–50% of total weight as non-edible part, which comprises gut (visceral), head, bone, skin, scales, fin, and flesh remaining on the bone (Fig. 7.1). However, shellfishes, mainly crustaceans processing, produce up to 80% of raw material as non-edible parts, which include head, shell (carapace), viscera, and appendages (Bhaskar et al. 2011; Suresh and Prabhu 2013). The major crustacean such as shrimps/prawn, crabs, and lobsters constitutes shellfish species that processed commercially. India is one of the leading processed shrimp-exporting countries and annually generates >1 lakh tonnes of solid by-products from shrimp processing (Bhaskar et al. 2011: Suresh and Prabhu 2013). Although no detailed estimate of non-edible by-products is available, a calculation (based on the amount of processed seafood for human consumption, 63 MT in 2014) indicated that between 31.5 and 50.4 MT (equivalent to 50–80% of processed seafood for human consumption) with an average of 40.95 MT are produced in 2014, globally. It was reported that about 20 MT of seafood by-products are produced worldwide by the various seafood processing operation (Guerard 2007; Bhaskar et al. 2011; Suresh and Prabhu 2013).

Fig. 7.1 Finfish processing by-products

Various kinds and quantities of by-product/discard are produced throughout the supply chain of seafood, beginning with harvesting and finishing with consumption (Ghosh et al. 2016). Seafood by-product/discard are formed through bycatch, onboard processing, other processing, transport, storage, retailers, and consumers (Sharp and Mariojouls 2012). In addition, huge amount of seafood is discarded during wild catching (fishing operation) as bycatch or accidental catching species. This problem has extensively studied, and despite environmental and business guidelines, there is still no effective solution to bycatch waste (Ghosh et al. 2016). It is projected that globally about 17.9–39.5 MT of whole seafood is discarded every year by commercial fishing operations (Ghosh et al. 2016). Also, a significant amount of the total catch from aquaculture is disposed of each year. In reality, not only does the by-product/discard disposal have a high cost, but also it has a significant environmental impact.

Concerning seafood processing, there are different terms such as by-product, coproduct, fish waste, fish offal, fish visceral mass, fish discards that are applied to describe the non-edible body part of seafood. According to Guerard (2007) and Sachindra and Mahendrakar (2015), 'by-product' specifies something that is not regarded as an ordinary saleable product but can use after treatment and the term 'waste' refers to products that cannot be used for feed or food but has to be composted or destroyed. Further, observed in the scientific literature that the most frequently used and appropriate term to describe the non-edible portion generated during seafood processing is 'by-product.' Hence, throughout this chapter, 'by-product/discard' are considered in a broad sense that comprises above-mentioned different terms. The composition and percentage of these non-edible parts in the total volume of by-products vary widely with the type of seafood and method of processing (Lee 2011; Bhaskar et al. 2011; Suresh and Prabhu 2013; Sachindra and Mahendrakar 2015; Ghosh et al. 2016). The major non-edible parts of finfish processing include head, gut (visceral), skin including scales, bones, frame, and flesh attached to the bone of processed fishes, while shellfish by-products, especially from crustaceans, principally comprises head (cephalothorax) and shell (carapaces) and also discolored and imperfectly processed whole animal.

The disposal route for seafood by-product/discard is not as straightforward as grains and other crops produced. It is because the disposal of seafood by-product/discard involves stringent hygiene, safety, and management of environmental issues during its disposal, and in many cases, its disposal is regulated by government organizations (Ghosh et al. 2016). Further, to reduce the disposal of a large quantity of seafood by-product/discard formed worldwide, some alternative strategies have developed to improve commercial value of these biomaterials. In general, two different approaches such as mass conversion/transformation and sorting/segregation have developed to increase the market value of seafood by-product/discard (Linder et al. 1995). In mass conversion/transformation, the seafood by-product/discard are converted into a single product. The converted seafood by-product/discard products comprise fish meal, fish oil, fertilizers, and hydrolysates such as protein hydrolysate. On the other hand, sorting/segregation involves utilizing various by-product/discard (seafood body parts) such as bones, visceral, and fins separately to enhance their economic value (Ghosh et al. 2016). Furthermore, sorting/segregation allows the extraction of commercially significant components from seafood by-product/discard such as peptides, oil and lipid, pigments, flavors, enzymes, chitin, collagen and gelatin, vitamins, minerals (Figs. 7.2 and 7.3). Widespread recognition and adoption of both methods could lead to significant reductions in by-products going to landfill or ocean dumping and reduce the damaging impact of seafood by-product/discard on the environment (Suresh and Prabhu 2013; Ghosh et al. 2016). At present, most seafood by-product/discard about 59% of all seafood go to landfill or dumping in the sea, and only about 39% are reused or incinerated (Jespersen et al. 2000; Ghaly et al. 2013). The traditional methods (landfill or dumping in the sea) of disposal of seafood by-product/discard not only create environmental pollutions but also losing various valuable components.

Moreover, it was observed that seafood by-product/discard processing could be a hard business in many nations owing to some problematic issues such as hygiene, safety, and environmental hazards (Ghosh et al. 2016). The most important factor that any commercial operation needs to consider is the economic viability of seafood by-product/discard processing (Hwang and Hansen 1998). It is noted that current seafood by-product/discard management in India is found to be inefficient and non-profitable (except in the case of chitin and chitosan production from crustacean by-products). It is highlighting the need for efficient by-product/

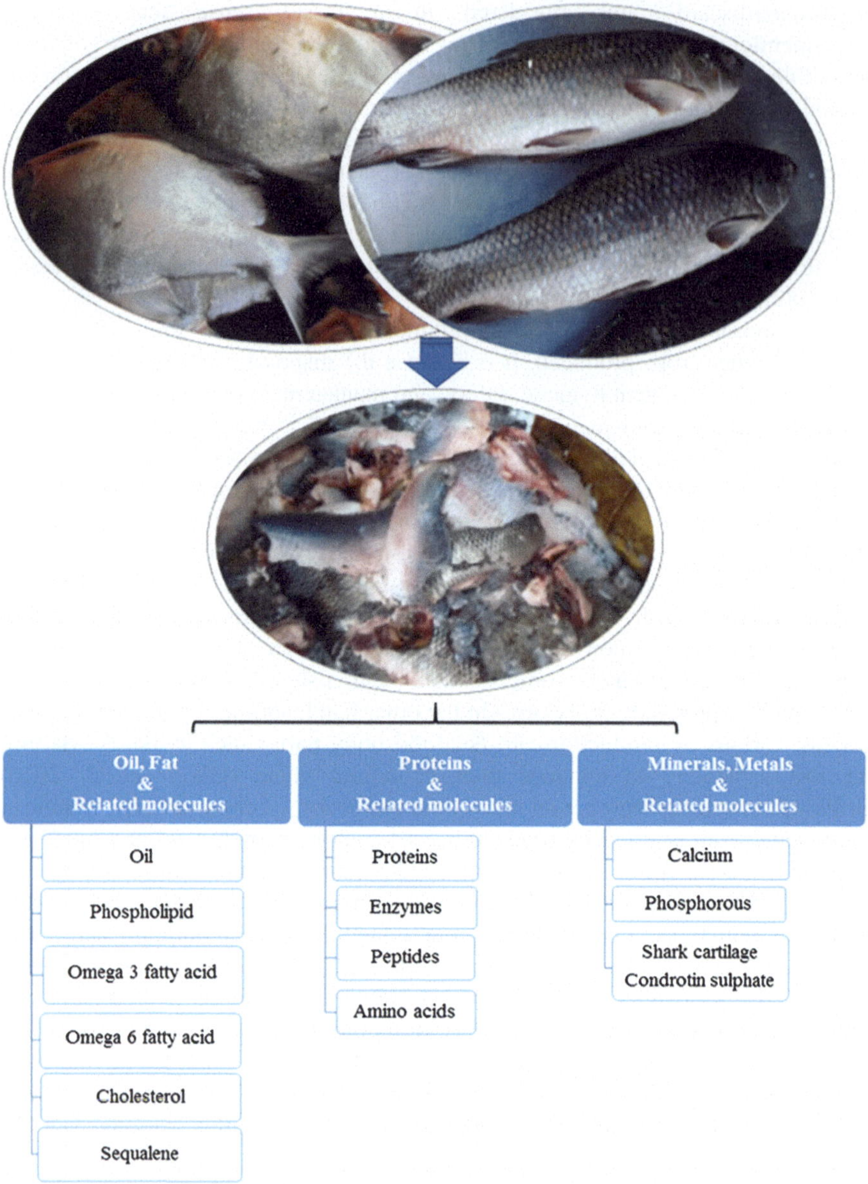

Fig. 7.2 Finfish processing by-products and possible valuable bioactive components

discard reduction strategies and by-product/discard utilization that can provide viable and profitable options for seafood by-product/discard processing. As stated by Ghosh et al. (2016), there are some good economic and environmental reasons to process seafood by-product/discard and produce valuable products. However,

Fig. 7.3 Shellfish specifically shrimp processing by-products and possible valuable bioactive components

additional research is required to make effective and efficient methods of handling seafood by-product/discard at an economically viable and commercially acceptable scale with as minimal ecological issues as possible.

In many developed countries, the chief drivers for food by-product/waste management strategies are government legislation concerning to safety, handling of hazardous waste materials, and the environmental influence of the business working practices, while, in emerging countries, factors such as food type, processing facilities, storage facilities, transport, and even climatic conditions are the primary drivers in food waste management strategies (Rutten 2013; Ghosh et al. 2016). Further, in the case of seafood by-product/discard utilization, the drivers are health safety and hygiene risks associated with the processing of these materials. Therefore, by-product/discard processing approaches must be optimized to encourage production efficiency and cost-effectiveness so that the final products are competitive in the marketplace (Ghosh et al. 2016). In a report, Ghosh et al. (2016) stated consumer is the most important factor for a food waste utilization strategy that targets to produce value-added products; further, without consumer acceptance of food waste reduction approaches, no sustainable, eco-friendly food waste utilization and management strategy can succeed.

Table 7.1 Major seafood processing by-product/discard (values shown in the table are compiled from various authors mentioned somewhere else in the text)

Seafood (100%)	Type of seafood	By-product/discard	% of by-product/ discards (w/w)
Finfish			
	Boney, cartilage, other finfishes	Gut	15–20
		Skin	1–3
		Head	15–20
		Trimming	2–5
		Bones	10–15
Shellfish			
Crustacean	Shrimp	Shell and head	60–80
	Crab	Shell and viscera	60–70
	Lobster	Shell and head	Up to 60
	Crayfish	Shell and head	Up to 85
	Krill	Shell and head	70–75
Mollusca	Oyster, mussel, clam, scallop, etc.	Shell and various non-edible portion	60–80
	Cuttlefish, squid, octopus, etc.	Ink bag, gladius, other non-edible organs	10–35

7.3 Treatment of Seafood By-Products by Mass Conversion/Transformation

7.3.1 Seafood By-Products/Discards into Fish Meal or Fish Oil

Fish meal is a natural feed component used in diets for farmed fish and crustaceans and as a nutrition supplement of pigs and poultry, as well as in pet food. Converting seafood by-product/discard into fish meal has been progressively increasing in recent years with many nations transforming their seafood by-product/discard using cost-effective reprocessing technologies (Shepherd 2012; Ghosh et al. 2016). In spite of the recycling costs connected with processing seafood by-product/discard into fish meal, fish meal's price as a feedstock for aquaculture has to balance the recycling costs. At present, there are only about ten major nations converting seafood by-product/discard into fish meal products, that is, Canada, Japan, Chile, Denmark, Iceland, Mexico, Norway, Thailand, Russian Federation, and the USA (Ghosh et al. 2016). However, these countries on average are only using around 25% of their seafood by-product/discard to produce fish meal products (Shepherd 2012). An estimate of about 15–25% of the global fish meal is currently produced from seafood processing by-product/discard (Shepherd 2012; World Bank 2013). Worldwide production of fish meal in 2030 is projected to be around 7.6 MT (World Bank 2013). It is reported that India increases imports over the period

(2010–30) and becomes the third largest net importer of fish meal in 2030, after China and Southeast Asia (World Bank 2013). Fish meal usage over a period of 50 years (1960–2010) reveals its increased use in aquaculture, while its use in both pig and poultry feeds has declined (Gunasekera et al. 2002; Ghosh et al. 2016). The demand for fish meal and fish oil likely becomes higher, given the accelerated expansion of the global aquaculture and slowness of the global capture fisheries that supply their constituents (World Bank 2013). Many of the species used to manufacture fish meal and fish oil are small, bony, not very palatable, or unfamiliar to consumers. However, globally in 2009, about 25% of fish meal and fish oil were produced from by-products, and this proportion has been rising steadily in recent years, by about 1–2% per annum. Conventionally, fish meal factories have often been located alongside canning and freezing plant to process and pack seafood for human consumption. Currently, aquaculture (63%) is the biggest consumer of the fish meal followed by the pig (25%), poultry (8%), and other animal (45) farmings (Chamberlain 2011; Ghaly et al. 2013).

Fish oil is the primary natural source of the long-chain omega-3 polyunsaturated fatty acids (eicosapentaenoic acid, EPA, and docosahexaenoic acid, DHA). There is an increasing market for fish oil as a human nutritional supplement in the form of oil capsules and as a food additive (Ghaly et al. 2013; Suresh et al. 2015; FAO 2016).

7.3.2 Converting Seafood By-Products/Discards into Fish Sauce

Fish sauce is a traditional fermented fish product, which is used as an important source of protein in several Southeast Asian countries and also consumed in Europe and North America. For example, such products are *nam-pla* in Thailand, *nuoc-mam* in Vietnam and Cambodia, *patis* in the Philippines, *ngapi* in Burma (Myanmar), and *yoshoku* in Japan (Suresh et al. 2015). Fish sauce is a clear liquid product, and it is produced from small pelagic fish or seafood processing by-products by the hydrolysis of fish protein in the presence of a high salt concentration (Salt fermentation). It is prepared by adding 20–30% marine salt to fish (small and commercially unimportant fish) such as anchovies, sardines, or round scad and permitting them to hydrolyze in closed containers (Shahidi and Kamil 2001). For complete hydrolysis of fish tissue and flavor development, the fermentations were carried out for 3 months to 2 years at 30–35 °C. During incubation, endogenous halotolerant enzymes present in the muscle gradually degrade the fish tissue in order to form a clear, amber-colored liquid with a high content of free amino acids and excellent flavor (Shahidi and Kamil 2001; Suresh et al. 2015). Due to the high salt content, in general, the growth of bacteria is negligible. It can be used as a condiment on vegetable dishes and is very nutritious due to the presence of essential amino acids (Ghaly et al. 2013).

7.3.3 *Converting Seafood By-Products/Discards into Fish Silage*

Fish silage is an excellent protein source with high nutritional properties for animal feeding. Fish silage can be made from spoiled fish, subutilized species, and seafood by-product/discard. It is a liquid product prepared from whole seafood or their processing by-product/discard that are liquefied by the activity of endogenous enzymes in them with added acid. During liquefaction, seafood proteins are broken into soluble units (amino acids), while acid accelerates the enzyme hydrolysis as well as prevents bacterial spoilage (Disney et al. 1977; Vidotti et al. 2003; Ghaly et al. 2013; Mahendrakar and Rathina 2015).

Lactic acid fermentation is also used in the preparation of fish silage, in which minced whole seafood or their by-product/discard are mixed with sugar or molasses. Further, a lactic acid fermentation is subjected by inoculating with a culture of *Lactobacillus* sp. The inoculated material incubated at a particular temperature for a desired period is a sealed vessel (Ghaly et al. 2013). According to Gildberg (1992), many bioactive products including peptone, oil, and pepsin can be obtained from fish silage.

7.3.4 *Production and Utilization of Seafood Protein*

Seafood processing by-products are rich sources of functionally active and nutritive marine proteins with various potential applications (Okada et al. 2008). Proteins from seafood by-products are recovered mainly in the form of protein hydrolysates, which are derived from proteins as peptides of various sizes after chemical and enzymatic hydrolysis as well as lactic acid fermentation (Guerard 2007; Kristinsson and Rasco 2000; Suresh and Prabhu 2013). Protein hydrolysate with specific functionalities and biological properties can be prepared by using specific enzymes and hydrolysis conditions (Kristinsson 2007; Sachindra et al. 2011). The protein hydrolysates have a broad range of potential applications including use of flavor enhancers, functional ingredients, or nutritional additives to foods of low protein quality (Sachindra et al. 2011). Recently, protein hydrolysates extracted from marine by-products have become popular in the food industry due to its high biofunctional action (Cordova-Murueta et al. 2007). Marine protein hydrolysates are widely used as feed for a variety of farmed animals as well as cultured food fish. Marine protein hydrolysates are rich sources of biologically active peptides, with considerable potential in pharmacology and as growth-stimulating agents in animal feed (Gildberg and Stenberg 2001).

7.4 Chemical Extraction of Seafood Protein

The most common extraction method used for the fish proteins is the solvent extraction process. A standard protocol for the solvent extraction of proteins has reported by Sikorski and Naczk (1981). The whole fish is first minced, and the protein is extracted with isopropanol. After grinding, the supernatant is collected and extracted three times. The first extraction is carried out at 20–30 °C for 50 min in isopropanol. The second extraction is performed at 75 °C for 90 min with isopropanol. The third extraction is carried at 75 °C for 70 min with azeotropic isopropanol. The final supernatant fraction is collected, dried, milled, and screened to separate out bone particles. Hermansson et al. (1971) reported that the fish protein concentrate could also produce at a temperature of 50 °C, but it will have lower emulsifying properties and poor solubility. The disadvantages of this method are reduced functionality, off-flavors, the high manufacture cost, and traces of the solvent in the final product, making it commercially unsuccessful. Batista (1999) reported on the extraction of proteins from hake and monkfish wastes using a chemical method. The extraction was carried out with HCl in the acid phase and with NaOH and Ca $(OH)_2$ in the alkaline phase. Further, Nurdiyana et al. (2008) optimized the extraction process of proteins from freeze-dried fish waste using response surface methodology.

The enzymatic processing of seafood by-product/discards could be helpful in producing a wide range of food ingredients and other products for a variety of applications (Kim and Park 2006; Suresh et al. 2015). Hydrolysis of the protein of seafood by-products can be achieved by both the digestive enzymes of fish itself and addition of enzymes of external sources (Rustad 2003). Most commercial proteases used for seafood protein hydrolysate preparation include those from microbial sources such as Alcalase, Neutrase, Flavourzyme, Umamizyme, Protamex; from plant sources such as papain, bromelain, ficin; and from animal sources such as pepsin, trypsin, chymotrypsin (Guerard 2007). Alcalase, an alkaline enzyme produced from *Bacillus licheniformis* and developed by Novozymes, has reported to be one of the most useful enzymes to hydrolysis fish proteins (Aspmo et al. 2005; Bhaskar et al. 2007a). By-products from several finfish and shellfish species have exploited for the commercial production of protein hydrolysate, with the most common source being fillet by-products (Gbogouri et al. 2004; Mukhi and Novikov 2001; Bhaskar et al. 2007a). Arctic and Antarctic krill has also reported as sources of protein and protein hydrolysate (Moren et al. 2006; Sountama et al. 2007). In addition, lactic fermentation has attempted to produce protein hydrolysates from processing by-products of shellfish such as crayfish (Bautista et al. 2001) and shrimp (Lopez-Cervantes et al. 2006; Bhaskar et al. 2007a, b; Bueno-Solano et al. 2009).

7.4.1 Treatment of Seafood By-Products by Sorting Methods

Seafood processing by-products are potential sources of some valuable components with a broad range of possible applications in the field of food, feed, biomedicine, fine chemical, and pharmaceuticals (Suresh and Prabhu 2013; Suresh et al. 2015; Sachindra and Mahendrakar 2015). As mentioned above, sorting method of seafood treatment enables the extraction of commercially significant components from by-product/discard such as specific protein, the bioactive peptide, oil and lipid, pigments, flavors, enzymes, chitin, collagen and gelatin, vitamins, minerals. However, the composition of these components varies according to the type of species, sex, age, and nutritional status, time of year and health, as well as a method of processing. The valuable components from by-products of finfish processing include bioactive peptides, enzymes, minerals, collagen, lipid, and long-chain omega-3 polyunsaturated fatty acids (n-3 or ω-3 PUFAs), while shellfish processing by-products especially from crustacean processing are the primary sources of chitin and chitosan, pigment, bioactive peptides, minerals (as calcium), etc. Recovery of these valuable components from seafood by-products is projected as a possible interim solution for the treatment of solid by-products of seafood processing since it provides additional profit for the processing plant and minimizes environmental hazard. Seafood industries across the world adopt different types of processing methods for treating their by-product/discard, and generally, they are grouped into different categories such as chemical, mechanical, and biological processing. However, in the past few decades, majority of the processing methods are of chemical and mechanical, due to the lack of adequate knowledge in the biological processing methods. Presently, due to the advancements in biotechnology, biological process is gaining much importance as sustainable processes and is on the edge of replacing the chemicals and mechanical processing of seafood by-products. It is well recognized that the biological processes are eco-friendly and sustainable, as well as provide the possibility to recover additional useful components than target component from the raw materials (Rustad 2003; Guerard 2007; Bhaskar et al. 2011; Suresh and Prabhu 2013). But, to date, very little has been done to fully develop and commercialize these types of products (Gupta et al. 1997).

7.4.2 Oil, Lipids, and Fatty Acids from Seafood Processing By-Products

The potential beneficial effects of fish oils on human health care are well documented (Berge and Barnathan 2005; Venugopal 2009). The fish oil consists of two primary polyunsaturated fatty acids, EPA and DHA, and is classified as omega-3 fatty acids. Marine oils are attractive from the nutritional point of view because they are thought to deliver specific physiological functions against thrombosis, cholesterol buildup, and allergies (Kimoto et al. 1994). The lipid-based compounds that can be obtained from seafood processing by-products include oils, polyunsaturated fatty acid (omega-3 fatty acids, n-PUFAs), phospholipids, squalene, vitamins, cholesterol (Amit et al. 2010). The occurrence of especially EPA (20:5 n-3) and DHA (22:6 n-3) in fish oils is responsible for the health benefit (Shahidi and Alasalvar 2011). Fish oils are still considered to be the least expensive natural source of n-PUFAs and the n-3PUFA content, and EPA/DHA ratios in marine oil tend to vary with season and species in both quality and quantity (Guerard 2007; Shahidi 2007a; Amit et al. 2010; Ghaly et al. 2013). Also, liver oil is a potential source of vitamin A and squalene (Shahidi and Alasalvar 2011). Various methods have attempted to extract oil from seafood by-product/discard including chemical, enzymatic, and microbial fermentation.

Fish oil can be extracted from fish and fish waste chemically using Soxhlet or Goldfisch, Folch, Bligh, and Dryer methods and acid digestion method with various solvents such as diethyl ether, petroleum ether, chloroform/methanol, and hexane (Shahidi 2003). Folch et al. (1957) described a chloroform/methanol lipid extraction procedure in which 1.5 g of fish tissue is mixed with 30 mL of 2:1 chloroform/ methanol. Aryee and Simpson (2000) and Zhong et al. (2007) extracted lipids from seafood by-product/discard by Bligh and Dyer methods using a different solvent mixture. Xiao (2010) extracted lipids from seafood by-product/discards using acid digestion method.

The enzymatic processes utilized for the oil recovery from seafood by-products can be classified into two categories such as (i) protease-catalyzed hydrolysis and (ii) lipase-catalyzed hydrolysis and esterification (Guerard 2007; Wanasundara 2011; Suresh et al. 2015). Protease-catalyzed hydrolysis method is used to extract the polyunsaturated fatty acids from the starting material. It is documented that oil from seafood processing by-product/discard with enzymatic hydrolysis is similar to that produced by conventional methods (Linder et al. 2002, 2005; Dumay et al. 2004). To concentrate n-3 fatty acids, the lipase-catalyzed hydrolysis and esterification of existing oils can be applied (Shahidi and Wanasundara 1998; Shahidi and Kamil 2001; Guerard 2007). It is reported that all of the enzymatic processes used for the production of n-PUFA concentrates are safer and more efficient than any other methods, since the enzymatic reactions, regardless of whether they involve hydrolysis or esterification, can conduct under ambient temperatures, normal pressure, and nitrogen-protected environment (Wanasundara 2011).

7.4.3 Enzymes from Seafood By-Products

Since the marine organisms adapted to drivers' extreme environmental conditions (high salt concentration, low or high temperature, and high pressure), enzymes from these organisms may differ from those of terrestrial organisms in their characteristic (Shahidi and Kamil 2001; Venugopal 2009; Sachindra et al. 2011). Hence, enzymes derived from marine sources have received much attention for their potential applications in various fields (Suresh and Prabhu 2013). It is stated that extraction of the enzyme from seafood by-products could help in the environmental and ethical anxieties surrounding the discards and improve the bottom line for seafood industries wishing to exploit new technologies and markets (Morrissey and Okada 2007). It is well documented that seafood processing by-products particularly viscera can serve as one of the cost-effective and viable sources for marine-derived enzymes. Some strategies such as ensilage, membrane technology, ohmic heating, precipitation, aqueous two-phase systems, and chromatography have applied to isolate enzymes economically from processing by-product/discard of seafood (Morrissey and Okada 2007; Sachindra et al. 2011; Suresh and Prabhu 2013). Recovery of enzymes from seafood by-products primarily depends on the localization of the target enzymes such as extracellular (located in the extracellular fluid) or intracellular (located inside the cell) (Morrissey and Okada 2007). Extractable enzymes in the seafood processing by-products include extracellular gastric proteinases, digestive proteinases, lipases, chitinases, phospholipases, transglutaminases, polyphenol oxidase. Among the various enzymes from seafood by-products, digestive proteolytic enzymes have received considerable interest past two decades owing to the availability of raw materials such as viscera and their high rate of enzymatic activity (Morrissey and Okada 2007; Sachindra et al. 2011). Overall, marine proteases possess particular characteristic properties such as higher catalytic efficiency at low temperature and greater stability at border pH range (Simpson 2000). These unusual catalytic activities of marine proteases have been used for their applications in many food processings (Vecchi and Coppes 1996; Swapna et al. 2011). In addition to digestive enzymes, a number of commercially significant enzymes have been isolated from seafood by-products such as chitinases, lipases, alkaline phosphatase, hyaluronidase, and β-acetyl-D-glucosaminidase (Fange et al. 1979; Matsumiya et al. 2002; Gutowska et al. 2004; Sachindra et al. 2011; Suresh and Prabhu 2013).

7.4.4 *Collagen and Gelatin from Seafood Processing By-Products*

Collagen is one of the major structural proteins in animal body and constitutes about 25% of total protein, while gelatin is a class of protein fractions that have no existence in nature, but are derived from the parent protein collagen by denaturation (Sachindra et al. 2011; Suresh and Prabhu 2013; Pal et al. 2015). The functional and structural features of collagen create it a desirable target for biomedical and food applications. The current healthcare demands depend on the collagen-based food, cosmetics, pharmaceutical, and other biomedical products. The international market regarding collagen was >3 billion US$ in 2015. Precisely, Asia-Pacific would be evolving as the most promising region for the international collagen market by 2022. Further, a Compound Annual Growth Rate concerning collagen is predicted to be >7% from 2016 to 2023. Regarding its commercial application, collagen has mainly bovine and porcine origins, which have been a matter of concern in the last years. In fact, partially due to religious constraints related to avoidance of porcine and bovine products and the recent outbreak of bovine spongiform encephalopathy (BSE) in bovines, other collagen sources are being discussed (Regenstein and Zhou 2007; Pal et al. 2015). In this regard, the use of collagen with marine origin is being considered highly attractive by the industry as an important alternative source. With this, the aquatic sources, especially seafood processing by-products, are considering as a promising source of safe and realistic collagens (Pal and Suresh 2016). Collagen can be isolated from skin, scale, bone, cartilage, swim bladder, and fins of finfish processing by-products and various by-products of shellfish such as cuttlefish, octopus, squid, jellyfish, and sea urchin (Regenstein and Zhou 2007; Suseela et al. 2015; Pal and Suresh 2016, 2017a, b).

Extensive research has carried on isolation of different collagens (salt-soluble collagen, acid-soluble collagen, and pepsin-soluble collagen) and gelatin from processing by-products of seafood using different strategies (Nagai et al. 2001; Nagai and Suzuki 2002; Arsesen and Gildberg 2002, 2007; Liu et al. 2010; Suseela et al. 2015; Pal and Suresh 2016, 2017a, b; Pal et al. 2015; 2017). In general, most of the collagen produced is converted to gelatin, owing to its wider industrial uses. Gelatin is usually manufactured by the acid treatment process (for type A gelatin) or the alkaline treatment process (type B gelatin) (Zhou and Regenstein 2004). In addition to food and pharmaceutical applications, collagen and gelatin are used in cosmetics, photographic industry, paper manufacturing, printing processes, and various other fields (Regenstein and Zhou 2007; Suseela et al. 2015).

7.4.5 Flavor from Seafood Processing By-Products

The complex flavor of seafood is composed of equally important taste and aroma-active components (Kim and Cadwallader 2011; Suresh and Prabhu 2013). Volatile aroma constituents are the key to flavor perception, and in seafood, aromas can form via several mechanisms including enzymatic conversion of lipid and other precursors to volatile aroma constituents, auto-oxidative degradation of free fatty acid to volatile aroma components, and thermal decomposition of precursors during processing and cooking (Josephson 1991). Extraction of flavors from seafood by-products is considered as an important method of valorization of seafood processing by-product/discard (Suresh and Prabhu 2013). The production of natural seafood flavor extracts from the seafood by-products has been a commercial practice in France and Japan (Lee 2007). These seafood flavors are being used in seafood sauces, chowders, soups, bisques, instant noodles, snacks, surimi-based products such as crab and shrimp analogues, and cereal-based extrusion products. Flavor extracts derived from seafood by-products are very popular items, particularly in Asian markets, and good quality seafood flavors are in high demand (Lee 2007). Processing by-products from finfish such as anchovies, salmon, cod, tuna and from shellfish such as clams, crab, squid, scallops, shrimp, krill, oyster, and lobster are excellent raw materials for the extraction of food flavors. In addition to the flavor extracts from seafood materials, the extracts of fermented fish (fish sauce) are also used as umami-giving seafood flavor (Lee 2007; Suresh and Prabhu 2013).

7.4.6 Pigment from Seafood By-Products

The chief pigments found in the by-products of seafood processing are the carotenoids. Shellfish by-product/discard, especially from crustaceans such as shrimp, lobster, crabs, crayfish and krill, are the important sources of natural carotenoids mainly astaxanthin (Sachindra et al. 2005, 2007, 2011; Sawmya and Sachindra 2015). The carotenoids extracted from crustacean by-product/discard have potential commercial application as a source of pigmentation in aquaculture. Also, to use in aquaculture feeds, carotenoids are used as colorants for food, drugs, and cosmetics (Simpson 2007; Sachindra et al. 2011). It is estimated that the global market for carotenoid pigments is about 935 million US dollar (Fraser and Bramley 2004). The carotenoid materials have also been incorporated in poultry feed to impart desirable coloration to egg yolks of various poultry animals (Suresh and Prabhu 2013). Some studies have been reported to recover the carotenoids from crustacean processing by-product/discard using different methods both chemical and biological. To extract carotenoids as stable carotenoprotein complex, biological methods using enzymatic techniques have been evaluated by various researchers (Simpson and Haard 1985; Cano-Lopez et al. 1987; Klomklao et al. 2009). It has been reported that biological approaches either enzymatic or fermentative methods are more desirable for the isolation of carotenoids from crustacean by-products as carotenoprotein (Sachindra et al. 2011). Recovery of carotenoids from silage of shrimp by-products by lactic acid fermentation followed by extraction with the solvent mixture and refined oil has attempted (Sachindra et al. 2007).

7.4.7 Chitin, Chitosan, and Their Derivatives from Seafood Processing By-Products

Chitin and its derivative chitosan are fascinating polysaccharides with unique properties that offer a wide variety of biological and industrial applications. They are widely used in textile, cosmetics, pharmaceuticals, agrochemicals, biomedicine, food, and environmental applications (Tharanathan and Kittur 2003; Synowiecki and Al-Khateeb 2003; Nidheeesh and Suresh 2015). Owing to the biodegradability, non-toxicity, and biocompatibility, chitin and chitosan have found several applications in food and related industries (Shahidi 2007b; Venugopal 2009; Nidheeesh and Suresh 2015).

Chitin is an amino polysaccharide of β-1,4-linked N-acetyl-D-glucosamine and is considered as the second most abundant and renewable biopolymer on earth, next only to cellulose and available to the extent of over ten giga tonnes annually (Tharanathan and Kittur 2003; Shahidi 2007b). Crustacean's by-product/discard from shrimp, crab, lobster, krill processing are currently the primary source for industrial production of chitin (Synowiecki and Al-Khateeb 2003; Suresh and Prabhu 2013). Commercial chitin preparation from shellfish by-product/discard involves removal of minerals using a strong acid (demineralization) and proteins using strong alkali (deproteinization) and pigments (decolouration) (Shimahara and Takiguchi 1988; Synowiecki and Al-Khateeb 2003; Nidheesh and Suresh 2015a, b). However, chemical chitin preparation process is extremely hazardous, energy-consuming, and damaging to the environment by high concentrations of strong acid and alkali (Healy et al. 2003). Alternative methods of chitin extraction from crustaceans by-products using biological processes such as enzymatic and fermentation have attempted (Simpson and Haard 1985; Rao et al. 2000; Shirai et al. 2001; Chaussard and Domard 2004; Bhaskar et al. 2007b). Chitosan is a linear copolymer of β-1, 4-linked GlcNAc and D-glucosamine (GlcN) derived from crustacean chitin by *N*-deacetylation. Most manufacturers produce chitosan, instead of chitin from crustacean by-products, owing to the variety of biological and industrial applications of chitosan than chitin. Chitosan is produced industrially by deacetylation of chitin extracted from crustacean shell by-products by treatment with a concentrated alkaline solution (up to 40%) and at very high temperature (up to 140 °C). To produce chitosan with consistent quality and to avoid the extreme alkali deacetylation process, an enzymatic method using microbial chitin deacetylase has attempted. However, at present, there is no commercial process available for the production of chitin deacetylase and its use in chitosan production (Suresh et al. 2011).

N-acetyl chitooligosaccharides (*N*-acetyl COSs) and chitooligosaccharides (COSs) are a hydrolytic derivative of chitin and chitosan, respectively. Chitin and chitosan oligomers have extensive uses in medicine, food, agriculture, and biotechnology owing to its water-soluble and biofunctional properties such as antimicrobial, antitumour, and immune-enhancing effects (Harish and Tharanathan 2007; Suresh et al. 2011; Nidheesh and Suresh 2015a, b). They are produced by partial hydrolysis of chitin and chitosan or crude crustacean processing chitinous by-products using acid or enzymes (Suresh et al. 2011). These oligomers may be more advantageous than polymer chitin and chitosan (Guerard 2007; Suresh and Prabhu 2013).

N-acetyl-D-glucosamine and D-glucosamine are the primary components of chitin and chitosan. These molecules have attracted much attention due to its potential therapeutic and health benefits (Venugopal 2009; Suresh and Prabhu 2013; Nidheesh and Suresh 2015b). Commonly, N-acetyl-D-glucosamine is produced by acid hydrolysis of chitin with concentrated HCl at high temperature (>80 °C) (Kuk et al. 2005). Biological methods (enzymatic and fermentative) have attempted for the production of *N*-acetyl-D-glucosamine from crustacean by-product/discard (Aloise et al. 1996; Sashiwa et al. 2002; Sukwattanasinitt et al. 2002; Matsumiya 2004; Kuk et al. 2005; Jung et al. 2007; San-Lang et al. 2009; Suresh and Kumar 2012; Suresh 2012).

7.4.8 Calcium, Other Minerals, Vitamin, Etc., from Seafood Processing By-Products

Calcium is an essential element required for numerous physiological functions in the human body including the strengthening of teeth and bones, nerve function, and many enzymatic reactions that require calcium as a cofactor (Kim and Jung 2007). Deficiency in calcium will cause hypocalcemia symptoms and other serious calcium malnutrition diseases such as osteoporosis. Phosphorous is the second major component of bone and teeth, is also important as a major regulator of energy metabolisms, and also plays a structural role in the formation of nucleic acids (Kim and Jung 2007). Fish bone contained a significant part of seafood by-product/discard and is considered as prospective sources of minerals, especially calcium. The inorganic mineral portion of fish bone (69.46% on dry basis) was mainly composed of 59.69% of calcium and 35.81% of phosphorus with the mole ratio of Ca/P of 1.67 (Kim and Jung 2007). Calcium powder prepared from the fish bone also contains traces of Cu, Mn, Zn, Se, and Fe in addition to calcium and phosphorus (Venugopal 2006; Suresh and Prabhu 2013). Enzymatic modification and fortification of bones from processing by-products of different seafood species toward use as a functionally active calcium supplement or fortifier and other minerals have attempted (Kim and Jung 2007).

Marine fish oils are rich sources of several vitamins such as A, D, and E (Suresh and Prabhu 2013). Vitamin A is concentrated mostly in fish liver oils (Venugopal 2006). Halibut and cod liver oils are rich sources of vitamins like A and D. Even though halibut and cod liver oils are rich in n-3 PUFA, they are used primarily as a vitamin A and vitamin D source. It was reported that sardine oil contains vitamins A and D with an average of 125 μg/g oil (Venugopal 2006; Suresh and Prabhu 2013).

Squalene is a natural polyunsaturated hydrocarbon (C30H50) and is important for the synthesis of steroid hormones, cholesterol, and vitamin D in the human body (Suresh and Prabhu 2013). It has several therapeutic potentials such as antioxidant, stimulate immune function, oxygenate the system, etc. (Kelly 1999). It is more commonly used in personal care products as a natural skin softener and lubricant (Shahidi and Alasalvar 2011). Squalene is one of the main components of shark liver oil, and its concentration (50–80%) is varying with species (Michael and Peter 1995). Squalene and foods containing squalene like shark liver are consumed widely as health foods (Zdenka and Slavomira 2007).

Shark cartilage is a kind of flexible connective tissue present in the animal's skeletal system. Shark cartilage has received significant commercial attention as therapeutic agent based on the incorrect notion that sharks do not get cancer. It is prepared from the fins, heads, and skeleton of shark (Shahidi and Alasalvar 2011; Suresh and Prabhu 2013). Chondroitin sulfate is a glycosaminoglycan, which has interactive roles in cell–cell recognition and cell growth as well as a crucial component of human cartilage that is essential for joint health (Uebelhart et al. 1998; Thomas et al. 2009). It is the principal bioactive component of shark cartilage and ideally obtained from shark cartilage. In finished shark cartilage, the content of chondroitin sulfate is about 21.3% (Sim et al. 2007).

7.4.9 *Bioactive Peptides and Amino Acid from Seafood Processing By-Products*

Amino acids can produce by hydrolyzing seafood proteins. Chemical (acid or alkali) and biological (enzymatic) methods are the most commonly used for the hydrolysis of seafood proteins. Microwave-induced hydrolysis of protein is also reported. The objective of the hydrolysis method is to liberate amino acids and recover them without changing their properties. The variables affecting the hydrolysis of proteins are time, temperature, hydrolysis agent, and additives, and these factors affect the quality and yield. Acid hydrolysis is the most commonly used process for the hydrolysis of proteins. The process itself is very harsh and hard to control but is still the preferred method for hydrolyzing proteins (Ghaly et al. 2013). Acid hydrolysis is performed by using HCl and in some cases with H_2SO_4. Hydrolysis of proteins can be carried out using NaOH, KOH, or BOH. The alkaline treatment is primarily utilized for the determination of tryptophan. It is also applied to the samples which have a higher percentage of carbohydrates as in the case of foods and formulation of pharmaceutical solutions which have a higher percentage of monosaccharides. The main drawback of this approach is that serine, threonine, arginine, and cysteine are destroyed, and all other amino acids are racemized (Ghaly et al. 2013).

Kim and Wijesekara (2010) reported that enzymatic hydrolysis of seafood proteins with enzymes such as alcalase, pronase, collagenase, pepsin, papain, protamex, bromelain, chymotrypsin, and trypsin allows for the preparation of bioactive peptides made up of a specific length of amino acids. The main advantage of enzymatic hydrolysis of proteins is that it allows quantification of asparagine and glutamine and other sensitive residues, which are usually destroyed by acid and alkali hydrolysis, and does not cause any racemization during digestion (Ghaly et al. 2013).

Food-derived bioactive peptides have potential regulatory roles in the human bodies beyond its core functions as nutrient sources (Hartmann and Meisel 2007). These peptides may be present in food or can release from different dietary proteins during gastrointestinal digestion, food processing, or fermentation processes (Venugopal 2006; Suresh and Prabhu 2013). Seafood proteins have gained much importance as an attractive source of bioactive peptides, due to the abundance of raw materials in the form of processing by-product/discard and underutilized species (Guerard 2007; Samaranayaka and Li-Chan 2011). Peptides can produce by hydrolyzing seafood proteins. Chemical (acid and alkali) and biological (enzymatic and fermentative) methods are most commonly used for the hydrolysis of seafood proteins. Enzymatic hydrolysis is the most common approach of producing bioactive peptides from seafood protein (Samaranayaka and Li-Chan 2011). It has been recognized that seafood by-product/discard are rich sources of various active peptides, specialty polypeptide, and other molecules such as angiotensin-converting enzyme (ACE) inhibitory peptides (Vercruysse et al. 2005; Murry and Fitz-Gerald 2007; Liu et al. 2011), protamine (a naturally occurring cationic polypeptide with 30–65 amino acid residues) (Samaranayaka and Li-Chan 2011), elastin (a cross-linked protein in the extracellular matrix that provides elasticity for many tissues), squalamine (a cationic water-soluble steroid) (Losse 2007).

7.4.10 Biodiesel and Biogas Form Seafood Processing By-Products

Biodiesel and biogas have drawn greater attention recently as they are made from renewable resources and giving many environmental and sustainability benefits (Suresh and Prabhu 2013; Ghosh et al. 2016). Fish oil and other seafood by-product/discard are considered as a potential source of renewable biofuels such as biodiesel and biogas (Zhang and El-Mashad 2007). Features of fish oil are determined by many factors, including species of origin, the season, and the unit operations used by the processors. In general, fish oils do not have significant differences among species in caloric content, but they do differ in the content of essential fatty acids. Many studies have been carried out to evaluate the potential of using seafood by-product/discard for biodiesel and biogas production (Zhang and El-Mashad 2007). Different methods are applied for converting oils and fats into biodiesel, such as micro-emulsions, thermal cracking, and the most commonly transesterification (Zhang and El-Mashad 2007). Enzymatic processes using lipases are also reported for converting oils and fats into biodiesel. However, application of enzymatic catalysis on an industrial scale may not be feasible because of the high cost associated with enzymes (Fukuda et al. 2001; Zhang and El-Mashad 2007). Also, many attempts have been made for methane production from seafood by-product/discard by different approaches (Rodenhizer and Boardman 1999; Achour et al. 2000; Mdhandete et al. 2004).

7.5 Aquatic Plant-Based By-Products

Aquatic plant foods such as algae have been used for both human and animal nutrition for thousands of years. Many aquatic plants are very rich in protein and are a highly nutritious food that can offer many beneficial advantages as a food supplement as well as having significant medicinal properties (Venugopal 2006; Kumar et al. 2008; Evans and Critchley 2014; Ghosh et al. 2016). Marine seaweeds such as kelp, Irish moss, and laver are important food sources in many Asian countries like Japan, China, and Korea, and production and consumption of edible aquatic plants had a long tradition in these countries (Senanayake et al. 2011; Suresh and Prabhu 2013; Ghosh et al. 2016). This long-standing practice has resulted in the widespread incorporation of aquatic plants into the global food supply. Instead of just trusting on marine capture, presently >95.5% of the total global production of aquatic plants is supplied by aquaculture (Ghosh et al. 2016). This compares to around 0.44 MT of marine capture and about 12 MT being produced by aquaculture in 2014 (FAO 2016). World aquaculture production aquatic plants were contributing 27.3 MT with the value of US$ 5.6 billion in 2014 (FAO 2016).

Reports have shown that the majority of aquaculture production, around 9 MT, was destined for human consumption. Phycocolloids were extracted from the remaining aquatic plants to be used as nutritional supplements in various forms of farm animal and aquaculture feedstock (Tacon and Metian 2013). So far, there has been very little data reported in the literature, and by-product/discard levels formed from aquatic plant food industries remain relatively unknown. Similarly, the treatment of wastes produced during processing remains largely unknown (Ghosh et al. 2016). Many attempts have been made to extract various bioactive compounds from seaweed (Pal et al. 2014). Bioactive substances from seaweeds currently receive more attention from the pharmaceutical companies for the drug development as well as the researchers. Seaweeds are receiving more attention from scientists because of its phenomenon bioactive compounds and its properties like antiviral, antitumor, anti-inflammatory, and antilipidemic and may more properties (Kumar et al. 2008; Pal et al. 2014).

7.6 Conclusion

Commercial seafood processing generates huge amount of non-edible by-product/discard every year, and it demands an efficient disposal method to minimize environmental issues. Thus, utilization of seafood by-product/discard for the development of high-value materials would be an effective approach for the sustainable use of unexploited seafood resources while minimizing environmental hazards. Various approaches have attempted by seafood industries to minimize the disposal of seafood by-products. Conventionally, methods of utilization of seafood by-product/discard include the mass transformation into a single product such as fish meal, fish oil, fertilizer/manure, fish silage, fish sauce, and protein hydrolysates. Although various kinds of products have been developed by mass transformation of seafood by-product/discard, their commercial value is comparatively small and thus discourages the use of seafood by-product/discard by employing these approaches in the different parts of the world. Alternatively, seafood by-product/discard can be utilized for the production of various high-value-added products such as oil, amino acids, minerals, enzymes, bioactive peptides, chitin, chitosan, collagen, and gelatin. Recently, due to the development of biotechnology, identification of biologically active products and their potential application in growing fields such as functional food, health care, nutraceutical, cosmetic has brought a new perception to seafood processing by-product/discard. Thus, comprehensive investigations on identified potential compounds/molecules and development of commercially acceptable and economically viable as well as sustainable processing methods for isolation and extraction will facilitate a successful journey of seafood processing by-product/discard in food and biomedical fields.

Despite advances in fisheries and aquaculture, as mentioned above, current seafood by-product/discard management in India is found to be inefficient and non-profitable (except in the case of chitin and chitosan production from crustacean by-products), thus highlighting the need for an efficient by-product/discard reduction strategies and by-product/discard utilization that can provide viable and profitable options for seafood by-product/discard processing.

References

Achour M, Khelifi O, Bouazizi I, Hamdi M (2000) Design of an integrated bioprocess for the treatment of tuna processing liquid effluents. Process Biochem 35:1013–1017

Aloise PA, Lumme M, Haynes CA (1996) *N*-acetyl-D-glucosamine production from chitin waste using chitinase from *Serratia marcescens*. In: Muzzarelli RAA, Chitin Enzy (eds), vol 2. European Chitin Society, Grottammare, pp 581–594

Amit KR, Swapna HC, Bhaskar N, Halami PM, Sachindra NM (2010) Effect of fermentation ensilaging on recovery of oil from fresh water fish viscera. Enzyme Microb Technol 46:9–13

Arsesen JA, Gildberg A (2002) Preparation and characterization of gelatin from the skin of harp seal (*Phoca groendlandica*). Bioresour Technol 82:191–194

Arsesen JA, Gildberg A (2007) Extraction and characterization of gelatin from the Atlantic salmon (*Salmo salar*) skin. Bioresour Technol 82:191–194

Aryee AN, Simpson BK (2000) Comparative studies on the yield and quality of solvent-extracted oil from salmon skin. J Food Eng 92:353–358

Aspmo SI, Horn SJ, Eijsink VG (2005) Enzymatic hydrolysis of Atlantic cod (*Gadus morhua* L.) viscera. Process Biochem 40:1957–1966

Batista I (1999) Recovery of proteins from fish waste products by alkaline extraction. Eur Food Res Technol 210:84–88

Bautista J, Jover M, Gutierrez JF, Corpas R et al (2001) Preparation of crayfish chitin by in situ lactic acid production. Process Biochem 37:229–234

Berge A, Barnathan G (2005) Fatty acids from lipids of marine organisms: molecular biodiversity, roles as biomarkers, biologically active compounds, and economical aspects. In: Le Gal Y,

Ulber R (eds) Marine biotech, advances in biochemical engineering/biotechnology, vol 96. Springer, Berlin, Heidelberg, pp 49–126

Bhaskar N, Suresh PV, Sakhare PZ, Sachindra NM (2007a) Shrimp biowaste fermentation with *Pediococcus acidolactici* CFR2182: optimization of fermentation conditions by response surface methodology and effect of optimized conditions on deproteination/demineralization and carotenoid recovery. Enzyme Microbiol Technol 40:1427–1434

Bhaskar N, Suresh PV, Sakhare PZ, Sachindra NM (2007b) Shrimp biowaste fermentation with *Pediococcus acidolactici* CFR2182: optimization of fermentation conditions by response surface methodology and effect of optimized conditions on deproteination/demineralization and carotenoid recovery. Enzyme Microbiol Technol 40(5):1427–1434

Bhaskar N, Sachindra NM, Suresh PV, Mahendrakar NS (2011) Microbial reclamation of fish industry by-products. In: Montet D, Ray RC (eds) Aquaculture microbiology and biotechnology. Science Publishers, Eneld, New Hampshire, pp 249–276

Bueno-Solano C, Lopez-Cervantes J, Campas-Baypoli ON, Lauterio-Garcia R, Adan-Bante NP, Sanchez-Machado DI (2009) Chemical and biological characteristics of protein hydrolysates from fermented shrimp by-products. Food Chem 112(3):671–675

Cano-Lopez A, Simpson BK, Haard NF (1987) Extraction of carotenoprotein from shrimp process waste with the aid of trypsin from Atlantic cod. J Food Sci 52:503–506

Chamberlain A (2011) Fishmeal and fish oil-the facts, fgures, trends, and IFFO's responsibility supply standard. International Fishmeal and Fish Oil Organization. http://www.seafish.org/media/publications/SeafishFishmealandFishOilFactsandFigures_201612.pdf. Accessed 31 Oct 2013

Chaussard G, Domard A (2004) New aspects of the extraction of chitin from squid pens. Biomacromol 5:559–564

Cordova-Murueta JH, Navarrete-del-Toro MA, Garcia-Carreno FL (2007) Concentrates of fish protein from by catch species produced by various drying processes. Food Chem 100:705–711

Disney GJ, Tatterson IN, Ollen J (1977) Recent development in fish silage. In: Proceedings of the conference on the handling processing and marketing of tropical fish, 1976. London Tropical Products Institute, London, pp 321–340

Dumay J, Barthomeuf C, Bergea JP (2004) How enzymes may be helpful for upgrading fish by-products: enhancement of fat extraction. Aquat Food Prod T 13(2):69–84

Evans FD, Critchley AT (2014) Seaweeds for animal production use. J Appl Phycol 26(2):891–899

Fange R, Lundblad G, Lind J, Slettengren K (1979) Chitinolytic enzymes in the digestive system of marine fishes. Mar Biol 53:317–321

FAO (2012a) The state of world fisheries and aquaculture. Fisheries and Aquaculture Department, Food and Agriculture Organization of the United Nations, Rome

FAO (2012b) Fisheries and aquaculture proceeding 22. Food and Agriculture Organization of the United Nations, Rome

FAO (2016) The state of world fisheries and aquaculture 2016. Contributing to food security and nutrition for all. Food and Agriculture Organization of the United Nations, Rome

Folch J, Less M, Stanley HS (1957) A simple method for the isolation and purification of total lipids from animal tissues. J Bio Chem 226:497–509

Fraser PD, Bramley PM (2004) The biosynthesis and nutritional uses of carotenoids. Pro Lipid Res 43:228–265

Fukuda H, Kondo A, Noda H (2001) Biodiesel fuel production by trans-esterification of oils. J Biosci Bioeng 92(5):405–416

Gbogouri GA, Linder M, Fanni J, Parmentier M (2004) Influence of hydrolysis degree on the functional properties of salmon by-products hydrolysates. J Food Sci 69:615–622

Ghaly AE, Ramakrishnan VV, Brooks MS, Budge SM, Dave D (2013) Fish processing wastes as a potential source of proteins, amino acids and oils: a critical review. J Microbio Biochem T 5:107–129

Ghosh PR, Fawcett D, Sharma SB, Poinern GEJ (2016) Progress towards sustainable utilisation and management of food wastes in the global economy. Int J Food Sci. https://doi.org/10.1155/2016/3563478

Gildberg A (1992) Recovery of proteinases and protein hydrolysates from fish viscera. Bioresour Technol 39:271–276

Gildberg A, Stenberg E (2001) A new process for advanced utilization of shrimp waste. Process Biochem 36:809–812

Guerard F (2007) Enzymatic methods for marine by-products recovery. In: Shahidi F (ed) Maximizing the value of marine by-products. Woodhead publishing Ltd., CRC Press, Cambridge, pp 107–136

Gunasekera RM, Turoczy NJ, De Silva SS, Gavine F, Gooley GJ (2002) An evaluation of the suitability of selected waste products in feeds for three fish species. J Aquat Food Prod T 11 (1):57–78

Gupta RK, Chang AC, Griffn P, Rivera R, Guo YY et al (1997) Determination of protein loading in biodegradable polymer microspheres containing tetanus toxoid. Vaccine 15:672–678

Gutowska MA, Drazen JC, Robison BH (2004) Digestive chitinolytic activity in marine fishes of Monterey Bay, California. Comp Biochem Physiol A Physiol 139:351–358

Harish PKV, Tharanathan RN (2007) Chitin/chitosan: modifications and their unlimited application potential-An overview. Trends Food Sci Technol 18:117–131

Hartmann R, Meisel H (2007) Food derived peptides with biological activity: from research to food applications. Curr Opin Biotechnol 18:63–169

Healy MG, Romo CR, Bustos R (2003) Bioconversion of marine crustacean shell waste. Resour Conserv Recy 11(1–4):139–147

Hermansson AM, Sivik B, Skjoldebrand C (1971) Factors affecting solubility, foaming and swelling of fsh protein concentrate. Lebensm Wiss Technol 4:201–204

Hwang S, Hansen CL (1998) Formation of organic acids and ammonia during acidogenesis of trout-processing wastewater. Tran ASAE 41(1):151–156

Jespersen C, Christiansen K, Hummelmose B (2000) Cleaner production assessment in fish processing. United Nations Environmental Program and Danish Environmental Protection Agency, Denmark, pp 1–99

Josephson DB (1991) Seafood. In: Maarse H (ed) Volatile compounds in food and beverages. Marcel Dekker Inc, New York, pp 179–202

Jung WJ, Souleimanov A, Park RD, Smith DL (2007) Enzymatic production of *N*-acetyl chitooligosaccharides by crude enzyme derived from *Paenibacillus illinosensis* KJA-424. Carbohydr Polym 67:256–259

Kelly GS (1999) Squalene and its potential clinical uses. Altern Med Rev 4(1):29–36

Kim JS, Park JW (2006) Mince from seafood processing by-products and surimi as food ingredients. In: Shahidi F (ed) Maximizing the value of marine by-products. Woodhead Publishing, Cambridge, London, pp 246–255

Kim SK, Jung VK (2007) Fish and bone as a calcium source. In: Shahidi F (ed) Maximizing the value of marine by-products. Woodhead Publishing, Cambridge, pp 328–339

Kim SK, Wijesekara I (2010) Development and biological activities of marine deprived bioactive peptides: a review. J Funct Foods 2:1–9

Kim H, Cadwallader KR (2011) Instrument analysis of seafood flavour. In: Alasalvar C, Shahidi F, Miyashita K, Wanasundara U (eds) Handbook of seafood quality, safety and health application. Wiley-Blackell, UK, pp 50–62

Kimoto H, Endo Y, Fujimoto K (1994) Influence of interesterification on the oxidative stability of marine oil triacylglycerols. J Am Oil Chem Soc 71:469–473

Klomklao S, Benjakul S, Visessanguan W, Kishimura H, Simpson BK (2009) Extraction of carotenoprotein from black tiger shrimp shells with the aid of bluefish trypsin. J Food Biochem 33(2):201–217

Kristinsson HG (2007) Aquatic food protein hydrolysates. In: Shahidi F (ed) Maximizing the value of marine by-products. Woodhead Publishing, Cambridge, pp 229–248

Kristinsson HG, Rasco BA (2000) Fish protein hydrolysates: production, biochemical and functional properties. Crit Rev Food Sci Nutr 32:1–39

Kuk JH, Jung WJ, Jo GH, Kim YC, Kim KY, Park D (2005) Production of *N*-acetyl-β-D-glucosamine from chitin by *Aeromonas* sp. GJ-18 crude enzyme. Adv Appl Microbiol 68:384–389

Kumar CS, Ganesan P, Suresh PV, Bhaskar N (2008) Seaweeds as a source of nutritionally beneficial compounds—a review. J Food Sci Technol 45:1–13

Lee CM (2007) Seafood flavor from processing by-products. In: Shahidi F (ed) Maximizing the value of marine by-products. Woodhead Publishing, Cambridge, pp 304–327

Lee CM (2011) Fish mince: cryostabilization and product formulation. In: Alasalvar C, Shahidi F, Miyashita K, Wanasundara U (eds) Handbook of seafood quality, safety and health application. Wiley-Blackell, UK, pp 156–170

Linder M, Fanni J, Parmentier M, Sergent M, Phan-Than-Luu R (1995) Protein recovery from veal bones by enzymatic hydrolysis. J Food Sci 60:949–952

Linder M, Matouba E, Fanni J, Parmentier M (2002) Enrichment of salmon oil with n-3 PUFA by lipolysis, filtration and enzyme re-esterification. Eur J Lipid Sci Technol 104:455–462

Linder M, Fanni J, Parmentier M (2005) Proteolytic extraction of salmon oil and PUFA concentration by lipases. Mar Biotechnol 15:70–76

Liu Z, Oliveira ACM, Su YC (2010) Purification and characterization of pepsin-solubilized collagen from skin and connective tissue of giant red sea cucumber (*Parastichopus californicus*). J Agric Food Chem 58:1270–1274

Liu Z, Chen D, Su YC, Zeng MY (2011) Optimization of hydrolysis conditions for the production of the angiotensin-I converting enzyme inhibitory peptides from sea Cucumber collagen hydrolysates. J Aquat Food Prod T 20(2):222–232

Lopez-Cervantes J, Sanchez-Machado DI, Rosas-Rodriguez JA (2006) Analysis of free amino acid in fermented shrimp waste by high-performance liquid chromatography. J Chromatogr A 1105:106–110

Losse J (2007) Using marine by-products in pharmaceutical, medical, and cosmetic products. In: Shahidi F (ed) Maximizing the value of marine by-products. Woodhead Publishing, Cambridge, pp 450–456

Mahendrakar NM, Rathina R (2015) Fish byproduct silage. In: Sachindra NM, Mahendraker NS (eds) Fish processing by-products: quality assessment and applications. Studium Press LLC, Houston, pp 299–345

Matsumiya M (2004) Enzymatic production of *N*-Acetyl-D-Glucosamine using crude enzyme from the liver of squids. Food Sci Technol Res 10(3):296–299

Matsumiya M, Kouji M, Atsushi M (2002) Characterization of 38 kDa and 42 kDa chitinase isozymes from the liver of Japanese common squid *Todarodes pacificus*. Fish Sci 68(3):603–609

Mdhandete A, Kivaisi A, Rubindamayugi M, Mattiasson B (2004) Anaerobic batch co-digestion of sisal pulp and fish wastes. Bioresour Technol 95:19–24

Michael JB, Peter DN (1995) Lipid, fatty acid and squalene composition of liver oil from six species of deep-sea sharks collected in southern Australian waters. Comp Biochem Physiol B-Biochem Mol Biol 110(1):267–275

Moren M, Suontama J, Hemre GI, Karlsen O, Olsen RE, Mundheim H, Julshamn K (2006) Element concentrations in meals from Krill and Amphipods—possible alternative protein sources in complete diets for farmed fish. Aquaculture 261:174–181

Morrissey MT, Okada T (2007) Marine enzymes from seafood by-products. In: Shahidi F (ed) Maximizing the value of marine by-products. Woodhead Publishing, Cambridge, pp 374–389

Mukhi VA, Novikov VY (2001) Enzymatic hydrolysis of proteins from crustaceans of the Barents Sea. Appl Biochem Microbiol 37:538–542

Murry BA, Fitz-Gerald RJ (2007) Angiotensin converting enzyme inhibitory peptides derived from food protein: biochemistry, bioactivity and production. Curr Pharm Des 53:773–791

Nagai T, Suzuki N (2002) Preparation and partial characterization of collagen from paper nautilus (*Argonauta argo*, Linnaeus) outer skin. Food Chem 76:149–153

Nagai T, Yamashita E, Taniguch K, Kanamori N, Suzuki N (2001) Isolation and characterization of collagen from the outer skin waste material of Cuttlefish (*Sepia lycidas*). Food Chem 72:425–429

Nidheesh T, Suresh PV (2014) Recent developments in chitosanase research and its biotechnological applications: a review. Food Chem 150:392–399

Nidheesh T, Suresh PV (2015a) Optimization of conditions for isolation of high quality chitin from shrimp processing raw byproducts using response surface methodology and its characterization. J Food Sci Technol 52(6):3812–3823

Nidheesh T, Suresh PV (2015b) Functional polysaccharides from shellfish by-products. In: Sachindra NM, Mahendraker NS (eds) Fish processing by-products: quality assessment and applications. Studium Press LLC, Houston, pp 299–345

Nurdiyana H, Mazlina MKS, Fadhilah MSN (2008) Optimization of protein extraction from freeze dried fish waste using response surface methodology. Int J Eng Technol 5:48–56

Okada T, Hosokawa M, Ono S, Miyashita K (2008) Enzymatic production of marine-derived protein hydrolysates and their bioactive peptides for use in foods and nutraceuticals. In: Hou CT, Shaw JF (eds) Biocatalysis and bioenergy. Wiley, Hoboken, pp 491–519

Pal GK, Suresh PV (2016) Sustainable valorisation of seafood by-products: recovery of collagen and development of collagen-based novel functional food ingredients. Innov Food Sci Eme Tech 37:201–215

Pal GK, Suresh PV (2017a) Physico-chemical characteristics and fibril-forming capacity of carp swim bladder collagens and exploration of their potential bioactive peptides by in silico approaches. Int J Biol Macromole 101:304–313

Pal GK, Suresh PV (2017b) Comparative assessment of physicochemical characteristics and fibril formation capacity of thermostable carp scales collagen. Mater Sci Eng C 70:32–40

Pal A, Kamthania MC, Kumar A (2014) Bioactive compounds and properties of seaweeds-a review. Open Access Libr J. https://doi.org/10.4236/oalib.1100752

Pal GK, Nidheesh T, Suresh PV (2015) Comparative study on characteristics and in vitro fibril formation ability of acid and pepsin soluble collagen from the skin of catla (*Catla catla*) and rohu (*Labeo rohita*). Food Res Int 76:804–812

Pal GK, Nidheesh T, Govindaraju K, Jyoti Suresh PV (2017) Enzymatic extraction and characterisation of a thermostable collagen from swim bladder of rohu (*Labeo raita*). J Sci Food Agr 97:1451–1458

Rao MS, Munoz J, Stevens WF (2000) Critical factors in chitin production by fermentation of shrimp biowaste. Appl Microbio Biotech 54:808–813

Regenstein JM, Zhou P (2007) Collagen and gelatin from marine by-products. In: Shahidi F (ed) Maximizing the value of marine by-products. Woodhead publishing Ltd., Cambridge, pp 279–299

Rodenhizer JS, Boardman GD (1999) Collection, analysis, and utilization of biogas from anaerobic treatment of crab processing waters. J Aqua Food Prod T 8(2):59–67

Rustad T (2003) Utilisation of marine by-products. EJEAF Chc 2:458–463

Rutten MM (2013) What economic theory tells us about the impacts of reducing food losses and/or waste: implications for research, policy and practice. Agric Food Secur 2(1):2–13

Sachindra NM, Mahendrakar NS (2015) Fishery by-products -an overview. In: Sachindra NM, Mahendrakar NS (ed) Fish processing by-products: quality assessment and applications. Studium Press LLC, Houston, USA

Sachindra NM, Bhaskar N, Mahendrakar NS (2005) Carotenoids in different component of Indian shrimps. J Sci Food Agri 85:167–172

Sachindra NM, Bhaskar N, Siddegowda GS et al (2007) Recovery of carotenoids from ensilaged shrimp waste. Biores Technol 98:1642–1646

Sachindra NM, Bhaskar N, Hosokawa M et al (2011) Value addition of seafood processing discards. In: Alasalvar C, Shahidi F, Miyashita K, Wanasundara U (eds) Handbook of seafood quality, safety and health application. Wiley-Blackell, UK, pp 390–398

Samaranayaka AGP, Li-Chan ECY (2011) Bioactive peptides from seafood and their health effects. In: Alasalvar C, Shahidi F, Miyashita K, Wanasundara U (eds) Handbook of seafood quality, safety and health application. Wiley-Blackell, UK, pp 485–493

San-Lang W, Jen-Yi L, Tzu-Wen L, Kao-Cheng L (2009) Conversion of squid pen by using Serratia sp. TKU020 fermentation for the production of enzymes, antioxidants, and N-acetyl chitooligosaccharides. Process Biochem 44(8):854–886

Sashiwa H, Fugishima S, Yamono N et al (2002) Production of *N*-acetyl-D-glucosamine from α-chitin by crude enzymes from *Aeromonas hydrophila* H-2330. Carbohydr Res 337:761–763

Sawmya R, Sachindra NM (2015) Carotenoid from fishery resources. In: Sachindra NM, Mahendraker NS (eds) Fish processing by-products: quality assessment and applications. Studium Press LLC, Houston, pp 273–298

Senanayake SPJN, Ahmed N, Fichtali J (2011) Neutraceutical and bioactive from marine algae. In: Alasalvar C, Shahidi F, Miyashita K, Wanasundara U (eds) Handbook of seafood quality, safety and health application. Wiley- Blackell UK, pp 456–463

Shahidi F (2003) Extraction and measurement of total lipids. Current protocols in food analytical chemistry

Shahidi F (2007) Marine oils from seafood waste. In: Shahidi F (ed) Maximizing the value of marine by-products. Woodhead publishing Ltd., Cambridge, pp 258–278

Shahidi F (2007) Chitin and chitosan from marine by-products. In: Shahidi F (ed) Maximizing the value of marine by-products. Woodhead publishing Ltd., Cambridge, pp 340–373

Shahidi F, Wanasundara U (1998) Omega-3 fatty acid concentrates: nutritional aspects and production technologies. Trends Food SciTech 9:230–240

Shahidi F, Kamil J (2001) Enzymes from fish and aquatic invertebrates and their application in the food industry. Trends Food SciTech 12:435–464

Shahidi F, Alasalvar C (2011) Marine oils and other marine nutraceuticals. In: Alasalvar C, Shahidi F, Miyashita K, Wanasundara U (eds) Handbook of seafood quality, safety and health application. Wiley-Blackell, UK, pp 444–454

Sharp M, Mariojouls CC (2012) Waste not, want not: better utilization of fish waste in the Pacific. SPC Fish Newslett 138:44–48

Shepherd J (2012) Aquaculture: are the criticisms justified? Feeding fish to fish. World Agriculture 3(2):11–18

Shimahara K, Takiguchi Y (1988) Preparation of crustacean chitin. In: Wood WA, Kellogg ST (eds) Methods in enzymology. Biomass. Part B. Lignin, pectin, and chitin. Academic Press, London, pp 417–423

Shirai K, Guerrero I, Huerta S (2001) Effect of initial glucose concentration and inoculation level of lactic acid bacteria in shrimp waste ensilation. Enz Micro Technol 28:446–452

Sikorski ZE, Naczk M (1981) Modification of technological properties of fish protein concentrates. Crit Rev Food Sci Nutr 14:201–230

Sim JS, A-Rang I, Seung MC, Hae JJ et al. (2007) Evaluation of chondroitin sulfate in shark cartilage powder as a dietary supplement: raw materials and finished products. Food Chem 101 (20):532–539

Simpson BK (2000) Digestive proteinases from marine animals. In: Haard NF, Simpson BK (eds) Seafood enzymes: utilization and influence on postharvest seafood quality. Marcel Dekker, New York, pp p191–124

Simpson BK (2007) Pigments from by-products of seafood processing. In: Shahidi F (ed) Maximizing the value of marine by-products. Woodhead publishing Ltd., Cambridge, pp 413–432

Simpson BK, Haard NF (1985) The use of proteolytic enzymes to extract carotenoproteins from shrimp wastes. J Appl Biochem 7:212–222

Sountama J, Kiessling A, Melle W et al (2007) Protein from Northern krill (*Thysanoessa inermis*), Antarctic krill (*Euphausia superba*) and the Arctic amphipod (*Themisto libellula*) can partially replace fish meal in diets to Atlantic salmon (*Salmo salar*) without affecting product quality. Aqua Nutri 13:50–58

Sukwattanasinitt M, Zhu H, Sashiwa H et al (2002) Utilization of commercial non-chitinase enzymes from fungi for preparation of 2-acetamido-2-deoxy-D-glucose from β-chitin. Carbohydr Res 337:133–137

Suresh PV (2012) Biodegradation of shrimp processing bio-waste and concomitant production of chitinase enzyme and *N*-acetyl-D-glucosamine by marine bacteria: production and process optimization. World J Microbiol Biotechnol 28:2945–2962

Suresh PV, Kumar PKA (2012) Enhanced degradation of α-chitin materials prepared from shrimp processing byproduct and production of *N*-acetyl-D-glucosamine by thermoactive chitinases from soil mesophilic fungi. Biodegradation 23:597–607

Suresh PV, Prabhu GN (2013) Seafood. In: Chandrasekaran M (ed) Valorization of food processing byproducts. CRC Press, Boca Raton, pp 685–736

Suresh PV, Kumar PKA, Sachindra NM (2011) Thermoactive β-*N*-acetylhexosaminidase production by a soil isolate of *Penicillium monoverticillium* CFR 2 under solid state fermentation: parameter optimization and application for *N*-acetyl chitooligosaccharides preparation from chitin. World J Microbiol Biotechnol 27:1435–1447

Suresh PV, Nidheesh T, Pal GK (2015) Enzymes in seafood processing. In: Chandrasekaran M (ed) Enzymes in food and beverage processing. CRC Press, Boca Raton, pp 353–377

Suseela M, Ninan G, Hema GS, Shiny K Lakshman PT (2015) Fish collagen and gelatin. In: Sachindra NM, NS Mahendraker (eds) Fish processing by-products: quality assessment and applications. Studium Press, LLC, Houston, pp 173–235

Swapna CH, Amit KR, Sachindra MN et al (2011) Seafood enzymes and their potential industrial application. In: Alasalvar C, Shahidi F, Miyashita K, Wanasundara U (eds) Handbook of seafood quality, safety and health application. Wiley-Blackell, UK, pp 522–532

Synowiecki J, Al-Khateeb NA (2003) Production, properties, and some new applications of chitin and its derivatives. Crit Rev Food Sci Nutr 43:145–171

Tacon AGJ, Metian M (2013) Fish matters: importance of aquatic foods in human nutrition and global food supply. Rev Fish Sci 21(1):22–38

Tharanathan RN, Kittur FS (2003) Chitin-the undisputed biomolecule of great potential. Crit Rev Food Sci Nutr 43(1):61–87

Thomas CV, William S, Juliana E (2009) A Review of evidence-based medicine for glucosamine and chondroitin sulfate use in knee osteoarthritis. Arthroscopy: J Arthros Relat Surg 25(1):86–94

Uebelhart D, Thonar EJ, Delmas PD et al (1998) Effects of oral chondroitin sulfate on the progression of kneedosteoarthritis: a pilot study. Osteoarthr Cartil 6(1):39–46

Vecchi SD, Coppes Z (1996) Marine fish digestive proteases-relevance to food industry and the South-west Atlantic region-a review. J Food Biochem 20:193–214

Venugopal V (2006) Seafood processing: adding value through quick freezing retortable packing and cook-chilling. Taylor and Francis, CRC Press, Boca Raton

Venugopal V (2009) Marine products for healthcare: functional and bioactive nutraceutical compounds from the ocean. Taylor and Francis, CRC Press, Boca Raton

Vercruysse L, John VC, Guy S (2005) ACE Inhibitory peptides derived from enzymatic hydrolysates of animal muscle protein: a review. J Agric Food Chem 53(21):8106–8115

Vidotti RM, Viegas EMM, Carneiro DJ (2003) Amino acid composition of processed fish silage using different raw materials. Anim Feed Sci Technol 105:199–204

Wanasundara U (2011) Preparative and industrial-scale isolation and purification of omega-3 polyunsaturated fatty acids from marine sources. In: Alasalvar C, Shahidi F, Miyashita K, Wanasundara U (eds) Handbook of seafood quality, safety and health application. Wiley-Blackell, UK, pp 464–475

World Bank (2013) FISH TO 2030, Prospects for fisheries and aquaculture, WORLD BANK Report number 83177-GLB. Agriculture and environmental services discussion paper 03. The World Bank, Washington

Xiao L (2010) Evaluation of extraction methods for recovery of fatty acids from marine products. Masters dissertation, University of Bergen

Zdenka K, Slavomira V (2007) Occurrence of squalene and cholesterol in various species of Czech freshwater fish. Czech J Food Sci 25(4):195–201

Zhang R, El-Mashad HM (2007) Bio-diesel and bio-gas production from seafood processing by-products. In: Shahidi F (ed) Maximizing the value of marine by-products. Woodhead publishing Ltd., Cambridge, pp 460–485

Zhong Y, Madhujith T, Mahfouz, N Shahidi F (2007) Compositional characteristics of muscle and visceral oil from steelhead trout and their oxidative stability. Food Chem 104:602–608

Zhou P, Regenstein JM (2004) Optimization of extraction conditions for Pollock skin gelatin. J Food Sci 69:393–398

Chapter 8
Bioeconomy and Biorefinery: Valorization of Hemicellulose from Lignocellulosic Biomass and Potential Use of Avocado Residues as a Promising Resource of Bioproducts

Anely A. Lara-Flores, Rafael G. Araújo, Rosa M. Rodríguez-Jasso, Mario Aguedo, Cristóbal N. Aguilar, Heather L. Trajano and Héctor A. Ruiz

Abstract Biorefineries of second generation (2G) are receiving more attention nowadays as an option for the development of bioeconomy all over the world. One of the main pretreatments utilized in this type of facilities for the conversion of lignocellulosic biomass is the use of hydrothermal processing using only water/steam as catalyst under different forms of heating (steam, electric heating jackets, or microwave radiation) at different temperatures. Currently, biorefineries are focused on obtaining feedstocks to produce biofuels, but the current position of these on the market shows that the new biorefineries must be integrated systems and so there is a need to focus on the valorization of the whole coproducts. One of them is hemicellulose, from which for instance oligomers could be derived and used in different areas as pharmaceutical products, food ingredients, fuels, chemicals, and bioplastics. In Mexico, avocado represents an important source for agro-industrial residues. These residues are in a process of valorization under the biorefinery concept, to obtain different types of bioproducts. This chapter describes the concepts usually utilized for the definition and understanding of biorefinery, especially the utilization

A. A. Lara-Flores · R. G. Araújo · R. M. Rodríguez-Jasso · C. N. Aguilar · H. A. Ruiz (✉)
Biorefinery Group, Food Research Department, Faculty of Chemistry Sciences,
Autonomous University of Coahuila, 25280 Saltillo, Coahuila, Mexico
e-mail: hector_ruiz_leza@uadec.edu.mx

A. A. Lara-Flores · R. G. Araújo · R. M. Rodríguez-Jasso · H. A. Ruiz
Cluster of Bioalcohols, Mexican Centre for Innovation in Bioenergy (Cemie-Bio),
Saltillo, Mexico

M. Aguedo
Laboratory of Biological and Industrial Chemistry, Gembloux Agro-Bio
Tech—University of Liège, Passage des Déportés 2, 5030 Gembloux, Belgium

H. L. Trajano
Department of Chemical and Biological Engineering, University of British Columbia,
2360 East Mall, Vancouver, British Columbia V6T 1Z3, Canada

R. R. Singhania et al. (eds.), *Waste to Wealth*, Energy, Environment,
and Sustainability, https://doi.org/10.1007/978-981-10-7431-8_8

of hydrothermal pretreatments. It focuses also on linking the concept of bioeconomy with biorefineries, and it introduces the utilization of avocado residues as an example of a Mexican residue with potential importance in the global market.

Keywords Autohydrolysis · Oligomers · Biomass · Hydrothermal processing Lignocellulosic biomass · Severity factor

8.1 Introduction

For years, the population of the world was concerned about commercial relationships and economic development with few concerns about what this represented for the environment. Now, major problems have been emerging from this: Climate change, global warming, and hunger are correlated with the depletion of natural resources and environmental contaminations. Due to overpopulation, the list of challenges for the upcoming years is increasing. This potentially propagates instability of the economic models and of the relation between countries. The main challenge for the society today is a sustainable way of living, meaning the implementation of economic changes such as bio-based industries, to mitigate pollution and damages that have been caused to the environment since industrial revolution.

The Rio Declaration on Environment and Development about 20 years ago was the first step to change this chaotic scenario for the years to come. During this meeting, the matter for social, scientific, and technological development for several countries was discussed. One of the strategies that impact in all the fields is the valorization of agro-industrial residues (Morais and Bogel-Lukasik 2013). This valorization comprehends the implementation of biorefineries. A biorefinery implies the conversion of biomass by diverse technologies to obtain a broad spectrum of products from the processing of a bio-based feedstock. Originally biorefineries are related to the production of bioethanol. However, to implement new biorefineries, the organization must be able to be further optimized and the production of biofuels and their derivatives alone is not enough. The integration of high value-added products in the process of production has become essential to make biorefineries sustainable for the investors and institutions involved (Ruiz et al. 2017a; Ruiz 2017b).

8.2 Bioeconomy

The twenty-first century has been in a period of economic transition, and the classic industrial activities are now endangered due to the abusive use of the natural resources. Internet and communications have brought major economic changes: expanding the perspectives of markets, decreasing market tariffs, and increasing international economic activity. Business also had a different change of command:

In 1950, the exportations were mainly about agriculture and manufacture; by 2004, agriculture had a reduction of 7%, manufacture reduced to 59%, and services increase up to 20% (Gilpin and Gilpin 2000; Dicken 2003). This is a clear example that business is in constant modification. The way of making business has changed and today the focus is on delivering information and merchandises, with the associated negative effects being the increased need for transportation, fuel, food, water as well as other elements for living.

World population is in deep need for solutions, to recover the balance with nature and develop in sustainable ways with high ranges of profitability and minimal damages. Taking this into account, the bioeconomy concept is developed with the idea of economy based on a sustainable profile.

But what is bioeconomy? As defined by the European Commission, one of the promising developers of this concept, bioeconomy comprises of the production of renewable resources from a biological source and the conversion of these in the form of foods, feed, bio-based products or bioenergy (https://ec.europa.eu/research/bioeconomy/index.cfm). The main feedstock utilized for this end is biomass; this is the point where biorefinery is widely appreciated in bioeconomy. Biorefinery is described as the generation of products, from different biomass, using different technologies in a combined form (Ruiz et al. 2013a). Bioeconomy involves the development of modern, viable, competitive, and powerful societies to produce food, materials, or energy as the result of a smart, sustainable, and inclusive growth by including areas of economy that rely on the use of renewable sources (Styhre and Sundgren 2011).

Franck Dumeignil mentioned that the European Union is working on redefining the concept of biorefinery in the countries that conform this political and economic union as EuroBioRef, a new highly integrated, diversified, and sustainable concept that involves all the biomass sector stakeholders. The potential of all the fractions issued from the various types of biomass is used to yield as high a value-added product as possible in sustainable and economical way (Dumeignil 2012). The concept previously mentioned unites the concepts of biorefinery and bioeconomy.

The economical aspect has defined for decades if an initial business goes ahead or is dismissed. It is an important task to take care of the world resources and exploit them in an intelligent sustainable form. Professor Lene Lange from the Technical University of Denmark (DTU) pointed out that "For the sake of the climate we need to use nature's own degradable and renewable resources to substitute the fossil resources and synthetics that we rely on today" (http://nordicway.org/2015/06/keyroleforbiorefineriesinthecircularbioeconomy/#.WJkF1hjmG8o). In 2011, humanity reached 7 billion persons. The number keeps increasing every day and by 2025 is expected to reach 8 billion (https://esa.un.org/unpd/wpp/DataQuery) and 9.8 billion persons by 2050 (https://esa.un.org/unpd/wpp/Publications/Files/WPP2017_KeyFindings.pdf). Based on this demographic information, it is crucial for the actual governments to anticipate the impact in terms of resources, pollution, and environmental perspectives. This growth represents an overuse of land and oceans just to produce feed. Added to this, water supplies, raw materials, and services need to be considered urgently. The utilization of residues, sidestreams, and wastes is not executed to their full potential at the present day, that is, where the use of biorefineries will be essential for a proper management of residues. Therefore, biorefineries appear vital to bioeconomy.

8.3 Integrated Biorefinery

New technologies to obtain compounds from a single feedstock are one of the new trends of investigation in research groups around the world. Biorefineries are an analogous and environmentally friendly form of oil refineries (Aguilar et al. 2017). In oil refineries, the process is designed to obtain a wide spectrum of compounds from petroleum by different processes. The initial concept of biorefineries was first implemented in the 1990s'. Until 1997, biorefinery integrated the concepts of sustainability, environmental consciousness, and the use of green technologies for this process (Maity 2015). To perform its operations, biorefinery uses different types of biomass waste (agricultural wastes, algae, organic residues to name a few) as raw material. Figure 8.1 represents the analogy between oil refineries and biorefineries, and some products that are obtained.

Biorefinery products are subdivided into two broad categories: material products and energy products. Energy products include biohydrogen, bioethanol, biomethane, biodiesel, pellets, lignin, charcoal (Cherubini and Jungmeier 2010; Aguilar-Reynosa et al. 2017a). Material products comprise of high value-added

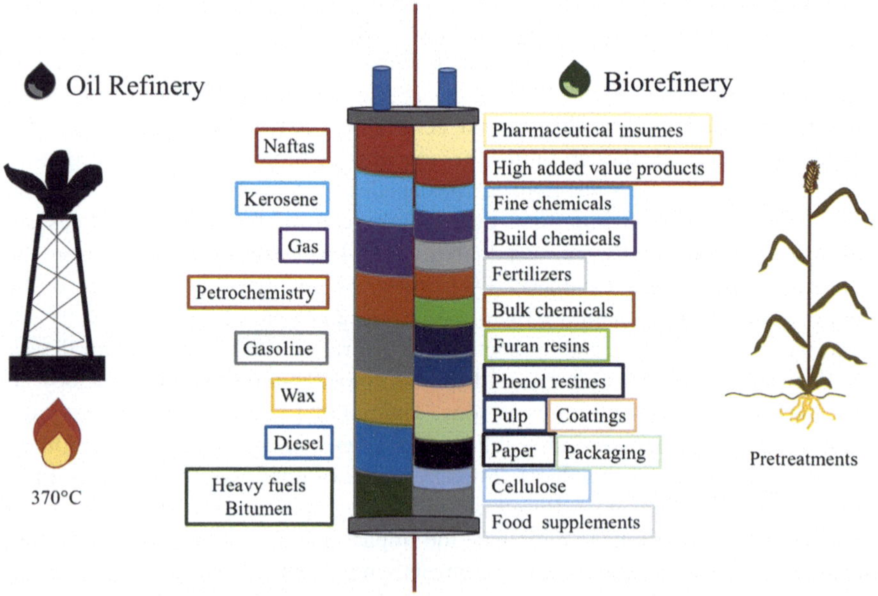

Fig. 8.1 Comparison between traditional oil refineries and biorefineries

Table 8.1 Biorefineries established in the world and products obtained by the processes established

Company	Country	Products
Lenzing AG	Austria	Cellulose-based textiles, fibers, and polymers
Zellstoff Stenday	Germany	Pulp production, biofuel, electricity
Ensyn	Canada	Liquid fuels and chemical from biomass and agricultural residues
Cristal Union	France	Alcohol, ethanol, natural, antioxidant products, dehydrated products
Abengoa Bioenergy	USA	First- and second-generation bioethanol, sugar, electricity, and animal feed
Neste Oil	USA	Renewable fuels, lubricants, diesel fuel, base oils
Pacific Ethanol	USA	Biofuel refineries, animal feed, yeasts
Beta Renewables	Italy	Bioethanol, energy, aromatics, theraphatic acids, phenols, N-butanol, isobutanol, butanediol, fatty alcohols, ethylene glycol, lactic acid, green diesel, succinic acid
Champolis	Finland	Ethanol, paper, board, packaging, hygiene products, acetic acid, furfural
Borregard	Finland	Vanillin, contrast media, cellulose, intermediates for pharmaceutical products

products, build chemicals, fine chemicals, bulk chemicals, organic acids, polymers, furan resins, phenol resins, pulp, paper, cellulose, food supplements for animal or human consumption, fertilizers, coatings, packaging, and pharmaceuticals (Hansen and Plackett 2008; Bergeron and Carrier 2012). Some examples of chemicals obtained from hemicellulose are ethanol, furfural, xylitol, 2,3-butanediol, lactic acid, citric acid, aspartic acid, succinic acid, fumaric acid, xylo-oligosaccharides, and xylitol (Bajpai 2015). Table 8.1 presents a summarized view of companies that take into consideration the biorefinery concept in its operations. A beneficial aspect of integrated biorefineries is the reduction, reutilization, and recycling of biomass that is involved in the process. By using biomass, the emissions of CO_2 are substantially reduced, since the CO_2 produced during the biorefinery process is captured again by growing biomass leading to a continuous cycle; with this, the complete process is recycled with focus on the reduction of waste, emissions, and the sustainability of the process (Kitani et al. 1989).

8.4 Biorefinery Generations

Different generations of biorefineries have been developed, each of them being particularly different from the others due to the feedstock utilized (Nigam and Singh 2011).

First-generation biorefineries use mainly grain foods as raw materials; some examples of these are as follows: corn, wheat, barley, sugarcane, sorghum, and sunflower (Azad et al. 2015). Despite being commercially applied in different parts of the world, mainly in Europe, the USA, and Brazil, it is still unpliable in other countries with limited resources for farming (Morais and Bogel-Lukasik 2013). This biorefinery has a controversial history of approval due to the use of food resources and the exploitation of land for agriculture (Dutta et al. 2014). However, the benefits of these biorefineries lay on the engineering and recovery aspects. Biofuels obtained from first-generation biorefineries could be blended with petroleum. Fuels could also be used in facilities already constructed without the risk of corrosion, transported in existing pipelines, and used in flexible fuel vehicles (Naik et al. 2010). Flexible fuel vehicles are those in which ethanol could be used in mixture with gasoline in up to 10 or 15% (https://www.fueleconomy.gov/feg/flextech.shtml).

The second-generation biorefinery takes a more sustainable approach, due to the use of residues from feedstock instead of the cereals, also known as lignocellulosic materials. Some examples of these are straws, forages, bagasse, stover, municipal solid waste, animal fat, etc., wastes that can be converted into either fuels or high value-added products (Geddes et al. 2011; Azad et al. 2015; Zanuso et al. 2017). Examples of the agricultural wastes used are cotton seed, forestry residues, switchgrass, wood wastes, and biomass residues from corn, rice, cane, agave to name a few (Dutta et al. 2014). These biorefineries coexist with first generation, but currently are not widely established due to the special requirements in the infrastructure for the processing of the feedstock utilized. Despite of this, second-generation biorefineries have been the best option for the development of a society as more sustainable and economically viable, without affecting other critical points such as the use of land to produce crops and competition for food resources (Dutta et al. 2014; Azad et al. 2015).

First- and second-generation biorefineries are gaining territory in business all over the world, with the principle of producing products, already established and new ones, with diverse characteristics to adapt the requirements of the Customers.

Third-generation biorefinery is based on the use of aquatic biomass, in the form of algae, as feedstock. This biomass is characterized by having a high content of oils, proteins, and carbohydrates (Martín and Grossmann 2012; Ruiz et al. 2013a; Ruiz et al. 2015a, b; Cervantes-Cisneros et al. 2017). The classification of this biomass is constituted by three groups: microalgae, cyanobacteria, and macroalgae. Microalgae correspond to prokaryotic or eukaryotic microorganisms able to accumulate biomass. Cyanobacteria also known as blue-green algae are prokaryotic microorganisms that have a metabolism similar to macroalgae. Finally, macroalgae, eukaryotic photosynthesizing organisms, are divided into green, brown, or red algae (Das 2015). For the better understanding of each generation and which feedstocks are considered, Fig. 8.2 summarizes the raw materials previously mentioned.

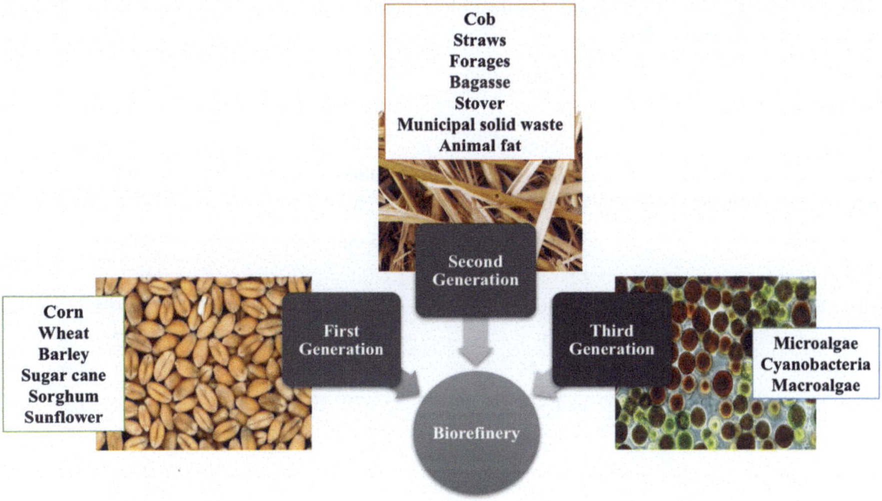

Fig. 8.2 Biorefinery generations first to third classified by feedstocks utilized

8.5 Lignocellulosic Materials as Feedstock for Second-Generation Biorefineries

Lignocellulosic biomass is an example of renewable materials rich in cellulose, hemicellulose, and lignin, which are polymeric compounds (Ruiz et al. 2013a). The sources of biomass are mainly agricultural residues, and the average composition of lignocellulosic materials is presented in Table 8.2. The main composition of these materials is cellulose (38–50%), followed by hemicellulose (23–32%) and lignin (15–25%) (Chávez-Sifontes and Domine 2013). Depending on the type of materials, the amounts of these three components are variable, due to the species, but also to seasonal variations and environment characteristics of cultivation.

8.5.1 Cellulose

Cellulose is the main polymer of natural origin, present in all lignocellulosic materials and thus highly available. In its native form, it is a crystalline and insoluble homopolysaccharide, composed of anhydrous glucose linked by β-1,4-glycosidic bonds. Cellulose could be obtained by the use of acidic or alkaline solutions (Saini et al. 2015). Cellulose has been used in the production of paper and textiles for hundreds of years; in recent times, the study on this polymer revealed its potential to be used as a substitute of plastic and for the obtainment of bioethanol (Siró and Plackett 2010; Michelin et al. 2015).

Table 8.2 Percentages of cellulose, hemicellulose, and lignin found in different lignocellulosic materials on a dry matter base

Raw Material	Cellulose (%)	Hemicellulose (%)	Lignin (%)
Corn cobs	38.8	33–36.4	13.1–18
Corn stover	34.32–36.5	20.11–31.3	11.9–13.55
Wheat bran (unmilled)	39.06	10.68	4.98
Wheat bran (milled)	38.99	10.91	5.08
Wheat straw	33–38	26–32	17–19
Coffee husks	43	7	9
Rice straw	28–36	23–28	12–14
Rice husk	33.43	20.99	18.25
Sugarcane	41.1–45	22.7–25.8	19.1–31.4
Sugarcane bagasse	34.1–49	15.79–29.6	19.4–27.2
Barley straw	31–45	27–38	14–19
Sorghum straw	32	24	13
Sweet sorghum bagasse	34–45	18–28	14–22
Sorghum stalks	27	25	11
Sunflower stalks	33.8	20.2–24.27	14.6–19.9
Cotton	80–95	5–20	–
Grasses	25–40	25–50	10–30
Switchgrass	41.2–32.97	25.95–31.1	17.34–19.1
Alfalfa stems	24.7	14.7	14.9
Hardwoods	45	30	20
Hardwood barks	22–40	20–38	30–55
Aspen	43.8	18	20.8
Hybrid poplar	48.95	21.73	23.25
Eucalyptus	44.6	21.4	30.1
Softwoods	42	27	28
Softwood barks	18–38	15–33	30–60
Pinus radiata	45.3	22.5	26.8
Spruce	43.8	20.8	28.83
Newspaper waste	60.3	16.4	12.4
Recycled paper	60.8	14.2	8.4
Chemical pulp	60–80	20–30	2–10

Taken and adapted from Gírio et al. (2010), Ruiz et al. (2013a, b), Kumar et al. (2015)

8.5.2 Hemicellulose

Hemicellulose is a hetero-polysaccharide, branched by pentoses or hexoses mainly composed by D-xylose, L-arabinose, D-mannose, and glucose. The link between the monomeric units is a bond between acetate groups, ester linkages, and hydroxyl groups in the sugars. These units form glucoronoxylans (GX), galactoglucomannans

(GGMs), arabinoglucuronoxylans (AGXs), arabinoxylans (AXs), and xyloglucans (XGs). Those structures form covalent bonds with lignin and hydrogen bonds with cellulose in the plant, and they constitute a physical barrier that surrounds and protects the cellulose (Scheller and Ulvskov 2010). Because of these characteristics, hemicellulose is less stable and degrades easily in thermal pretreatments than cellulose or lignin.

8.5.3 Lignin

Lignin is a complex structure that works as a web between cellulose and hemicellulose (Le Floch et al. 2015). Its native structure is still a theme for debate since it is usually modified by the extraction methodologies. Lignin is a polymer constituted by phenylpropanoid structures: p-coumaryl alcohol, coniferyl alcohol, and sinapyl alcohol. These polymerized units form complex biopolymers that increase the strength of the plant cell wall. Lignin is a hydrophobic structure which is very resistant to degradation by chemicals or biological agents (Carrot and Carrot 2007; Potumarthi et al. 2013).

8.6 Oligomers and Potential Uses in the Biorefinery Context

Traditionally, second-generation biorefineries were only related to the obtainment of biofuels from cellulose with some side products during the process such as lactic acid or glycerol. However, the process itself develops the production of other high value-added products such as hemicellulose oligomers. These oligomers are defined as carbohydrates with high or low molecular weights, and such compounds are available in natural sources such as honey, fruits, or vegetables. They may be obtained from lignocellulosic materials; for example, as the result of the pretreatment of lignocellulosic materials in a hydrothermal pretreatment, oligomers are solubilized in the liquid fraction recovered (Moura et al. 2015; Singh et al. 2015). Oligomers could be used in different areas with significant economic value, i.e., materials, pharmaceuticals, foods, paper products, agriculture, textiles, building blocks, and fuels by obtaining the monomeric sugars from oligomers. Food and pharmaceutical industries are particularly important: These compounds may be used in the development of novel products or as ingredients in prebiotics, fortified foods, sweetener with low caloric effect, or symbiotic products. Several pharmaceutical uses and benefits are reported in the literature: treatment and prevention of several diseases such as infections in the gastrointestinal tract, constipation, colon cancer, hepatic encephalopathy, osteoporosis, otitis, skin and hair disorders, antibiotic therapy, absorption of minerals, immunomodulatory effects, lipid reduction, production of short chain fatty acids, and antioxidant properties (Gullón et al. 2009; Samanta et al. 2015). Due to the increased commercial value of these high value-added products in the USA, Europe, and Japan, the current applications of these products are still in the constant development, and for now, these oligomers are the raising stars in the biorefinery development.

8.7 Pretreatments for Lignocellulosic Materials to Solubilize Hemicellulose

The process for first- and second-generation biorefineries is structured already: It begins with the obtainment of raw materials, and it follows with the pretreatment. For second-generation biorefinery, the pretreatment begins with delignification, then the hydrolysis of cellulose and hemicellulose to obtain sugars. Pretreatments can be physical, chemical, physicochemical, or biological. Physical pretreatments include comminution, extrusion, and ultrasound. Chemical pretreatments comprise of ionic liquids, inorganic or organic acids, alkalis, solvents, or a mixture of them such as organosolv (solvents and organic acids). Physicochemical pretreatments are usually related to hydrothermal pretreatments including different forms of heating, i.e., conventional, microwave, or steam explosion. Biological pretreatments correspond to the use of enzymes from brown, white, or soft rot fungi (Mosier et al. 2005; Agbor et al. 2011; Sarkar et al. 2012; Ruiz et al. 2013a; Zheng et al. 2014). Figure 8.3 summarizes the pretreatments for second-generation biorefineries. Pretreatments are reviewed in detail in the following sections.

8.7.1 Biological Pretreatment

Microorganisms such as fungi produce enzymes that are an alternative of pretreatment for lignocellulosic materials with an environmentally friendly approach. A great variety of brown, soft, or white rot fungi are used to hydrolyze the components of different types of feedstocks. Brown fungi such as *Serpula lacrymans, Coniophora puteana, Meruliporia incrassate, Laetoporeus sulphureus,* and *G. trabeum* are used for the pretreatment of wood. Soft rot fungi cause degradation by erotion of the substrate; some examples of these fungi are the ones compelled in the genera *Daldinia, Hypoxylon,* and *Xylaria.* Unfortunately, brown and soft rot fungi consume almost immediately cellulose and hemicellulose during the invading process. For white rot fungi that are mostly applied, this group has shown effectiveness in the conversion of several lignocellulosic materials; thus by a delignification process almost without consuming the cellulose and hemicellulose present, some examples of these fungi include *Phanerochaete chrysosporium, Ceriporia lacerate, Cyathus stercoreus, Ceriporiopsis subvermispora, Pycnoporus*

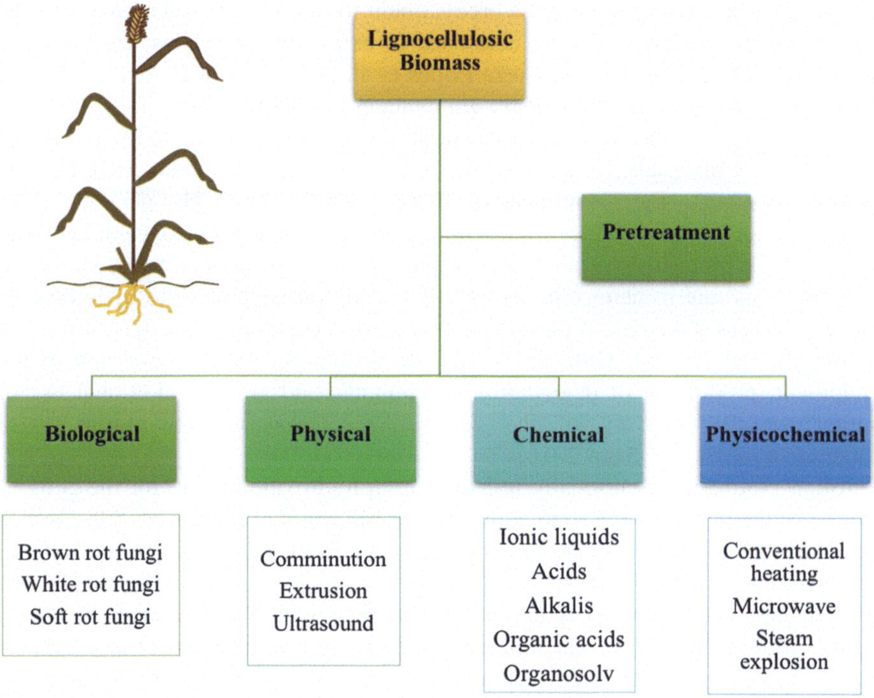

Fig. 8.3 Pretreatment performed in second-generation biorefinery

cinnabarinus, Pleurotus ostreatus, or *P. chrysosporium*. However, the pretreatment by biological agents in larger scale still seems uncertain, due to the high operational cost detached from the sterile conditions required, the long times of procedure for the pretreatment process, and also the hydrolysis of cellulose and hemicellulose in some cases (Narayanaswamy et al. 2013; Larran et al. 2015; Maurya et al. 2015; Rouches et al. 2016a, b; Sindhu et al. 2016).

8.7.2 Chemical Pretreatments

Chemical pretreatments are characterized using catalyst to treat the lignocellulosic materials. Some of the chemical pretreatments used are as follows: ionic liquids, inorganic or organic acids, alkalis, or solvents.

Ionic liquids (ILs) are a group of organic salts only available in liquid form at temperatures under 100 °C; ILs under these temperatures are usually called room-temperature ILs. ILs are usually nonvolatiles, nonflammables, stable, and can also be easily modified by changing the structure of anions and cations in the salts; the first ones are usually organic, and the anions could be either inorganic or organic. Besides its use in biomass pretreatment to dissolve the components of the lignocellulosic materials, ILs are also used in extractions, catalysis, electrochemistry, and organic synthesis (da Costa Lopes et al. 2013; Luo et al. 2013; Abe and Ohno 2014; Wang et al. 2014; Perez-Pimienta et al. 2016).

In acid pretreatments, different concentrations of acid are added to hydrolyze the lignocellulosic materials. Commonly, the acids used are H_2SO_4, SO_2, HCl, H_3PO_4; these acids lead to the obtainment of aliphatic carboxylic acids, phenolic compounds, furans, or oligomers from cellulose or hemicellulose. Concentrated acid allows a high production of monomers of cellulose at low temperatures. However, the hydrolysis rate in hemicellulose is faster, and consequently, the hydrolysis in one step increases the risk of producing degradation compounds such as furfural or hydroxymethyl furfural. Other associated disadvantages are the corrosion of the equipment, production of degradation compounds, and also an additional use of energy for the recovery of the acid utilized (Jönsson and Martín 2016; Rabemanolontsoa and Saka 2016).

Alkali pretreatments normally utilize sodium hydroxide mainly for the delignification of the material, and so to allow the solubilization of polysaccharides. In this procedure, the temperatures are low, but the solubilization times are larger, ranging between hours or even days. The pretreatment is successful in lignocellulosic materials such as hardwood or agricultural residues with lower contents of lignin, besides sodium hydroxide, calcium hydroxide (lime pretreatment) at low temperatures showed a good performance as a pretreatment agent solubilizing hemicellulose and lignin, it is also the least expensive per kilogram, also it is possible to recover the calcium with carbon dioxide (Kumar et al. 2009; Agbor et al. 2011; Karimi 2015).

The use of organic acids represents a solution for the problems mentioned above such as corrosion and degradation compounds. In these pretreatments, maleic, succinic, oxalic, monocarboxylic acids can be used. These acids are better for the pretreatment of plant materials with high percentages of cellulose and low amounts of hemicellulose (Rabemanolontsoa and Saka 2016).

The organosolv process uses solvents such as ethanol, methanol, acetone, or ethylene glycol to extract lignin and so to increase the pore volume and surface area available in cellulose. When compared with other chemical pretreatments, it has many advantages: (i) easy recovery of the solvent, (ii) environmentally friendly, and (iii) efficient lignin recovery as a high-quality product. But the detrimental point is that the operational costs of using solvents in big amounts hinder the development of the process (Sun et al. 2016).

8.7.3 Physical Pretreatments

Physical pretreatments include extrusion, ultrasound, milling, or heating. In these pretreatments, the effect is directly on the material; no other corrosive intermediaries are added to the process.

Comminution of the lignocellulosic materials is achieved after milling or chipping by using hammer mills, choppers, chippers, shredders, or disk mills, in order to reduce the particle size and to increase the available specific area in the material (Agbor et al. 2011). Both processes are commonly used to reduce the crystallinity of cellulose. Crystallinity of cellulose has an important impact on the physical properties in the fibers and on the chemical performance (Krässig 1993). Milling produces a particle size of 0.2–2 mm, and it is recognized as inexpensive and also as extremely effective for a posterior application of pretreatments (Karunanithy et al. 2012). For the chipping process, the size of the material is between 10 and 30 mm and the mechanical comminution is conditioned by the characteristic of the lignocellulosic material LCM (Sun and Cheng 2002; Cuevas et al. 2015).

Extrusion is another form of physical process used for lignocellulosic pretreatment; in this process, the LCM is passed through an extruder barrel, a high shear is exerted by the screw of the material, also high pressure and temperature are produced by the mechanical action, and this causes defribration, fibrillation, and shortening of the fibers. The speed, temperature (between 40–200 °C), screw speed/profile, and pressure are believed to cause an important effect on the structure of the material, as it is assumed that the correlation between all of them is the key of the complexity in the process (Karunanithy et al. 2012; Karunanithy and Muthukumarappan 2013; Liu et al. 2013).

Ultrasounds are ultrasonic radiations used to break the polymeric network contained in the lignocellulosic biomass by disrupting the cell wall structure and to increase the specific surface area. Ultrasounds lead to cavitation in liquids, producing bubbles that collapse with the structure of the LCM produced by hydroxyl radicals and hydrogen atoms. This facilitates the dissolution of cellulose and compounds contained in the LCM (García et al. 2011; Zheng et al. 2014).

8.8 Physicochemical Pretreatments

Another form of pretreatments is when a chemical agent and a physical factor, normally temperature and pressure, are used to treat the LCM. The main example of this type of pretreatment is the hydrothermal processing; this type of pretreatment can be heating by conduction–convection, microwave, or steam explosion (Cherubini and Jungmeier 2010; Menon and Rao 2012).

8.8.1 Hydrothermal Pretreatment and Heating Sources for the Obtainment of Hemicellulose Oligomers

Hydrothermal pretreatments are also referred to as: autohydrolysis, hydrothermal treatment, hot compressed water, hydrothermolysis, liquid hot water, aquasolve process, aqueous processing, or pressure-cooking in water (Ruiz et al. 2013a; Ruiz et al. 2017a). In these types of pretreatments, the use of acids or bases is replaced with water at elevated temperature and pressure. Often, this type of pretreatment is used to enhance the efficiency at enzymatic hydrolysis (Tekin et al. 2014; Cuevas et al. 2015) but recent works claim that the use of this pretreatment is enough to obtain high value-added products. Figure 8.4 shows a scheme of a physicochemical pretreatment, specifically hydrothermal pretreatment performed in agricultural residues and the products delivered by the pretreatment.

Hydrothermal pretreatments are an excellent alternative to the use of chemicals for the pretreatment due to the unique characteristics of water: It is an environmentally friendly solvent, it does not generate toxic compounds, it is relatively inexpensive and readily available, and it could be adapted to different types of biomass without a need for an initial drying operation. From an operational perspective, there are no major corrosion issues due to the media pH and economic studies have shown that the use of water is advantageous compared to alternative technologies (Nigam and Pandey 2009; Tekin et al. 2014; Aguilar-Reynosa et al. 2017b). Another benefit of hydrothermal pretreatments is the production of low-lignin solids that can be enzymatically hydrolyzed to produce bioethanol or another biofuels (Shafiei et al. 2015).

Hydrothermal processes can be conducted under isothermal or non-isothermal conditions. Under isothermal conditions, after the desired temperature is reached a residence time is applied during which the temperature is maintained. Under non-isothermal conditions, the desired temperature is reached and then the system is immediately rapidly cooled to reduce the amount of energy used (Ruiz et al. 2015a, b).

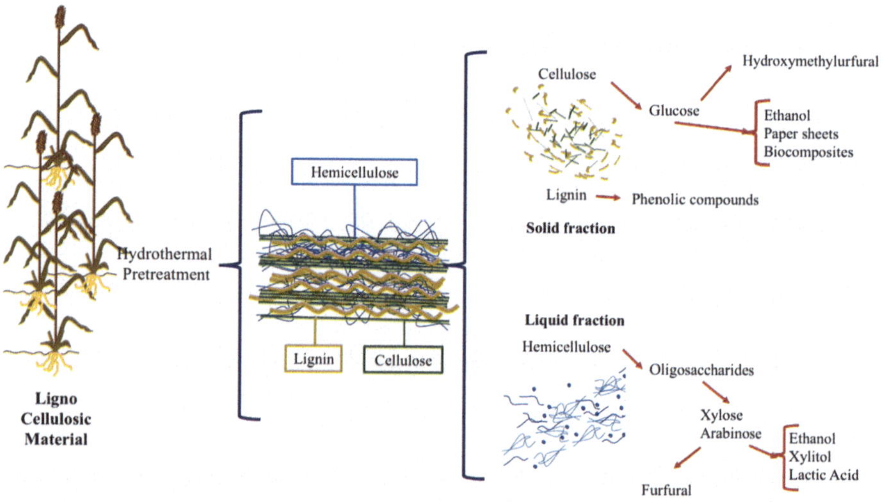

Fig. 8.4 General representation of physicochemical pretreatment and products

The hydrolysis process (Fig. 8.5) occurs at temperatures from 150 to 230 °C and begins with the weakening of hydrogen bonds; this allows the autoionization of water to produce a catalyst in the form of acidic hydronium ions (H_3O^+) and hydroxide ions (OH^-). The hydronium ions initiate the formation of acetic acid and other organic acids via the hydration of acetyl groups and uronic acids in hemicellulose (Ruiz et al. 2013a; da Silva and Chandel 2014). These acids cause the chains of hemicellulose to progressively break down producing soluble products, mainly in the form of oligomers. Cellulose and lignin remain in the solid fraction without, however the fractions showed considerable alteration in structure such as degree of polymerization and crystallinity, but giving a solid ready for enzymatic hydrolysis (Gullón et al. 2006). Hydrothermal pretreatments are usually performed in a conventional form in different types of reactors for a conventional form of heating, but recently in another form of heating such as microwave or steam explosion systems.

Recent hydrothermal pretreatment studies have employed novel forms of heating such as microwave to achieve the solubilization of hemicellulose in the form of xylo-oligosaccharides which are summarized in the following lines (Aguilar-Reynosa et al. 2017a, b).

Xiao et al. (2015) conducted autohydrolysis of bamboo culms under non-isothermal conditions to produce xylo-oligosaccharides with a degree of polymerization of 2–6 units; the experiment was performed at temperatures from 152 to 208 °C, with reaction times of 1.72–58.28 min. Xiao et al. (2015) found the optimum process conditions to be 182 °C for 31 min; this produced a xylo-oligosaccharide yield of 36.4% with scavenging activity equivalent to the founded in commercial antioxidants for superoxide and hydroxyl radicals. Moniz et al. (2014) used hydrothermal pretreatment to recover xylo-oligosaccharides from rice straw; the highest concentration of xylo-oligosaccharides (12.86 g L^{-1} of xylo-oligosaccharides with 3.95 g L^{-1} of gluco-oligosaccharides) was achieved by non-isothermal treatment at 210 °C. Another successful example of

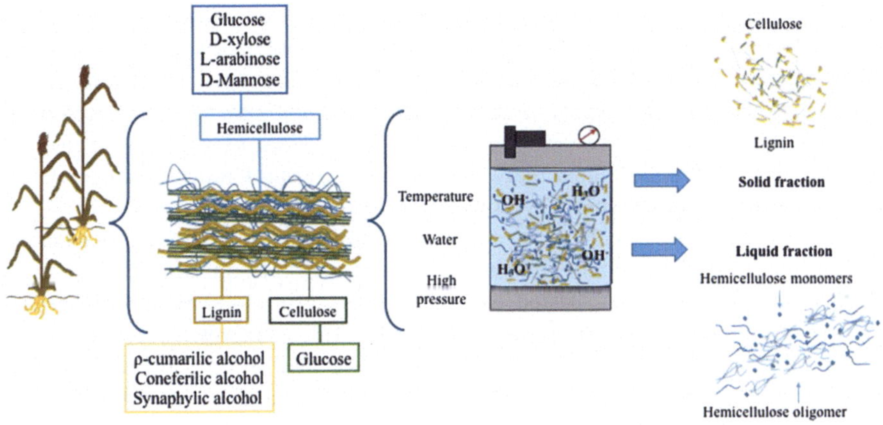

Fig. 8.5 Autohydrolysis process in lignocellulosic materials

xylo-oligosaccharide production by hydrothermal pretreatment was reported by Nabarlatz et al. (2007). In this study, a 10 L batch reactor at 179 °C for 23 min was used, in a ratio of 1–6 obtaining 17.4 g/100 kg of dry almond shells, this corresponds to the 57.0% of the potential material (Nabarlatz et al. 2007). Based on these studies, it is correct to infer that the recovery of oligomers with commercial value is achievable through hydrothermal pretreatment in a variety of configurations.

Microwaves are electromagnetic waves with frequencies between 0.3 and 300 GHz and can be used as an alternative form of heating for the solubilization of biologically active compounds in a selective, efficient, and in an environmentally friendly manner avoiding the excessive use of energy in traditional forms of heating. The heating provided by the microwave is achieved by two mechanisms: dipolar polarization and ionic conduction. When the irradiation takes place, the ions or dipoles in the reaction media are aligned with an electric field created by a magnetron; this electric field is in oscillation; thus, as the dipoles and ions realign with the field, there is a loss of energy that is expressed as heat. The amount of heat produced is determined by the polar properties of the solvent as well as the material properties of the material. This helps to promote autohydrolysis in the lignocellulosic material, and by the effective heating in the material, degradation compounds are decreased in the media, and the energetic consume of energy decreases (Zheng et al. 2014; Aguedo et al. 2015; Aguilar-Reynosa et al. 2017a).

The most relevant solvent property to determine the performance of pretreatments by microwave is the dissipation factor (tan δ); this is the energy absorbed from the microwave and converted into heat. Water's dissipation factor, 1.570×10^{-4}, is much lower than other solvents for solubilization such as ethanol or methanol, but it can work as a "green" solvent that could convert efficiently microwave energy into heat by also using the water available in the feedstocks (Delazar et al. 2012).

Tsubaki et al. (2010) applied microwave-assisted hydrolysis to extract polysaccharides and phenolic compounds from pickled and native *Prunus mume* stone. In this study, powdered stone was suspended in distilled water at temperatures of 110, 140, 170, 200, or 230 °C for 2 min. The results show that solubilization from native and pickled stone was successful at 230 °C with 37.7 and 55.9% of the total solubilization, respectively. The total yield of polysaccharides was 48% from native stone and 60.8% from pickled stone. Phenolic compounds were also recovered from 84.1 to 97.9% of the total raw material at 200–230 °C. These experiments are a clear example of the amount of sugar that can be recovered from lignocellulosic material without a subsequent enzymatic treatment (Tsubaki et al. 2010).

Microwave heating has also been combined with another pretreatment to enhance the solubilization yield. Wang and Lu (2013) obtained xylo-oligosaccharides from wheat bran by microwave-assisted enzymatic hydrolysis (1400 W for 120 s). To improve the production of xylo-oligosaccharides, three microwave pretreatments were performed, and then a subsequent enzymatic hydrolysis was conducted in series resulting in the recovery of 3.2 g of xylo-oligosaccharides from 50 g of wheat bran powder; this is compared with just an enzymatic pretreatment (Wang and Lu 2013).

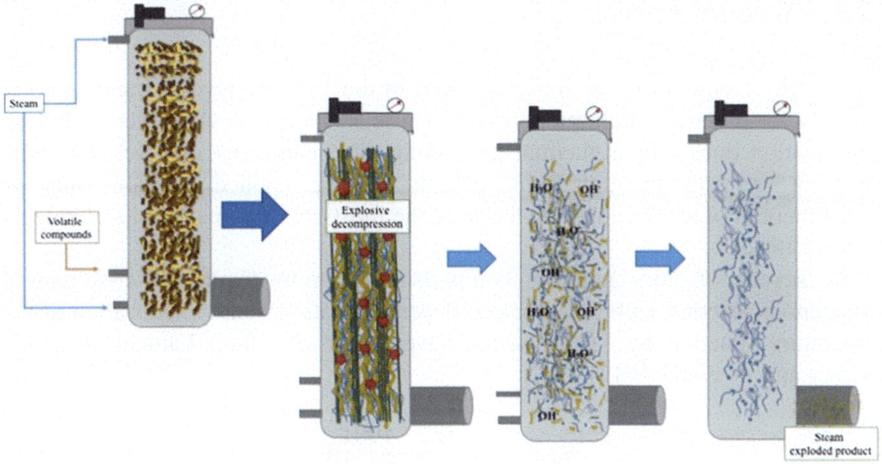

Fig. 8.6 General scheme of different steam explosion pretreatments in lignocellulosic materials

Another hydrothermal pretreatment for the solubilization of hemicellulose is steam explosion; this is the primary heating system used for pretreatment at the pilot plant and industrial scale due to the volumes of production. This approach is characterized by using high-pressure steam and a rapid decompression of the system to create an explosion that fractionates the material (Fig. 8.6). Typical temperatures for this process are between 160 and 260 °C at pressures from 0.7 to 4.8 MPa for reaction times ranging from seconds to minutes. Steam pretreatments have been successfully applied to a wide range of residues including hardwoods (aspen, white birch, red maple) or agricultural residues without adding a catalyst to the media. In the case of softwoods (jack pine, Japan cedar, white spruce), a catalyst such as H_2SO_4, CO_2, or SO_2 is recommended to be added (Agbor et al. 2011). In terms of profitably, the steam explosion pretreatments is effective for a ton of material 0.5–1.0 of steam is consumed, also the pretreatment condition allows the separation of cellulose, hemicellulose and lignin into different value-added product with commercial value, also the production of degradation compounds could be reduced under optimal process conditions, also there are reports of pilot plants using this pretreatment in Sweden, Italy, and Canada (Chen 2005).

8.9 Important Parameters in Hydrothermal Processing

To control and analyze in a better form, a hydrothermal pretreatment variables must be considered such as severity factor, particle size, and viscosity; these are reviewed in the following points.

8.9.1 Severity Factor

The severity factor (R_0) was first introduced in the 1987 by Overend and Chornet primarily for the use in pulp processing but was adapted to describe hemicellulose solubilization during hydrothermal processing. The parameter captures the interaction of temperature, time, and pH; it can be applied to isothermal or non-isothermal processes, making it one of the most versatile mathematical models in recent history.

The severity factor facilitates comparison of pretreatments by assuming a homogenous reaction with an Arrhenius dependent rate law and linearization of the temperature function by Taylor series (Overend et al. 1987; Lam et al. 2013; Chornet and Overend 2017).

$$R_0 = \int_0^t \exp\left[\frac{T - 100}{14.75}\right] \mathrm{d}t \tag{8.1}$$

In subsequent work by Overend and Chornet, the severity factor was used to describe scaling-up pretreatment from batch conditions ($R_{0,\ \mathrm{Batch}}$) to a continuous process ($R_{0,\ \mathrm{Continuous}}$) (Montané et al. 1998):

$$\log R_{0,\,\mathrm{Batch}} = 1.50 \times \left(\log R_{0,\mathrm{Continuous}} - 1\right) \tag{8.2}$$

Romaní et al. (2014) adapted the severity factor to describe a non-isothermal process by considering the heating and cooling periods. In the model, $t_{\max}$ is the time needed to achieve the highest temperature in the system, and t_f is the total time required for heating and cooling. *T' (t)* and *T (t)* are the temperature profiles for the heating and cooling, and ω is an empirical parameter (Romaní et al. 2014).

$$\log R_0 = \log\left[R_{0\,\mathrm{Heating}} + R_{0\,\mathrm{Cooling}}\right] \tag{8.3}$$

$$\log R_0 = \left[\int_0^{t_{\mathrm{MAX}}} \frac{T(t) - 100}{\omega}\right] \mathrm{d}t + \int_{t_{\mathrm{MAX}}}^{t_F} \frac{T'(t) - 100}{\omega} \mathrm{d}t \tag{8.4}$$

During the hydrothermal pretreatment, there is a decrease in the pH due to the ionization of water and the formation of organic acids. This can be accounted for by using the combined severity parameter (Overend et al. 1987; Galbe and Zacchi 2007)

$$\mathrm{CS}_{(\text{combined severity})} = \log R_0 - \mathrm{pH} \tag{8.5}$$

8.9.2 Particle Size

Particle size represents a point of interest for the successful recovery of components from lignocellulosic materials. Particle size is closely tied to heat transfer. For very large particles, the exterior may begin to overcook resulting in the production of degradation products, while the interior may not efficiently react at all. However, large particle sizes may be required to achieve processing goals. For example, Ballesteros et al. (2000) reported the effect of the particle size during steam explosion pretreatment. They used particles sizes of softwood in the range from 2–5, 5–8 to 8–12 mm at 190–210 °C for 4–8 min; the highest recovery of cellulose and hemicellulose from the pretreatment was obtained using the 8–12 mm as well as enzymatic hydrolysis. In this study, the use of very small particles was found to be detrimental (Ballesteros et al. 2000).

Ruiz et al. (2011) reported the effect of particle size, temperature, and time on the production of fermentable products from wheat straw. Particle sizes (1, 0.5, 0.3, 0.15 mm) were mixed in different frequency blends; after this, the hydrothermal process was performed at different temperatures (160, 180, 200 °C) and residence times (10, 30, 60 min). It was found that the size distribution did not affect the extent of sugar degradation during pretreatment but did affect the total sugar recovery; the highest glucose yield was found to be 21.1% at 160 °C and 10 min using a blend containing primarily 0.3–0.5 mm particles, while the highest xylose yield, 49.3%, was achieved using a blend containing primarily 0.15–0.3 mm particles and small amounts of 0.5–1 mm particles at 180 °C and 10 min (Ruiz et al. 2011).

8.9.3 Viscosity

The severity factor may also be related to the viscosity of the system. Yanagida et al. (2010) reported that the severity factor could be used to predict viscosity. Dewatered sewage sludge was heated to different temperatures (150, 175, 200, 225 °C) in a tube-type reactor from 0 to 75 h, and then the viscosity (μ) was measured. The severity factor was calculated according to Eq. 8.6, and then two correlations were developed to predict viscosity as a function of severity factor. Equation 8.6 is valid for $R_0 < 2.0 \times 10^7$, and Eq. 8.7 is valid for $R_0 > 2.0 \times 10^7$ (Overend and Chornet 1987; Yanagida et al. 2010).

$$\mu_A = 2.755 \times 10^5 \times R_0^{-0.8250} \tag{8.6}$$

$$\mu_B = 0.2611 \times \exp(-1.655 \times 10^{-7}) \times \left(R_0 - 2.0 \times 10^7\right) \tag{8.7}$$

8.10 Avocado as a Valuable Agro-industrial Residue in Mexico

8.10.1 Avocado

Avocado (*Persea americana* Mill) is a tropical and subtropical fruit native to Mexico and Central America; the first evidence of avocados was dated from the presence of avocado seeds in the Coxcatlan Cave, Tehuacan Valley Puebla, Mexico (Zafar and Sidhu 2011), and there is evidence of their consumption in Mexico for the last 10,000 years (Gutiérrez-Contreras et al. 2010). Avocado has a high worldwide acceptance, wide distribution and marketing, and with many benefits for human health (SIAP 2015; Chel-Guerrero et al. 2016; López-Cobo et al. 2016). In 2014, global avocado production exceeded 5 million tons with more 5.47×10^5 hectares dedicated to avocado production and Mexico, the largest avocado producer, produced 1.5 million tonnes (FAO 2014). Other key avocado producers are Israel, the USA, Colombia, and Dominican Republic, but avocado production is adaptable a variety of tropical regions (FAO 2014; Chel-Guerrero et al. 2016). Currently, there are more than 500 varieties of avocado but a very few are marketed or produced due problems such as production time, quality of the fruit in terms of protein and fat content, and susceptibility to damage during transportation (Yahia and Woolf 2011). The Hass variety is the most produced and dominates the worldwide market due the quality of the fruit, productivity, ease to commercial management, and consistent availability (Rodríguez-Carpena et al. 2011a, b; Cowan and Wolstenholme 2016).

8.10.2 Avocado Residues

Avocado is an unusual fruit with two stages of maturation, one on the tree named physiological maturation where the avocado is ready for harvest but not for consumption and can be kept on the tree for many months, and the normal maturation that starts two days after harvest (Blakey et al. 2009; Yahia and Woolf 2011; Alvarez et al. 2012). Avocado processing to guacamole and oil produces residues composed of seeds and peels from the fresh fruit. The peels represent 21–30% of fresh fruit and the seed represents 13–16%, depending on the variety of avocado (Wang et al. 2010; Rodríguez-Carpena et al. 2011a; López-Cobo et al. 2016; Perea-Moreno et al. 2016). The use of avocado residues is not well-developed, and for this reason, they are discarded thus generating problems of accumulation and contamination (Camberos et al. 2013). Table 8.3 shows the average composition of avocado seeds and peels from Hass, Utz, Booth 8, Panchoy, and Shupte varieties. The carbohydrate composition was studied in Drymifolia and Fuerte seeds and was determined to be: 1.2–1.9% sucrose, 4.1–4.5% arabinose, 4.5–7.3% total sugars, and 57.2–59.9% starch on a dry basis (Dominguez et al. 2014).

Table 8.3 Chemical composition of lyophilized avocado seeds and peels, average of Hass, Utz, Booth 8, Panchoy, and Shupte varieties (% dry matter basis) according Bressani et al. (2009)

	Avocado seeds (%)	Avocado peels (%)
Moisture before lyophilization	58.82	67.17
Moisture	8.75	10.09
Carbohydrates	62.82	27.98
Proteins	4.98	4.93
Lipids	5.72	5.91
Crude fiber	3.43	48.3
Ashes	3.43	4.19

Avocado residues are a potential source of bioactive compounds that can be used in the alimentary, cosmetic, or pharmaceutical industries. For example, the antioxidant capacity is several fold greater than that of blueberries, which are known for high antioxidant capacity (Ayala-Zavala et al. 2011; Agnieszka et al. 2012; Gómez et al. 2014; López-Cobo et al. 2016). The bioactive compounds are produced by the plant's secondary metabolism(Guzmán-Rodríguez et al. 2013). Avocado residues are rich in many polyphenolic compounds such as catechin and polymeric compounds such as proanthocyanidins (Soong and Barlow 2004; Agnieszka et al. 2012; López-Cobo et al. 2016). Recent reports about phenolic compounds in avocado residues demonstrated the presence of quercetin glycosides, catechin, caffeoylquinic acid, coumaroylquinic acid, procyanidin dimers of type A and B, and procyanidin trimers of type A (Agnieszka et al. 2012; López-Cobo et al. 2016). These compounds are associated with the bioactivities reported for avocado residues such as larvicidal, antifungal, antimicrobial, antioxidant, antiprotozoal, antidiabetic, antihypertensive, hypocholesterolemic, antimycobacterial activities, and inhibition of lipid and protein oxidation (Morton 1987; Leite et al. 2009; Yasir et al. 2010; Dabas et al. 2013; Jiménez-Arellanes et al. 2013). These activities and others yet to be discovered demonstrate the importance and need to recover avocado residues and make value-added products.

8.10.3 Bioproducts Recuperation and Applications

Several disadvantages are associated with the use of agro-industrial biomass such as avocado residues; these disadvantages include low energy density, high moisture content, and costs associated with transportation and storage. Preliminary results on the use of thermal processes with avocado seeds indicate significant high potential to commercially produce charcoal and liquid fuel by torrefaction and pyrolysis, and also the study showed that incorporation of avocado seeds into a porous media reactor with methane/air mixtures alters the process temperature, flame propagation velocity, and gas composition and increases the concentration of syngas

(Dominguez et al. 2014). By one estimate in the processing of this fruit, avocado seeds could generate more than 3.7 million MWh (Perea-Moreno et al. 2016).

Rodrigues et al. (2011) demonstrated the use of avocado seeds to produce activated carbon to remove phenol compounds from wastewater with high adsorption capacity. The seed fibrous residues were studied by Barbosa-Martín et al. (2016) and reported yields of extraction about 45 and 48% (w/w), with two methods of extraction. Table 8.2 summarizes the composition of this fibrous residue as an initial point for a further valorization of the residues (Table 8.4).

The value-added compounds with bioactivity are primarily recovered from plant and fruit waste. Currently, there are a variety of methods to obtain these compounds but technological and economic viability differs given the kind or conditions of the residue, the compounds that will be obtained, and the final application of recovered compounds. Methods commonly used for value-added compounds extraction are solid-liquid extraction, ultrasound-assisted extraction, microwave-assisted extraction, supercritical fluid extraction, pressurized solvent extraction, pulsed electric field extraction, and enzyme-assisted extraction (Galanakis 2012; Chan et al. 2014; Dietrich et al. 2016). Hydrothermal processes such as microwave-assisted and ultrasound-assisted extraction are considered green and safe technologies for the extraction of compounds such as polyphenols and polysaccharides such as pectin and starch (Galanakis 2012). Segovia et al. (2016) demonstrated that temperature and ultrasound power were more effective than conventional extractions on the extraction of polyphenols from avocado seed; a linear relationship between total polyphenol content and antioxidant capacity was found when using water as the solvent.

Microwave-assisted extraction is the most common alternative to traditional extraction methods to recover secondary metabolites from plants for use as essential oils, antioxidants, aromas, food colorants, pharmaceutical, and nutraceutical compounds (Zhang et al. 2011; Routray and Orsat 2012). Microwave-assisted extraction is associated with short processing times higher extraction yields, and reduced solvent consumption. The technology is used at industrial scales to obtain active compounds, mainly antioxidants, oils rich in carotenoids, or polyphenol extracts from plants or residues and is sometimes combined with other techniques such as ultrasound and pulsed electric field (Zhang et al. 2011; Chemat and Cravotto 2013).

Table 8.4 Total, insoluble, and cellulose in avocado seed fiber residues extracted with two methods (A and B) (g/100 g sample) according (Adapted and modified from Barbosa-Martín et al. 2016)

Component	Method A (NaHSO3)	Method B (NaHSO3, NaCl and trishydroxymethyl-aminomethane)
Total dietary fiber	47.84 ± 0.78	47.41 ± 0.14
Hemicellulose	19.81 ± 0.18	18.45 ± 0.22
Cellulose	7.64 ± 0.28	7.74 ± 0.25

The major compounds of interest with added value in avocado residues are the polyphenol compounds, pigments, fibers, and starch (Lacerda et al. 2015; Barbosa-Martín et al. 2016). Currently, there are no reports about the use of innovative methods of extraction to recover value-added compounds from avocado residues and only a few reports using solid-liquid extraction to obtain polyphenolic compounds (Rodríguez-Carpena et al. 2011a; Agnieszka et al. 2012; Ramos-Jerz et al. 2013; Gómez et al. 2014; López-Cobo et al. 2016). There are some reports about starch extraction from avocado seed but only with conventional technology and a low yield of 20% (Builders et al. 2010; Lacerda et al. 2014; Chel-Guerrero et al. 2016).

Another vision is the production of bioethanol from seeds as the seeds contain a high quantity of polysaccharides, starch, and hemicellulose that could be converted into sugars via hydrothermal pretreatment and then subjected to fermentation (Davis et al. 2006; Zhang et al. 2013; Tanimura et al. 2015; Yang et al. 2015).

Additional study is needed to determine the effect of hydrothermal processing to produce oligomers for use in food, chemical, and pharmaceutical industries (Ruiz et al. 2013a, b).

Avocado residues can be a new source of biocompounds and biomaterials with a large variety of applications; the following diagram (Fig. 8.7) summarizes the potential valorization routes.

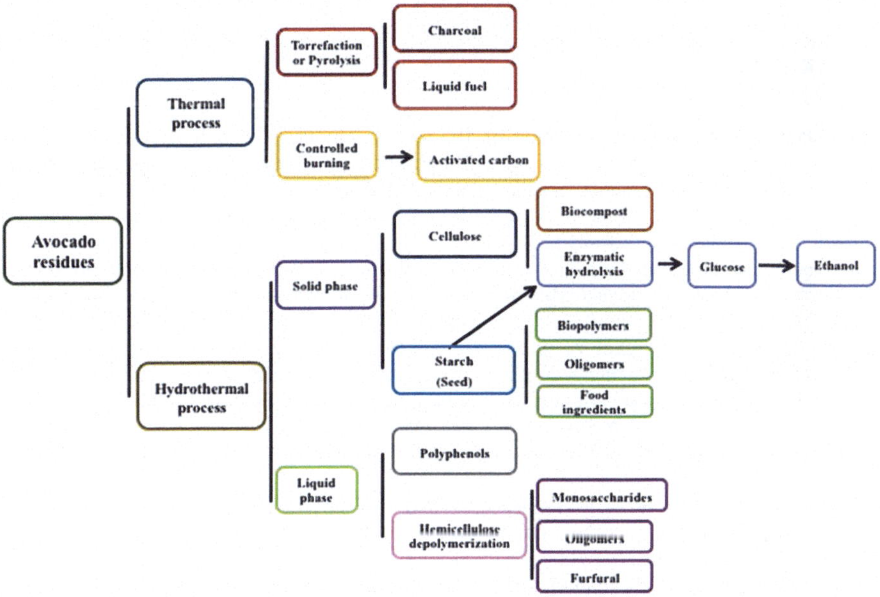

Fig. 8.7 General scheme of products and processes of products from avocado residues

8.11 Conclusions and Future Remarks

The development of second-generation biorefineries to upgrade lignocellulosic materials by obtaining high value-added products and fuels along with the implementation of hydrothermal pretreatments is promising. Hydrothermal pretreatment can help overcome the challenges associated with conversion of lignocellulosic materials. Biorefineries represent a new business model for the twenty-first century, one in which social, environmental, and economic benefits are incorporated. Avocado residues have a number of characteristics that make them an extremely interesting biorefinery feedstock, and further study is needed. Phenolic compounds have a wide variety of biological applications as functional food, nutraceuticals, or cosmetics, and the fibrous content could be used for renewable fuel. A biorefinery designed to promote the integrated use of avocado residues would be a significant technological advance.

Acknowledgements Financial support is gratefully acknowledged from the Energy Sustainability Fund 2014–05 (CONACYT-SENER), Mexican Centre for Innovation in Bioenergy (Cemie-Bio), Cluster of Bioalcohols (Ref. 249564). We gratefully acknowledge support for this research by the Mexican Science and Technology Council (CONACYT, Mexico) for the Infrastructure Project—INFR201601 (Ref. 269461) and Basic Science Project -2015–01 (Ref. 254808). The authors Anely Lara and Rafael Araújo thank Mexican Science and Technology Council (CONACYT, Mexico) for Master and PhD fellowship support.

References

Abe M, Ohno H (2014) Solubilization of Biomass Components with Ionic Liquids Toward Biomass Energy Conversions. In: Fang Z, Smith LR Jr, Qi X (eds) Production of Biofuels and Chemicals with Ionic Liquids. Springer, Dordrecht p, pp 29–59

Agbor VB, Cicek N, Sparling R et al (2011) Biomass pretreatment: fundamentals toward application. Biotechnol Adv 29:675–685

Agnieszka K, Karamac M, Estrella I et al (2012) Phenolic compound profiles and antioxidant capacity of persea americana Mill. peels and seeds of two varieties. J Agric Food Chem 60:4613–4619

Aguedo M, Ruiz HA, Richel A (2015) Non-alkaline solubilization of arabinoxylans from destarched wheat bran using hydrothermal microwave processing and comparison with the hydrolysis by an endoxylanase. Chem Eng Process Process Intensif 96:72–82

Aguilar-Reynosa A, Romaní A, Rodríguez-Jasso RM et al (2017a) Microwave heating processing as alternative of pretreatment in second-generation biorefinery: an overview. Energy Convers Manage 136:50–65

Aguilar-Reynosa A, Romaní A, Rodríguez-Jasso RM et al (2017b) Comparison of microwave and conduction-convection heating autohydrolysis pretreatment for bioethanol production. Bioresour Technol 243:273–283

Alvarez LD, Moreno AO, Ochoa FG (2012) Avocado. In: Siddiq M (ed) Tropical and subtropical fruits: postharvest physiology, processing and packaging. Wiley-Blackwell, Oxford, pp 437–454

Ayala-Zavala JF, Vega-Vega V, Rosas-Domínguez C et al (2011) Agro-industrial potential of exotic fruit byproducts as a source of food additives. Food Res Int 44:1866–1874

Azad AK, Rasul MG, Khan MMK et al (2015) Prospect of biofuels as an alternative transport fuel in Australia. Renew Sustain Energy Rev 43:331–351

Bajpai P (2015) Pulp and paper industry chemicals. Elsevier, USA

Ballesteros I, Oliva JM, Navarro AA et al (2000) Effect of chip size on steam explosion pretreatment of softwood. Appl Biochem Biotechnol 84:97–110

Barbosa-Martín E, Chel-Guerrero L, González-Mondragón E et al (2016) Chemical and technological properties of avocado (Persea americana Mill.) seed fibrous residues. Food Bioprod Process 100:457–463

Bergeron C, Carrier DJ (2012) Ramaswamy S Biorefinery co-products: phytochemicals, primary metabolites and value-added biomass processing. Wiley, Noida

Blakey RJ, Bower JP, Bertling I (2009) Influence of water and ABA supply on the ripening pattern of avocado (Persea americana Mill.) fruit and the prediction of water content using Near Infrared Spectroscopy. Postharvest Biol Technol 53:72–76

Bressani R, Rodas B, Ruiz AS (2009) La Composición Química, Capacidad Antioxidativa y Valor Nutritivo de la Semilla de Variedades de Aguacate. Final Report of the Project FODECYT 02–2006 (National Science and Technology Fund), Universidad del Valle, Guatemala http://glifos.concyt.gob.gt/digital/fodecyt/fodecyt%202006.02.pdf Accessed 15 Apr 2017

Builders PF, Nnurum A, Mbah CC et al (2010) The physicochemical and binder properties of starch from Persea americana Miller (Lauraceae). Starch/Stärke 62:309–320

Camberos E, Velázquez M, Fernández JM, Rodríguez S (2013) Acute toxicity and genotoxic activity of avocado seed extract (Persea americana Mill., c.v. Hass). Sci World J 2013:1–4

Carrot SPJM, Carrot MMLR (2007) Lignin from natural adsorbent to activated carbon: a review. Bioresour Technol 98:2301–2312

Cervantes-Cisneros DE, Arguello-Esparza D, Cabello-Galindo A et al (2017) Hydrothermal process for extraction of macroalgae high value-added compounds. In: Ruiz HA, Thomsen MH, Trajano HL (eds) Hydrothermal processing in biorefineries. Springer, Switzerland, pp 161–181

Chan CH, Yusoff R, Ngoh GC (2014) Modeling and kinetics study of conventional and assisted batch solvent extraction. Chem Eng Res Des 92:1169–1186

Chávez-Sifontes M, Domine ME (2013) Lignin, structure and applications: depolymerization methods for obtaining aromatic derivatives of industrial interest. Av en Ciencias e Ingenería 4:15–46

Chel-Guerrero L, Barbosa-Martín E, Martínez-Antonio A et al (2016) Some physicochemical and rheological properties of starch isolated from avocado seeds. Int J Biol Macromol 86:302–308

Chemat F, Cravotto G (2013) Microwave-assisted Extraction for Bioactive Compounds. Springer, Boston

Chen HZ (2005) Biotechnology of lignocellulose. Springer, Dordrecht (2013)

Cherubini F, Jungmeier G (2010) LCA of a biorefinery concept producing bioethanol, bioenergy, and chemicals from switchgrass. Int J Life Cycle Assess 15:53–66

Chornet E, Overend RP (2017) How the severity factor in biomass hydrolysis came about. In: Ruiz HA, Thomsen MH, Trajano HL (eds) Hydrothermal processing in biorefineries. Springer, Switzerland, pp 1–3

Cowan AK, Wolstenholme BN (2016) Avocado. In: Caballero B, Finglas MP, Toldrá F (eds) Encyclopedia of Food and Health. Academic Press, Oxford, pp 294–300

Cuevas M, García JF, Hodaifa G et al (2015) Oligosaccharides and sugars production from olive stones by autohydrolysis and enzymatic hydrolysis. Ind Crops Prod 70:100–106

da Costa Lopes AM, João KG, Morais ARC et al (2013) Ionic liquids as a tool for lignocellulosic biomass fractionation. Sustain Chem Process 1:1–3

da Silva SS, Chandel AK (2014) Biofuels in Brazil: Fundamental Aspects, Recent Developments, and Future Perspectives. Springer, Cham

Dabas D, Shegog RM, Ziegler GR et al (2013) Avocado (Persea americana) seed as a source of bioactive phytochemicals. Curr Pharm Des 19:6133–6140

Das D (2015) Algal Biorefinery: An Integrated Approach. Springer, Cham

Davis L, Rogers P, Pearce J et al (2006) Evaluation of Zymomonas-based ethanol production from a hydrolysed waste starch stream. Biomass Bioenerg 30:809–814

Delazar A, Nahar L, Hamedeyazdan S et al (2012) Microwave-Assisted Extraction in Natural Products Isolation. Nat Prod Isol 864:89–115

Department of Energy Fuel Economy U.S. Ethanol https://www.fueleconomy.gov/feg/ethanol.shtml. Accessed 12 May 2017 (2017)

Dicken P (ed) (2003) Global shift: Reshaping the global economic map in the 21st century sixth edition: mapping the changing contours of the world economy. Sage, London

Dietrich T, Velasco MCV, Echeverría PJ et al (2016) Crop and Plant Biomass as Valuable Material for BBB. Alternatives for Valorization of Green Wastes. In: Poltronieri P, D'Urso OF (eds) Biotransformation of Agricultural Waste and By-products: The Food, Feed, Fibre, Fuel (4F) Economy. Elsevier, Cambridge, pp 1–20

Dominguez MP, Araus K, Bonert P et al (2014) The avocado and its waste: an approach of fuel potential/application. In: Lefebvre G, Jiménez G, Cabañas B (eds) Environment, energy and climate change II: energies from new resources and the climate change. Springer International Publishing, Switzerland, pp 199–223

Dumeignil F (2012) A new concept of biorefinery comes into operation: the EuroBioRef concept. In: Aresta M, Dibenedetto A, Dumeignil F (eds) Biorefinery: from biomass to chemicals and fuels, 1st edn. Walter de Gruyter, Germany, pp 1–17

Dutta K, Daverey A, Lin JG (2014) Evolution retrospective for alternative fuels: First to fourth generation. Renew Energy 69:114–122

EC European Comission Research and Innovation Bioeconomy https://ec.europa.eu/research/bioeconomy/index.cfm. Accessed 23 Apr 2017 (2017)

FAO Food and Agriculture Organization of the United Nations. Agriculture Database (2014) http://www.fao.org/faostat/en/#data/QC. Accessed 10 Apr 2017

Galanakis CM (2012) Recovery of high added-value components from food wastes: Conventional, emerging technologies and commercialized applications. Trends Food Sci Technol 26:68–87

Galbe M, Zacchi G (2007) Pretreatment of lignocellulosic materials for efficient bioethanol production. In: Olsson L (ed) Biofuels. Advances in Biochemical Engineering/Biotechnology Springer, Berlin, pp 41–65

García A, Alriols MG, Llano-Ponte R et al (2011) Ultrasound-assisted fractionation of the lignocellulosic material. Bioresour Technol 102:6326–6330

Geddes CC, Nieves IU, Ingram LO (2011) Advances in ethanol production. Curr Opin Biotechnol 22:312–319

Gilpin R, Gilpin JM (2000) The challenge of global capitalism: The world economy in the 21st century. Princeton University Press Princeton, New Jersey

Gírio FM, Fonseca C, Carvalheiro F et al (2010) hemicelluloses for fuel ethanol: a review. Bioresour Technol 101:4775–4800

Gómez F, Sánchez S, Iradi M et al (2014) Avocado seeds: extraction optimization and possible use as antioxidant in food. Antioxidants 3:439–454

Green Growth Web Magazine The Nordic Way Key role for biorefineries in the circular bioeconomy. http://nordicway.org/2015/06/key-role-for-biorefineries-in-the-circular-bioeconomy/#.WUiHJ8bmH-k Accessed 20 Apr 2017 (2017)

Gullón P, Domínguez H, Parajo JC (2006) Advances in the manufacture, purification and applications of xylo-oligosaccharides as food additives and nutraceuticals. Process Biochem 41:1913–1923

Gullón P, Pereiro G, Alonso JL et al (2009) Aqueous pretreatment of agricultural wastes: characterization of soluble reaction products. Bioresour Technol 100:5840–5845

Gutiérrez-Contreras M, Lara-Chávez MBN, Guillén-Andrade H et al (2010) Agroecología de la franja aguacatera en Michoacán. México. Interciencia 35:647–653

Guzmán-Rodríguez JJ, López-Gómez R, Suárez-Rodríguez LM et al. (2013) Antibacterial activity of defensin PaDef from avocado fruit (Persea americana var. drymifolia) expressed in endothelial cells against escherichia coli and staphylococcus aureus. Biomed Res Int 2013:1–9

Hansen NML, Plackett D (2008) Sustainable films and coatings from hemicelluloses: a review. Biomacromol 9:1493–1505

Jiménez-Arellanes A, Luna-Herrera J, Ruiz-Nicolás R et al (2013) Antiprotozoal and antimycobacterial activities of Persea americana seeds. BMC Complement Altern Med 13 (109):1–5

Jönsson LJ, Martín C (2016) Pretreatment of lignocellulose: formation of inhibitory by-products and strategies for minimizing their effects. Bioresour Technol 199:103–112

Karimi K (2015) Lignocellulose-based bioproducts. Springer, Switzerland

Karunanithy C, Muthukumarappan K (2013) Thermo-mechanical pretreatment of feedstocks. In: Gu T (ed) Green Biomass Pretreatment for Biofuels Production. Springer, Dordrecht, pp 31–65

Karunanithy C, Muthukumarappan K, Gibbons WR (2012) Extrusion pretreatment of pine wood chips. Appl Biochem Biotechnol 167:81–99

Kitani O, Hall CW, Wagener K (1989) Biomass handbook. Gordon and Breach Science Publishers, New York

Krässig HA (1993) Cellulose: structure, accessibility and reactivity. Taylor and Francis, London

Kumar P, Barrett DM, Delwiche MJ et al (2009) Methods for pretreatment of lignocellulosic biomass for efficient hydrolysis and biofuel production. Ind Eng Chem Res 48:3713–3729

Lacerda LG, Colman TAD, Bauab T et al (2014) Thermal, structural and rheological properties of starch from avocado seeds (Persea americana, Miller) modified with standard sodium hypochlorite solutions. J Therm Anal Calorim 115:1893–1899

Lacerda LG, Da Silva Carvalho Filho MA, Bauab T et al (2015) The effects of heat-moisture treatment on avocado starch granules: thermoanalytical and structural analysis. J Therm Anal Calorim 120:387–393

Lam PS, Tooyserkani Z, Naimi LJ et al (2013) Pretreatment and pelletization of woody biomass. In: Fang Z (ed) Pretreatment techniques for biofuels and biorefineries. Springer-Verlag, Berlin Heildinberg, pp 93–116

Larran A, Jozami E, Vicario L et al (2015) Evaluation of biological pretreatments to increase the efficiency of the saccharification process using Spartina argentinensis as a biomass resource. Bioresour Technol 194:320–325

Le Floch A, Jourdes M, Teissedre P-L (2015) Polysaccharides and lignin from oak wood used in cooperage: composition, interest, assays: a review. Carbohydr Res 417:94–102

Leite JJG, Brito EHS, Cordeiro RA et al (2009) Chemical composition, toxicity and larvicidal and antifungal activities of Persea americana (avocado) seed extracts. Rev Soc Bras Med Trop 42:110–113

Liu C, Van Der Heide E, Wang H et al (2013) Alkaline twin-screw extrusion pretreatment for fermentable sugar production. Biotechnol Biofuels 6:97–108

López-Cobo A, Gómez-Caravaca AM, Pasini F et al (2016) HPLC-DAD-ESI-QTOF-MS and HPLC-FLD-MS as valuable tools for the determination of phenolic and other polar compounds in the edible part and by-products of avocado. LWT—Food Sci Technol 73:505–513

Luo J, Cai M, Gu T (2013) Pretreatment of lignocellulosic biomass using green ionic liquids. In: Gu T (ed) Green biomass pretreatment for biofuels production. Springer, Dordrecht p, pp 127–153

Maity SK (2015) Opportunities recent trends, and challenges of integrated biorefinery: part I. Renew Sustain Energy Rev 43:1427–1445

Martín M, Grossmann IE (2012) Optimal Synthesis of Sustainable Biorefineries. In: Stuart PR, El-Halwagi MM (eds) Integrated biorefineries design analysis and optimization. USA, CRC, pp 325–347

Maurya DP, Singla A, Negi S (2015) An overview of key pretreatment processes for biological conversion of lignocellulosic biomass to bioethanol. 3 Biotech 5:597–609

Menon V, Rao M (2012) Trends in bioconversion of lignocellulose: biofuels, platform chemicals and biorefinery concept. Prog Energy Combust Sci 38:522–550

Michelin M, Ruiz HA, Silva DP et al (2015) Cellulose from lignocellulosic waste. In: Ramawat KG, Mérillon JM (eds) Polysaccharides: bioactivity and biotechnology. Springer, Switzerland, pp 475–511

Moniz P, Pereira H, Duarte LC et al (2014) Hydrothermal production and gel filtration purification of xylo-oligosaccharides from rice straw. Ind Crops Prod 62:460–465
Montané D, Overend RP, Chornet E (1998) Kinetic models for non-homogeneous complex systems with a time-dependent rate constant. Can J Chem Eng 76:58–68
Morais AR, Bogel-Lukasik R (2013) Green chemistry and the biorefinery concept. Sustain Chem Process 1:18
Morton JF (1987) Fruits of warm climates, creative resource systems Inc, Winterville
Mosier N, Wyman C, Dale B et al (2005) Features of promising technologies for pretreatment of lignocellulosic biomass. Bioresour Technol 96:673–686
Moura FA, Macagnan FT, Silva LP (2015) Oligosaccharide production by hydrolysis of polysaccharides: a review. Int J Food Sci Technol 50:275–281
Nabarlatz D, Torras C, Garcia-Valls R et al (2007) Purification of xylo-oligosaccharides from almond shells by ultrafiltration. Sep Purif Technol 53:235–243
Naik SN, Goud VV, Rout PK et al (2010) Production of first and second generation biofuels: a comprehensive review. Renew Sustain Energy Rev 14:578–597
Narayanaswamy N, Dheeran P, Verma S et al (2013) Biological pretreatment of lignocellulosic biomass for enzymatic saccharification. In: Fang Z (ed) Pretreatment techniques for biofuels and biorefineries. Springer-Verlag, Berlin Heidelberg p, pp 3–34
Nigam PS-N, Pandey A (2009) Biotechnology for agro-industrial residues utilisation: utilisation of agro-residues. Springer, Dordrecht
Nigam PS, Singh A (2011) Production of liquid biofuels from renewable resources. Prog Energy Combust Sci 37:52–68
Overend RP, Chornet E (1987) Fractionation of lignocellulosic by steam-aqueous pretreatments. Philos Trans R Soc A 321:523–536
Overend RP, Chornet E, JA Gascoigne (1987) Fractionation of lignocellulosics by steam-aqueous pretreatments [and discussion]. Philos Trans R Soc London A Math Phys Eng Sci 321:523–536
Perea-Moreno AJ, Aguilera-Ureña MJ, Manzano-Agugliaro F (2016) Fuel properties of avocado stone. Fuel 186:358–364
Perez-Pimienta JA, Flores-Gómez CA, Ruiz HA et al (2016) Evaluation of agave bagasse recalcitrance using AFEXTM, autohydrolysis, and ionic liquid pretreatments. Bioresour Technol 211:216–223
Potumarthi R, Baadhe RR, Bhattacharya S (2013) Fermentable sugars from lignocellulosic biomass: technical challenges. In: Gupta KV, Tuohy GM (eds) Biofuel technologies: recent developments. Springer, Berlin Heidelberg, pp 3–27
Rabemanolontsoa H, Saka S (2016) Various pretreatments of lignocellulosics. Bioresour Technol 199:83–91
Ramos-Jerz MDR, Villanueva S, Jerz G et al. (2013) Persea americana Mill. seed: fractionation, characterization, and effects on human keratinocytes and fibroblasts. Evidence-based Complement Altern Med
Rodrigues LA, da Silva MLCP, Alvarez-Mendes MO et al (2011) Phenol removal from aqueous solution by activated carbon produced from avocado kernel seeds. Chem Eng J 174:49–57
Rodríguez-Carpena JG, Morcuende D, Andrade MJ et al (2011a) Avocado (Persea americana Mill.) phenolics, in vitro antioxidant and antimicrobial activities, and inhibition of lipid and protein oxidation in porcine patties. J Agric Food Chem 59:5625–5635
Rodríguez-Carpena JG, Morcuende D, Estévez M (2011b) Avocado by-products as inhibitors of color deterioration and lipid and protein oxidation in raw porcine patties subjected to chilled storage. Meat Sci 89:166–173
Romaní A, Ruiz HA, Pereira FB et al (2014) Integrated approach for effective bioethanol production using whole slurry from autohydrolyzed Eucalyptus globulus wood at high-solid loadings. Fuel 135:482–491
Rouches E, Herpoël-gimbert I, Steyer JP et al (2016a) Improvement of anaerobic degradation by white-rot fungi pretreatment of lignocellulosic biomass: a review. Renew Sustain Energy Rev 59:179–198

Rouches E, Zhou S, Steyer JP et al (2016b) White-Rot Fungi pretreatment of lignocellulosic biomass for anaerobic digestion: Impact of glucose supplementation. Process Biochem 51:1784–1792

Routray W, Orsat V (2012) Microwave-assisted extraction of flavonoids: a review. Food Bioprocess Technol 5:409–424

Ruiz HA, Ruzene DS, Silva DP et al (2011) Evaluation of a hydrothermal process for pretreatment of wheat straw—effect of particle size and process conditions. J Chem Technol Biotechnol 86:88–94

Ruiz HA, Rodríguez-Jasso RM, Fernandes BD et al (2013a) Hydrothermal processing, as an alternative for upgrading agriculture residues and marine biomass according to the biorefinery concept: A review. Renew Sustain Energy Rev 21:35–51

Ruiz HA, Romaní A, Michelin M et al. (2013b) A importância dos pré-tratamentos no conceito das biorrefinarias. In: Biotecnologia. Sociedade Portuguesa de Biotecnologia. http://www.cienciaviva.pt/img/upload/Boletim3.pdf#page=5. Accesed 24 Apr 2017

Ruiz HA, Rodríguez-Jasso RM, Aguedo M et al (2015a) Hydrothermal pretreatments of macroalgal biomass for biorefineries. In: Prokop A, Bajpai R, Zappi M (eds) Algal Biorefineries. Springer, Cham p, pp 467–491

Ruiz HA, Parajó JC, Teixeira JA (2015b) Biorefinery strategies for macroalgae-based in bioethanol production. In: Sharma UC, Prasad R, Sivakumar S (eds) Energy science and technology, Energy management, vol 12. Studium Press LLC, Houston p, pp 349–361

Ruiz HA, Thomsen MH, Trajano HL (2017a) Hydrothermal processing in biorefineries: production of bioethanol and high added-value compounds of second and third generation biomass. Springer Nature, Cham

Ruiz HA (2017b) High pressure processing for the valorization of biomass. In: Łukasik RM (ed) High Pressure Technologies in Biomass Conversion, The Royal Society of Chemistry, Croydon, UK, pp vii–viii

Saini JK, Saini R, Tewari L (2015) Lignocellulosic agriculture wastes as biomass feedstocks for second-generation bioethanol production: concepts and recent developments. 3. Biotech 5:337–353

Samanta AK, Jayapal N, Jayaram C et al (2015) Xylooligosaccharides as prebiotics from agricultural by-products: production and applications. Bioact Carbohydrates Diet Fibre 5:62–71

Sarkar N, Ghosh SK, Bannerjee S et al (2012) Bioethanol production from agricultural wastes: an overview. Renew Energy 37:19–27

Scheller HV, Ulvskov P (2010) Hemicelluloses. Annu Rev Plant Biol 61:263–289

Segovia F, Corral-Pérez JJ, Al, ajano MP (2016) Avocado Seed: modeling extraction of bioactive compunds. Ind Crops Prod 85:213–220

Shafiei M, Kumar R Karimi K (2015) Pretreatment of lignocellulosic biomass. In: Karimi K (ed) Lignocellulose-based bioproducts. Springer, Cham p, pp 85–154

SIAP Servicio de Información Agroalimentaria y Pesquera (2015) http://infosiap.siap.gob.mx/aagricola_siap_gb/ientidad/index.jsp. Accessed 21 May 2017

Sindhu R, Binod P, Pandey A (2016) Biological pretreatment of lignocellulosic biomass—an overview. Bioresour Technol 199:76–82

Singh RD, Banerjee J, Arora A et al (2015) Prebiotic potential of oligosaccharides: a focus on xylan derived oligosaccharides. Bioact Carbohydrates Diet Fibre 5:19–30

Siró I, Plackett D (2010) Microfibrillated cellulose and new nanocomposite materials: a review. Cellulose 17:459–494

Soong YY, Barlow PJ (2004) Antioxidant activity and phenolic content of selected fruit seeds. Food Chem 88:411–417

Styhre A, Sundgren M (2011) Venturing into the Bioeconomy: Professions, Innovation. Identity. Palgrave Macmillan, London

Sun Y, Cheng J (2002) Hydrolysis of lignocellulosic materials for ethanol production: a review. Bioresour Technol 83:1–11

Sun S, Sun S, Cao X, Sun R (2016) The role of pretreatment in improving the enzymatic hydrolysis of lignocellulosic materials. Bioresour Technol 199:49–58

Tanimura A, Kikukawa M, Yamaguchi S et al (2015) Direct ethanol production from starch using a natural isolate, Scheffersomyces shehatae: toward consolidated bioprocessing. Sci Rep 5:9593

Tekin K, Karagöz S, Bektaş SA et al (2014) Review of hydrothermal biomass processing. Renew Sustain Energy Rev 40:673–687

Tsubaki S, Ozaki Y, Azuma J (2010) Microwave-assisted autohydrolysis of Prunus mume stone for extraction of polysaccharides and phenolic compounds. J Food Sci 75:152–159

UN United Nations (2017) World population prospects https://esa.un.org/unpd/wpp/DataQuer. Accessed 14 Apr 2017

Wang TH, Lu S (2013) Production of xylooligosaccharide from wheat bran by microwave assisted enzymatic hydrolysis. Food Chem 138:1531–1535

Wang W, Bostic TR, Gu L (2010) Antioxidant capacities, procyanidins and pigments in avocados of different strains and cultivars. Food Chem 122:1193–1198

Wang H, Gurau G, Rogers RD et al (2014) Dissolution of biomass using ionic liquids. In: Zhang S, Wang J, Zhau Q, Zhou Q (eds) Structures and interactions of ionic liquids. Springer, Berlin Heidelberg, pp 79–105

Xiao X, Wang C-Z, Bian J et al (2015) Optimization of bamboo autohydrolysis for the production of xylo-oligosaccharides using response surface methodology. RSC Adv 5:106219–106226

Yahia EM, Woolf AB (2011) Avocado (Persea americana Mill.). In: Yahia EM (ed) Postharvest biology and technology of tropical and subtropical fruits. Woodhead Publishing Limited, Sawston, pp 125–186

Yanagida T, Fujimoto S, Minowa T (2010) Application of the severity parameter for predicting viscosity during hydrothermal processing of dewatered sewage sludge for a commercial PFBC plant. Bioresour Technol 101:2043–2045

Yang M, Kuittinen S, Zhang J et al (2015) Co-fermentation of hemicellulose and starch from barley straw and grain for efficient pentoses utilization in acetone-butanol-ethanol production. Bioresour Technol 179:128–135

Yasir M, Das S, Kharya M (2010) The phytochemical and pharmacological profile of Persea americana Mill. Pharmacogn Rev 4:77

Zafar T, Sidhu JS (2011) Avocado: production, quality, and major processed products. In: Sinha N (ed) Handbook of vegetables and vegetable processing. Wiley-Blackwell, Oxford, pp 525–543

Zanuso E, Lara-Flores AA, Aguilar DL et al (2017) Kinetic modeling, operational conditions, and biorefinery products from hemicellulose: depolymerization and solubilization during hydrothermal processing. In: Ruiz HA, Thomsen MH, Trajano HL (eds) Hydrothermal processing in biorefineries. Springer International Publishing, Switzerland, pp 141–160

Zhang HF, Yang XH, Wang Y (2011) Microwave assisted extraction of secondary metabolites from plants: current status and future directions. Trends Food Sci Technol 22:672–688

Zhang P, Chen C, Shen Y et al (2013) Starch saccharification and fermentation of uncooked sweet potato roots for fuel ethanol production. Bioresour Technol 128:835–838

Zheng Y, Zhao J, Xu F, Li Y (2014) Pretreatment of lignocellulosic biomass for enhanced biogas production. Prog Energy Combust Sci 42:35–53

Chapter 9
Land Applications of Biochar: An Emerging Area

Anil Kumar Patel

Abstract In recent years, there has been increasing interest on land application of biochar for improved carbon sequestration, pollutants removal, and soil amelioration. The biomass conversion into biochar and subsequent land application of biochar significantly stabilizes the ecosystem via GHG emission reduction and carbon sequestration, thus leading to the climate change mitigation. Biochar properties (e.g., surface area, microporosity, and pH) significantly improve the soil physiochemical (e.g., water-holding capacity, O_2 content, moisture level, nutrient adsorption/desorption, pollutants immobilization), and biological properties (e.g., microbial abundance and activity) improves the soil health. Current research mainly aims to exploit biochar to recover nutrients from waste matters and utilize the resulting nutrient-enriched biochar as a source of micronutrients especially in nutrient-depleting soils to sustain the crop productivity. This chapter compiles the recent advances of biochar in land application, focusing important physiochemical attributes and mechanisms pertinent to soil amelioration and plant growth promotion. Moreover, biochar application rate and methods of land applications are also outlined.

Keywords Biochar · Adsorption · Nutrient · Agriculture · Environment
Pollutant

9.1 Introduction

There are several challenges toward developing sustainable agriculture, improving the poor economies in rural areas, and decreasing climate change, e.g., soil, water and atmospheric pollutions, desertification, soil eruption, and nutrient leaching. Increasingly, adaption of agrochemical usage in agriculture practices since green

A. K. Patel (✉)
Department of Molecular Biosciences and Bioengineering,
University of Hawaii at Manoa, Honolulu, HI 96822, USA
e-mail: patelani@hawaii.edu; anilkpatel22@gmail.com

R. R. Singhania et al. (eds.), *Waste to Wealth*, Energy, Environment,
and Sustainability, https://doi.org/10.1007/978-981-10-7431-8_9

revolution and non-regulated disposal of industrial wastes into environment has considerably increased the persistence of organic and heavy metal contaminations in the food chain and the surrounding environment. It has been raised a serious public concern for the protection of the environment and human health (WHO 2017). These challenges have driven researchers to investigate an effective solution to mitigate the rising greenhouse gas (GHG) emissions in atmosphere, remove the organic and inorganic contaminants from environment, prevent the transfer of toxicity of pollutants from land to crops, and improve the health of the soil. In the recent year, application of biochar has been extensively examined to cope up these challenges. Land application of biochar to improve soil fertility, however, is not a novel concept, and it has been used as anthropogenic soils or Hortic Anthrosols (IUSS 2006) in ancient age, in which carbon content has been elevated from charcoal residues and termed as '*Terra Preta.*'

Biochar is coproduced from thermochemical (e.g., pyrolysis, gasification) and hydrothermal processes from various biomass, e.g., agriculture residues, forest residues, energy crops, animal manures, waste sludge, at temperature range of 250–850 °C under O_2-limiting condition (Kambo and Dutta 2015; Lehmann et al. 2009). Biochar has attractive characteristics for soil amelioration, e.g., high surface area, microporosity, water-holding capacity, cation exchange capacity (Brewer and Brown 2012; Mohanty et al. 2013). Biochar has been more popular for remediation applications in recent years and emerged as cost-effective substitute of activated carbon to remove numerous organic pollutants such as agrochemicals (e.g., pesticides, herbicides, insecticides), antibiotics/drugs, industrial chemicals including polycyclic aromatic hydrocarbons (PAHs), volatile organic compounds (VOCs), cationic aromatic dyes, and a series of inorganic pollutants from soil, aqueous and gaseous phases (Ahmad et al. 2014). The mechanisms for organic pollutant removal are sorption through hydrophobic, electrophilic, electrostatic attraction/repulsion via π-π electron donor–acceptor (H-bonding), partitioning, and chemical transformation (Tong et al. 2011; Ahmad et al. 2014), however, for inorganic pollutants, ion exchange, surface complexation, precipitation, and cationic and anionic interactions between pollutants and active functional groups (−COOH, R–OH, and −OH) on the biochar surface (Lu et al. 2012). These interactions also have large implications to reduce pollution and metal toxicity of contaminated land and later in soil health improvement through biological activities.

Biochar application as soil additive is an emerging area for environmental and agricultural research mainly to remove various pollutants from environment and to sustain agronomic production as well as nutrient quality of the grains. Soil amendment of biochar improves physiochemical and biological properties of the soil such as aeration, moisture content, water-holding capacity (WHC), cationic exchange capacity (CEC), microbial proliferation, pollutant immobilization and degradation, heavy metal toxicity reduction, carbon sequestration (Ameloot et al. 2013; Gul et al. 2015). These biochar characteristics finally contribute to soil carbon sequestration (Malghani et al. 2013), greenhouse gas (GHG) emission reduction (Stewart et al. 2013) and thus contribute to an overall improvement in soil health and agricultural productivity (Zhang et al. 2013). Land application of biochar also restricts heavy metals and herbicides mobility into the crop (Bolan et al. 2014). Deficiency of micronutrients in the soil and crops has been raised, which steadily reduced the crop yield across worldwide (David et al. 2016). Researchers are foreseeing nutrient-enriched biochar could play a critically important role to supplement micronutrient to the nutrient-deficient soil to sustain crop yield.

This chapter focuses on various aspects of biochar for land applications and soil amelioration and does not cover the details of biochar for removal of organic and inorganic pollutants from soil, water, and gaseous media, for which readers may refer to the recent publications (Ahmad et al. 2016). The focus of this chapter is to address how the biochar characteristics play a crucial role to alter the physiochemical properties of soil for enhancement of crop yields and lead to changes in microbial abundance, community structure, and activities, moreover, also demonstrating the land application methods and application rate of biochar.

9.2 Biochar Mechanism in Soil Amelioration and Plant Growth Promotion

In the recent years, the mechanism of biochar to improve soil fertility has been greatly studied (Woolf et al. 2010). Various attractive properties (e.g., pH, surface area, microporosity, O/C and H/C ratios, biodegradation, WHC, CEC) of biochar are explored in several environmental and agriculture applications. These biochar characteristics critically affect the physiochemical characteristics of the soil. Figure 9.1 shows various biochar characteristics, exploited in suitable applications during last decades. Biochar effect has been examined on various soils through pot and field studies. In the tropical soil, biochar amendment has been attributed to prevent nutrient leaching from the soil (Steiner et al. 2008), whereas in arid and semiarid soil (Mediterranean area), where nutrients are poor due to low organic concentration, soil amendment of biochar enhanced the total soil organic matter (SOM) (Lehmann et al. 2009). Moreover, biochar found to reduce the rate of SOM mineralization and works as slow-release fertilizer (Thies and Rilling 2009).

The pH attributed as the most governing factor for the regulation of soil properties (Ameloot et al. 2013; Stewart et al. 2013). Ash content of the biochar attributed to play a major role for resulting biochar pH, which also provides inorganic minerals to plant (Mengel and Kirkby 2001). The effect of alkaline biochar for the enrichment of soil quality and crop productivity has been evident in acidic soils mainly due to biochar-mediated pH enhancement and buffering capacity of biochar–soil mixture (Curtin and Trolove 2013). Biochar blending with other solid soil amendments, e.g., organic fertilizers, manure, compost, or lime, has also been tested prior to the soil application. Biochar exhibited significant reduction of nutrient leaching when amended with compost–sand mixtures (Iqbal et al. 2015), sandy soils (Yao et al. 2012) and swine manure (Laird et al. 2010) prior to land

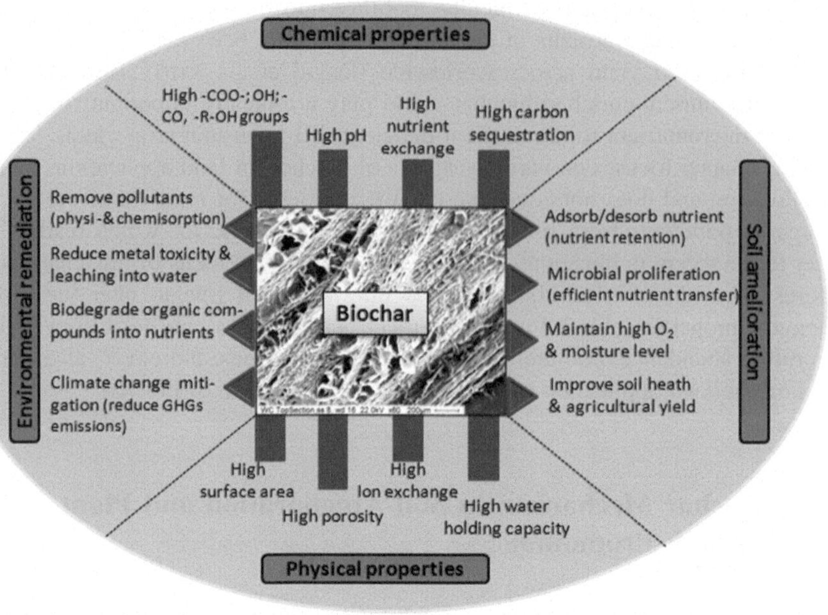

Fig. 9.1 Biochar characteristics and utilization for various applications

application in midwestern USA. Role of biochar pH in charge distribution of soil improves cation exchange capacity of the soil through binding of cationic groups of soil organic matters (SOMs). Hence, CEC of biochar-amended soil is highly dependent on SOM content in sandy soil (Basso et al. 2013). Surface charge density of biochar also enhances the CEC of the soil and thus facilitates cations retention, which plays crucial role to control soil pH (Liang et al. 2006).

Biochar has unique ability to adsorb various nutrients via adsorption on its porous surface and interaction through surface functional groups (COOH, OH, ROH, etc.). Interactions of nutrients with these radical groups prevent leaching process, and hence, it provides prolonged benefits to the soil by the mechanism of slow nutrient diffusion (Major et al. 2009; Novak et al. 2009). These functional groups enable biochar to adsorb other organic molecules and associated nutrients such as N, P, K, Mg, Ca, S which is readily available to plant uptake with the release of acidic legends by the plants to solubilize them prior to uptake. Mineralization process through various microorganisms also facilitates solubilization of bound complex organic compounds at biochar surface into bioavailable minerals to the plant. Moreover, biochar also contains various nutrients, and nutrient content is highly depended on feedstock type. For example, animal manure-derived biochar (at 400 °C) contained high N (3.2%) and P (6.1%) (Tsai et al. 2012), whereas agriculture residue (*Arundo donax*)-derived biochar pyrolyzed at 400 °C comprised low N (0.69%) and P (0.13%) contents (Zheng et al. 2013). Moreover, poultry litter-derived biochar at 350 °C had high ash content of 30.7% (Cantrell et al. 2012) and then ash content (1.5%) of pinewood chip-derived biochar at 350 °C (Spokas et al. 2011).

One of the possible mechanisms for yield improvement in biochar-amended soil may be the augmentation of soil water-holding capacity (Jeffery et al. 2011). Due to high total porosity of biochar, it can hold water in small pores and thus increase WHC, which may be increased over 22% (Peake et al. 2014). Moreover, it contributes pore-mediated water infiltration from the ground surface to the topsoil after heavy rain (Asai et al. 2009).

Apart from the direct implication of various biochar characteristics in soil amelioration and plant growth promotion, indirect mechanisms have also been reported. Biochar can also modify microbial and nutritional status of the soil at rhizosphere through changing the soil physical properties (e.g., porosity, bulk density, particle size distribution, CEC, WHC). Due to optimum moisture content, nutrients availability, and unique porous structure for microbial habitation, biochar promotes the microbial proliferation in the rhizosphere; thus, increased microbial activity improves the O_2 content or aeration of the soil (Gul et al. 2015). Increased O_2 content greatly supports oxidation or biodegradation of complex organic compounds into bioavailable minerals to improve the nutrient contents of the soil. Due to alteration of these soil parameters, the fertility of soils is significantly improved, which are attributed to increase in nutrient availability, water use efficiency, and crop productivity. A study revealed that land application of rice husk biochar has increased soil pore structure parameters by 20% and shear strength as well as decreased soil swelling parameters by >10% (Lu et al. 2014).

Quorum sensing is known mechanism of interaction between soil microorganisms and plant–microbe (among intra- and interspecies) for signaling, cell–cell detection. Biochar derived from wood is recognized to hinder cell interaction between gram-negative soil bacteria through N-(3-oxododeca-noyl)-L-homoserine lactone. Moreover, hindrance process further increased with biochar produced at higher pyrolysis temperature (700 °C) as compared to biochar produced at lower temperature (300 °C) (Masiello et al. 2013). Through these mechanisms, biochar elicits a collection of effects on gene expression dependent on intercellular signaling, connecting with the parameter used for biochar preparation so that it can be tuned to control microbial-dependent soil processes, such as nitrogen fixation and pest attack. This ability of biochar was attributed to increasing surface area of biochar, which increases with increasing pyrolysis temperature. Soil-amended biochar interacts with several organic and inorganic substances; therefore, interruption magnitude of microbial signaling depends on free space on biochar surfaces, where it can be adsorbed.

9.3 Factor Affecting Biochar Characteristics in Land Application

Each biochar exhibits different physiochemical properties such as surface area, microporosity, pH, macro- and micronutrients compositions. These variations in biochar characteristics resulted mainly due to feedstock type and pyrolysis temperature used for biochar production. However, pH is another affecting factor, which is correlated with biochar and resulted after pyrolysis of carbonaceous biomass. At higher pyrolysis temperature, resulting biochar mainly has higher pH. During pyrolysis, feedstock is thermally broken down and forming three major products: biochar, bio-oil, and non-condensable gases, e.g., CO, CO_2, CH_4, and H_2 (Ahmad et al. 2012; Suliman et al. 2016). At high pyrolysis temperature (>500 °C), removal of C, H, O elements into gases and other volatile compounds results in changing O/C and H/C ratios in biochar (Brewer et al. 2012). The O/C and H/C ratio in biochar correlates directly with aromaticity, biodegradability, and polarity, which greatly affects its physiochemical properties and thus the application such as pollutant removal, organic matter immobilization, and microbial proliferation (Crombie et al. 2013). Biochar produced from seaweeds, animal manure, and agro-residues has higher pH, and they are richer in nutrients and less in nitrogen and carbon contents than that of the lignocellulosic rich feedstocks, e.g., wood chips, wood bark (Brewer and Brown 2012; Novak et al. 2013). Moreover, at high pyrolysis temperature, the effect of feedstock has been evident, and biochar derived from woody biomass and crop residues has higher surface area compared to that of the solid municipal wastes- and animal manure-derived biochars (Ahmad et al. 2014). Furthermore, biochar produced at higher temperature exhibits lower H/C and O/C ratios than that at a lower temperature, indicating a gradual increase in aromaticity and decrease in polarity with increasing temperature (Suliman et al. 2016). A fully carbonized biochar produced at a higher pyrolysis temperature (>500 °C) has greater surface area and hydrophobicity and exhibits more affinity for organic pollutants. On the other hand, partly carbonized biochar produced at a lower pyrolysis temperature (<500 °C) contains a higher content of dissolved organic carbon and O-bearing functional groups and hence is more appropriate for removal of inorganic pollutants (Ahmad et al. 2014). These biochar attributes have high implications toward remediation of polluted agriculture lands based on the types of pollutants.

Biochar pH has significant role in removal of polar organic and charged inorganic pollutants from the soil. Alkaline pH of biochar positively correlated with higher pyrolysis temperature and feedstock types is used for its production. Wood-derived biochar results in more alkaline than biochar derived from agriculture residues and animal manures. Moreover, the presence of anionic radical groups (OH, COOH, ROH), silicates, carbonates, and bicarbonates on biochar surface also affects the biochar pH, which bind H^+ ions from soil microenvironment and reduce its concentration, thus increasing the soil pH (Brewer and Brown 2012). The degree of ionization of soil pollutants is affected through biochar pH, and thus, application rate is highly crucial for the regulation of soil pH. At higher pH, the H-bonding occurs between anionic organic pollutants and COOH or OH group of biochar. At lower and neutral pH, π-π electron donor–acceptor interactions, as well as cation exchange, are the dominant mechanisms between biochar and anionic organic pollutants (Teixido et al. 2011; Vithanage et al. 2014). The metal removal efficiency is also mainly dependent on pH (Lima et al. 2010; Uchimiya et al. 2011). High adsorption efficiency of these biochars is attributed to have high C and O contents as COOH groups, high O/C molar ratio, and polarity index (Bogusz et al. 2015). Biochar pH also has great role in nutrient desorption process. From study of Zheng et al. (2013) revealed that the release of PO_4^{3-} and NH_4^+ were pH-dependent and their release were decreased with increasing pH between pH 2–7, whereas the release of K^+ and NO_3^- was not depended on biochar pH.

9.4 Benefits of Biochar Application

There has been significant interest on land application of biochar for soil amelioration and plant nutrition in the recent years. Numerous characteristics of biochar found to be relevant to use biochar not only for soil remediation but also as carrier substrate to provide nutrients for long time in the soil. Biochar enhances soil quality by improving water permeability, cation exchange capacity (CEC), water-holding capacity (WHC), mineral accumulation, mineralization and release, improving fertility thus facilitate in improved crop productivity. Biochar offers numerous environmental benefits such as climate change mitigation, nutrient cycling, bioremediation.

9.4.1 Biochar Role in Climate Change Mitigation

Biochar has numerous unique properties found relevant to play role in climate change mitigation. It is non-biodegradable carbonaceous material that carries several nutrients and also adsorbs numerous organic pollutants from the soil. It persists thousands of years in the soil and supports microbial proliferation and activity. Microbial activities play crucial role in nutrient cycling in rhizospheric soil, thus facilitating slow nutrient release (Lehmann 2007). Retention of nutrients in soil leads to significant reduction in fertilizer application, thereby reducing the process costs and chemical fertilizer-related net GHG emissions during its production, transportation, and application and thus contributing in climate change mitigation and toward more sustainable agronomy (Wang et al. 2015). Widespread use of N fertilizers in agriculture is responsible for two-thirds of global N_2O emissions. N_2O is the most important ozone-depleting compound emitted to the atmosphere with an increased concentration of 270 ppb by volume (ppbv) in the preindustrial era to

324 ppbv by the year 2011 (Cayuela et al. 2014). Agriculture is the main sector of these emissions, largely due to the extensive consumption of synthetic fertilizers (Laird et al. 2011). Reducing N fertilizer inputs may help for widening the amount of N assimilation by crops with better N management that can be attained by biochar usage. Biochar can decrease emissions of N_2O and CH_4 from soil via biotic and abiotic mechanisms (van Zweiten et al. 2009). Biochar has potential to store carbon in a stable form for long duration and acts as atmospheric CO_2 sink in terrestrial ecosystems by reducing N_2O and CH_4 emissions (Rutherford et al. 2012; IBI 2013). Decrease in N_2O emissions up to 54% can be achieved by the application of biochar in agricultural land (Cayuela et al. 2014). With sustainable concept of biochar usage, emissions of N_2O and CH_4 can be circumvented by adapting pyrolysis of waste biomass.

Carbon sequestration could be another effective strategy for climate change mitigation. The CO_2 captured by plants during photosynthesis would eventually return to the atmosphere during decomposition or burning and available for photosynthesis, hence completing the carbon cycle (Lal et al. 2007; Lehmann and Joseph 2015). By converting this biomass into biochar, the CO_2 absorbed by photosynthesis would no longer be released and bound to the final structure of the biochar; moreover, it can significantly help to control the climate change, thus remitting the threat of global warming (Lal et al. 2007; Lehmann and Joseph 2015). It has been estimated that 373 metric ton (Mt) yr^{-1} of biochar production can sequester CO_2 about 0.55 billion Mt yr^{-1}, which would be equivalent to 1.5% of total annual CO_2 emissions (Windeatt et al. 2014).

9.4.2 *Energy Production*

Biochar is coproduced with bio-oils and biogas during thermochemical of hydrothermal treatment of biomass. It could be an alternative to fossil energy with low CO_2 emissions (Ahmad et al. 2014). Biochar production can be tailored to suit its potential application while generating energy. A critical control of the thermochemical process and conditions utilized for biochar production will vary the yield of biofuel, especially syngas or bio-oils (Ahmad et al. 2014).

Syngas is a mixture of H_2 and CO, CH_4, CO_2, H_2O and several other low molecular weight volatile organic compounds. Syngas can be used to generate the heat and electricity needed to run the pyrolyzer (Laird et al. 2011). Bio-oils coproduced with biochar are acidic in nature and are not stable, resulting as non-suitable transportation fuel. However, it can be converted into synthetic transportation fuels through Fischer–Tropsch (FT) catalytic synthesis (Laird et al. 2011). Constraints as a high capital investment required to establish a FT facility and a relatively low-carbon conversion efficiency of FT refineries are the major challenges in the coproduction of bio-oil with biochar (Laird et al. 2011).

9.4.3 Biochar as Potential Soil Amendment

Application of biochar as soil additive ameliorates the amended soil and elicits several positive responses to the plants. Soil amendment of biochar improves physical (e.g., O_2 content, moisture level, water-holding capacity, nutrient sorption, pollutants immobilization), chemical (e.g., carbon sequestration and slow nutrient release), and biological (e.g., microbial proliferation, diversity and activity) properties of soils (Gul et al. 2015). The physiochemical properties of biochar prevent nutrient leaching from soil and provide microbial habitat at porous surface by immobilizing organic substances, whereas biological activities result in mineralization of biochar as well as other recalcitrant soil organic substances, which promotes phytoavailability and efficient uptake of nutrients by the plants (Ameloot et al. 2013; Gul et al. 2015). Moreover, it facilitates nutrient bioavailability by increasing soil's WHC and the adsorption of compounds for better CEC (Verheijen et al. 2010), increasing soil pH to buffer against soil acidity (Van Zwieten et al. 2010), remediation of pollutants, and plant–microbial interactions in the soil. However, its response varies with soil type, feedstock quality, pyrolysis condition, application rate, and type of crop (Lehmann and Joseph 2015).

Most of biochars carry basic pH; therefore, they find large application as liming agent to neutralize acidic soils. Improved soil pH of acidic soil and facilitate bioavailability of the minerals Ca^{2+}, Mg^{2+}, and K^{+} to promote nutrient retention for microbial and plants growth (Laird et al. 2011). The liming effect varies based on feedstock type and pyrolysis temperature. Paper mill waste-derived biochar pyrolyzed at 550 °C has liming effect approx 30% of $CaCO_3$ due to its basic pH (van Zweiten et al. 2010). This pH-based mechanism has high implication to alter the pollutant properties where biochar finds wider remediation applications in the soil system. Alkaline biochar reduces bioavailability of pollutants by reducing their solubility; hence, their accumulation is prevented by plants. In contrast, biochar blending in neutral or alkaline soils has limited effect; for example, phosphorus (P) availability is affected in the presence of alkaline metals, and they inhibit Ca-driven P uptake (DeLuca et al. 2015). Blending of paper mill waste-derived biochar in these soils altered the effect of alkaline metals by precipitation and surface complexation and makes P bioavailable to plants; however, blending in acidic soil facilitates phosphate binding with free cations, e.g., Mg^{2+}, Ca^{2+}, Al^{3+}, and Fe^{3+}, and hence, P is slowly released to plant uptake (Xu et al. 2014a).

The C/N ratio of biochar greatly affects its inorganic mineralization and longevity in the soil. C/N ratio less than 20 is desirable for soil amendment; however, higher ratio promotes the inorganic N immobilization and results in N deficiencies in plants (Uras et al. 2012). Higher pyrolysis temperatures (>500 °C) derived for lignocellulosic biochars usually have more C/N ratio than the biochars produced at lower temperature (Purakayastha et al. 2016). Biosolids and manure-derived biochars contain lower C/N ratio, but they are rich in ash content as compare to the lignocellulosic biochars (Gonzaga et al. 2017). Ash content is a rich source of inorganic minerals, and hence, soil blending with biosolids and manure-derived biochars improved the agricultural yields. High ash containing biochar usually contains lower amount of recalcitrant carbon. Hence, biochar from biomass with higher ash content could be a beneficial nutrient source, which may lead to a possible reduction in fertilizer application rate (Gonjaga et al. 2017).

Biochar has great capacity to raise net nitrification in acidic soils, and biochar blending in this soil improves nitrification by sorption of phenolic compounds, which normally suppresses nitrification. Biochar application promotes the nitrifying bacteria and suppresses ammonia-oxidizing bacteria (DeLuca et al. 2015). Biochar blending improves the biological N_2 fixation and NH_3/NH_4^+ adsorption and also greatly affects the associated processes such as N-leaching, ammonification, nitrification–denitrification, NH_3 volatilization, and nitrous oxide (N_2O) emissions (Wang et al. 2015; Bruun et al. 2012). Nitrogen loss reduces the fertilizers potential and economic value of the end product, thus also causing the environmental pollution (Chen et al. 2015). For stimulation of nitrification, biochar derived from Douglas-fir, pinewood, bark, brazilian pepperwood, and peanut hull at pyrolysis temperature 300–800 °C was greatly exploited (Gundale and DeLuca 2006; Yao et al. 2012).

The effect of biochar varies with the soil types. Soil texture has high correlation with soil organic matter retention. Fine textured soils are able to retain more carbon and nitrogen than those of coarse textured soils at equal SOM, which is correlated with the role of clay for better protection of organic carbon (Ganjegunte et al. 2009). In this context, biochar would be more effective for controlling nutrient dynamics in coarse-grained soils. Physical structure and surface property of biochar provide a framework for building a slow-release fertilizer. Biochar is rich in nutrients and also binds nutrients from fertilizer, thus it not only retains the fertilizer's nutrition in the soil but also works as slow-release fertilizer (Major et al. 2009). Moreover, ash content of biochar has numerous cations, which are accessible as dissolved salts and thus readily available to the plant (Glaser et al. 2002). Cao and Harris (2010) used dairy manure as a feedstock to develop biochar-based slow-release phosphorus fertilizer. Moreover, NPK slow-release biochar fertilizer has been developed by Day et al. (2005).

Recent studies have addressed declining drift of nutrient deficiency in crops and in cropland (David et al. 2016; Emmalea 2016). At present, most of the provided fertilizers contain several nutrients together for plant growth and its application is considered as very important preventative measure to retain the crop yield in the near future. Plants absorb nutrients in its inorganic forms such as NH_4^+, $H_2PO_4^-$, NO_3^{2-}, SO_4^{2-}. Thus, they must be readily present in these forms prior to meet the crop nutrient requirements (Brady and Weil 2013). These are usually very less abundant in agriculture soils than that of organically bound forms (Bohn et al. 1986). Current studies on biochar system are mainly focused on remediation applications. Recent trends are shown in biochar research for nutrient recovery from waste and application of resulting biochar for plant nutrition. Figure 9.2 shows the recent trend of biochar application for plant nutrition. Little information is available

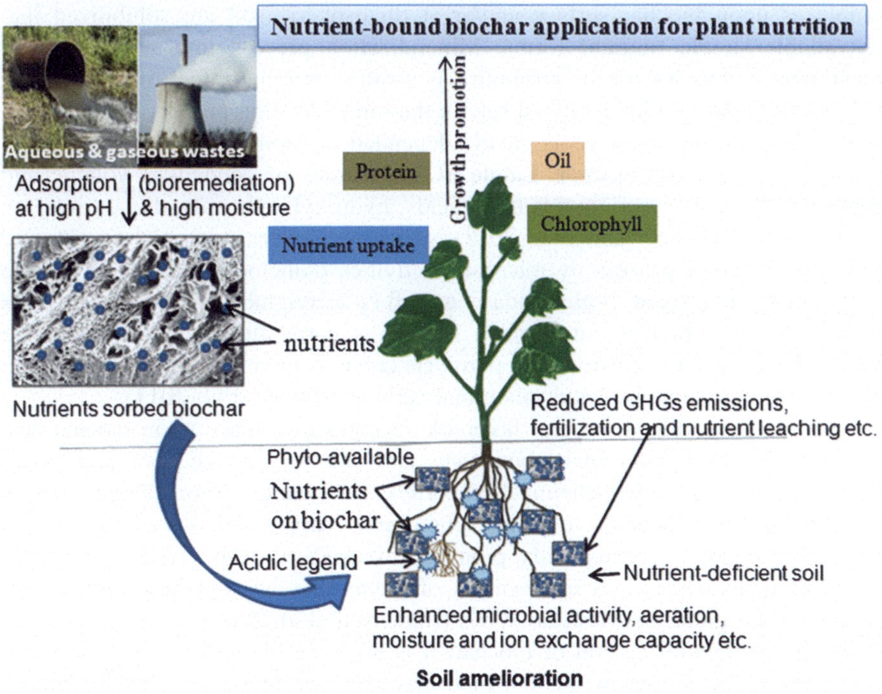

Fig. 9.2 Biochar use in nutrient recovery from waste and its application as fertilizer

on the land application and dynamics of nutrient transformations, e.g., N, Ca, Mg, K, and P by biochars (DeLuca et al. 2015; Laird et al. 2011), Recently, H_2S-laden biochar research is being examined to facilitate SO_4^{2-} availability and their effect on plant growth in pot experiments (Kanjanarong et al. 2017). H_2S-adsorbed biochar contains S on its surface in SO_4^{2-} and other forms of mineral precipitates (Na, K)$_2SO_4$ (Xu et al. 2014b), which are bioavailable and can be easily up taken by the plant roots. Biochar derived from rice hull is used for H_2S removal from gaseous stream. Such biochar with adsorbed S may have great potential for application as soil amendment, especially on S-depleted soils to provide the S for plant health.

9.4.4 Biochar Mineralization

From various studies, biochars have been examined as source of bioavailable nutrients for plant nutrition. Biochar is a rich source of plant nutrients, e.g., P, K, Mg, Ca, Fe, Cu, Na, Zn, and Cl (Crombie et al. 2015; Spokas et al. 2012). Biochar mineralization is a collective process of biotic/abiotic by which insoluble organic complexes upon biochar surface and/or at rhizospheric soil are solubilized into bioavailable simple nutrient forms. Mineralization process improves after soil amendment of biochar and is attributed as positive priming effect of the biochar. Soil microorganisms play a crucial role in the mineralization process (Novak et al. 2010). Mineralization process is greatly depended on labile simple organic matter (SOM) content of the biochar. Labile SOM is dregs of condensed volatile and semi-volatile organic compounds (e.g., acids, hydroxyls, phenols, diols, triols) on biochar surface (Graber and Elad 2013), which accelerates microbial growth and thus mineralization process by microbial activities (Ameloot et al. 2013; Wardle et al. 2008). Improved β-glucosidase and β-N-acetylglucosaminidase activities confirm the microbial activity, which has direct correlation with sum of labile matters. Biochar obtained from fast pyrolysis contains more labile organic fraction (up to 40%) and supports better microbial activity (Bailey et al. 2011).

After soil amendment, several favorable changes take place upon biochar surface, which affect microbial abundance and activity in the soil and result improvement in soil physiochemical properties and organic carbon. These changes facilitate numerous benefits to plants, importantly suppress toxicity of pesticides and allelochemicals, consequently promote rhizobacteria followed by arbuscular and other mycorrhizae and intervention of plant–microbe signaling, and protect microbes from grazers (Qiu et al. 2009; Kothamasi et al. 2006).

Due to recalcitrant carbon of biochar, it is not very likely for better microbial growth on it. But longer incubation time may facilitate biphasic biochar degradation. Primarily, labile SOM is important for initial microbial growth and biochar mineralization; however, for further mineralization, more microbial colonization is required which can access recalcitrant biochar components. Several fungal saprophytes can utilize these components by secreting several extracellular enzymes, e.g., lignin peroxidases, laccases, manganese peroxidase, and phenoloxidase; moreover, reactive phenoxy and peroxy radicals are also required to degrade the aromatic structures (Hockaday 2006; Atkinson et al. 2010). Fungi basidiomycetes and ascomycetes (Hofrichter et al. 1999) are able to utilize biochar carbons. Moreover, *Mycobacterium* sp. and *Beauveriasul furescens* are also known to utilize aromatic biochar compounds (Novak et al. 2010). Due to these microbial activities, biochar mineralization is likely to be changed. Fundamental benefits of biochar application on environment and plants are summarized in Table 9.1.

9.5 Biochar Characteristics and Land Application Methods

9.5.1 Biochar Forms and Application Rate of Biochar

Biochar application in soil has been reported for several agronomic benefits such as increased seed germination (30%), shoot heights (24%), biomass production (13%),

Table 9.1 Positive effects of soil application of biochar on environment and plants

Benefits	Description	References
Influence soil pH	pH of biochar is affected with the pyrolysis temperature, feedstock types, and biochar application rate; it is apparent in acidic and alkaline soils	Liu and Zhang (2012); Chintala et al. (2014)
Provide microbial habitats	Biochar contains nutrients and is also adsorbed from surrounding, thus facilitating microbial growth and activity	Steiner et al. (2008); Jaafar et al. (2014)
Increase soil cation exchange capacity (CEC)	Biochar increases CEC of soil through COOH, OH, and ROH groups on its surface, and the efficiency increases depending on pH, $O_{2,}$ and moisture content of the soil	Cheng et al. (2006, 2008)
Priming effect	Mineralization of biochar improves with soil amendment, and it is attributed as positive priming effect of the biochar. Soil microorganisms play a crucial role in mineralization process	Novak et al. (2010); Kuzyakov et al. (2000)
Influence N cycle	Soil amendment of biochar affects microbial processes, increases ammonification, and reduces denitrification	Yanai et al. (2007)
Reduce soil salinity	Biochar adsorbs salts and mitigates salt stress to the plants, and this property explains biochar application to mitigate salinity in agricultural soils	Thomas et al. (2013); Lashari et al. (2015)
Increase earthworm activity	Increased growth of earthworms has been recorded in soils amended with biochar than the soils with no biochar addition	Topoliantz and Ponge (2005); Van et al. (2006)
Increase arbuscular mycorrhizae activity	Biochar interferes with the soil microbes and plant–fungus signaling, detoxifies the allelochemicals, and protects refugia from fungal grazers. Due to such alteration of soil properties, biochar indirectly affects positively onarbuscular mycorrhizae growth and activity	Warnock et al. (2007); Solaiman et al. (2010)
Remove pollutants and reduce toxicity	Biochar removes series of organic and inorganic pollutants from soil. It also immobilize several toxic heavy metals and reduce their mobility to the plant	Ahmad et al. (2014)
Reducing greenhouse gas emissions	Biochar decreases emission of N_2O and CH_4 due to high surface area and microbial activity from agricultural soil and greatly helps in climate change mitigation	Rondon et al. (2005); Feng et al. (2012)

(continued)

Table 9.1 (continued)

Benefits	Description	References
Carbon sequestration	Instead of burning biomass, its use for biochar production in O_2-limiting condition reduces CO_2 emission notably and carbon is bound in biochar structure. Hence, biochar amendment in soil increases the soil carbon	Van Zwieten et al. (2010)
Influence soil organic matter (SOM)	Due to high surface area, it adsorbs SOM and prevents their runoff, and it is influenced by the combinations of soil, climate, and management factors	Marschner et al. (2008)
Improve soil water-holding capacity (WHC) and aeration	Soil blending of biochar significantly affects the soil water retention and aeration through high surface area and physiochemical interaction via radical groups found on its surface, and WHC also depends on the soil type	Sohi et al. (2009); Cheng et al. (2006)

and crop yields (up to 200%) on highly degraded sites (Kimetu et al. 2008). Therefore, there has been significant interest recently spiked for biochar land application; however, application rate, application method, and size of biochar application were not precisely addressed to guide the farmers. Moreover, these parameters also differ with the type of soil and soil organic content.

One of the challenges for biochar field application is linked with right particle size to improve soil moisture retention. Biochar has been used as soil additive in pellets and dust forms using numerous applications methods: topdressing, deep banding, blending, layering (more discussion is in the next section). Biochar in pellets is appropriate for field application and transportation, and it has been used in nursery planting with peat moss at a ratio of 1:1; however, other ratios (e.g., 0.5,0.75. 1.25, 1.5) have exhibited lower yields (Dumroese et al. 2011). Moreover, pelleted biochar needs mechanical tool to spread in the soil, especially pneumatic systems, which are designed for granular material, e.g., seeds and granular fertilizer (Kilicken and Guner 2006).

Land application of biochar in dust form is more challenging task. Biochar in dust form has been negatively attributed for transportation and soil application due to particle's median aerodynamic diameter about 10 μm (US Bureau of Mines 2008). However, after moisturization of dusted biochar, 10–50% (depends on soil condition) has been applied in soil using topdressing and layering methods (Tom 2008). Land application of biochar is smooth with the larger particles, whereas management of fine dust biochar is tricky during wind and rain incidence (Gómez-Rey et al. 2012). Fine biochar is lost during soil application, which is unaffordable (Major 2010). To avoid such loss in farming, biochar has mixed with compost, manures, and organic or chemical fertilizers (Tom 2008).

Application rate of biochar is very important to get maximum benefits and avoid detrimental effect of over dosage. Biochar application rate 0.5–5 kg m^{-2} and proper nutrient management have resulted in better productivity (Tom 2008). The wheat yield has been increased by 58% by altering soil pH 4.5–6.0 through 76.5 t ha^{-1} of biochar application (Collins 2008). Radish yield increased from 42 to 96% when dosage of poultry litter biochar increased from 10–50.5 t ha^{-1}, with respect to non-amended control (Chan et al. 2008). Likewise, 51 and 109% maize yield increased with 10 and 50.5 t ha^{-1} biochar application, respectively (Van Zwieten 2007). The yield of maize grain increased by 150 and 98% with the application rate of 15 and 20 t ha^{-1} of biochar, respectively (Uzoma et al. 2011). However, decrease in grain yield by 23.3, 10, and 26.7% is also recorded with the biochar application rate of 4, 8, and 16 t ha^{-1}, respectively (Asai et al. 2009). Decreased crop yield in this study was attributed to high-volatile matter and toxic substance of biochar, which may have reduced nutrient uptake as well as plant growth. Thus, it is noticeable, and enhancement of crop yield and plant growth may be based on the properties of biochar and soil, moreover biochar suitability for land application also depends on soil type. Hence, it is very crucial to characterize the biochar and understand the mechanisms, which may induce changes on soil after biochar application. Apart from the application of native biochar (as above studies), some studies have also been carried out utilizing fortified biochar with single or multiple nutrients. A brief report upon the effect of sulfur-enriched biochar application (42.5–171 mg kg^{-1} soil) derived from digested dairy manure (at 850 °C) elicited better corn growth response (about 31–49%) than unamended control in pot studies (Zhang et al. 2016). In these studies, biochar has been treated with H_2S and SO_2 gases at high moisture (>80% wt) and pH (>7.0) to adsorb sulfur on biochar surface.

It has been recorded that crop yields enhanced significantly when biochar is applied together with other inorganic or organic fertilizers (Glaser et al. 2002). Improvement in beans yield is recorded with biochar–fertilizer blend, compared to biochar alone (Van Zwieten 2007). Dosage of 11.25 t ha^{-1} of forest wood biochar and mineral fertilizer has improved rice and sorghum yield by 2 times; however, 1.4 times stove yield improved when biochar blended with compost or fertilizer (Steiner et al. 2007). Table 9.2 shows effect of various biochar applications on crop yield improvement. All these studies mainly are carried on pot experiment; however, results from field application of biochar are limited to understand the efficacy of biochar for soil amelioration and enhancement of crop productivity. Soil amendment precision also needs numerous field data with respect to biochar characteristics (e.g., pH, ash content), soil types, and crops varieties; these information would be crucial to fill the knowledge gaps prior to exploit biochar on field. Deleterious effect of biochar due to organic and inorganic contamination must be evaluated. Apart from contaminants, their detrimental effect also arises due to the application mode. A potential detrimental effect of biochar includes contaminant, for instance PAHs, heavy metals, and dioxins; moreover, negative effects also raised due to high rate of biochar application (Verheijen et al. 2010).

Table 9.2 Effect of biochar on crop yields in different soil

Biochar type	Application rate (t ha^{-1})	Soil type	Crop	Yield rise over control (%)	References
Hardwood	19	Midwestern mollisol	Maize	10	Rogovska et al. (2014)
Wheat straw	40	Upland red soil	Rapeseed	36	Liu et al. (2014)
Dairy manure (pot studies)	0.7 (171 mg/kg)	Professional Potting soil	Maize	49	Zhang et al. (2016)
Black carbon	20	Oxisol	Maize	28	Major et al. (2010)
Wheat straw	40	Paddy soil	Rice	14	Zhang et al. (2010)
Eucalyptus wood	30	Inceptisol	Rice	294	Noguera et al. (2010)
Cow manure	15	Sandy soil	Maize	150	Uzoma et al. (2011)
Greenwaste	100	Alfisol	Radish	266	Chan et al. (2008)
Peanut hull + fir bark	76.5	Quincy sand soil	Wheat	58	Collins (2008)
Orchard pruning	22	Sandy clay loam	Grape	20	Genesio et al. (2015)
Poultry litter	50.5	Alfisol	Radish	96	Chan et al. (2008)
Paper mill sludge	50.5	Ferrosol	Maize	109	Van Zwieten et al. (2008)
Forest residue	137	Xanthic Ferrasol	Cowpea	100	Glaser et al. (2002)
Forest residue + fertilizer	11.25	Xanthic Ferrasol	Rice and sorghum	2	Steiner et al. (2007)
Charcoal	0.5	Delhi soil	Moong	22	Glaser et al. (2002)

9.5.2 *Methods of Land Application*

In conventional cropping method, biochar application in the soil can be managed jointly with lime. Lime is frequently applied as a fine solid and homogeneously mixed into soil. This method can be implemented using traditional farm machinery and added into routine field operations. These approaches give lower cost input to use biochar as soil conditioner. Following four methods are well accepted under conventional cropping:

9.5.3 Broadcast and Incorporate Method

It is most popular and practically adapted method for biochar application in the soil (Steiner et al. 2007; Asai et al. 2009). Broadcasting is manually used on small scales; however, larger scale is carried out using lime/solid manure spreaders or broadcast seeders. Land application of moisturized biochar is appropriate with manure spreaders than that of the lime spreaders. Biochar can be mixed in soil by several plowing methods at any scale, including animal draft, hand hoes, rotary hoes, disk harrows, chisels etc. (Blackwell et al. 2009).

9.5.4 Deep Banded Method

It is similar to routine banding operation of seeds and fertilizers and well adopted in mechanized agriculture. Such method involves biochar addition in a narrow band via tools or hands that forms hole without disturbing the entire soil surface, and biochar is placed inside the soil. Biochar is added into 50–100 mm broad and 200–600 mm spacing depth, compatible with the particular cropping system. For deep banding, usually pneumatic system is used in industrialized agriculture. Biochar is applied up to 1 t ha^{-1} for whole field with 100 mm bands and 300 mm spacing by belt-driven feed from supply hoppers. This method is applicable when crop is established. Wheat yield was significantly improved by adopting banding biochar application method in Western Australia (Blackwell et al. 2007).

9.5.5 Blending with Solid Manures

Biochar can be blended with other solid soil amendments, e.g., organic fertilizers, organic manure, compost, or lime prior to land application. Biochar has potential to adsorb and shield nutrient leaching and facilitate prolonged slow release of these nutrients into the soil (Major et al. 2009; Novak et al. 2009). Such blending may save several field operations and improve manure efficiency. Combined effect of biochar and N-P_2O_5-K_2O (NPK) fertilizer at proportion of 20 t and 60–175–120 kg ha^{-1} has shown better yield in three soybean varieties as compared to control with no biochar and fertilizer (Mete et al. 2015).

9.5.6 Blending with Liquid Manures

Biochar can also be mixed with liquid manures and can be applied as slurry. Fine biochars would likely be best suited to this type of application through existing application equipment, which may solve dust application drawbacks. Biochar can also be mixed with manure at holding ponds and potentially may reduce the gaseous nitrogen losses. This method is normally being practiced prior to soil application (Yanai et al. 2007). Main aim of such application includes the odor reduction (Kleegberg et al. 2005), P retention in liquid manure, and dust reduction (Lehmann 2007). The application rate of biochar 25, 50, and 150 g kg^{-1} of local fertilizers, compost blend, liquid compost, and lake sediment is used in the potting medium (Carter et al. 2013).

9.5.7 Case-Specific Application Methods

9.5.7.1 Subsurface Banding

Poultry manure has been applied to perennial grassland, adopting the subsurface banding method (Sistani et al. 2009). It was accomplished using the prototype equipments to mix biochar into the soil, which has significantly reduced the leaching of manure from the soil during raining. This method is attributed to be effective to reduce potential losses of biochar via wind and water erosion and may be suitable on slopes even for fine biochar materials. The large pieces of litter or biochar are crushed into fine powder, and then, it was gravity fed into furrows opened by equipment disks; furthermore, the furrows were closed with wheels and cover the biochar material with the soil.

9.5.7.2 Surface Application or Topdressing

This application method is suitable for conditioning matter to spread on no-till cropping system, pasture system, and perennial vegetation. It needs proper pretreatment of biochar prior to land application to ensure the health risks (Vassilev et al. 2013) and losses. Therefore, usually manures, compost, and synthetic fertilizer are mixed with biochar to restore fertility in perennial systems. Biochar can be top-dressed on perennial plants upon spaces between plants manually by using disk or rotating hammer spreaders. Such practices on perennials were made effectively without significant losses of biochar in England (Gathorne-Hardy et al. 2008); moreover, biochar mixing was facilitated by earthworms into the soil. Activities such as subsequent leaf fall, macro-faunal action, and water infiltration may help to add biochar into the topsoil.

9.5.7.3 Localized Applications

In this method, biochar can be applied in radial ditches that would spread and utilized from the base of established trees in the later phase. Biochar could also be placed into the soil using "air excavation apparatus." Instead of removing soil around the tree roots, it can be placed into several small ditches between the trees. Such treatment has been practiced for old trees in Japan (Japan Biochar Association). This method would be suitable when the field soil is harsh due to high or low pH; biochar placement in radial ditches can improve from resulting unstable trees with poor root structures. Nursery plants grown with biochar soil in the field could benefit from biochar on early stage, but roots will usually grow outside the biochar-amended area in a while.

9.5.7.4 Broadcast and Uniform Mixing with New Topsoil Applications

Biochar can be blended with sand, compost, new topsoil, or turf grass prior to use in various fields especially in the landscape. Spreader can be applied to mix biochar to the whole area mechanically using hand or draft animals during primary and secondary tillage. Offset disk plow generally provides better mixing. The influence of deep tillage on the soil structure varies based on types of tillage and soil type (Coulouma et al. 2006). This may be used in bulk applications in topdressing soil in flowerbeds. In important places such as golf court and sport turfgrass, moreover, places where vital facilities are desirable such as decompression and quick drainage in artificial rooting zones. Blending biochar with sand can facilitate the fast drainage and resistance to sand compaction while increasing moisture retention and availability to turf. The benefits of uniform topsoil mixing mainly increase the soil fertility, improvement in CEC and WHC, adsorption of leachable herbicides, and reduction of GHG emissions.

9.5.7.5 Future Perspectives

There has been growing interest on biochar application in soil since green revolution for contaminants removal and heavy metal toxicity reduction. Current interest of biochar application in soil is mainly focused to examine its efficacy as slow-release fertilizer to improve single or multiple nutrient transformations to the plants. The effect of biochar application on improvement of soil quality and crop yields mainly in alkaline and nutrient-rich soils is still unclear. Research is needed to standardize biochar characteristics based on different feedstocks and pyrolysis conditions prior to gain insights for their specified applications such as contaminant adsorption and desorption of nutrients via mineralization. Moreover, further research is required for biochar surface chemistry regarding interaction with soil constituents, especially micronutrients' binding, preferred forms of binding on biochar, and their exchange mechanisms. Most importantly, these researches are being carried out independently with limited parameters; however, it needs to cover wider range of each parameter and preferably in collaborative manners to fill the information gaps. These approaches will help to find complete information about the characteristics of biochar pertinent to targeted applications and would be reliable to replicate into the fields. Methods of application are also important to extract maximum benefits from biochar; therefore, more research should be centered to develop effective method and the timing of application with respect to plant stage or cropping season to elicit better response.

Role of microbial communities for plant growth especially for nutrient transformation in biochar-amended soil has not been well characterized. Moreover, correlation of microbial community with respect to biochar properties (e.g., pH, particle size, microporosity, nutrient content) has not been examined. With growing challenge of nutrient deficiency in soil, biochar application for scrubbing nutrients from wastes could open up new opportunity for application of resulting nutrient-bound biochar as a source of micronutrient in nutrient-depleted soils.

Apart from the biochar applicability in environmental and agriculture sectors, industrial domains also need deep insights for specified biochar applications to economies their processes and attain the waste management standards to deliver sustainable technology to the society. In situ removal of key toxic compounds, e.g., furfural, HMF, H_2S, SO_2, NH_3, CO_2, Cl, Pb, Cr, is required in several industries, e.g., biofuel, oil and mining, pharmaceutical, leather, textiles, during primary and later stage of their processes; therefore, research directions must be set, focusing these red areas to exploit biochar technology effectively. Moreover, utilized biochar for above-mentioned applications can also be explored for soil amelioration and plant growth promotion. Various studies have shown that numerous toxic compounds have underwent significant structural changes after biochar binding and converted into bioavailable nutrients. Biochar associated with microbial activities would be another key factor in chemical transformation of toxic compound into nutrients during weathering process of biochar in later stage of the soil applications.

9.6 Conclusion

Recent trend of biochar application has been aimed not only in remediation of various organic and inorganic pollutants in soil system, but also as soil amendment and source of nutrients for plant growth promotion. Recent interest on biochar research significantly spiked toward biochar-mediated single or multiple nutrients recovery from waste matters and application of resulting nutrient-enriched biochar for plant nutrition. Notably, biomass conversion into biochar and its subsequent land application has other series of benefits toward ecosystem sustainability via GHG emission reduction and carbon sequestration, thus leading to climate change mitigation. Various dissolved organic compounds and simple organic compounds adsorbed on biochar surface provide a favorable microbial habitation, thus affecting microbial abundance, microbial activity, and mineralization process in soil system. The interaction between biochar and soil improves soil physiochemical properties, facilitates nutrient bioavailability by increasing water-holding capacity, adsorbs inorganic compounds to improve cation exchange capacity, increases soil pH to buffer against soil acidity, detoxifies the allelochemicals, etc. These changes bring the favorable environment for microbial proliferations and activities, which significantly improve mineralization and nutrient transformation to the plants and thus the agricultural productivity.

Acknowledgements The study is being supported partly by the HATCH grant from the College of Tropical Agriculture and Human Resources (CTAHR), University of Hawaii at Manoa (UHM), and Coordenação de Aperfeiçoamento de Pessoal de Nível Superior, Ministério da Educação (Capes/MEC), Brazil.

References

Ahmad M, Lee SS, Dou X, Mohan D, Sung JK, Yang JE, Ok YS (2012) Effects of pyrolysis temperature on soybean stover-and peanut shell-derived biochar properties and TCE adsorption in water. Bioresour Technol 118:536–544

Ahmad M, Ok YS, Rajapaksha AU, Lim JE, Kim BY, Ahn JH, Lee YH, Al-Wabel MI, Lee SE, Lee SS (2016) Lead and copper immobilization in a shooting range soil using soybean stover-and pine needle-derived biochars: chemical, microbial and spectroscopic assessments. J Hazard Mater 301:179–186

Ahmad M, Rajapaksha AU, Lim JE, Zhang M, Bolan N, Mohan D, Ok YS (2014) Biochar as a sorbent for contaminant management in soil and water: a review. Chemosphere 99:19–33

Ameloot N, Graber E, Verheijen F, De Neve S (2013) Interactions between biochar stability and soil organisms: review and research needs. Eur J Soil Sci 64:379–390

Asai H, Samson BK, Stephan HM, Songyikhangsuthor K, Homma K, Kiyono Y, Inoue Y, Shiraiwa T, Horie T (2009) Biochar amendment techniques for upland rice production in Northern Laos 1. Soil physical properties, leaf SPAD and grain yield. Field Crop Res 111:81–84

Atkinson CJ, Fitzgerald JD, Hipps NA (2010) Potential mechanisms for achieving agricultural benefits from biochar application to temperate soils: a review. Plant Soil 337:1–18

Bailey VL, Fansler SJ, Smith JL, Bolton H (2011) Reconciling apparent variability in effects of biochar amendment on soil enzyme activities by assay optimization. Soil Biol Biochem 43:296–301

Basso AS, Miguez FE, Laird DA, Horton R, Westgate M (2013) Assessing potential of biochar for increasing water-holding capacity of sandy soils. Glob Change Biol Bioenerg 5:132–143

Blackwell P, Riethmuller G, Collins M (2009) Biochar application to soil (Chapter 12). In: Lehmann J, Joseph S (eds) Biochar for environmental management: science and technology. Earthscan, London, UK, p 207

Blackwell P, Shea S, Storer P, Kerkmans M, Stanley I (2007) Improving wheat production with deep banded oil mallee charcoal in Western Australia. In: Talk given at the international agrichar conference 2007, Terrigal, NSW Australia, 27 April–2 May 2007

Bogusz A, Oleszczuk P, Dobrowolski R (2015) Application of laboratory prepared and commercially available biochars to adsorption of cadmium, copper and zinc ions from water. Bioresour Technol 196:540–549

Bohn HL, Barrow NJ, Rajan SS, Parfitt RL (1986) Reactions of inorganic sulfur in soils. In: Tabatabai MA (ed.) Sulfur in agriculture, agronomic monogram. CSSA, ISSSA, Madison, WI, 27, 233–249

Bolan N, Kunhikrishnan A, Thangarajan R, Kumpiene J, Park J, Makino T, Kirkham MB, Scheckel K (2014) Remediation of heavy metal(loid)s contaminated soils—to mobilize or to immobilize. J Hazard Mater 266:141–166

Brady NC, Weil RR (2013) The nature and properties of soils, 14th edn. Prentice Hall, Upper Saddle River, New Jersey, USA

Brewer CE, Brown RC (2012) Biochar. In: Sayigh A (ed) Comprehensive renewable energy. Elsevier, Oxford, pp 357–384

Brewer CE, Hu YY, Schmidt-Rohr K, Loynachan TE, Laird DA, Brown RC (2012) Extent of pyrolysis impacts on fast pyrolysis biochar properties. J Environ Qual 41(4):1115–1122

Bruun EW, Ambus P, Egsgaard H, Hauggaard-Nielsen H (2012) Effects of slow and fast pyrolysis biochar on soil C and N turnover dynamics. Soil Biol Biochem 46:73–79

Cantrell KB, Hunt PG, Uchimiya M, Novak JM, Ro KS (2012) Impact of pyrolysis temperature and manure source on physicochemical characteristics of biochar. Bioresour Technol 107:419–428

Cao X, Harris W (2010) Properties of dairy-manure-derived biochar pertinent to its potential use in remediation. Bioresour Technol 101:5222–5228

Carter S, Shackley S, Sohi S, Suy TB, Haefele S (2013) -. Agronomy 3:404–418

Cayuela ML, Van Zwieten L, Singh BP, Jeffery S, Roig A, Sánchez-Monedero MA (2014) Biochar's role in mitigating soil nitrous oxide emissions: a review and meta-analysis. Agr Ecosyst Environ 191:5–16

Chan KY, Van Zwieten L, Meszaros I, Downie A, Joseph S (2008) Using poultry litter biochars as soil amendments. Aus J Soil Res 46(5):437–444

Chen G, Chen Y, Zhao G, Cheng W, Guo S, Zhang H, Shi W (2015) Do high nitrogen use efficiency rice cultivars reduce nitrogen losses from paddy fields?. Agric Ecosyst Environ 209:26–33

Cheng CH, Lehmann J, Engelhard MH (2008) Natural oxidation of black carbon in soils: changes in molecular form and surface charge along a climosequence. Geochim Cosmochim Acta 72:1598–1610

Cheng CH, Lehmann J, Thies JE, Burton SD, Engelhard MH (2006) Oxidation of black carbon by biotic and abiotic processes. Org Geochem 37:1477–1488

Chintala R, Mollinedo J, Schumacher TE, Malo DD, Julson JL (2014) Effect of biochar on chemical properties of acidic soil. Arch Agron Soil Sci 60:393–404

Collins H (2008) Use of biochar from the pyrolysis of waste organic material as a soil amendment: laboratory and greenhouse analyses. A quarterly progress report prepared for the biochar project

Coulouma G, Boizard H, Trotoux G, Lagacherie P, Richard G (2006) Effect of deep tillage for vineyard establishment on soil structure: a case study in southern France. Soil Tillage Res 88:132–143

Crombie K, Mašek O, Cross A, Sohi S (2015) Biochar–synergies and trade-offs between soil enhancing properties and C sequestration potential. GCB Bioenerg 7(5):1161–1175

Crombie K, Mašek O, Sohi SP, Brownsort P, Cross A (2013) The effect of pyrolysis conditions on biochar stability as determined by three methods. GCB Bioenerg 5(2):122–131

Curtin D, Trolove S (2013) Predicting pH buffering capacity of New Zealand soils from organic matter content and mineral characteristics. Soil Res 51:494–502

David MB, Gentry LE, Mitchell CA (2016) Riverine response of sulfate to declining atmospheric sulfur deposition in agricultural watersheds. J Environ Qual 45:1313–1319

Day D, Evans RJ, Lee JW, Reicosky D (2005) Economical CO_2, SOx, and NOx capture from fossil-fuel utilization with combined renewable hydrogen production and large-scale carbon sequestration. Energy 30:2558–2579

DeLuca TH, Gundale MJ, MacKenzie MD, Jones DL (2015) Biochar effects on soil nutrient transformations. Biochar for environmental management: science. In: Technology and Implementation. Taylor and Francis, New York, USA, pp 421–454

Dumroese RK, Heiskanen J, Englund K, Tervahauta A (2011) Pelleted biochar: chemical and physical properties show potential use as a substrate in container nurseries. Biomass Bioenerg 35:2018–2027

Emmalea E (2016) Sulfur deficiency on field corn, https://extension.udel.edu/weeklycropupdate/?p=9161. Last accessed 2017/03/04

Feng YZ, Xu YP, Yu YC, Xie ZB, Lin XG (2012) Mechanisms of biochar decreasing methane emission from Chinese paddy soils. Soil Biol Biochem 46:80–88

Ganjegunte GK, Wick AF, Stahl P, Vance GF (2009) Accumulation and composition of total organic carbon in reclaimed coal mine lands. Land Degrad Dev 20:156–175

Gathorne-Hardy A, Knight J, Woods J (2008) Surface application of biochar to pasture—changes in yield, diversity, forage quality, and its incorporation into the soil. In: Poster presented at the 2nd Annual meeting of the International Biochar Initiative (IBI), Newcastle, UK, 8–10 Sept 2008

Genesio L, Miglietta F, Baronti S, Vaccari FP (2015) Biochar increases vineyard productivity without affecting grape quality: results from a four years field experiment in Tuscany. Agric Ecosyst Environ 201:20–25

Glaser B, Lehmann J, Zech W (2002) Ameliorating physical and chemical properties of highly weathered soils in the tropics with charcoal—a review. Biol Fertil Soils 35:219–230

Gómez-Rey MX, Madeira M, Coutinho J (2012) Wood ash effects on nutrient dynamics and soil properties under Mediterranean climate. Ann For Sci 69:569–579

Gonzaga MIS, Mackowiak CL, Comerford NB, Molin EFV, Shirley JP, Guimaraes DV (2017) Pyrolysis methods impact biosolids-derived biochar composition, maize growth and nutrition. Soil Til. Res. 165:59–65

Graber ER, Elad Y (2013) Biochar impact on plant resistance to disease. CRC Press, Boca Raton

Gul S, Whalen JK, Thomas BW, Sachdeva V, Deng H (2015) Physico-chemical properties and microbial responses in biochar-amended soils: mechanisms and future directions. Agr Ecosyst Environ 206:46–59

Gundale MJ, DeLuca TH (2006) Temperature and source material influence ecological attributes of Ponderosa pine and Douglas-fir charcoal. For Ecol Manag 231:86–93

Hockaday WC (2006) The organic geochemistry of charcoal black carbon in the soils of the University of Michigan Biological Station. Ohio State University, Ohio, Columbus

Hofrichter M, Ziegenhagen D, Sorge S, Ullrich R, Bublitz F, Fritsche W (1999) Degradation of lignite (low-rank coal) by ligninolytic basidiomycetes and their manganese peroxidase system. Appl Microbiol Biotechnol 52:78–84

International Biochar Initiative (IBI) (2013) Standardized product definition and product testing guidelines for biochar that is used in soil. http://www.biochar-international.org/. Last accessed 2017/05/04

Iqbal H, Garcia-Perez M, Flury M (2015) Effect of biochar on leaching of organic carbon, nitrogen, and phosphorus from compost in bioretention systems. Sci Total Environ 521–522:37–45

IUSS Working Group WRB (2006) World reference base for soil resources, 2nd edn. World soil resources report no 103, UN Food and Agriculture Organization, Rome, pp 128

Jaafar NM, Clode PL, Abbott LK (2014) Microscopy observations of habitable space in biochar for colonization by fungal hyphae from soil. J Integr Agr 13:483–490

Jeffery S, Verheijen FGA, Van Der Velde M, Bastos AC (2011) A quantitative review of the effects of biochar application to soils on crop productivity using meta-analysis. Agric Ecosyst Environ 144:175–187

Kambo HS, Dutta A (2015) A comparative review of biochar and hydrochar in terms of production, physico-chemical properties and applications. Renew Sust Energ Rev 45:359–378

Kanjanarong J, Giri BS, Oliveira FR, Boonsawang P, Chaiprapat S, Singh RS, Balakrishna A, Jaisi DP, Khanal SK (2017) Removal of hydrogen sulfide generated during anaerobic treatment of sulfate-laden wastewater using biochar: evaluation of efficiency and mechanisms. Bioresour Technol 234:115–121

Kilicken A, Guner M (2006) Pneumatic conveying characteristics of cotton seeds. Biosyst Eng 95:537–546

Kimetu JM, Lehmann J, Ngoze SO, Mugendi DN, Kinyangi J, Riha SJ, Verchot L, Recha JW, Pell AN (2008) Reversibility of soil productivity decline with organic matter of differing quality along a degradationgradient. Ecosystems 11:726–739

Kleegberg KK, Schlegemilch M, Strees J, Stcinhart H, Stegmann R (2005) Odour abatement strategy for a sustainable odour management. In: Proceedings of the tenth international waste management and landfill symposium, Sardinia

Kothamasi D, Kothamasi S, Bhattacharyya A, Kuhad RC, Babu CR (2006) Arbuscular mycorrhizae and phosphate solubilising bacteria of the rhizosphere of the mangrove ecosystem of great Nicobar Island, India. Biol Fert Soils 42:358–361

Kuzyakov Y, Friedel JK, Stahr K (2000) Review of mechanisms and quantification of priming effects. Soil Biol Biochem 32:1485–1498

Laird D, Fleming P, Wang B, Horton R, Karlen D (2010) Biochar impact on nutrient leaching from a Midwestern agricultural soil. Geoderma 158(3):436–442

Laird DA, Rogovska NP, Garcia-Perez M, Collins HP, Streubel JD, Smith M (2011) Pyrolysis and biochar–opportunities for distributed production and soil quality enhancement. In: Sustainable alternative fuel feedstock opportunities, challenges and roadmaps for six US regions. Proceedings of the sustainable feedstocks for advanced biofuels workshop, SWCS publisher, Atlanta, GA, pp 257–281

Lal R, Follett RF, Stewart BA, Kimble JM (2007) Soil carbon sequestration to mitigate climate change and advance food security. Soil Sci 172(12):943–956

Lashari MS, Ye YX, Ji HS, Li LQ, Kibue GW, Lu HF, Zheng JF, Pan GX (2015) Biochar-manure compost in conjunction with pyroligneous solution alleviated salt stress and improved leaf bioactivity of maize in a saline soil from Central China: a two-year field experiment. J Sci Food Agr 95:1321–1327

Lehmann J (2007) A handful of carbon. Nature 447:143–144

Lehmann J, Czimczik C, Laird D, Sohi S (2009) Stability of biochar in soil. Biochar Environ Manag: Sci Technol:183–206

Lehmann J, Joseph S (eds) (2015) Biochar for environmental management: science, technology and implementation. Taylor and Francis Group, Routledge, USA

Liang B, Lehmann J, Solomon D, Kinyangi J, Grossman J, O'Neill B, Skjemstad JO, Thies J, Luizão FJ, Petersen J, Neves EG (2006) Black carbon increases cation exchange capacity in soils. Soil Sci Soc Am J 70

Lima IM, Boateng AA, Klasson KT (2010) Physicochemical and adsorptive properties of fast-pyrolysis bio-chars and their steam activated counterparts. J Chem Technol Biotechnol 85 (11):1515–1521

Liu XH, Zhang XC (2012) Effect of biochar on pH of alkaline soils in the loess plateau: results from incubation experiments. Int J Agr Biol 14:745–750

Liu Z, Chen X, Jing Y, Li Q, Zhang J, Huang Q (2014) Effects of biochar amendment on rapeseed and sweet potato yields and water stable aggregate in upland red soil. CATENA 123:45–51

Lu H, Zhang W, Yang Y, Huang X, Wang S, Qiu R (2012) Relative distribution of Pb^{2+} sorption mechanisms by sludge-derived biochar. Water Res 46:854–862

Lu SG, Sun FF, Zong YT (2014) Effect of rice husk biochar and coal fly ash on some physical properties of expansive clayey soil (Vertisol). CATENA 114:37–44

Major J (2010) Guidelines on practical aspects of biochar application to field soil in various soil management systems. 1–23

Major J, Rondon M, Molina D, Riha SJ, Lehmann J (2010) Maize yield and nutrition during 4 years after biochar application to a Colombian savanna oxisol. Plant Soil 333:117–128

Major J, Steiner C, Downie A, Lehmann J (2009) Biochar effects on nutrient leaching (Chapter 15). In: Lehmann J, Joseph S (eds), Biochar for environmental management: science and technology. Earthscan, London, UK, p 271

Malghani S, Gleixner G, Trumbore SE (2013) Chars produced by slow pyrolysis and hydrothermal carbonization vary in carbon sequestration potential and greenhouse gases emissions. Soil Biol Biochem 62:137–146

Masiello CA, Gao YCX, Liu S, Cheng H-Y, Bennett MR, Rudgers JA, Wagner DS, Zygourakis K, Silberg JJ (2013) Biochar and microbial signaling: production conditions determine effects on microbial communication. Environ Sci Technol 47:11496–11503

Marschner B, Brodowski S, Dreves A, Gleixner G, Gude A, Grootes PM, Hamer U, Heim A, Jandl G, Ji R, Kaiser K, Kalbitz K, Kramer C, Leinweber P, Rethemeyer J, SchÅNaffer A, Schmidt MWI, Schwark L, Wiesenberg GLB (2008) How relevant is recalcitrance for the stabilization of organic matter in soils. J Plant Nutr Soil Sci 171:91–110

Mengel K, Kirkby EA (2001) Principles of plant nutrition, 5th edn. Kluwer Academic Publishers, Dordrecht

Mete FJ, Mia S, Dijkstra FA, Abuyusuf M, Iqbal Hossain ASM (2015) Synergistic effects of biochar and NPK fertilizer on soybean yield in an alkaline soil. Pedosphere 25(5):713–719

Mohanty P, Nanda S, Pant KK, Naik S, Kozinski JA, Dalai AK (2013) Evaluation of the physiochemical development of biochars obtained from pyrolysis of wheat straw, timothy grass and pinewood: Effects of heating rate. J Anal Appl Pyrol 104:485–493

Noguera D, Rondón M, Laossi KR, Hoyos V, Lavelle P, de Carvalho MHC, Barot S (2010) Contrasted effect of biochar and earthworms on rice growth and resource allocation in different soils. Soil Biol Biochem 42:1017–1027

Novak JM, Busscher WJ, Laird DL, Ahmedna M, Watts DW, Niandou MAS (2009) Impact of biochar amendment on fertility of a southeastern coastal plain soil. Soil Sci 174:105–112

Novak JM, Busscher WJ, Watts DW, Laird DA, Ahmedna MA, Niandou MAS (2010) Short-term CO2 mineralization after additions of biochar and switchgrass to a Typic Kandiudult. Geoderma 154:281–288

Novak JM, Cantrell KB, Watts DW (2013) Compositional and thermal evaluation of lignocellulosic and poultry litter chars via high and low temperature pyrolysis. Bioenerg Res 6:114–130

Peake LR, Reid BJ, Tang X (2014) Quantifying the influence of biochar on the physical and hydrological properties of dissimilar soils. Geoderma 235:182–190

Purakayastha TJ, Das KC, Gaskin J, Harris K, Smith JL, Kumari S (2016) Effect of pyrolysis temperatures on stability and priming effects of C3 and C4 biochars applied to two different soils. Soil Til Res 155:107–115

Qiu Y, Zheng Z, Zhou Z, Sheng GD (2009) Effectiveness and mechanisms of dye adsorption on a straw-based biochar. Bioresour Technol 100:5348–5351

Rogovska N, Laird DA, Rathke SJ, Karlen DL (2014) Biochar impact on Midwestern Mollisols and maize nutrient availability. Geoderma 230:340–347

Rondon M, Ramirez JA, Lehmann J (2005) Greenhouse gas emissions decrease with charcoal additions to tropical soils. In USDA (ed) Proceedings of the third USDA symposium on greenhouse gases and carbon sequestration in agriculture and forestry. USDA, Baltimore, p 208

Rutherford DW, Wershaw RL, Rostad CE, Kelly CN (2012) Effect of formation conditions on biochars: compositional and structural properties of cellulose, lignin, and pine biochars. Biomass and Bioenerg 46:693–701

Sistani KR, Torbert HA, Way TR, Bolster CH, Porte DH, Warren JG (2009) Boiler litter application a d runoff timing effect on nutrient and Escherichia coli losses from tall fescue pasture. J Environ Qual 38:1216–1223

Sohi S, Lopez-Capel E, Krull E, Bol R (2009) Biochar, climate change and soil: a review to guide future research. CSIRO Land and Water Science Report 05/09. CSIRO, Canberra

Solaiman ZM, Blackwell P, Abbott LK, Storer P (2010) Direct and residual effect of biochar application on mycorrhizal root colonisation, growth and nutrition of wheat. Soil Res 48: 546–554

Spokas KA, Cantrell KB, Novak JM, Archer DW, Ippolito JA, Collins HP, Lentz RD (2012) Biochar: a synthesis of its agronomic impact beyond carbon sequestration. J Environ Qual 41 (4):973–989

Spokas KA, Novak JM, Stewart CE, Cantrell KB, Uchimiya M, DuSaire MG, Ro KS (2011) Qualitative analysis of volatile organic compounds on biochar. Chemosphere 85:869–882

Steiner C, Glaser B, Teixeira WG, Lehmann J, Blum WEH, Zech W (2008) Nitrogen retention and plant uptake on a highly weathered central Amazonian Ferralsol amended with compost and charcoal. J Plant Nutr Soil Sci 171:893–899

Steiner C, Teixeira WG, Lehmann J, Nehls T, Vasconcelos JL, Blum WEH, Zech W (2007) Long term effects of manure, charcoal and mineral fertilization on crop production and fertility on a highly weathered Central Amazonian upland soil. Plant Soil 291:275–290

Stewart CE, Zheng J, Botte J, Cotrufo F (2013) Co-generated fast pyrolysis biochar mitigates greenhouse gas emissions and increases carbon sequestration in temperate soils. Glob Change Biol Bioenerg 5:153–164
Suliman W, Harsh JB, Abu-Lail NI, Fortuna AM, Dallmeyer I, Garcia-Perez M (2016) Influence of feedstock source and pyrolysis temperature on biochar bulk and surface properties. Biomass Bioenerg 84:37–48
Teixidó M, Pignatello JJ, Beltrán JL, Granados M, Peccia J (2011) Speciation of the ionizable antibiotic sulfamethazine on black carbon (biochar). Environ Sci Technol 45:10020–10027
Thies E, Rilling MC (2009) Characteristics of biochar: biological properties. In: Lehmann J, Joseph S (eds) Biochar for environmental management: science and technology. London, UK, Earthscan, pp 85–106
Thomas SC, Frye S, Gale N, Garmon M, Launchbury R, Machado N, Melamed S, Murray J, Petroff A, Winsborough C (2013) Biochar mitigates negative effects of salt additions on two herbaceous plant species. J Environ Manage 129:62–68
Tom M (2008) Biochar white paper, the international biochar initiative 2007, biochar: a soil amendment that combats global warming and improves agricultural sustainability and environmental impacts. http://biochar.bioenergylists.org/ibiaboutbiochar. Last accessed 2017/05/12
Tong SJ, Li JY, Yuan JH, Xu RK (2011) Adsorption of Cu(II) by biochars generated from three crop straws. Chem Eng J 172:828–834
Topoliantz S, Ponge JF (2005) Charcoal consumption and casting activity by Pontoscolex corethrurus (Glossoscolecidae). Appl Soil Ecol 28:217–224
Tsai WT, Liu SC, Chen HR, Chang YM, Tsai YL (2012) Textural and chemical properties of swine-manure-derived biochar pertinent to its potential use as a soil amendment. Chemosphere 89:198–203
Uchimiya M, Klasson KT, Wartelle LH, Lima IM (2011) Influence of soil properties on heavy metal sequestration by biochar amendment: 1. Copper sorption isotherms and the release of cations. Chemosphere 82:1431–1437
Uras U, Carrier M, Hardie AG, Knoetze JH (2012) Physico-chemical characterization of biochars from vacuum pyrolysis of South African agricultural wastes for application as soil amendments. J Anal Appl Pyrol 98:207–213
US Bureau of Mines (2008) Dust and its control, in dust control handbook for mineral processing, US Department of the interior, Chapter 1. www.osha.gov/SLTC/silicacrystalline/dust/chapter_1.html. Last accessed 2016/12/04 (2008)
Uzoma KC, Inoue M, Andry H, Fujimaki H, Zahoor A, Nishihara E (2011) Effect of cow manure biochar on maize productivity under sandy soil condition. Soil Use Manag 27:205–212
Van DTT, Mui NT, Ledin I (2006) Effect of processing foliage of *Acacia mangium* and inclusion of bamboo charcoal in the diet on performance of growing goats. Anim Feed Sci Tech 130:242–256
Van Zwieten L, Kimber S, Morris S, Downie A, Berger E, Rust J, Scheer C (2010) Influence of biochars on flux of N2O and CO2 from Ferrosol. Soil Res. 48:555–568
Van Zwieten L, Singh B, Joseph S, Cowic A, Chan K (2009) Biochar and emissions of non-CO_2 greenhouse gases from soil. In: Lehmann J, Joseph S (eds) Biochar for environmental management. Sci. Technol. Earthscan, London, pp 227–249
Van Zwieten L (2007) Research confirms biochar in soils boosts crop yields (last cited Dec 2016). http://biopact.com/2007/06/research-confirms-biochar-in-soils.htmlS
Van Zwieten L, Meszaros I, Downie A, Chan YK, Joseph S (2008) Soil health: can the cane industry use a bit of black magic. In: Australian Canegrower 17
Vassilev SV, Baxter D, Andersen LK, Vassileva CG (2013) An overview of the composition and application of biomass ash: part 2. Potential utilization, technological and ecological advantages and challenges. Fuel 105:19–39
Verheijen F, Jeffery S, Bastos AC, van der Velde, M. Diafas, I (2010) Biochar application to soils: a critical scientific review of effects on soil properties, processes and functions. EUR 24099 EN, Office for the Official Publications of the European Communities. Luxembourg

Vithanage M, Rajapaksha AU, Tang X, Bruhn ST, Kim KH, Lee SE, Ok YS (2014) Sorption and transport of sulfamethazine in agricultural soils amended with invasive-plant-derived biochar. J Environ Manage 141:95–103

Wang K, He C, You S, Liu W, Wang W, Zhang R, Ren N (2015) Transformation of organic matters in animal wastes during composting. J Hazard Mater 300:745–753

Wardle DA, Nilsson MC, Zackrisson O (2008) Response to comment on fire-derived charcoal causes loss of forest humus. Science 321:1295d

Warnock DD, Lehmann J, Kuyper TW, Rillig MC (2007) Mycorrhizal responses to biochar in soil–concepts and mechanisms. Plant Soil 300:9–20

Windeatt JH, Ross AB, Williams PT, Forster PM, Nahil MA, Singh S (2014) Characteristics of biochars from crop residues: potential for carbon sequestration and soil amendment. J Environ Manage 146:189–197

Woolf D, Amonette JE, Street-Perrott FA, Lehmann J, Joseph S (2010) Sustainable biochar to mitigate global climate change. Nature Commun 1:56

World Health Organization (2017) Don't pollute my future!, The impact of the environment on children's health. Geneva, Licence: CC BY-NC-SA 3.0 IGO

Xu G, Sun J, Shao H, Chang SX (2014a) Biochar had effects on phosphorus sorption and desorption in three soils with differing acidity. Ecol Eng 62:54–60

Xu X, Cao X, Zhao L, Sun T (2014b) Comparison of sewage sludge and pig manure derived biochars for hydrogen sulfide removal. Chemosphere 111:296–303

Yanai Y, Toyota K, Okazaki M (2007) Effects of charcoal addition on N_2O emissions from soil resulting from rewetting air-dried soil in short-term laboratory experiments. Soil Sci Plant Nutr 53:181–188

Yao Y, Gao B, Zhang M, Inyang M, Zimmerman AR (2012) Effect of biochar amendment on sorption and leaching of nitrate, ammonium, and phosphate in a sandy soil. Chemosphere 89 (11):1467–1471

Zhang A, Cui L, Pan G, Li L, Hussain Q, Zhang X, Zheng J, Crowley D (2010) Effect of biochar amendment on yield and methane and nitrous oxide emissions from a rice paddy from Tai Lake Plain. China Agric Ecosyst Environ 139:469–475

Zhang K, Cheng X, Dang H, Ye C, Zhang Y, Zhang Q (2013) Linking litter production, quality and decomposition to vegetation succession following agricultural abandonment. Soil Biol Biochem 57:803–813

Zhang H, Voroney RP, Price GW, Andrew J, White AJ (2016) Sulfur-enriched biochar as a potential soil amendment and fertilizer. Soil Res. http://dx.doi.org/10.1071/SR15256

Zheng H, Wang Z, Deng X, Zhao J, Luo Y, Novak J, Herbert S, Xing B (2013) Characteristics and nutrient values of biochars produced from giant reed at different temperatures. Bioresour Technol 130:463–471

Chapter 10
Vermicomposting: A Green Technology for Organic Waste Management

Kavita Sharma and V. K. Garg

Abstract Population growth, urbanization, industrialization, intensification of agriculture and food production have considerably contributed to solid waste generation in recent times. To dispose this burgeoning solid waste, destructive techniques such as landfilling and incineration are used. These methods wipe out various nutrients present in the solid waste which otherwise can be recycled using other methods. Solid waste is heterogeneous in nature, and any single method is not sufficient for its management. But Non-toxic fraction of the solid wastes can be used as feedstock for various biological processes to recover or produce value-added products from solid wastes. Such biological processes include biomethanation, composting and vermicomposting. Among these, vermicomposting has been reported as a practicable, economical and swift technique for proficient management of the solid wastes. In this process, earthworms convert compostable fraction of the solid wastes into stabilized, finely divided peat-like material called vermicompost that can be used as manure in agricultural fields to improve soil health. Different waste residues like animal excreta, agricultural residues, domestic waste, sewage sludge, industrial wastes etc. have been used as earthworm feedstock in various research trials. In this chapter, solid wastes, vermicomposting process, various feedstocks, vermicompost quality etc. have been discussed in detail.

Keywords Vermicomposting · Waste management · Green technology Maturity

K. Sharma · V. K. Garg
Department of Environmental Science and Engineering, Guru Jambheshwar University of Science and Technology, Hisar 125001, India

V. K. Garg (✉)
Centre for Environmental Sciences and Technology, Central University of Punjab, Bathinda 151001, India
e-mail: vinodkgarg@yahoo.com

R. R. Singhania et al. (eds.), *Waste to Wealth*, Energy, Environment, and Sustainability, https://doi.org/10.1007/978-981-10-7431-8_10

10.1 Introduction

Solid waste from agricultural, domestic and industrial sectors contributes to environmental pollution and may cause health hazards. Various studies have reported that solid waste generation is alarming and should be dealt immediately to avoid further environmental degradation (Lim et al. 2016; Soobhany et al. 2015). These studies conclude that solid waste generation, at global level, is likely to be increased in near future. However, quantity of waste generation may differs from country to country depending on population, industrialisation and economic growth. Problems associated with waste generation are enormous and demand for sustainable and decentralized waste management (Kumar et al. 2017).

Globally, current solid waste disposal methods are not eco-friendly and have higher operating and capital costs. Unscientific and improper disposal methods release large amounts of carbon dioxide in the atmosphere leading to global warming and climate change. According to WHO (2000) in the developing countries, 10% population is severely affected, with intestinal worms, due to improper waste and excreta management. Still, unscientific landfilling and open dumping of solid wastes are commonly practised for waste disposal. These practices are a source of water pollution, air pollution and soil pollution in different ways. In developing countries, significant quantities of solid waste are disposed by these ways which affect human health with increased incidence of diseases (Joshi and Ahmed 2016). Other solid waste disposal techniques like incineration, gasification, digestion are also not sustainable. So it is essential to use sustainable waste disposal method or some suitable alternative for waste recycling and its conversion into valuable products.

One of the most suitable biological solid waste recycling methods is vermicomposting. This is a bio-oxidation process used for waste stabilization exploiting mutual interactions between earthworms and microorganisms. Due to high nutrient content and biodegradable nature, organic fraction of solid wastes can be an ideal and suitable feedstock for the earthworms. Various authors have reported the use of different solid wastes including animal dung (Garg et al. 2006); food waste (Sharma and Garg 2017); municipal solid waste (Soobhany et al. 2017); weed (Hussain et al. 2016a, b, c); agricultural waste (Pigatin et al. 2016) and sewage sludge (He et al. 2016) as feedstock for earthworms. Different species of earthworms have been utilized for the vermicomposting depending on the climatic conditions and their ecological niche. *Eisenia fetida*; *Eisenia andrei*; *Eudrilus eugeniae* and *Perionyx excavates* are commonly employed worm species for vermicomposting (Sahariah et al. 2015). Vermicomposting can convert the organic fraction of non-toxic solid wastes into valuable vermicompost. Vermicompost is porous, homogenised, humified, nutrient-rich manure that is a good quality plant growth promoter. In recent years, farmers are attracted to vermicompost as a sustainable alternative to chemical fertilizer. The reasons for the adoption of vermicompost by farmers may be increased demand of organic food products and consumer awareness to food quality. Vermicompost as fertilizer maintains soil health and fertility, removes adverse effect of agro-chemicals and also economical to farmers (Lim et al. 2016). This present chapter focuses on the use of vermicomposting for organic waste management, process of vermicomposting and also presents an overview of different parameters used to evaluate vermicompost maturity and stability.

10.2 Solid Waste Generation

Solid wastes are unwanted organic and inorganic materials, which are not of immediate use, generated from various anthropogenic activities. Solid waste can be classified in several types based on the origin, toxicity and content (Fig. 10.1).

Solid waste generation at global level is expected to be approximately 27 billion tonnes per year by 2050. In Asia, China and India will be the major contributors to burgeoning solid waste generation (Modak et al. 2010). In India also, waste generation is likely to increase by 5% year^{-1} attributed to increase in population, industries and lifestyles. Per capita solid waste generation in India may be 0.7 kg/person/day by 2025 which is 4–6 times elevated as compared to 1999 (Kumar et al. 2017).

10.3 Composition of Organic Fraction of Solid Waste

Organic waste forms largest fraction (46%) of solid waste at global level (Hoornweg and Bhada-Tata 2012). In India, about 41% portion of solid waste is of organic nature and 19% is of recyclable materials. The organic waste mainly includes:

- food and vegetable waste
- yard waste
- human and animal excreta

Classification of solid waste

Based on origin -	**Based on contents-**	**Based on toxicity-**
• Industrial • Domestic • Construction • Commercial or institutional	• **Organic** • Glass • Metal • Plastic	• Toxic • Non-toxic • Infectious • Radioactive

Fig. 10.1 Classification of solid waste

- municipal waste (organic residue)
- leaf litter
- sewage sludge
- weeds
- agricultural and agro-industrial waste

It is evident from available literature that solid wastes are a promising resource (feedstock) for bioenergy and nutrients in agriculture.

10.4 Current Waste Management Systems

Solid waste management is one among the major issues that affects both developed and developing economies. Waste management practices used in a nation are largely dependent on its climatic conditions, socio-economic status and political issues. Other factors which influence waste management system include population growth, consumption pattern and technology (Lim et al. 2016). Currently, several methods/techniques are in use for the waste management. A comparative overview of various methods used for the organic solid waste management is encapsulated in Table 10.1.

Inappropriate and unscientific disposal of waste can lead to environmental, economic and health issues. So it is essential to use appropriate waste disposal method for waste recycling and its conversion into valuable products. Biological methods like composting and vermicomposting involve utilization of organic waste and convert it into valuable manure (Lim et al. 2016). Vermicomposting is preferred over composting due to certain good qualities in the final product such as higher nutrients, lesser heavy metals and pathogens.

10.5 Vermicomposting: A Green Technology for Organic Waste Management

Vermicomposting is one among the most suitable green technologies that have long been in use for the organic waste management. Vermicomposting has increasingly been used for recycling of different types of organic waste from urban, industrial and agricultural sites (Reinecke et al. 1992).

Vermicomposting is the process of transformation of organic waste into stable form through the mutual action of earthworms and microorganisms. Earthworms modify organic waste physico-chemically and enhance aeration and decomposition of organic substances. However, microorganisms present in the earthworm guts help in biochemical degradation of waste (Aira et al. 2007). This is a biological

Table 10.1 Various waste management methods prevalent worldwide

Treatment method	Type	Process	Advantage/disadvantages
Open dumping	Physical	It is indiscriminate and uncontrolled dumping of solid waste on land	No environmental and health concerns, uneconomical use of the available space, produces unpleasant odours. It is not a sustainable and eco-friendly practice. It is susceptible to open burning and also acts as disease vectors
Landfilling a Operated b Sanitary	Physical	Oldest and most prevalent form of solid waste disposal	Promotes greenhouse gas effect, require proper maintenance and continuous care. Leachate cause groundwater contamination
Pyrolysis	Thermo-chemical	Thermal degradation of waste in the absence of air to produce syngas, pyrolysis oil or solid (char, mainly ash and carbon) takes place between temperature 400–1000 °C	Reducing and avoiding corrosion and emissions by retaining alkali and heavy metals
Gasification	Thermo-chemical	This is the process of converting organic waste into CO, H_2 and CO_2 at higher temperatures (1000–1400 °C) in a controlled amount of oxygen	
Incineration	Thermo-chemical	Combustion of wastes under controlled conditions at 850 °C in an enclosed structure and at last waste is converted to carbon dioxide, water and non-combustible materials with solid residue state called incinerator bottom ash (IBA)	It is suitable for non-biodegradable waste with low moisture content and reduced waste volume of up to 90%. Also offers recovery of energy Ash produced from incineration contains hazardous substances. High capital, technical and operation costs. Less positive energy balance
Anaerobic digestion	Biological	Biological decomposition of organic materials in the absence of oxygen to produce a biogas and digestate	Biogas and slurry are produced during the treatment process
Composting	Biological	Aerobic degradation of organic wastes through microorganisms to produce an end product 'compost'	Risk of heavy metals in the compost, may have chances of pathogens, compost is coarser textured
Vermi composting	Biological	Bio-oxidation process for stabilization of organic waste using earthworms to produce 'vermicompost'	odourless, cost efficient, free of toxic waste and its resultant is a valuable end product

Source (Kumar et al. 2017; Lim and Wu 2016; Singh et al. 2011; Patidar et al. 2014)

process which occurs under mesophilic conditions further aided by biochemical action of microorganisms (Dominguez and Edwards 2010). Earthworms are most critical for the process as they do aeration and fragmentation of the substrate. This fragmentation increases surface area of the waste which provides more substrate to microorganisms for microbial activity and decomposition (Dominguez and Edwards 2011). Vermicompost is the end product of vermicomposting process which is a sustainable source of macronutrients and micronutrients, humic substances, hormones and enzymes (Ravindran and Jonathan 2016). Vermicompost is an efficient growth promoter for plants as it contains available nutrients. Vermicompost has low toxicity, pathogens and heavy metals and also protects plants against pests and diseases.

Earthworms are terrestrial invertebrates representing major animal biomass involved in various activities in soil such as ploughing, nutrient turnover decomposition, stabilization. At present, more than 4000 species of earthworms are known at global level (Edwards and Bohlen 1996). The classification of earthworms on the basis of their natural habitat and physiology is given in Table 10.2.

Table 10.2 Ecological categories of earthworms

Category	Biological characteristics	Habitat	Feeding habit	Examples
Anecic	– Large size – Dorsally pigmented – Low reproductive rate	Live in permanent burrows in deep soil layers and bury organic matter, forms vertical burrows	Phytogeophagous, excrete organo-mineral faeces	*Lumbricus terrestris*, *Lumbricus polyphemus* and *Aporrectodea longa*
Endogeic ○ Polyhumic ○ Mesohumic ○ Oligohumic	– Small to large in size – Medium life cycle – Moderate reproduction rate	Lives in top soil, made horizontal burrows	Geophagous, excrete organo-mineral faeces	*Aporrectodea caliginosa*, *Aporrectodea trapezoides*, *Aporrectodea rosea*, *Pontoscolex corethrurus*, *Allolobophora chlorotica*
Epigeic	– Small size – Short life cycle – High reproduction rate	Lives in superficial soil surface, litter dwellers	Phytophagous, excrete holorganic faecal pellets	*Eisenia fetida*, *Eudrilus eugeniae*, *Perionyx excavatus* *Lumbricus rubellus*, *Bimastus eiseni*, *Dendrobaena veneta*

Source (Dominguez and Edwards 2011; Edwards and Bohlen 1996; Lavelle et al. 1992; Edwards 1998)

Epigeic earthworms are extensively used for the vermicomposting of solid wastes. *E. eugeniae*, *E. fetida* and *P. excavates* have great potential as waste decomposers. Some of the desirable characteristics of earthworm species which make them suitable for decomposition are as follows:

- higher rate of waste consumption
- digestion and assimilation of organic matter
- tolerance to a wide range of environmental factors
- short life cycle and high reproductive rate
- endurance and resistance to handling

10.6 Process of Vermicomposting

Main biotic component of vermicomposting process is earthworms assisted by the microorganisms (bacteria, fungi, actinomycetes). Vermicomposting process involves different steps starting from waste collection to vermicompost production. First requirement for vermicomposting is suitable organic waste that should be non-toxic and biodegradable. Second step is pre-treatment as some of the raw materials can directly not be used as worm feed. They need to be mixed with some suitable organic materials before use in vermicomposting. Pre-treatment is followed by pre-composting to make the feed palatable to worms and remove foul odour. Period of pre-composting is variable from 7 to 30 days. Then, earthworms are introduced into the semi-composted waste mixture and left for the vermicomposting. During this period, earthworm processes organic waste and convert feed into vermicompost. Processing of organic waste via vermicomposting occurs in two major phases:

(a) *Gut-associated process or active phase or direct phase*—During the active phase, earthworms modify physical properties of waste by various metabolic activities like ingestion, digestion and assimilation (Lores et al. 2006).
(b) *Cast-associated process or maturation phase or indirect phase*—These changes takes place during earthworms movement toward fresh layers of organic waste. This process is mainly taking over by the microorganisms and involves vermicompost preparation. Finally, earthworms are harvested and vermicompost is stored for further use as organic manure.

10.6.1 Factors Affecting Vermicomposting

Vermicomposting depends on several abiotic factors including feedstock, bedding material, pH, temperature, moisture, aeration and also on biotic factors such as earthworm species used. Table 10.3 encapsulated optimum range/quality of the factors affecting vermicomposting process.

Feedstock—Success of vermicomposting process largely depends on physico-chemical properties of the waste used as feedstock for the earthworms. Several types of organic wastes have been used as feedstock for vermicomposting such as animal dung, sewage sludge, fruit and vegetable waste, paper waste, agro-waste. But for successful vermicomposting, suitable feedstock is crucial which would increase efficiency of worms (Yadav and Garg 2011).

Feeding rate—Various factors which affect feeding rate include organic content moisture, particle size of the feedstock, type of feed, pre-treatment and method of preparation (Wright 1972).

Bedding material—Material which provides suitable habitat to earthworm is bedding material such as animal dung, crop residues, wood chips, saw dust, paper waste. Suitable bedding material must have high absorbency and C/N ratio with excellent bulking potential.

pH—Generally, feeds having pH in the range of 5–9 are acceptable to worms (Edward 1998), but 7.5–8.0 pH is considered optimum.

Moisture—Moisture in the range of 60–80% is basic requirement for vermicomposting. Excessive moisture combined with poor aeration may render anaerobic conditions, and low moisture content can delay sexual development of earthworms.

Temperature—The optimum temperature range for the vermicomposting process is 15–28 °C. Higher temperatures may lead to the reduction of oxygen level and may result into mortality. Low temperatures reduce metabolic activities of worms which cease reproduction (Hand et al. 1988). However, tolerances and preferences range of temperature vary from species to species.

Stocking density—Ndgewa et al. (2000) reported that for vermicomposting, stocking density of 1.60 kg worms/m^2 is optimum. Furthermore, stocking density

Table 10.3 Optimum range of various process parameters for vermicomposting

Parameters	Quality/optimum range	Source
Bedding material	• Good bulking potential	Edwards and Bohlen (1996), Edwards (1998), Ndgewa et al. (2000), Ali et al. (2015)
	• High absorbency	
	• High C/N ratio	
Moisture	60–80%	
Temperature	10–35 °C	
pH	4.2–8.0	
Stocking density	1.60 kg worms/m^2	
Microbes	• Diverse type	
	• Present in earthworm gut	
	• Helps in biochemical degradation of waste	

of earthworms is influenced by feedstock quality and quantity, temperature, pH, moisture etc. (Lee 1985).

C/N ratio—C/N ratio of feedstock in the range of 30 is consider optimum for vermicomposting, but it is evident from the literature that organic wastes having higher C/N ratio are also used as feed by the worms (Edwards and Bohlen 1996; Pramanik and Chung 2011).

10.7 Types of Vermicomposting Process

Researchers have proposed several variants of the vermicomposting process to improve the efficiency of process starting from simple vermibin to large cement rings. These vermicomposting systems are divided into continuous flow system, batch mode system and wedge system. Vermicomposting process is divided into two categories on the basis of scale, i.e. domestic scale and commercial scale. A domestic system of vermicomposting requires a vermibin (container), bedding material (bedding material), worms and appropriate process conditions. On the basis of technology, vermicomposting process has been categorized into three types: low technology, medium technology and high technology systems. Windrow and batch systems are considered as low technology system required large land area and more labour. Any type of container is suitable for the batch vermicomposting. Medium and high technology vermicomposting consist of physically operated or fully automated continuous flow systems. These are economical due to low operating cost, and they can run for several years without any extra efforts (Edwards 1998).

10.8 Vermicomposting of Organic Substrates

For vermicomposting, organic waste can be utilized in two ways: (i) as feedstock for earthworms and (ii) as bulking agent. Animal excreta has been cited as most favourable organic substrate for vermicomposting and considered as nutrient-rich feedstock for earthworms (Garg et al. 2006). Several environmental problems are associated with the disposal of raw animal manure such as foul odour, pathogens, greenhouse gas emission, nitrogen losses and water pollution. Conversion of animal manure into vermicompost provides it fertilizer value and helps in reduction of environmental pollution. Vermicomposting potential of animal excreta is given in Table 10.4.

A number of solid wastes alone or after mixing with suitable organic amendment have successfully been converted into vermicompost. Benefits associated with the use of bulking agents in earthworm feed during vermicomposting are as follows:

Table 10.4 Animal excreta as organic substrate for vermicomposting

Sr. No.	Animal waste	Earthworm species	Refs.
1.	Cow dung	*Eisenia fetida*	Aira et al. (2011)
2.	Cow dung	*Perionyx ceylanensis*	Karmegam and Daniel (2009)
3.	Cow dung	*Eisenia fetida*	Contreras-Ramos et al. (2005)
4.	Cattle manure	*Eisenia Andrei*	Lazcano et al. (2008)
5.	Cattle manure	*Eisenia andrei*	Elvira et al. (1998)
6.	Goat manure	*Eisenia fetida*	Loh et al. (2005)
7.	Rabbit manure	*Eisenia fetida*	Molina et al. (2013)
8.	Rabbit manure	*Eisenia fetida*	Gomez Brandon et al. (2013)
9.	Pig manure	*Eisenia fetida*	Aira et al. (2007)
10.	Cow manure and poultry droppings	*Eisenia fetida*	Lv et al. (2016)
11.	Cattle solids, pig solids, horse solids and turkey waste	*Perionyx excavatus*	Edwards et al. (1998)
12.	Cow dung and biogas plant slurry	*Eisenia fetida*	Yadav et al. (2013)
13.	Cow, sheep, pig and chicken Wastes	*Eudrilus Eugeniae*	Coulibaly and Zoro Bi (2010)
14.	Cow manure and poultry droppings	*Metaphire posthuma*	Bisht et al. (2007)
15.	Cow, buffalo, horse, donkey, sheep, goat and camel wastes	*Eisenia fetida*	Garg et al. (2006)
16.	Pig dung, poultry dung, rabbit dung, cattle dung, sheep dung	*Eisenia fetida*	Vodounnou et al. (2016)
17.	Lion, hippopotamus, elephant, horse, rhino	*Eisenia fetida*	Perez Godinez et al. (2017)

- makes the waste more palatable to earthworms
- reduces the negative effects of toxic compounds present in waste such as sewage sludge
- provides acceptable chemical composition to the waste and hence enhances the efficiency of the process
- reduces the moisture content of waste
- acts as microbial inoculums
- initial C/N ratio of the feed stock is improved.

Mixing of cow dung with industrial wastes enhances vermicomposting process by providing nutrients and microorganisms (Mupambwa et al. 2016). Some of the toxic and complex wastes such as sewage sludge are mixed with some organic waste to reduce the negative effects of various contaminants (Garg et al. 2012).

Some organic wastes like industrial sludges have higher moisture content so need some bulking material to reduce the moisture content. Due to higher carbon content, lignocellulosic biomass usually mixed with such waste having low carbon content. Waste with low C/N ratio, e.g. tannery wastes is mixed with leaf litter (Castillo et al. 2013). Various studies have inferred that physico-chemical properties of feedstock significantly affect the fecundity of the earthworms (Xie et al. 2016).

Although animal excreta is commonly used organic amendment, but vegetable waste, fruit waste, crop residues (straw, husk) and paper waste have also been used as organic amendments as well as feedstock in vermicomposting process (Lim et al. 2012). Recently, use of matured compost and vermicompost (Huang et al. 2014) as organic amendment has been suggested because compost provides initial habitat for earthworms and also act as microbial inoculums (Castillo et al. 2013). Animal manure-based vermicomposting is easy, but vermicomposting of phytomass containing feedstock is complicated as it require pre-composting, supplementation and has low vermicomposting efficiency. To overcome this, high-rate vermicomposting technique to use phytomass as feedstock without any pre-composting or supplementation has been developed (Abbasi et al. 2015). In high-rate vermicomposting, the rate of feedstock decomposition is 3–4 times faster than usual.

Management of variety of organic waste using composting and vermicomposting offers recycling of nutrients and can be beneficial to agriculture. Table 10.5 gives an account of vermicomposting studies carried out for the stabilization of diverse organic wastes.

10.9 Modifications in the Organic Waste After Vermicomposting

To assess maturity and stability of vermicompost, any single parameter is not sufficient; so it is better to evaluate it on the basis of different parameters. These parameters may be physical, chemical or biological (Bernal et al. 2009). In addition to these, vermicompost maturity can also be evaluated using Fourier transform infrared spectroscopy (FTIR), gas chromatography–mass spectrometry (GC-MS), thermogravimetric analysis (TGA), differential scanning calorimetry analysis (DSC) and scanning electron microscope (SEM). These parameters are useful to compare the quality of vermicompost prepared from different feedstocks and also measure organic matter degradation. Figure 10.2 presents various properties of vermicompost studied by different authors to evaluate its maturity and quality. Modifications in waste characteristics after vermicomposting of different types of organic waste have been summarised in Table 10.6.

Table 10.5 Organic waste management using vermicomposting

Sr. No.	Organic Waste	Earthworm species	Bulking material	Worm stocking density	Duration	Pre-composting	Conclusion	Reference
1	Household wastes, biological and chemical sludge	*Eisenia fetida*	Rice straw	400 adult worm per reactor	70 days	20 days	• C/N ratio from vermicompost household solid waste, biological and chemical sludge was 16.5, 14.5, and 15, respectively • Nitrogen and phosphate increased, however, TOC and heavy metal concentrations decreased with time • Earthworm tissues biologically accumulate heavy metals	Amouei et al. (2017)
2	Municipal solid waste	*Eudrilus eugeniae*	Cow dung	1.60 kg worms per m^2	70 days	14 days	• C/N ratios of the vermicompost were in the range of 15.0–16.2 • FTIR revealed greater reduction in readily degradable materials in vermicomposts as compared to composts • Residual mass after thermal degradation was slightly higher for vermicomposts than composts as indicated from TG. DSC curves also revealed intense degradation of complex and larger compounds to simpler compounds after vermicomposting • SEM images of vermicompost reflected a flakier and disintegrated matrix than compost	Soobhany et al. (2017)

(continued)

Table 10.5 (continued)

Sr. No.	Organic Waste	Earthworm species	Bulking material	Worm stocking density	Duration	Pre-composting	Conclusion	Reference
3	Food and vegetable processing waste	*Eisenia fetida*	Buffalo manure	20 worm per kg	90 days	21 days	• Three different combinations vegetable waste and buffalo waste were prepared • Highest growth and reproduction of earthworm was achieved in 100% buffalo waste • Vermicomposting enhanced nitrogen, total available phosphate and total potassium content • Buffalo waste could be suitable alternative to cattle dung	Sharma and Garg (2017)
4	Horse manure, apple pomace, grape pomace and digestate (50% manure slurry + 40% corn silage + 10% haylage)	*Eisenia andrei*	–	200 worm per reactor	240 days		• Study evaluated vermicompost characteristics based on 120 days old layer and 240 days old layer in vermicreactor • Maximum biomass of earthworms was in 120 d-old-layer. After 240 days, microbial biomass activity decreased due to decrease in the earthworm activity indicating a high degree of stabilization. Enzyme activities differ according to the age of layers and type of waste • Germination index increases after vermicomposting and higher with apple pomace and digestate as compared to horse manure and grape pomace	Sanchez et al. (2017)

(continued)

Table 10.5 (continued)

Sr. No.	Organic Waste	Earthworm species	Bulking material	Worm stocking density	Duration	Pre-composting	Conclusion	Reference
5	Cow manure and wheat residues	*Eisenia fetida*	–	20 adult worm per kg	60 days	–	• Urease activity is suitable indicator of vermicompost maturity and waste stabilization during the process of vermicomposting • Urease activity was highly correlated with the time of vermicomposting resulted into $r = 0.97$for cattle manure and $r = 0.99$ for wheat waste • Urease activity showed significant correlations with the C:N ratio	Sudkolai and Nourbakhsh (2017)
6	Pig dung, poultry dung, rabbit dung, cattle dung, sheep dung and vegetal compost	*Eisenia fetida*	Wheat straw		90 days	14 days	• Highest worm production and growth rate was obtained with cow dung followed by pig dung, however, earthworm growth decreased in vegetable compost • Maximum earthworm growth rate was found on 90th day • Growth and worm production depend on the biochemical quality of the feedstocks	Vodounnou et al. (2016)
7	Sawdust, boxwood leaves and cardboard	*Eisenia fetida*	Compost (MSW)	100 g worms per kg	100 days	–	• MSW and carbonaceous materials in different proportions, viz. 50:50, 70:30, 85:15 and 100:0 were vermicomposted for 100 days • 75-days vermicomposting is sufficient for vermicompost maturity • Phosphorous, nitrogen and pH were higher in vermicompost	Alidadi et al. (2016)

(continued)

Table 10.5 (continued)

Sr. No.	Organic Waste	Earthworm species	Bulking material	Worm stocking density	Duration	Pre-composting	Conclusion	Reference
8	Orange peel and filter cake	*Eisenia fetida*	Cattle manure	500 worms box	135 days	6 week	• Vermicomposts have higher macronutrients and micronutrients • Filter cake and orange peel + cattle manure were better as compared to vermicomposts prepared using cattle manure only • 2:1 orange peel + cattle manure had the highest N content at the end of the vermicomposting process • A positive correlation exhibited between Mg and K and P and Ca	Pigatin et al. (2016)
9	*Salvinia natans*	*Eisenia fetida*	Cattle manure and saw dust	100 worms per 2.5 kg	45 days	–	• Total concentration of heavy metals (Zn, Cu, Mn, Fe, Cr, Pb, Cd, Ni) increased, however, water soluble and plant available heavy metals reduced in the final vermicompost • Toxicity characteristics leaching procedure (TCLP) tests confirmed suitability of vermicompost for agriculture	Singh and Kalamdhad (2016)
10	Cashew leaf litter	*Perionyx excavates*	Cow dung, Horse dung, Sheep dung	34–36 worms per kg	60 days	1 week	• Cashew leaf litter mixed with cow dung at 2:2 ratio was found to best in terms of vermicompost properties • Vermicompost produced had lower pH, organic carbon, C:N ratio, C:P ratio, lignin, cellulose, hemicellulose and phenol content, however, higher NPK,	Parthasarathi et al. (2016)

(continued)

Table 10.5 (continued)

Sr. No.	Organic Waste	Earthworm species	Bulking material	Worm stocking density	Duration	Pre-composting	Conclusion	Reference
							dehydrogenase and humic acid content as compared to waste and compost • Reduction in the lignocellusic and phenol content is due to the combined action of gut lignocellulolytic microflora and earthworm duting vermicomposting process	
11	Municipal solid waste	*Metaphire posthuma and Eisenia fetida*	Cow dung	10 worms per kg substrate	60 days	–	• Comparison of vermicomposting efficiency of *Metaphire posthuma* and *Eisenia fetida* using MSW and cow dung as feedstock • *M. posthuma* can be utilized as a successful candidate for vermiconversion of toxic wastes • *Vermicomposting* resulted in higher bioavailability of N, P, K and Fe	Sahariah et al. (2015)
12	Sludge, tannery waste, saw dust	*Eisenia fetida*	Cattle manure	500 worms per pot	135 days	1 week	• Vermicomposting process affects humic acid biostimulant properties	Scagalia et al. (2016)
13	Submerged and solid state fermented tannery waste.	*Eudrilus eugeniae*	Cow dung, leaf litter	50 worms per kg	25		• In vermicompost, phytohormones were maximum • C/N ratio of vermicomposts was in the range of 17.3–10.3. • Microbial count was very high	Ravindran and Jonathan (2016)

(continued)

Table 10.5 (continued)

Sr. No.	Organic Waste	Earthworm species	Bulking material	Worm stocking density	Duration	Pre-composting	Conclusion	Reference
14	Municipal green waste	*Eisenia fetida*	Cattle manure	52-kg live weight m^{-3}	18 week	1 week	• Organic matter decomposition was higher during composting as compared to vermicomposting • Vermicompost had a lower bulk density and greater total porosity and higher microbial biomass as compared to composts • Basal respiration and metabolic quotient were also higher for vermicompost	Haynes and (Haynes and Zhou 2016)
15	Fly ash	*Eisenia fetida*	Cow dung, paper waste	25 g worms per kg	10 week	1 week	• Study experimented on the effect of inoculation of effective microorganisms with earthworms during vermicomposting • EM + *E. fetida* treatment resulted higher weekly Olsen P release as compared to *E. fetida* and EM alone and control	Mupambwa et al. (2016)
16	Ipomoea	*Eisenia fetida*	–	50 worms per kg	30 days	–	• Total carbon contents reduced from 527.3 to 282.8 g/kg; total nitrogen contents increased from 20.2 to 28.5 g/kg • C/N ratio of ipomoea vermicompost was 9.9 • Spectroscopic analysis revealed transformation of weed into potent organic fertilizer	Hussain et al. (2016a)

(continued)

Table 10.5 (continued)

Sr. No.	Organic Waste	Earthworm species	Bulking material	Worm stocking density	Duration	Pre-composting	Conclusion	Reference
17	Sewage sludge	*Eisenia fetida*	Wheat straw + biochar	0.28 kg worms m^2	25 weeks	14 days	• Highest production of cocoons was observed after 4 weeks vermicomposting • Trace elements, ph and C/N ratio decreased after vermicomposting	Malinska et al. (2016)
18	Coconut husk	*Eudrilus eugeniae*	Poultry manure, pig slurry	500 worms	120 days	21 days	• Highest relative N and K recovery were observed for 20% feedstock substitution by pig slurry • Highest P recovery was recorded with poultry manure substitution • In the vermicompost, pH, microbial biomass carbon, macronutrients and micro nutrients were higher as compared to initial waste with different maturity indices	Swarnam et al. (2016)
19	Cow dung	*Eisenia fetida*	Poultry manure	25 worms per kg		90 days	• After vermicomposting, pH, total organic carbon and C/N ratio were reduced however EC and humic acid increased • Heavy metals stabilized	Lv et al. (2016)
20	Decanter cake	*Eudrilus eugeniae*	Rice straw	10 worms per 200 gm	4 weeks	2 weeks	• Four treatments with different ratio of decanter cake and rice straw (2:1, 1:1, 1:2, 1:3) were prepared • Two parts decanter cake and one part rice straw (w/w) found to be best among all the treatments	Lim and Wu (2016)

(continued)

Table 10.5 (continued)

Sr. No.	Organic Waste	Earthworm species	Bulking material	Worm stocking density	Duration	Pre-composting	Conclusion	Reference
21	Fresh fruit and vegetable wastes	*Eisenia fetida*	Vermicompost (cow dung + vegetable waste) and soil		60 days	48 h	• In the vermireactor, optimal loading was 30 g fresh fruit and vegetable waste/day • Banana peels found to be harmful for the survival of *Eisenia fetida* • Molecular biological approaches revealed that earthworms could broaden bacterial diversity in their products, with significant greater populations of actinobacteria and ammonia oxidizing bacteria. • Vermicomposting efficiency differs with the types and loadings of waste	Huang et al. (2016)
22	Waste paper	*Eisenia fetida*	Chicken manure	1.6 kg worms per m^2	7 weeks	20 days	• Main objective was to determine the optimum C/N ratio • Ideal C/N ratio for the waste mixture was 40 • Total N, P, and K concentrations increased however total carbon, C/N ratio, electrical conductivity (EC), and heavy metal content gradually decreased with time	Ravindran and Mnkeni (2016)
23	Sewage sludge	*Eisenia fetida and Eisenia andrei*	Soil	40 worms per bin	9 weeks	1.0 month	• Cd, Cu, Ni, and Pb found to be reduced whilst Zn concentration increased • *E. andrei* has higher capabilities to accumulate some metals • During the first six weeks of vermicomposting, riboflavin content decreased to some extent and after that restored till the end of the 9-week experiment	Rorat et al. (2016)

(continued)

Table 10.5 (continued)

Sr. No.	Organic Waste	Earthworm species	Bulking material	Worm stocking density	Duration	Pre-composting	Conclusion	Reference
24	Digestate, kitchen waste, sewage sludge	*Eisenia andrei*	Wheat straw, paper waste, garden biowaste	300 g worms tray	5 month	2 weeks	• Study compared the effects of the composting and the vermicomposting on the distribution of particles into three size fractions, i.e. more than 12 mm, between 12–5 mm and less than 5 mm • Vermicomposting was able to achieve the finer and more homogeneous final product compared to classical composting • Total NPK, extractable Phosphorus, N–NH_4^+ and N–NO_3 were higher in vermicompost compared to compost	Hanc and Dreslova (2016)
25	Sludge	*Eisenia fetida*	Cow dung, Swine manure	20 worms per 0.2 kg		60 days	• Sludge amended with 40% swine manure proved a great medium for the growth of *E. fetida* • Sludge amended with 40% cow dung can be a suitable medium for the fecundity of *E. fetida* • Finally, lower pH value, total organic carbon, ammonium nitrogen and C/N ratio were reported. • Total available phosphorous content was higher indicating optimal stability and maturity	Xie et al. (2016)

(continued)

Table 10.5 (continued)

Sr. No.	Organic Waste	Earthworm species	Bulking material	Worm stocking density	Duration	Pre-composting	Conclusion	Reference
26	Sewage sludge (fresh and composted)	*Eisenia andrei*	Wood chips	120 worms per 2.0 kg		112 days	• Highest number of mature earthworms, cocoons and hatchlings were obtained in vermicomposting treatment • Hydrolytic enzymes activities and microbial biomass (fungal and bacterial) decreased throughout the vermicomposting process • Combined composting-vermicomposing was most appropriate	Villar et al. (2016)
27	Lantana	*Eisenia fetida*	–	–	–	–	• C/N ratio reduced from 22.7 to 8.1 • Humification index from 8.38 to 2.03 • FTIR spectra revealed complete degradation of phenols and sesquiterpene lactones and formation of simple compounds • GC–MS revealed transformation of 24–86 constituents	Hussain et al. (2016b)
28	Salvinia	*Eisenia fetida*	–	–	–	–	• chemicals responsible for the allelopathy of salvinia are destroyed • Scanning electron microscopy shows marked disaggregation of the material in the vermicompost as compared to well-formed matrix of salvinia leaves • The C/N ratio of salvinia was reduced sharply (from 53.9 to 9.35)	Hussain et al. (2016c)

(continued)

Table 10.5 (continued)

Sr. No.	Organic Waste	Earthworm species	Bulking material	Worm stocking density	Duration	Pre-composting	Conclusion	Reference
29	Tomato plant debris + paper mill sludge	*Eisenia fetida*	–	1000 worms per 18 kg	6 month		• Characterize humic acids isolated from different waste mixtures before and after vermicomposting • Humic acid content increased by 15.9–16.2% • Vermicompost produced from tomato debris/paper mill sludge (2:1) recorded higher C content and C:N ratio • Humic acid from tomato debris/paper mill sludge (1:1) vermicompost showed a higher O content and O: C ratio	Gomez et al. (2015)
30	Filter cake	*Eisenia fetida*	Cattle manure	50 worms per kg	120 days	30 days	• Positive correlation of phosphatase activities with TOC, pH and WSP, however, negatively correlated with HA content • Nanopore volume found to be negatively correlated with phosphatase activities for filter cake but not for cattle manure • HA content of filter cake vermicompost was higher as compared to cattle manure vermicompost	Busato et al. (2016)
31	*Eichornia crasipes*	*Eudrilus eugeniae*	Cow dung	100 worms	40 days	30 days	• FTIR spectra reveal the presence of humic substance in compost and vermicompost • GC–MS analysis shows maximum level of Benzene propanoic acid (95.98%) and by 2-Propanone, 1-Phenyl-, OXIM (10.10%) from vermicompost	Kumar et al. (2015)

(continued)

Table 10.5 (continued)

Sr. No.	Organic Waste	Earthworm species	Bulking material	Worm stocking density	Duration	Pre-composting	Conclusion	Reference
32	Pelletized dewatered sludge (PDS)	*Bimastus parvus*	No	200 worms per reactor	60 days		• Fresh PDS with two pellet size 4.5 and 14.5 mm • Organic matter decreased by 31.7–39.9%. • 4.5 mm pelletized sludge could be more suitable for vermicomposting	Fu et al. (2015)
33	Agave bagasse	*Eisenia fetida*	Sewage sludge		45 days with fungi	45 days	• Degradation of lignin in the different treatments ranged between 4 and 63% • Degradation time of bagasse was lower (3 months) for vermicomposting as compared to the traditional composting process (6–8 months)	Moran-Salazar et al. (2016)
34	Disposable paper cups	*Eisenia fetida*	Cow dung slurry	0.15–0.2 g worms per tub	19 week	20 days	• Ratio of 1:1 (paper cups: cow dung) found to be the best on the basis of the physico-chemical parameters and the time taken for degradation • FTIR analysis revealed higher degradation of aliphatic, aromatic, carboxylic, and phenolic groups present in the paper cup wastes	Arumugam et al. (2015)
35	*Parthenium hysterophorus* Weed		Manure (cow, goat, poultry, swine) and Farm waste		60 days	20 days	• Four different treatments of parthenium mixed with farm wastes and animal manures • Use of organic supplements for the degradation of *Parthenium* helps in the reduction of weed biomass and its conversion to organic manure	Yadav (2015)

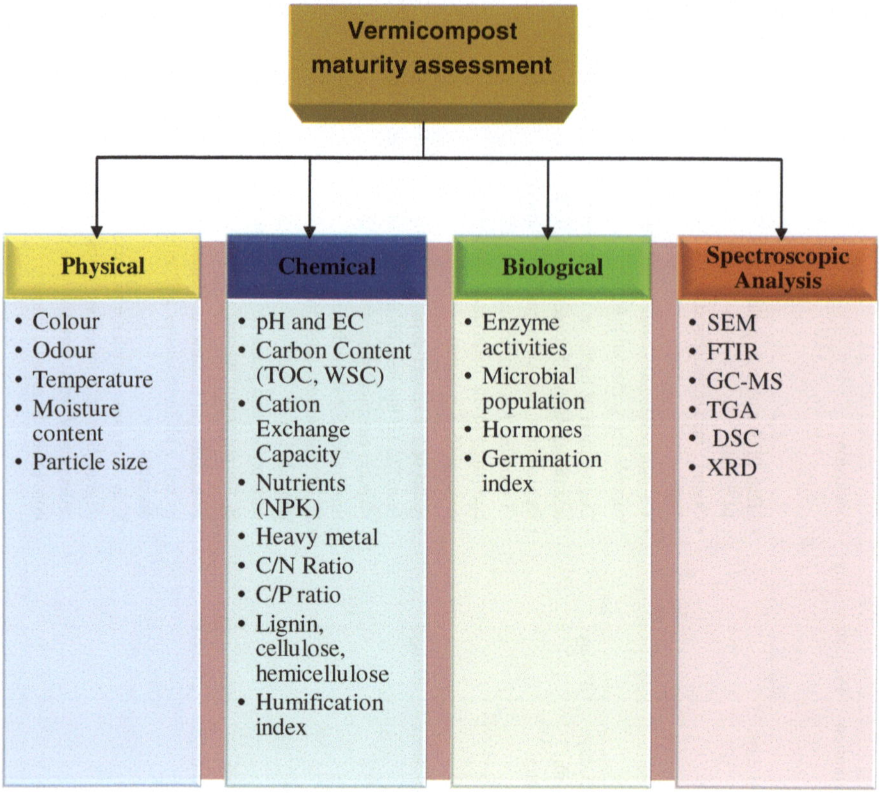

Fig. 10.2 Various parameters studied for vermicompost maturity and quality

10.9.1 pH and EC

pH and EC of organic matter play an important role in vermicomposting efficiency and also affect survival and growth of earthworms. Vermicomposting of different organic wastes produces different intermediate species, leading to pH shift either acidic or alkaline approaching neutrality. Lv et al. (2016) reported lowering in pH during vermicomposting of cattle dung and pig manure. Many other studies have also reported that pH of vermicompost is lesser as compared to feedstocks (Bhat et al. 2015; Parthasarathi et al. 2016). Contrary to this, Lim and Wu (2016) reported that pH (8.37–8.75) of the vermicompost was more feedstocks (decancter cake and rice straw). Singh and Kalamdhad (2016) have also reported an increase in the pH after vermicomposting of *Salvinia natans weed*. Regulatory factors of pH of vermicompost are mineralization of nitrogen and phosphorus, formation of ammonium ions and humic acids (Pramanik et al. 2007). Lowering of pH of feedstocks after

Table 10.6 Comparison of physico-chemical characteristics of organic waste before and after vermicomposting

Organic waste		Physico-chemical parameters							Refs.
		pH	EC (dS/m)	TOC (g/kg)	TKN (g/kg)	TK (g/kg)	TAP	C/N	
Citronella waste + cow dung	Initial	6.8	0.3	190	4.4	184	408	43	Deka et al. (2011)
	Final	6.21	0.61	134	18.2	736	666	7.4	
Cow manure + mushroom residues	Initial	8.42	1.31	369.7	10.53	12.41	18.2	35.36	Song et al. (2014)
	Final	7.57	2.98	257.6	23.03	15.43	15.51	11.32	
Pig manure + mushroom residues	Initial	8.66	1.49	388.5	11.54	9.91	10.21	33.83	
	Final	7.35	3.26	273.6	26.21	15.33	19.79	10.43	
Tomato plant debris + paper mill sludge (2:1)	Initial	7.8	5.3	283	17.6	16.1	2.8	24.9	Gomez et al. (2015)
	Final	9.3	2.6	179	14.2	12.6	3.7	13.1	
Tomato plant debris + paper mill sludge (1:1)	Initial	8.6	4	232	19.4	11.9	2.3	18.2	
	Final	8.9	2.2	163	15.3	10.6	3	9.6	
Press mud + cattle dung	Initial	6.70–8.35	3.23–4.13	421.0–462.8	13.4–19.7	–	–	21.92–34.53	Bhat et al. (2015)
	Final	6.91–7.18	4.5–5.83	323.2–389.4	20–24.8	–	–	14.18–19.37	
Sludge, swine manure, cow dung	Initial	7.48–8.67		411–448	2.98–3.09		2.88–5.68	14.5–18.8	Xie et al. (2016)
	Final	5.80–9.14		320–373	2.33–2.99		3.79–8.50	10.7–14.5	

(continued)

Table 10.6 (continued)

Organic waste		Physico-chemical parameters							Refs.
		pH	EC (dS/m)	TOC (g/kg)	TKN (g/kg)	TK (g/kg)	TAP	C/N	
Decanter cake and rice straw	Initial	6.13–6.97	2160–2590 (µS/cm)	–	–	12.64–15.69	1.37–191	20.54–34.60	Lim and Wu (2016)
	Final	8.37–8.75	1511–2215 (µS/cm)	–	–	22.45–3015	3.27–3.91	8.76–1178	
Fruit and vegetable waste	Initial	7.38–7.56	2.40–2.95	399.1–482.5	7.82–10.46	7.77–9.01	4.80–7.67	46.13–51.09	Sharma and Garg (2017)
	Final	6.55–7.29	3.38–3.94	235.4–340.3	15.46–20.43	12.04–12.75	7.74–11.74	11.35–22.02	
Wheat residue	Initial	7.5	1.56	372.7	6.27	–	–	59.5	Sudkolai and Nourbakhsh (2017)
	Final	7.2	1.99	323.7	20.1	–	–	16.2	
Cattle manure	Initial	7.9	2.9	177.4	11.3	–	–	15.5	
	Final	8.5	3.02	122.2	14.1	–	–	8.7	

vermicomposting is attributed to the production of carboxylic and phenolic groups in humic acids; however, production of ammonium ions causes increase in the pH. Electrical conductivity (EC) is measured to evaluate soluble salts in an organic amendment and used as an indicator to evaluate its applicability for agricultural purposes. Generally, EC is higher than feedstocks due to release of soluble salts (phosphate, ammonium and nitrate) and availability of mineral salts by organic matter decomposition (Dominguez and Edwards 2010). Some studies have reported reduction of EC after vermicomposting as compared to the initial feedstocks (Pramanik et al. 2007). The decrease may be due to the leaching of soluble salts, decomposition of organic acids and higher pH.

10.9.2 Organic Carbon

Total organic carbon (TOC) of feedstock is reduced during the vermicomposting process. Reduction in TOC is mainly attributed to carbon loss from initial feedstock through microbial respiration in the form of carbon dioxide. Earthworms' metabolic activities like excretion (ammonia and urea) and secretion (enzymes and mucus) promote microbial communities which further fasten the organic matter decomposition and production of carbon dioxide.

During vermicomposting, about half of the organic compounds present in the feedstock are converted to CO_2 and rest of the organic compounds are converted to more stable compounds. Total organic carbon reduction after vermicomposting was found to be higher in vermicompost as compared to compost (Song et al. 2014). Various studies have evaluated changes in TOC after vermicomposting which are given in Table 10.6.

10.9.3 Nutrient Content (NPK)

Vermicomposting improves nutrient status of the solid wastes that depends on several factors such as constituents of feedstocks, initial nitrogen content, C/N ratio (Garg et al. 2006). In the vermicompost prepared from coconut husk mixed with animal excreta, nitrogen content was 300% more. Significant increment of 841% was recorded for phosphorus content with pig slurry substitution (Swarnam et al. 2016). Parthasarathi et al. (2016) also reported enhanced levels of NPK in the vermicompost obtained from organic waste.

Hanc and Pliva (2013) have reported that when pre-composted kitchen waste amended with woodchips and paper waste subjected to vermicomposting resulted into increased N (15%), P (20%), K (18%), Ca (18%) and Mg (31%). Lim and Wu (2016) also reported higher nutrients content in the vermicompost and control as compared to feedstocks. Increase in nutrient content during vermicomposting has largely been attributed to loss of weight and organic matter decomposition. Nitrogen increment may also be due to loss of dry matter as carbon dioxide and nitrogen addition by earthworm mucus (Karmegam and Daniel 2009). Plant nutrients, present in feedstock, whilst passing through worm gut are converted into available from (Garg et al. 2012).

10.9.4 C/N Ratio

The C/N ratio is an important factor employed to characterize the quality and maturity of the vermicompost. C/N ratio depends on qualitative and quantitative nature of organic waste used as feedstock. Composting of organic waste by earthworms usually decreases C/N ratio with time due to losses of carbon as carbon dioxide that stabilizes at 15–20 (Golueke 1981). C/N ratio of vermicompost prepared from different organic wastes is given in Table 10.6.

10.9.5 Humification Ratio

Vermicomposts prepared from organic wastes have higher humic acid content as compared to the parent materials (Soobhany et al. 2015). Various activities of earthworms (feeding, burrowing and digestion) accelerate humification process and increase the humic acid content (Song et al. 2012, 2014). Song et al. (2014) reported that humic acid fraction carbon increased from 15.33 to 56.23 mg/g at the end of vermicomposting. Humification rate and index also increase in similar fashion. However, fulvic acid fraction carbon changed slightly from 33 to 34 mg/g. Fornes et al. (2012) found that humic acid content gradually increased but fulvic acid decreased during vermicomposting. Parthasarathi et al. (2016) also reported higher humic acid content in the vermicompost prepared from cashew leaf litter and animal excreta. Increase in humic acid contents after vermicomposting may be due to the microbial action and various gut-associated processes of earthworm.

Busato et al. (2016) reported linear increase in humic acid content after 120 days of vermicomposting. Humic acid of filter cake was 29.7% higher as compared to cattle manure which was increased by 116.8%. Gomez et al. (2015) found that after vermicomposting of organic waste (debris + sludge), humic acids increased by 15–16%. Further, the humic acid isolated from vermicompost has low C, H and N content and higher S, O, C:N, C:H and O:C ratios as compared to waste humic acid.

Humification index achieved during vermicomposting reflects transformation of initial waste into a highly humified and stabilized product. Most of the studies concluded that humification index value less than 5 confirms humification of waste and vermicompost maturity (Zbytniewski and Buszewski 2005). Hussain et al. (2016a, b) reported reduction in the humification index of ipomoea weed from 7.74 to 1.90 and lantana weed from 8.38 to 2.03 after vermicomposting.

10.9.6 Heavy Metals

Fate, translocation and compartmentization of heavy metals, during the vermicomposting process of various organic wastes, have been studied by several authors. Generally, total heavy metal quantity increases after vermicomposting due to reduction of organic matter and decrease in weight and volume of the feedstock (Sharma and Garg 2017; Malinska et al. 2016).

Singh and Kalamdhad (2016) during vermicomposting of *Salvinia* weed found that total concentration of heavy metals increased after vermicomposting. Sharma and Garg (2017) also reported that heavy metal content was higher in the vermicompost prepared from vegetable processing waste.

Some studies have found lesser heavy metal content in vermicompost than parent material that may be attributed to their bio-accumulation in earworm tissues (Soobhany et al. 2015). He et al. (2016) reported that Cu, Ni and Zn content of feedstock was reduced by 81.9, 60.2 and 38.5%, respectively, after vermicomposting. Earthworms have been found to accumulate heavy metals in their choragogen cells which result in reduction in heavy metal content present in the waste (Huges 1980). Vermicomposting process mainly reduces the bioavailable fraction of heavy metals (Singh and Kalamdhad 2013). Earthworms' activity and chemical nature of bulking agent play a synergistic role in sequestration of heavy metals during vermicomposting (He et al. 2016). Lv et al. (2016) studied the speciation and mobility of heavy metals during vermicomposting and inferred that total heavy metals content was higher in vermicompost than substrate (animal manure). However, mobile fraction of heavy metal reduced because vermicomposting retarded the migration and availability of heavy metals.

10.9.7 Spectroscopic Analysis

Spectroscopic properties reveal changes in the texture and structure of organic waste after vermicomposting. Most commonly used techniques are SEM, FTIR, TGA, GC-MS, UV-vis spectroscopy etc. (Lim and Wu 2016; Bhat et al. 2017). A SEM image provides vital information on surface morphology of a substrate and offers comparative study before and after vermicomposting. Most of the studies have revealed that vermicompost is more fragmented, porous with smaller particle size as compared to compost and initial feedstock (Hussain et al. 2016a, b; Kumar et al. 2014).

Kumar et al. (2014) used flower waste as feedstock for vermicomposting and compared the surface morphology, functional groups and chromatographic properties of compost, vermicompost and feedstock. Surface area of vermicompost exhibited single particles packed together to form aggregates and numerous surface irregularities which confirm compost maturity. Furthermore, protein and lignin matrix appeared disaggregated in vermicompost due to earthworms and microbes activities. SEM micrograph showed two distinctive morphologies between initial feedstock and vermicompost of rice straw amended palm oil mill effluent. Initial waste mixture contained long fibres whilst the vermicompost was fragmented and porous which revealed its suitability as an organic fertilizer (Lim et al. 2012).

Lim and Wu (2015) linked BET surface area with the vermicompost maturity and found that surface area of the vermicompost (1.2303 m^2/g) was higher as compared to initial feedstock (0.4882 m2/g) and control (1.0728 m^2/g). Rationale behind this is the earthworm activities during vermicomposting which increase the surface area of the organic wastes resulting into waste stabilization. Hussain et al. (2016a) subjected ipomoea weed to vermicomposting and evaluated vermicompost maturity on the basis of spectroscopic properties. SEM images of vermicompost reflected disaggregated materials which indicate progressive degradation of the weed. After vermicomposting, FTIR spectra revealed reduction in the alcoholic and phenolic content which indicates removal of allelopathic effect of weed. Disappearance of the peak at 1738 cm^{-1} and shortening of the peak at 1652 cm^{-1} after vermicomposting of weed revealed degradation of lignocellulose and lignin content. TGA analysis revealed that maximum mass loss in the vermicompost (53.6%) was lower than the maximum mass loss in the ipomoea (82.9%). GC-MS of vermicompost verified lower number of fatty acids, alcohols, alkanes, alkenes and nitrogenous compounds as compared to feedstock. This indicates conversion of complex compounds of ipomoea to simpler compounds in the vermicompost (Hussain et al. 2016c).

Soobhany et al. (2017) subjected municipal solid waste to composting and vermicomposting and characterized respective compost and vermicompost using spectroscopic, thermogravimetric and structural analysis. SEM images of compost exhibited packed fluffy and floc-like structure whilst vermicompost was more scattered and less fragmented. This reflected strong fragmentation and intra-structural degradation of municipal solid waste after vermicomposting. TGA of vermicompost showed lesser mass loss when compared to compost indicating its stability. FTIR spectra revealed flattening of bands between 2918–2639 cm^{-1} and missing peak between 1500 and 1600 cm^{-1} which may be due to decomposition of methylene groups and degradation of lignocellulose, respectively, in both compost and vermicompost (Hussain et al. 2016c).

10.10 New Perspectives in Vermicomposting Field

Significant research work has been undertaken in the field of vermicomposting especially for the waste management and sustainable agriculture. In recent years, few studies have reported the use of vermicompost in pollution control.

Vermifiltration technique involving earthworms and microorganisms has also been developed for wastewater treatment and a few such plants are in operation too (Sinha et al. 2008). Li et al. (2009) used vermifiltration to process wastewater using vermicompost prepared from swine manure. Authors have also worked to develop different form of vermifiltrations. Lourenc and Nunes (2017) carried out a study to compare single stage vermifiltration with four-stage vermifiltration. Results revealed that four-stage sequential vermifiltration significantly reduced biological oxygen demand, chemical oxygen demand, total suspended solid and ammonia (99.1%).

Vermicomposting can also be employed as cost-effective bioremediation technique for metal and dye removal from the waste. Vermicompost has been used as an adsorbent due to its greater capacity for cation exchange and larger surface area (Zhu et al. 2017; Dey et al. 2017). Zhu et al. (2017) investigated ability of cattle manure and cattle manure vermicompost to remove Pb^{2+} and Cd^{2+} from aqueous solution through a series of batch experiments. Results verified that vermicompost adsorbs heavy metal more effectively. High adsorbtion capacity of vermicompost also confirmed through SEM images.

Rorat et al. (2016) used earthworm for the vermiremediation of heavy metal and polycyclic aromatic hydrocarbons present in sewage sludge. Results revealed accumulation of heavy metal in the earthworms' body in the order of Cd > Cu > Zn > Ni > Cr > Pb. Polycyclic aromatic hydrocarbons content also reduced after vermicomposting in the sewage sludge.

For on-site organic waste management, a vermireactor fed with cow dung, kitchen waste and vegetable market waste has been developed to receive organic wastes up to 0.263 m^3 and was controlled similar to vermibins (Taeporamaysamai and Ratanatamskul 2016). Findings of the study suggest that this reactor could be monitored and operated on-site with greater hygienic. It was so designed that liquid vermicompost, solid vermicompost, vermiwash and earthworms can be harvested easily.

10.11 Environmental Impacts of Vermicomposting

The impacts of vermicomposting on the environment mainly in terms of greenhouse gas emissions have also been studied in yesteryears. Emission of CH_4, NH_3 and other volatile compounds during vermicomposting is lesser as compared to other waste disposal methods like landfilling, incineration and composting (Lim and Wu 2016; Lleo et al. 2013). Luth et al. (2011) also reported that earthworms reduce emissions of ammonia; hence, earthworm could be used to depress the greenhouse gases' production. The greenhouse gas emission can be diminished by adding mature compost, carbon-enriched bulking agent prior to composting or vermicomposting process (Lim and Wu 2016).

10.12 Conclusion

Modernization, urbanization and technological advances have lead to development but have significantly contributed solid waste generation. It is well documented that large quantities of organic wastes generated all over the world, especially in developing countries, cause environmental problems. Solid waste management has become one of the biggest challenge that call for some sustainable and green approach for its management. Over the last two decades, vermicomposting is favoured for the organic waste management diverting energy of waste towards agri-production and soil conservation. Several researchers have employed vermicomposting for conversion of different types of organic wastes into vermicompost and earthworm biomass. Use of vermicompost as organic manure is an alternative to chemical fertilizers since it favours microbial biodiversity and improves soil structure and fertility along with crop production.

References

Abbasi SA, Nayeem Shah M, Abbasi T (2015) Vermicomposting of phytomass: limitations of the past approaches and the emerging directions. J Clean Prod 93:103–114

Aira M, Monroy F, Dominguez J (2007) Earthworms strongly modify microbial biomass and activity triggering enzymatic activities during vermicomposting independently of the application rates of pig slurry. Sci Total Environ 385:252–261

Aira M, Gomez-Brandon M, Gonzalez-Porto P, Dominguez J (2011) Selective reduction of the pathogenic load of cow manure in an industrial-scale continuous-feeding vermireactor. Biores Technol 102:9633–9637

Ali U, Sajid N, Halid A, Riaz L, Rabbani MM, Syed JH, Malik RN (2015) A review on vermicomposting of organic wastes. Environ Progress Sustain Energy 34(4):1050–1062

Alidadi H, Hosseinzadeh A, Najafpoor AA, Esmaili H, Zanganeh J, Takabi MD, Piranloo FG (2016) Waste recycling by vermicomposting: maturity and quality assessment via dehydrogenase enzyme activity, lignin, water soluble carbon, nitrogen, phosphorous and other indicators. J Environ Manage 182:134–140

Amouei AI, Yousefi Z, Khosravi T (2017) Comparison of vermicompost characteristics produced from sewage sludge of wood and paper industry and household solid wastes. J Environ Health Sci Eng 15:5. https://doi.org/10.1186/s40201-017-0269-z

Arumugam K, Ganesan S, Muthunarayanan V, Vivek S, Sugumar S, Munusamy V (2015) Potentiality of *Eisenia fetida* to degrade disposable paper cups-an ecofriendly solution to solid waste pollution. Environ Sci Pollut Res 22:2868–2876

Bernal MP, Alburquerque J, Moral R (2009) Composting of animal manures and chemical criteria for compost maturity assessment. A review. Biores Technol 22:5444–5453

Bhat SA, Singh J, Vig AP (2016) Effect on growth of earthworm and chemical parameters during vermicomposting of pressmud sludge mixed with cattle dung mixture. In: International conference on solid waste management, 5IconSWM 2015. Procedia Environmental Sciences vol 35, pp 425–434

Bhat SA, Singh J, Vig AP (2017) Instrumental characterization of organic wastes for evaluation of vermicompost maturity. J Anal Sci Technol 8:2

Bisht R, Pandey H, Bharti D, Bisht SPS, Kaushal BR (2007) Reproductive potential of the earthworm *Metaphire posthuma* (Oligochaeta) in different food substrates. Trop Ecol 48 (1):107–114

Busato JG, Papa G, Canellas LP, Adani F, Oliveira AL, Leao TP (2016) Phosphatase activity and its relationship with physical and chemical parameters during vermicomposting of filter cake and cattle manure. J Sci Food Agric 96:1223–1230

Castillo JM, Romero E, Nogales R (2013) Dynamics of microbial communities related to biochemical parameters during vermicomposting and maturation of agro industrial lignocelluloses wastes. Biores Technol 146:345–354

Contreras-Ramos SM, Escamilla-Silva EM, Dendooven L (2005) Vermicomposting of biosolids with cow manure and oat straw. Biol Fertil Soils 41:190–198

Coulibaly SS, Zoro Bi IA (2010) Influence of animal wastes on growth and reproduction of the African earthworm species *Eudrilus eugeniae* (Oligochaeta). Eur J Soil Biol 46:225–229

Deka H, Deka S, Baruah C, Das J, Hoque S, Sarma H, Sarma N (2011) Vermicomposting potentiality of *Perionyx excavatus* for recycling of waste biomass of java citronella – an aromatic oil yielding plant. Biores Technol 102:11212–11217

Dey MD, Das S, Kumar R, Doleya R (2017) Bhattacharya, Mukhopadhyaya, R.: Vermiremoval of methylene blue using *Eisenia fetida*: a potential strategy for bioremediation of synthetic dye-containing effluents. Ecol Eng 106:200–208

Dominguez J, Edwards CA (2010) Relationships between composting and vermicomposting. In: Edwards CA, Arancon NQ, Sherman RL (eds) Vermiculture technology: earthworms, organic wastes, and environmental management. CRC Press, Boca Raton, FL pp 11–26

Dominguez J, Edwards CA (2011) Biology and ecology of earthworm species used for vermicomposting. In: Vermiculture technology-earthworms, organic wastes, and environmental management. CRC Press, Boca Raton, FL, pp 27–40

Edwards CA (1998) Earthworm ecology. St. Lucie Press, Boca Raton, p 389

Edwards CA, Bohlen PJ (1996) Biology and ecology of earthworms. Chapman and Hall, London

Edwards CA, Dominguez J, Neuhauser EF (1998) Growth and reproduction of *Perionyx excavatus* (Perr) (Megascolecidae) as factors in organic waste management. Biol Fertil Soils 27:155–161

Elvira C, Sampedro L, Benitez E, Nogales R (1998) Vermicomposting of sludges from paper mill and dairy industries with *Eisenia andrei*: a pilot-scale study. Biores Technol 63:205–211

Fornes F, Mendoza Hernandez D, Garcia-de-la-Fuente R, Abad M, Belda RM (2012) Composting versus vermicomposting: a comparative study of organic matter evolution through straight and combined processes. Biores Technol 118:296–305

Fu X, Huang K, Chen X, Li F, Cui G (2015) Feasibility of vermistabilization for fresh pelletized dewatered sludge with earthworms *Bimastus parvus*. Biores Technol 175:646–650

Garg VK, Yadav YK, Sheoran A, Chand S, Kausik P (2006) Livestock excreta management through vermicomposting using epigeic earthworm *Eisenia fetida*. Environmentalist 26: 269–276

Garg VK, Suthar S, Yadav A (2012) Management of food industry waste employing vermicomposting technology. Biores Technol 126:437–443

Golueke CG (1981) Principles of biological resource recovery. Biocycle 22(4):36–40

Gomez Brandon M, Lores M, Dominguez J (2013) Changes in chemical and microbiological properties of rabbit manure in a continuous-feeding vermicomposting system. Biores Technol 128:310–316

Gomez FMJ, Nogales R, Plante A, Plaza C, Fernandez MJ (2015) Application of a set of complementary techniques to understand how varying the proportion of two wastes affects humic acids produced by vermicomposting. Waste Manag 35:81–88

Hanc A, Dreslova M (2016) Effect of composting and vermicomposting on properties of particle size fractions. Biores Technol 217:186–189

Hanc A, Pliva P (2013) Vermicomposting technology as a tool for nutrient recovery from kitchen bio-waste. J Mater Cycles Waste Manage 15:431–439

Hand P, Hayes W, Frankland J, Satchell J (1988) Vermicomposting of cow slurry. Pedobiologia 31:199–209

Haynes RJ, Zhou YF (2016) Comparison of the chemical, physical and microbial properties of composts produced by conventional composting or vermicomposting using the same feedstocks. Environ Sci Pollut Res 23:10763–10772

He X, Zhang Y, Shen M, Zeng G, Zhou M, Li M (2016) Effect of vermicomposting on concentration and speciation of heavy metals in sewage sludge with additive materials. Bioresour Technol 218:867–873

Hoornweg D, Bhada-Tata P (2012) What a waste: a global review of solid waste management. World Bank, Washington, DC, USA

Huang K, Li F, Wei Y, Fu X, Chen X (2014) Effects of earthworms on physiochemical properties and microbial profiles during vermicomposting of fresh fruit and vegetable wastes. Biores Technol 170:45–52

Huang K, Xia H, Li F, Wei Y, Cui G, Fu X, Chen X (2016) Optimal growth condition of earthworms and their vermicompost features during recycling of five different fresh fruit and vegetable wastes. Environ Sci Pollut Res 23:13569–13575

Huges, H. B. N.: The Biology of Polluted Waters. University Press, Liverpool UK, pp 96–104 (1980)

Hussain N, Abbasi T, Abbasi SA (2016a) Transformation of toxic and allelopathic lantana into a benign organic fertilizer through vermicomposting. Spectrochim Acta Part A Mol Biomol Spectrosc 163:162–169

Hussain N, Abbasi T, Abbasi SA (2016b) Vermicomposting-mediated conversion of the toxic and allelopathic weed ipomoea into a potent fertilizer. Process Saf Environ Prot 103:97–106

Hussain N, Abbasi T, Abbasi SA (2016c) Vermicomposting transforms allelopathic parthenium into a benign organic fertilizer. J Environ Manage 180:180–189

Joshi R, Ahmed S (2016) Status and challenges of municipal solid waste management in India: a review. Cogent Environ Sci 2:1139434

Karmegam N, Daniel T (2009) Investigating efficiency of *Lampito mauritii* (Kinberg) and *Perionyx ceylanensis* michaelsen for vermicomposting of different types of organic substrates. Environmentalist 29(3):287–300

Kumar DS, Kumar PS, Rajendran NM, Kumar UV, Anbuganapathi G (2014) Evaluation of vermicompost maturity using scanning electron microscopy and paper chromatography analysis. J Agric Food Chem 62:2738–2741

Kumar MS, Rajiv P, Rajeshwari S, Venckatesh R (2015) Spectroscopic analysis of vermicompost for determination of nutritional quality. Spectrochim Acta Part A Mol Biomol Spectrosc 135:252–255

Kumar S, Smith SR, Fowler G, Velis C, Kumar SJ, Arya SR, Kumar R, Cheeseman C (2017) Challenges and opportunities associated with waste management in India. Roy Soc Open Sci 4:160764

Lavelle P, Melendez G, Pashanasi B, Schaefer R (1992) Nitrogen mineralization and reorganization in cast of the geophagous tropical earthworm *Pontoscolex corethrurus* (Glossoscolecidae). Biol Fertil Soils 14:49–53

Lazcano C, Gomez-Brandón M, Dominguez J (2008) Comparison of the effectiveness of composting and vermicomposting for the biological stabilization of cattle manure. Chemosphere 72:1013–1019

Lee KE (1985) Earthworms: their ecology and relationship with soil and land use. Academic Press, Sydney

Li YS, Xiao YQ, Qiu JP, Dai YQ, Robin P (2009) Continuous village sewage treatment by vermifiltration and activated sludge process. Water Sci Technol 60(11):3001–3010

Lim S, Wu TA (2015) Determination of maturity in the vermicompost produced from palm oil mill effluent using spectroscopy, structural characterization and thermogravimetric analysis. Ecol Eng 84:515–519

Lim SL, Wu TY (2016) Characterization of matured vermicompost derived from valorization of palm oil mill by product. J Agric Food Chem 64:1761–1769

Lim SL, Wu TY, Sim EYS, Lim PN, Clarke C (2012) Biotransformation of rice husk into organic fertilizer through vermicomposting. Ecol Eng 41:60–64

Lim SL, Lee LH, Wu TY (2016) Sustainability of using composting and vermicomposting technologies for organic solid waste biotransformation: recent overview, greenhouse gases emissions and economic analysis. J Clean Prod 111:262–278

Lleo T, Albacete E, Barrena R, Font X, Artola A, Sanchez A (2013) Home and vermicomposting as sustainable options for biowaste management. J Clean Prod 47:70–76

Loh TC, Lee YC, Liang JB, Tan D (2005) Vermicomposting of cattle and goat manures by *Eisenia foetida* and their growth and reproduction performance. Biores Technol 96:11–14

Lores M, Gomez Brandon M, Perez-Diaz D, Dominguez J (2006) Using FAME profiles for the characterization of animal wastes and vermicomposts. Soil Biol Biochem 38:2993–2996

Lourenc N, Nunes LM (2017) Optimization of a vermifiltration process for treating urban wastewater. Ecol Eng 100:138–146

Luth, Robin P, Germain P, Lecomte M, Landrain B, Li YS, Cluzeau D (2011) Earthworm effects on gaseous emissions during vermifiltration of pig fresh slurry. Biores Technol 102:3679–3686

Lv B, Xing M, Yang J (2016) Speciation and transformation of heavy metals during vermicomposting of animal manure. Biores Technol 209:397–401

Malinska K, Zabochnicka Swiateka M, Caceresb R, Marfa O (2016) The effect of precomposted sewage sludge mixture amended with biochar on the growth and reproduction of *Eisenia fetida* during laboratory vermicomposting. Ecol Eng 90:35–41

Modak P, Jiemian Y, Hongyuan Y, Mohanty CR (2010) Municipal solid waste management: turning waste into resources. In: Shanghai manual: a guide for sustainable urban development in the 21st century, pp 1–36

Molina MJ, Soriano MD, Ingelmo F, Llinares J (2013) Stabilisation of sewage sludge and vinasse bio-wastes by vermicomposting with rabbit manure using *Eisenia fetida*. Biores Technol 137:88–97

Moran-Salazar RG, Marino-Marmolejo EN, Rodriguez CJ, Vazquez GD, Contreras-Ramos SM (2016) Use of agave bagasse for production of an organic fertilizer by pretreatment with *Bjerkandera adusta* and vermicomposting with *Eisenia fetida*. Environ Technol 37(10): 1220–1231.

Mupambwa HA, Ravindran B, Mnkeni PNS (2016) Potential of effective micro-organisms and *Eisenia fetida* in enhancing vermi-degradation and nutrient release of fly ash incorporated into cow dung–paper waste mixture. Waste Manag 48:165–173

Ndgewa PM, Thompson SA, Das KC (2000) Effects of stocking density and feeding rate on vermicomposting of biosolids. Biores Technol 71:5–12

Parthasarathi K, Balamurugan M, Prashija KV, Jayanthi L, Basha SA (2016) Potential of *Perionyx excavatus* (Perrier) in lignocellulosic solid waste management and quality vermifertilizer production for soil health. Int J Recycl Org Waste Agric 5:65–86

Patidar A, Gupta R, Tiwari A (2014) Integrated composting and vermicomposting: a boon to industry for waste clearance. Int J Environ Waste Manage 13(3):274–290

Perez Godinez EA, Zarate JL, Hernandez JC, Aceves MB (2017) Growth and reproductive potential of *Eisenia foetida* (Sav) on various zoo animal dungs after two methods of pre-composting followed by vermicomposting. Waste Manag 64:67–78

Pigatin LBF, Atoloye IO, Obikoya OA, Borsato AV, Rezende MOO (2016) Chemical study of vermicomposted agroindustrial wastes. Int J Recycl Org Waste Agric 5:55–63

Pramanik P, Chung YR (2011) Changes in fungal population of fly ash and vinasse mixture during vermicomposting by *Eudrilus eugeniae* and *Eisenia fetida*: documentation of cellulase isozymes in vermicompost. Waste Manag 31:1169–1175

Pramanik P, Ghosh GK, Ghosal PK, Banik P (2007) Changes in organic- C, N, P and K and enzyme activities in vermicompost of biodegradable organic wastes under liming and microbial inoculants. Biores Technol 98:2485–2494

Ravindran B, Mnkeni PNS (2016) Bio-optimization of the carbon-to-nitrogen ratio for efficient vermicomposting of chicken manure and waste paper using *Eisenia fetida*. Environ Sci Pollut Res 23:16965–16976

Ravindran B, Jonathan JWC, Selvam A, Sekaran G (2016) Influence of microbial diversity and plant growth hormones in compost and vermicompost from fermented tannery waste. Biores Technol 217:200–204

Reinecke AJ, Viljoen SA, Saayman RJ (1992) The suitability of *Eudrilus eugeniae*, *Perionyx excavatus* and *Eisenia fetida* (Oligochaeta) for vermicomposting in Southern Africa in terms of their temperature requirements. Soil Biol Biochem 24:1295–1307

Rorat A, Suleiman H, Grobelak A, Grosser A, Kacprzak M, Płytycz B, Vandenbulcke F (2016) Interactions between sewage sludge-amended soil and earthworms—comparison between *Eisenia fetida* and *Eisenia andrei* composting species. Environ Sci Pollut Res 23:3026–3035

Sahariah B, Goswami L, Kim KH, Bhattachatyya P, Bhattacharya SS (2015) Metal remediation and biodegradation potential of earthworm species on municipal solid waste a parallel analysis between *Metaphire posthuma* and *Eisenia fetida*. Biores Technol 180:230–236

Sanchez GM, Tausnerova H, Hanc A, Tlustos P (2017) Stabilization of different starting materials through vermicomposting in a continuous-feeding system: changes in chemical and biological parameters. Ecol Eng 106:200–208

Scagalia B, Nunes RR, Rezende MOO, Tambone F, Adani F (2016) Investigating organic molecules responsible of auxin-like activity of humic acid fraction extracted from vermicompost. Sci Total Environ 562:289 295

Sharma K, Garg VK (2017) Management of food and vegetable processing waste spiked with buffalo waste using earthworms (*Eisenia fetida*). Environ Sci Pollut Res. https://doi.org/10.1007/s11356-017-8438-2

Singh J, Kalamdhad AS (2013) Assessment of bioavailability and leachability of heavy metals during rotary drum composting of green waste (Water hyacinth). Ecol Eng 52:59–69

Singh WR, Kalamdhad AS (2016) Transformation of nutrients and heavy metals during vermicomposting of the invasive green weed *Salvinia natans* using *Eisenia fetida*. Int J Recycl Org Waste Agric 5:205–220

Singh RP, Singh P, Araujoc SFM, Ibrahima HM, Sulaiman O (2011) Management of urban solid waste: vermicomposting a sustainable option. Resour Conserv Recycl 55:719–729

Sinha RK, Bharambe G, Chaudhari U (2008) Sewage treatment by vermifiltration with synchronous treatment of sludge by earthworms: a low-cost sustainable technology over conventional systems with potential for decentralization. Environmentalist 28:409–420

Song X, Liu M, Wu D, Qi L (2014) Ye, Jiao, J., Hu, F.: Heavy metal and nutrient changes during vermicomposting animal manure spiked with mushroom residues. Waste Manag 34:1977–1983

Soobhany N, Mohee R, Garg VK (2015) Comparative assessment of heavy metals content during the composting and vermicomposting of Municipal Solid Waste employing *Eudrilus eugeniae*. Waste Manag 39:130–145

Soobhany N, Gunasee S, Rago YP, Joyram H, Raghoo P, Mohee R, Garg VK (2017) Spectroscopic, thermogravimetric and structural characterization analyses for comparing municipal solid waste composts and vermicomposts stability and maturity. Bioresour Technol 326:11–19.

Sudkolai ST, Nourbakhsh F (2017) Urease activity as an index for assessing the maturity of cow manure and wheat residue vermicomposts. Waste Manag 64:63–66

Swarnam TP, Velmurugan A, Pandey SK, Roy SD (2016) Enhancing nutrient recovery and compost maturity of coconut husk by vermicomposting technology. Biores Technol 207:76–84

Taeporamaysamai O, Ratanatamskul C (2016) Co-composting of various organic substrates from municipal solid waste using an on-site prototype vermicomposting reactor. Int Biodeterior Biodegrad 113:357–366

Villar I, Alves D, Perez-Diaz D, Mato S (2016) Changes in microbial dynamics during vermicomposting of fresh and composted sewage sludge. Waste Manag 48:409–417

Vodounnou DSJV, Kpogue DNS, Tossavi CE, Mennsah GA, Fiogbe ED (2016) Effect of animal waste and vegetable compost on production and growth of earthworm (*Eisenia fetida*) during vermiculture. Int J Recycl Org Waste Agric 5:87–92

WHO (2000) Air quality guidelines for Europe. European series No. 91. World Health Organization Regional Office for Europe, Copenhagen

Wright M (1972) Factors' governing ingestion by the earthworm *Lubricus terrestris* (L.), with special reference to apple leaves. Annals Appl Biol 70:175–188

Xie D, Wu W, Hao X, Jiang D, Li X, Bai L (2016) Vermicomposting of sludge from animal wastewater treatment plant mixed with cow dung or swine manure using *Eisenia fetida*. Environ Sci Pollut Res. 23(8):7767–7775

Yadav HR (2015) Assessment of different organic supplements for degradation of *Parthenium hysterophorus* by vermitechnology. J Environ Health Sci Eng 13:44

Yadav A, Garg VK (2011) Recycling of organic wastes by employing *Eisenia fetida*. Biores Technol 102:2874–2880

Yadav A, Gupta R, Garg VK (2013) Organic manure production from cow dung and biogas plant slurry by vermicomposting under field conditions. Int J Recycl Org Waste Agric 2:21

Zbytniewski R, Buszewski B (2005) Characterization of natural organic matter (NOM) derived from sewage sludge compost. Part 1: chemical and spectroscopic properties. Biores Technol 96:471–478

Zhu W, Du W, Shen X, Zhang H, Ding Y (2017) Comparative adsorption of $Pb^{2+}$and Cd^{2+} by cow manure and its vermicompost. Environ Pollut 227:89–97

Chapter 11
Microbial Fuel Cell Technology for Bioelectricity Generation from Wastewaters

Surajbhan Sevda, Pranab Jyoti Sarma, Kaustubha Mohanty, T. R. Sreekrishnan and Deepak Pant

Abstract Bioelectrochemical system (BES) has made significant advancement in last decade indifferent modes for converting organic waste material into bioelectricity. In this chapter, we review the basic aspects of a microbial fuel cell (MFC) that produces bioelectricity through the biodegradation of organic waste present in wastewaters and study the influencing factors such as membranes in MFC, electrodes, pH and temperature. A classic MFC subsists of cathode and anode electrode chambers separated by a cation or anion exchange membrane. Various substrates such as synthetic wastewater, municipal wastewater, biowaste, and industrial wastewater rich in organic content can be treated in MFC for removal of organic content with bioelectricity generation simultaneously. In this chapter, the effect of various substrate (biowaste) and configuration on the MFC performance on bioelectricity generation is presented. Also, this chapter demonstrates the improvements made in MFCs in recent years with summarization of their advantages and possible future applications. Different key influencing factors affecting bioelectricity generation on MFCs are elaborated, and these key parameters are fully discussed.

Keywords Microbial fuel cell · Bioelectricity · Wastewater · Electrodes

S. Sevda (✉)
Department of Bioscience and Biotechnology, Indian Institute of Technology Guwahati, Guwahati, India
e-mail: sevdasuraj@iitg.ernet.in

P. Jyoti Sarma · K. Mohanty
Department of Chemical Engineering, Indian Institute of Technology Guwahati, Guwahati, India

T. R. Sreekrishnan
Department of Biochemical Engineering and Biotechnology, Indian Institute of Technology Delhi, New Delhi, India

D. Pant (✉)
Separation and Conversion Technology, Flemish Institute for Technological Research (VITO), Boeretang 200, 2400 Mol, Belgium
e-mail: deepak.pant@vito.be

R. R. Singhania et al. (eds.), *Waste to Wealth*, Energy, Environment, and Sustainability, https://doi.org/10.1007/978-981-10-7431-8_11

11.1 Introduction

Fossil fuels such as petroleum, coal, natural gas were formed millions of years ago by natural processes. These fossil fuels are extremely efficient sources to produce power. In the past several decades, human civilization has been utilizing these fossil fuels for power production and other activities such as heating of houses. However, these fuel sources are not infinite, since their underground levels are attenuating at a startling rate. In addition, many of these fossil fuels such as oil may no longer be available in coming decades. Hence, it is indispensable to look at different sources of energy for the continuation of the industrialized civilization at the present rate (Sevda et al. 2013a). Renewable energy is produced from natural processes which are continuously provisioned and cannot be exhausted soon unlike fossil fuels. This includes water, sunlight, geothermal heat, wind, tides, and various forms of biomass. Recently, the generation cost of renewable energy has come down extremely due to technological uplift and research activities in this field (Lund 2007).

MFCs represent a standard and rising technology where microbial catalytic reactions at the anode result in electric power generation from waste and renewable biomass (Inglesby et al. 2012; Rosenbaum et al. 2005). MFCs also assist in the bioremediation of specific pollutants and nutrients in wastewaters (Mathuriya and Yakhmi 2014). Recovery of heavy metals, decolourization of dyes, production of bioenergy such as biomethane, biohydrogen, and even biomass are other applications of MFCs (Mathuriya and Yakhmi 2014; Mohan et al. 2014). Thus, MFCs have the dual benefits of power generation and wastewater treatment by which the process becomes as a whole more eco-friendly and economically feasible (Logan et al. 2006).

The electrical energy production through MFC using microbes is one such sustainable and renewable accession which is now well thought-out to be most capable (Hao et al. 2007; Salgado 2009; Elmekawy et al. 2017; Sevda et al. 2013b). MFCs act as bioreactor where chemical energy is being transformed using electroactive bacteria as a biocatalyst through the oxidization of accessible biodegradable substrates. Therefore, the MFCs can be used for biological wastewater treatment and production of bioelectricity (He et al. 2017; Sevda and Sreekrishnan 2012; Mathuriya and Sharma 2009; Mohan et al. 2008). Wastes, especially wastewaters (when treated at source and having no transportation costs), are inexpensive and can be processed conventionally. Therefore, generation of electrical energy from such wastes using MFC can offer a reliable and cheap solution for addressing the environmental pollution and energy catastrophe issues in the near future. The important components of MFC are proton exchange membrane (PEM), electrolyte medium, anaerobic microbes (anodic chamber), anode and cathode electrode which are connected with an external resistances (Cheng et al. 2006a, b; Mathuriya 2014; Ma et al. 2017; Sevda et al. 2016). In a MFC, microbial community endures in the anodic chamber. The microbes generate protons and electrons by employing organic substrates as fuels to generate electrons (Rabaey and Verstraete 2005; Sonawane et al. 2017). These electrons are passed through electron transport chain by nicotinamide adenine dinucleotide (NADH). These electrons are further shifted to the final electron acceptor such as nitrate, oxygen and then translocated to the outer membrane proteins (Logan et al. 2006). The protons are transferred from anodic chamber by membrane to cathodic chamber. The produced electrons are transferred through an external electrical circuit, thereby producing bioelectricity (Salgado 2009). PEM helps to diffuse the produced protons to the cathodic chamber where it combines with the O_2 and electrons to form water. As oxygen inhibits electricity production, therefore, anaerobic stipulation is strictly maintained in the anode compartment. However, the cathode is exposed to oxygen to help in the formation of water.

11.2 From Microbial Fuel Cells to Bioelectrochemical Systems (BESs)

BES represents one of the modern approaches of generating electricity by using the inherent capacity of electroactive microorganisms. This represents one of the simple forms of energy generation process where microorganisms breakdown organic products and harness the chemical energy stored in it into electricity. This technology in recent times has gained much importance as any kind of biodegradable waste can be converted into electricity, in turn creating an environment which is green and healthy at a very reasonable cost especially when treatment takes place on-site of wastewater generation (no transport cost) and the cost of energy used can be generated on-site from MFCs. In BESs, the energy produced by microorganisms is harnessed by placing two electrodes, i.e., anode and/or cathode into the system where electrochemical reactions take place through microbial interactions with the electrodes leading to electricity generation.

The term “BES” in broader sense covers different domains based on the mode of operation and biocatalyst used and has many applications, viz. wastewater treatment and renewable energy recovery, generation of energy carriers, such as H_2 or CH_4, biosensors for remote devices, bioremediation of recalcitrant compounds, and production of valuable products, a few to name (Pant et al. 2012). As compared to conventional fuel cells, BESs are not energy-intensive and can utilize wide variety of organic substrates without generally requiring expensive catalysts. The BESs in its nascent stages have got the future potential to become a source of power for our coming generation. In recent times, a lot of research has been carried out for the development and use of various alternative materials to be used as electrodes, membrane separators, and catalysts along with innovation in system designs, and therefore, BESs are becoming a very promising technology for our future (Mohan et al. 2009).

11.2.1 Principle of Microbial Fuel Cell

MFC technologies are one of the latest ways for producing bioelectricity in situ from biodegradable substrate metabolism using various microbes without requiring any added supply of energy from outside, whereas a conventional electrochemical cell requires external source of energy supply for power generation (Imhoff et al. 2004). A typical MFC system consists of anode and a cathode compartments together either as single chamber or as two individual chambers. Figure 11.1 illustrates a two-chamber MFC system. On the anodic compartment of the fuel cell, the microbes use its cellular respiratory pathways to transfer the chemical energy stored in the organic substances into electrical energy. Under anaerobic conditions in the anodic chamber, the sugar substrates are converted into carbon dioxide, protons, and electrons (Logan 2008). These electrons, produced at the anode, flow through the external load toward cathode surface and get reduced at the cathode upon interaction with protons and oxygen. Hence, a potential difference is generated between the cathode and the anode, resulting in the generation of electrical power. The reactions that take place at the electrode surfaces are shown below:

$$\text{At Anode: } C_{12}H_{22}O_{11} + 13H_2O \rightarrow 12CO_2 + 48H^+ + 48e^- \tag{11.1}$$

$$\text{At Cathode: } 6O_2 + 24H^+ + 24e^- \rightarrow 12H_2O \tag{11.2}$$

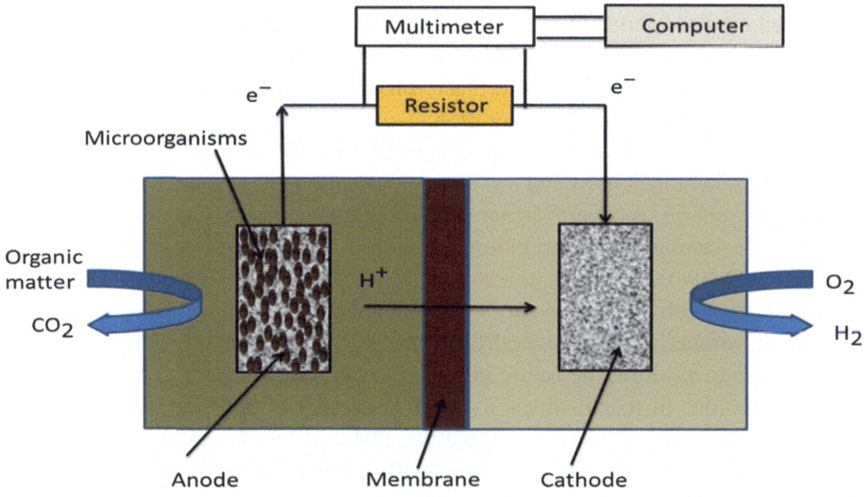

Fig. 11.1 Schematic diagram of a two chamber microbial fuel cells (MFCs) (PEM: proton exchange membrane, R: external resistance; M: multi-meter)

11.3 Use of Ion Permeable Membrane in Bioelectrochemical System

Membranes play a very crucial role in the BES. These membranes provide a support to the MFC for separation of anodic and cathodic chambers and also provide a route to transfer the protons from anodic chamber to the cathodic chamber. In BES system, mainly three different membranes are used. These membranes manufactured by different companies have different characteristics. These membranes such as Nafion membranes (DuPont Co., DE), CMI-7000 membranes (Membranes International Inc., NJ), and forward osmosis (FO) membranes (Hydration Technology Innovation, AZ) are used in different BESs. Bacterial processes can be combined with chemical processes, but are often segregated to separate chambers to reduce limiting factors, such as troublesome by product synthesis and competitive inhibition (Logan et al. 2006; Rabaey et al. 2009). In a single-chamber system, the anode and cathode, and relevant reaction components, are contained in a homogenous environment. Dual-chamber BESs are constructed with the cathode and anode chambers separated by a PEM membrane (Kim et al. 2007). The type of membrane used depends on the overall function of the BES and the desired components that are required in one chamber or another. Most BESs confide on microbial metabolism and the alteration of electrons on the anode electrode, while using either an air or water cathode. Regardless of the solution composition used within the cathode chamber, the use of membranes allows the two chambers to function simultaneously without mixing or contamination. Effectively separating the anode and cathode chambers can expand the system potential to include both reverse and forward configurations. Using both reverse and forward configurations involves combining a biologically driven reaction in one chamber, typically oxidation within the anode, with either a chemically driven reduction reaction or a biologically driven reduction reaction (Logan et al. 2006; Rabaey et al. 2009). Membrane performance and viability are mainly vulnerable on the level of interpenetration of the membrane and the pH of the solution to which the membrane is exposed. One challenge associated with selecting the ideal membrane for system efficiency is the supplementary charge-carrying chemicals, such as potassium (K^+), calcium (Ca^{2+}), and sodium (Na^+), that may be present due to the waste stream or bacterial medium utilized in the anode (Rozendalet al. 2006). The accumulation of protons, which can be induced by flux of alternative charge-carrying chemicals, can create a disadvantageous environment for microbes in the anode, which can dramatically affect the efficiency of the system.

11.3.1 Anion and Cation Exchange Membranes Used in BES System

PEMs and cation exchange membranes (CEMs) are charge-selective, nonpolar, ion exchange membranes that selectively transfer positively charged ions, like larger cation bodies and protons, from aside of the membrane to the other. The semi-permeable membranes selectively function by appropriate immobilized charged groups docking the polymer network to allow flux of oppositely charged ions while dispersing or blocking similarly charged ions from moving across the membrane. This conception follows the Donnan law, where charged ions that are incapable to move across a semi-permeable membrane, whether due to size stipulation or immobilization, discombobulate equilibrium of the solutions on either side of the membrane (Harnisch et al. 2007; Ng et al. 2013; Cheng et al. 2003).

Selectivity of a membrane can be compromised by the unavoidable permeability by neutral species, such as electrolyte salts, gases, and organic compounds. The flux of neutral species leads to the concentration gradient between the cathodic and anodic compartments.

Nafion-117 membranes (Dupont Co., DE) are encompassing of a chemically stabilized perfluorosulfonic acid (PFSA) polymer that allows for the selective transport of cations (Dupont 2017). PEMs, like Nafion-117 membranes, are well suited for sustaining autonomous cathode and anode chambers while allowing the flux of protons (H^+) between chambers. The selective transfer of protons from the anode can be used to abutment product synthesis in the cathode, such as water (H_2O) or methane (CH_4). Nafion-117 membranes are said to be chemically resistant, durable, stable in mildly acidic environments (pH 4–7). Pretreated membranes are applicable, but untreated, dry membranes are the most common form, as they can easily be cut to fit the custom dimensions of a system before treatment. Pretreatment of the membranes commences membrane expansion and assists with membrane selectivity.

CEMs function in the equal manner as PEMs, but require a less arduous pretreatment process. CMI-7000 is produced by Membranes International, Inc., which has a similar structural content to that of the Nafion-117. The polymer structures of the CMI-7000 are gel polystyrene cross-linked with divinylbenzene and with a sulfonic acid functional group (Membrane International 2017), and CMI-7000S membranes are chemically stable in strong acidic to weak basic environments, ranging from pH of 1.0 to 10.0. CMI-7000 membranes include a coated woven fabric 18 for stability, creating a rigidity that is not seen with the Nafion-117 membranes. There are very few manufacturers producing PEMs and CEMs, which may cause differences between membrane types. Major differences include actual selectivity versus reported selectivity, stability and durability, and longevity of membrane life. Nafion-117 membranes are reportedly used in a majority of previous and ongoing BES research and have been tested extensively.

11.3.2 Forward Osmosis (FO) Membranes

Generally, FO membranes are not used in MFC and BES research, but are periodically seen in water treatment technology development. FO membranes, exclusively produced by Hydration Technology Innovation (HTI), are only selective for water molecules (HTI 2017). The flux of water across the membrane is driven by osmotic gradient between a chamber, which naturally drives the flux of water in order to settle equal dissemination and concentration between the chambers separated by the FO membrane. Due to the driving force of the osmotic pressure gradient, FO technologies do not require energy input to trigger and maintain filtration. The main disadvantage of using FO membranes in BESs is the requirement to a draw solution in either the anode or the cathode, which limits the customizability of the system to one chamber (Qin et al. 2017).

11.3.3 Membrane Fouling

Membrane fouling is prevalent after membranes were used in functioning reactors with frequent treatments. Considering the configuration of the reactors, membrane fouling could have been due to bacterial degradation and growth. Each potential cause of membrane fouling may influence a reduction in system efficiency. While microbial growth may not be unfavorable to the membrane, it may alter the charge or osmotic gradients along the surface of the membrane, which may influence a change in selectivity. If fouling affects the selectivity of the membrane, unwanted ionic species may not be blocked from moving across the membrane. Additionally, microbial growth on the surface of the membrane may be an indicator that one microbial species from the consortia is benefiting from an accumulation of charged ions on the membrane surface.

11.4 Microbial Culture, Medium, and Electrode Material Used in MFCs

The selection and enrichment of the microbial source are significant for optimizing the performance of MFCs. The enrichment of anaerobic culture is carried out in the anodic chamber of MFC in closed circuit mode. In the higher power mode, exoelectrogenic bacterial community are developed, and this is similar to the anaerobic digestion where methanogen bacteria are enriched. Even the source of bacteria for MFC and anaerobic digestion is same but due to the process difference, enrichment of culture is different from each other. Generally, in MFCs mixed culture is used; normally, they are collected from the wastewater treatment plants. It has been observed that specific class of bacteria such as *Geobacter* is dominant in the anodic

Table 11.1 Comparison of MFC performances in systems that employed different inoculum, substrate, and designs

MFC configuration	Inoculum source	Substrate used in anodic chamber	Obtained power density ($\mu W/cm^2$)	References
Dual-chamber	Anaerobic sludge	Acetate 10 g L^{-1}	13.5	Passos et al. (2015)
Dual-chamber	Anaerobic sludge	Glucose 3 g L^{-1}	1.0	Yusoff et al. (2013)
Dual-chamber	Anaerobic sludge	Glucose 3 g L^{-1}	4.2	Yusoff et al. (2013)
Dual-chamber	Pure culture of *Shewanella oneidensis* MR-1	Lactate 1.6 g L^{-1}	1.0	Manohar and Mansfeld (2009)
Dual-chamber	Anaerobic sludge	Pretreated activated sludge	5.6	Xiao et al. (2011)
Single-chamber	Pre-acclimated bacteria from MFC	Acetate 1 g L^{-1}	240	Logan et al. (2008)
Single-chamber	Effluent from MFC	Fermented primary sludge with 15.5 g L^{-1}sCOD[a]	32	Yang et al. (2013)
Single-chamber	Effluent from MFC	Fermented primary sludge with 15.5 g L^{-1}sCOD[a]	103	Yang et al. (2013)

[a]Soluble COD (chemical oxygen demand)

chamber of MFCs (Li et al. 2013) . These bacteria donate their electron to the anode electrode. In mixed culture, specific class of Proteobacteria (e.g., *Geobacter, Shewanella, and Pseudomonas*) is found. Some researcher also used pure culture of *Geobacter*, but the MFC performance is greater with mixed culture Osman et al. 2010) (Table 11.1).

11.4.1 Electrode Characterization in Full MFC

Electrode characterization is the most important aspect to evaluate the performance of a working MFC. It helps us to understand the behavior of electrode and hence the potential of a MFC for electricity generation. Electrode output can be measured in terms of cell voltages as volts (V) against time, which can be evaluated directly from the potential difference between the anode and cathode electrode. The electrode potential of anode and cathode electrode is measured separately by the use of the reference electrode (Ag/AgCl). The power density is measured with reference to the surface of electrode, while volumetric power density is determined by the use of the volume of anodic chamber of MFC (Bard and Faulkner 2000).

11.4.2 Polarization Curve

For the effective analysis of MFCs, polarization curve is used, and this is also known as power curve. The easiest approach to obtain the polarization curve is through changing the external resistance from higher to lower (10,000 Ω to 1 Ω) in a 20-min time interval (Fig. 11.2). The voltage and current are measured with reference to change in the external voltage (Logan 2008).

Current density and power density can be calculated using Eqs. 11.3 and 11.4, respectively.

$$I = \frac{V}{\alpha R} \tag{11.3}$$

$$P = \frac{V^2}{\alpha R} \tag{11.4}$$

where α is the electrode area. Normally, anodic surface area is taken as effective surface area, as it is the surface where microbial biofilm is established and it is determined by the power production.

To understand the BES in detail, potentiostatic method of analysis can also be carried out. A potentiostat is normally operated in a three-electrode setup consisting of a working electrode (anode or cathode), a counter electrode, and reference electrode. Different electrochemical techniques are used with the use of potentiostat such as linear sweep voltammetry (LSV), cyclic voltammetry (CV), and electrochemical impedance spectrometry (EIS) (Dominguez-Benetton et al. 2012; Sevda et al. 2015).

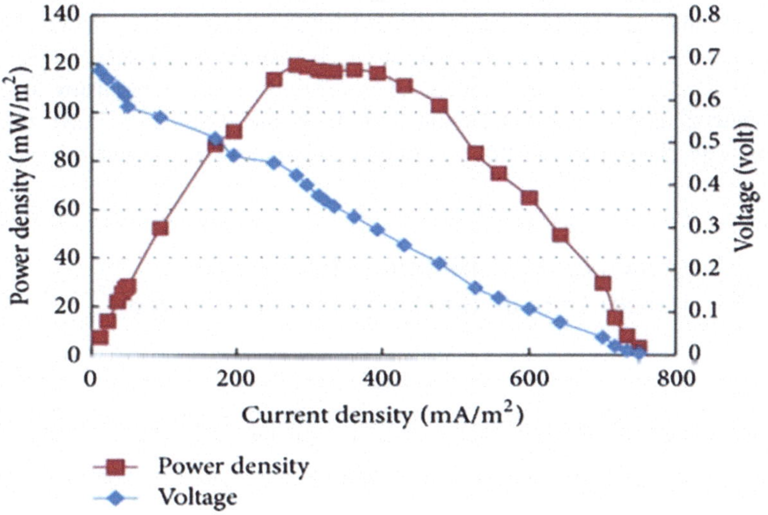

Fig. 11.2 A typical polarization curve. *Source* Logan (2008)

The voltammetric studies represent an important characterization steps to monitor the bacterial activity on the electrode surface with a potentiostat (Logan 2008; Logan et al. 2006). The individual electrode behavior MFC is done by use of a reference electrode and counter electrode (Logan 2008; Logan et al. 2006). Characterization of electrode potential in an electrochemical cell provides us insight into the capabilities of a microorganism to transfer electrons by directly measuring specific electrical potentials that allow electron release from the cell. Hence, electrode potential helps us to understand the development of biofilm and the power generation process (Sevda et al. 2014).

11.4.3 Microbial Community Analysis

Morphology of electrode is highly important for the performance consideration in a MFC. Electrode surface should be favorable for the growth of microorganisms and forms biofilm on its surface. Hence, morphological study of electrode is important to estimate the attachment of microbes to the electrode surface and can be studied by micrograph of scanning electron microscope. This in turns provides information on amount of electricity generation capacity of microbes and their adaptation to the internal MFC environment (Deng et al. 2010). Microbial community analysis of bacteria present on the electrodes reveals the bacteria phylum in an active MFC. Generally, the sample is collected from the surface of anode electrode area. In biocathode MFC, sample is collected from the cathode electrode surface. Molecular characterization of biofilm that develops on the anode surface over a period of time can be done by harvesting the biofilm from the anode surface and hence extracting the genomic DNA using specific DNA extraction tool. Genomic DNA can be extracted from specific regions of bacterial 16S rRNA (Jia et al. 2013). Microbial community analysis is performed by using denaturing gradient gel electrophoresis. In a study with electricity generation from microorganism grown on food waste, Jia et al. (2013) carried out phylogenetic classification of microorganisms to understand bacterial diversity at the anodic biofilm. In the study (Fig. 11.3), microorganisms were classified into different phylum, class, and genus levels.

11.5 Electroactive (EA) Biofilm: The Microbial Electrocatalysts of Bioelectrochemical Systems

In BES, microbes play a very important role by degrading the pollutant present in the wastewater. Microorganisms occupy almost every ecological niche on earth. Generally, in BES, a mixed culture is used, and this is enriched with anaerobic microbes; these microbes develop a biofilm on the anodic electrode surface.

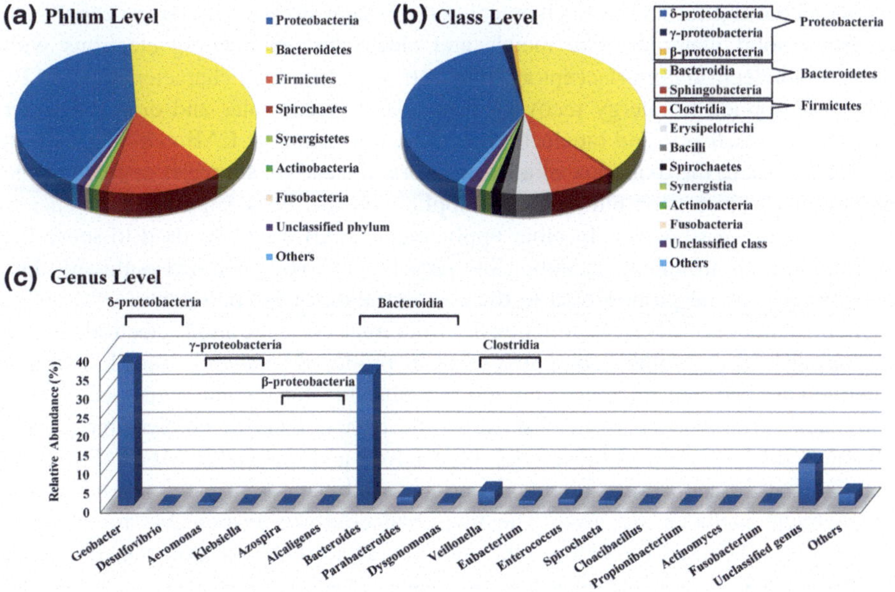

Fig. 11.3 Microbial community structure based on 454 pyrosequencing for a single-chamber MFCs fed with food wastewater. *Source* Jia et al. (2013)

In MFC, the developed biofilm is enriched with electroactive microbes that utilize specific pathways (Kim et al. 2006; Gottenbos et al. 1999; Davies et al. 1998). Biofilm developed on the anode electrode enhanced the direct electron transfer between anode electrode surface and microbes (Picioreanu et al. 2007). In the anodic chamber, microbes work as biocatalyst and it differs from the natural microbial process, since in this electrons are delivered to the anode electrode instead of another electron acceptor (Rabaey and Verstraete 2005). Still, the MFC studies are at the laboratory experimental stage, so biofilm behavior at higher scale needs to be checked. The biofilm plays an important role in the overall performance of the system (Davies et al. 1998). In MFCs, there are three different ways of electron transfer occurred: The first one is through use of external mediator, second one is through use of direct electron transfer (biofilm), and third one through use of nanowires (El-Naggar et al. 2010; Kalathil and Pant 2016). In this chapter, here we discuss the main mechanism of the electroactive biofilm formation on the anode electrode and electron transfer mechanism in the MFCs. Electroactive biofilm is capable of extracellular electron transfer and is therefore relevant to geochemistry, biocatalysis, and energy production. Although electroactive biofilm is widely applied in MFCs for renewable electricity production, where they transfer electrons to solid electrodes, the extracellular electron transfer chain is not completely understood. Electroactive biofilms (EABs) are microstructured communities composed of microbes that thrive at the solid/liquid interface. The biofilm is composed by the viable microbes and a matrix material that confers mechanical stability and increases biofilm resistance to chemical and physical stresses. EABs are capable of respiring solid materials, like metals and electrodes, exchanging electrons with these solid electron donor/acceptors. Because of their unique characteristics, EABs might be applied to energy recovery from waste, wastewater and drinking water treatment, biosensors, and bioelectrosynthesis. For example, EAB can be grown in fuel cells, where the oxidation of a reduced organic carbon source is coupled to the reduction of oxygen or nitrates. This application has been suggested for energy recovery from wastewater. In other applications, EAB might be used to sense the surrounding environment, including the presence of toxic chemical pollutant, thus generating a signal proportional to the concentration of the pollutant.

In mediator-less MFC, electron transfer from microbe to the anode electrode is also performed by use of soluble compounds that are produced by microbes and function as intermediate between microbes and anode electrode (Rabaey and Verstraete 2005). This lack of knowledge is an obstacle to the development of efficient bioelectrochemical devices, such as biosensors, biocatalyst, and innovative MFCs. For characterizing the electroactive microbes, various combinations of electrochemical and biochemical methods are used. The microbial consortia responsible for electron transfer are identified for higher efficiency in extracellular electron transfer. After identifying the responsible microbes, the best electroactive biofilm is grown on the anode electrode of MFCs. EABs may have applications in biosensors and MFCs. They are essentially a film of microorganisms able to supply electrons for electricity, but the exact mechanism of this electron transfer process remains poorly understood.

Both of these have structures in their outer membranes that allow for direct electron transfer between the bacterium and an electrode. Both of these have structures in their outer membranes that allow for direct electron transfer between itself and an electrode.

For *Shewanella*, researchers discovered that electrons are indirectly transferred in young biofilm (this produces more current), but that the process becomes mostly direct as the biofilm ages. In general, electron transfer is also very rapid in thin biofilm. However, the development of reliable EAB-based technologies requires a thorough understanding of charge transfer mechanism at the biofilm/electrode interface. Furthermore, the scale-up of EAB-based devices for industrial applications is usually accompanied by dramatic loss of power output, and this problem has prevented so far a large-scale implementation of EAB-based technology. *Geobacter* uses prevalently direct electron transfer through their outer membrane cytochromes, iron-containing proteins that can connect directly the bacterial cells with the electrode. *Shewanella* possess outer membrane cytochromes and is capable of direct electron transfer. However, it also produces redox compounds, namely quinones and flavins, (similar to vitamin B12) that get reduced at the bacterial surface and then in turn reduce the electrode, thereby producing a net electron transfer. The relative importance of direct vs. mediated electron transfer in EABs depends on various factors, including the electrode material, biofilm age, the potential at which electron transfer occurs. However, the electron transfer becomes mostly direct as the biofilm ages. In general, the mediated electron transfer is very rapid in thin biofilm, as the diffusion of microbial produced redox agent is not impeded by thick and dense biofilm. However, old biofilm such as those envisioned in practical technological applications will have a predominant direct electron transfer mode, thus producing less current than young biofilm (Read et al. 2010). This finding is relevant to the design of efficient EAB bioreactors, where the biofilm thickness should be kept at a minimum. EAB research requires the development of techniques capable of interrogating several aspects of EAB activity and structure.

A biofilm model suggests that the total number of electrons transferred to the anode electrode is proportional to the anode electrode surface area (Picioreanu et al. 2007). Also, when a biofilm is developed, with time it became thicker, this will not decrease the efficiency of the MFC when the electroactive bacteria present on the biofilm and the dead cells from the biofilm can be removed (Mah and O'Toole 2001). The use of mix culture has better advantage for biofilm development, as different microbes such as bacteria and archaea can form communities and it helps in enhancing the performance of a MFC (Hall-Stoodley et al. 2004). Biofilm formed on the anode has more quantity of *Geobacter* species family, because its metabolic pathway extracellularly reduces iron and metal oxide as the electron acceptor (Logan 2009). In comparison with the pure culture, mixed culture biofilm gave better performance in MFC because of the synergistic effect of many bacteria present such as *Pelotomaculum thermopropionicum, Methanothermobacter thermautotrophicus,* and *Geobacteraceae* (Gottenbos et al. 1999; Logan 2009; Kim et al. 2006; Lovely 2006).

11.6 Mechanisms of Electrons Transfer with Solid Electrodes

The real understanding of working of an MFC comes from the knowledge of association of microorganisms with the electrode materials (Kalathil et al. 2016). In case of bacteria, three mechanisms of electron transfer to a surface are known, viz. (a) presence of electrically conductive appendages known as microbial or bacterial nanowires produced by certain species of bacteria, viz. *Geobacter* and *Shewanella* (Pant et al. 2012); (b) direct electron transfer (DET) involving proteins located on cell surfaces; and (c) mediated electron transfer (MET) through use of small, redox reactive molecules that "shuttle" electrons from bacteria to the electrode surfaces by a diffusion-limited process. Figure 11.4 describes about some of the known mechanism of electron transfer in BES (Pham et al. 2009; Lovely 2008; Bond and Lovley 2003).

Mediators or electron shuttles are the substance that helps in the transfer of electrons produced by the microorganisms from inside the cell surfaces to the electrode surface. Based on nature, mediators are of two types, i.e., external and internal mediators. Certain chemicals, viz. neutral red, anthraquinone-2-6, disulfonate (AQDS), thionine, potassium ferricyanide, methyl-viologen, are used as mediator in MFCs. These compounds were known as exogenous mediators. However, for certain species like *Geobacter* bacteria, external mediator is not required. However, some bacteria like *Pseudomonas aeruginosa* produce their own

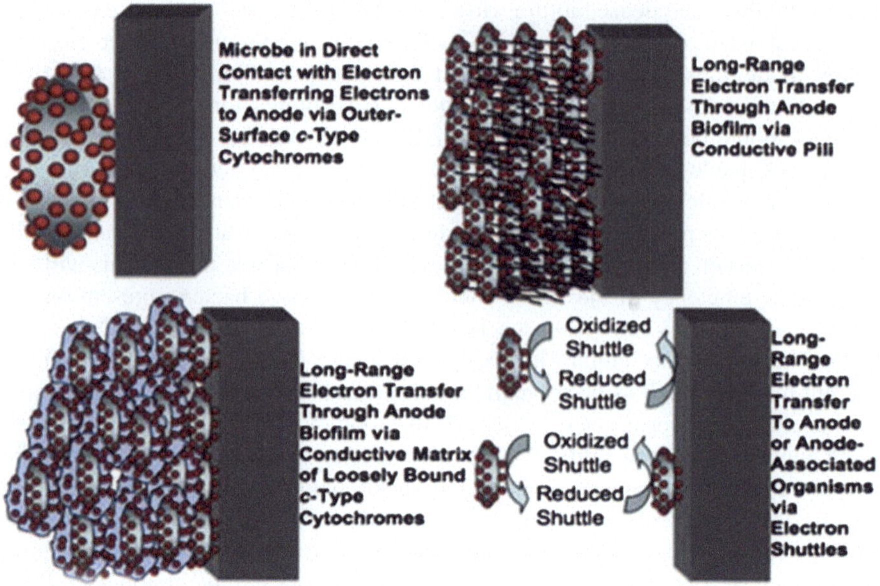

Fig. 11.4 Four different ways of electron transfer in BES: (i) exogenous external mediators, (ii) self-produced endogenous mediators and DET by (iii) single outer membrane cytochrome, and (iv) "nanowires." *Source* Lovely (2008)

mediator (phycocyanin), and the produced mediator shuttles electrons to the electrode from microbes (Logan 2008). Bacteria produce certain kind of conductive appendages which are known as "nanowires" (Gorby and Beveridge 2005). *Geobacter* and *Shewanella* species are having the appendages of nanowires. It is thought that these appendages are useful to establish connections between bacterial cells and with a surface carrying electron from the cell to a surface. Cytochromes are conventional electron transport protein that helps nanowires for transporting electrons from *G. sulfurreducens* to the electrode surface (Reguera et al. 2005).

Apart from conductive nanowires, there was also evidence of presence of certain surface blebs, i.e., protrusions on the microbial surface that do not exist as nanowires but certainly could be conductive points of contact (Logan 2008). *S. oneidensis* under anaerobically grown adhere to an iron surface with much greater forces than aerobically grown cells.

11.7 Anode and Cathode Materials

Electrodes are the most important components of MFC as it decides the performance and the cost of the system. The design of a suitable electrode is very useful, and it can help for scale-up studies (Rabaey et al. 2009). Protons and electrons produced by the bacteria in MFC travel from the point of production on the anode surface to cathode surface. So, electrode materials should have the characteristic of being highly electrically conductive, but they must also be cost-effective and non-corrosive. In addition, the surface of the material must favor bacterial attachment to form biofilm on its surface.

Therefore, the selection of electrode plays an important role in the development and commercialization of MFC. Among all the materials, carbon is known to be one of a trusted material for application as electrode. In MFC research, carbon products have been under use in various forms because of its affability in its size and shape, viz. carbon cloth, carbon paper, carbon fiber. The use of carbon-based materials has significantly improved the MFC research, and it drastically reduces the cost of MFC system (Govind 2015).

Anode surface is the heart of a MFC system as it favors the formation of microbial biofilm, and hence, a anode material must fulfill certain requirements; i.e., it should have high specific surface area (area per volume), highly conductive, non-corrosive, high porosity for biofilm development, non-fouling (i.e., bacteria do not fill it up) (Logan 2008). Usually, carbon material plays a major role as electrodes in most of MFC for its characteristics and cost-effectiveness. Carbon-based electrodes are popular among fabrication of MFC. Table 11.2 shows direct comparisons on the effects of various carbon electrodes on power generations considering size of electrode, MFC type, and bacterial source used in a system.

Unlike anode, cathode is one of the most important components in an MFC setup and its design represents a challenge in making MFC a useful and scalable technology. It is the surface of a cathode where tri-phase reaction of combining electrons, protons, and oxygen takes place resulting in generation of bioelectricity and producing water. Hence, surface of a cathode should favor the process of this reduction reaction (Logan 2008). In selecting a cathode material, various low-cost

Table 11.2 Comparison of various carbon-based anode materials and their performances

Electrode material	Size	Bacterial source	MFC type	Power output (mW/m^2)	References
Carbon brush	4 cm by 3 cm dia	Pre-accumulated bacteria from active MFC	Cube air cathode	2400	Logan et al. (2007)
Graphite plate	155 cm^2	*Shewanella oneidenis* (MR-1)	Two-chamber air cathode	1410	Dewan et al. (2008)
Activated carbon cloth	1.3 cm^2	*D. desulfuricans* strain	Single chamber air cathode	0.51	Zhao et al (2008)
Carbon Mesh	7 cm^2	Pre-accumulated bacteria from active MFC	Single-chamber cube air cathode	893	Wang et al. (2009)

Table 11.3 Comparison of various carbon-based cathode materials and their performances

Electrode material	Size (cm^2)	Catalyst	MFC Type	Power output (mW/m^2)	References
Carbon cloth with Nafion binder	7	Pt	Single-chamber cube air cathode	480	Cheng et al. (2006a, b)
Carbon fiber felt	36	Microbes	Cylindrical two-chamber aqueous air cathode without catalyst	315	Deng et al. (2010)
Carbon cloth PTFE binder	7	Pt	Single-chamber cube air cathode	360	Cheng et al. (2006a, b)
Graphite coating on ultrafiltration(UF) membrane	54	CoTMPP	Tubular single chamber	18	Zuo et al. (2007)

carbon-based materials are available but when the cost of catalyst (optional) if included may lead to increase in cost. The same materials that have been described above for use as anode have also been used as cathode. A catalyst is usually used for effective reduction reaction to take place, i.e., Pt or ferricyanide for oxygen reduction but not always needed. When catalyst is used, it is held on to the surface by a binder, which allows transfer of protons, electrons, and oxygen. Nafion is commonly used as binder due to its high proton conductivity and oxygen permeability. Other materials such as polytetrafluoroethylene (PTFE) suspension have also been used. Performance of the MFCs can also be improved by using air cathode and microorganisms as catalyst. Hence, surface of the electrode should allow growth of microorganisms for reduction reaction to carry out. This type of biocathode can substantially reduce the cost of a MFC. Table 11.3 shows the various cathode materials used in different MFCs.

11.8 MFC Implementation in Practical Applications

It is reported that domestic wastewater contains 9.3 times more energy that currently used in treatment of wastewater by conventional methods (Logan et al. 2006). The MFC can utilize this energy present in the wastewater; the produced bioelectricity can be saved and used for real-life applications. This energy can be used in the remote area as a biosensor. The MFCs can be worked at a lower temperature (15–25 °C), compared to conventional anaerobic treatment (37 °C needed). More energy can be produced by adding more MFCs in series or parallel mode. More wastewater can be treated in the continuous mode operation. The stack MFC provides a platform for treating more wastewater. The BES is also used for the production of useful product by use of carbon dioxide or acetate at the cathode. However, there is still a long way to go for scale-up studies and then at big scale. The key challenges are the better understanding of the electron transfer, electrode surface area, and stability of membrane for longer duration operation (Shukla et al. 2004). The MFCs can also be used as a sensor for determining chemical oxygen demand in synthetic and real wastewater. The synergy between the conventional system and MFC may be a better application in near future. This combined system may work better, and it utilized substrate in better way. Nonetheless, MFC cells may prove practical sooner for higher volume reactors.

11.9 Economic Analysis of Microbial Fuel Cell

The concept of electricity generation by electroactive bacteria though discovered in 1911 by MC Potter (Potter 1911), it took more than 100 years for this research to reach at the current stage and now it can be commercially used. Over the recent years, several small companies were started and this was summarized by Pant et al. (2011). Table 11.4 shows the leading companied for MFC prototype construction.

However, even these companies are still at pilot scale and a commercial prototype is yet to arrive. The first market application of MFCs was brought by an American company (Cambrian innovation), who came out with their treatment system called "EcoVolt," last year. This was a plant on truck kind of concept which also included bioelectricity generation (http://www.fuelcellstore.com/mudwatt-microbial-fuel-cell-kit). Very recently, a consulting company called market research came out with a detailed report called "Global Microbial Fuel Cell Market Research Report" (https://www.marketresearch.com/Micro-Market-Monitor-v4042/Global-Microbial-Fuel-Cells-MFC-9977629/). This included an overview of the MFC market globally and reflect the current market analysis and companies overview. Finally, the full commercial application was discussed by Trapero et al. (2017). They concluded that the implementation of MFC is a promising alternative to the use of classical aerated activated sludge, and it has potential economic benefits. The bottom line is that besides the sustainable and green nature of MFCs

Table 11.4 Details of companies based on MFC technology

Company	Country	Website
Trophos energy	USA	http://www.trophosenergy.com
Lebone	USA	http://www.lebone.org/
Emefcy	Israel	http://www.emefcy.com/
Int ActLAbLLC	USA	http://www.intactlabs.com/
Hy-SyEnce	USA	http://www.hy-syence.com/
Pant–e	The Netherlands	www.plant-e.com

Source Pant et al. (2011)

along with a reduced carbon footprint, their integration into a WWTP could reduce the operational cost because they diminish electricity consumption (Logan and Regan 2006).

11.10 Perspectives

Even though the electroactive nature of specific bacteria was discovered more than 100 years ago (Potter 1911), the research has been intensified in this field only in last 15 years. Initially, limited to proof-of-concept studies, the research explored the material aspect of MFCs also in the last decade. Significant advances have been made in the performance of these systems in terms of electric current and power obtained, the range of substrates explored (Pant et al. 2010; Pandey et al. 2016), electrode materials (both anode and cathode) (Govind 2015; Mashkour and Rahimnejad 2015), types of separators (Kokabian and Gude 2015), and design of the systems. Furthermore, the size of the systems has moved from small laboratory-scale reactors with mL volume to large pilots treating several 100 L of wastewater (Logan 2010). BES as a technology itself has moved beyond only wastewater treatment and energy recovery in the form of MFCs to encompass several other applications such as hydrogen production (MEC), desalination (MDC), and chemicals production (microbial electrosynthesis). More recently, companies are showing increased interest in the technology, and some of them are promising to come up with the commercial prototypes in near future. At the same time, though the field has grown tremendously over the years, there is still need for uniform standards to be adopted by thc research community in order to better compare the results coming from different laboratories. Some efforts have been made already in this regard (Patil et al. 2015). Given the research efforts being directed toward this technology, the future appears bright and a real practical application of BESs can be foreseen soon.

11.11 Conclusion

Alternative energy sources are being developed due to increased and heavy consumption of fossil fuels. Renewable energy is an eco-friendly alternative and includes options such as biomass and biowaste converted to energy and fuel. Resource recovery and organic waste recycle is a major concern nowadays allover the world. MFC is a novel technology with potential to generate energy/electricity from low-cost substrates such as wastewaters. At the current stage, the net power production is low in MFCs so it is difficult to use them in real-world applications. The efforts are being made to reduce the construction and operating cost to improve the overall performance of MFCs, and a commercial prototype can be expected in coming years.

References

Bard AJ, Faulkner LR (2000) Electrochemical methods: fundamentals and applications, 2nd edn. Wiley, New York

Barron B, Rosenbaum M, Balcerzak P, Angenent LT A Curriculum for high school science education: microbial Fuel Cell a living battery 1–89. http://angenent.bee.cornell.edu/MFCcurriculum.pdf

Bond DR, Lovley DR (2003) Electricity production by *Geobacter sulfurreducens* attached to electrodes. Appl Environ Microbiol 69:1548–1555

Cheng S, Liu H, Logan BE (2003) Power densities using different cathode catalysts (Pt and CoTMPP) and polymer binders (Nafion and PTFE) in single chamber MFCs. Environ Sci Technol 40:364–369

Cheng S, Liu H, Logan BE (2006a) Increased performance of single-chamber microbial fuel cells using an improved cathode structure. Electrochem Commun 8:489–494

Cheng S, Liu H, Logan BE (2006b) Increased power generation in a continuous flow MFC with advective flow through the porous anode and reduced electrode spacing. Environ Sci Technol 40:2426–2432

Davies DG, Parsek MR, Pearson JP, Iglewski BH, Costerton JW, Greenberg EP (1998) The involvement of cell-to-cell signals in the development of a bacterial biofilm. Science 280 (5361):295–298

Deng Q, Li XY, Zuo JE, Ling A, Logan BE (2010) Power generation using inactivated carbon fiber felt cathode in an upflow microbial fuel cell. J Power Sources 195:1130–1135

Dewan A, Beyenal H, Lewandowski Z (2008) Scaling up microbial fuel cells. Environ Sci Technol 42:7643–7648

Dominguez-Benetton X, Sevda S, Vanbroekhoven K, Pant D (2012) The accurate use of impedance analysis for the study of microbial electrochemical systems. Chem Soc Rev 41:7228–7246

DuPont (2017) DuPont Fuel Cells: DuPont™ Nafion PFSA Membranes. DuPont Technical Documents

Elmekawy A, Hegab HM, Losic D, Saint CP, Pant D (2017) Applications of graphene in microbial fuel cells: the gap between promise and reality. Renew Sustain Energy Rev 72:1389–1403

El-Naggar MY, Wanger G, Leung KM, Yuzvinsky TD, Southam G, Yang J, Lau WM, Nealson KH, Gorby YA (2010) Electrical transport along bacterial nanowires from Shewanella oneidensis MR-1. Proc Natl Acad Sci U S A 107:18127–18131

Gorby YA, Beveridge TJ (2005) Composition, reactivity and regulation of extracellular metal-reducting structures (nanowires) produced by dissimilatory metal reducing bacteria. Warrenton VA

Gottenbos B, van der Mei HC, Busscher HJ (1999) Models for studying initial adhesion and surface growth in biofilm formation on surfaces. Methods Enzymol 310:523–534

Govind AM (2015) Review on carbon electrodes in microbial fuel cell. Int Res J Eng Technol 02:424–427

Hall-Stoodley L, Costerton JW, Stoodley P (2004) Bacterial biofilms: from the Natural environment to infectious diseases. Nat Rev Microbiol. https://doi.org/10.1038/nrmicro821

Hao YE, Cheng S, Scott K, Logan BE (2007) Microbial fuel cell performance with non-Pt cathode catalysts. J Power Sources 171:275–281

Harnisch F, Schröder U, Scholz F (2007) The suitability of monopolar and bipolar ion exchange membranes as separators for biological fuel cells. Environl Sci Technol 42(5):1740–1746

He L, Du P, Chen Y, Lu H, Cheng X, Chang B et al (2017) Advances in microbial fuel cells for wastewater treatment. Renew Sustain Energy Rev 71:388–403

Hydration Technology Innovations (HTI) Forward Osmosis Solutions. HTI Water Technology Technical Document, URL: http://www.htiwater.com/technology/forward_osmosis/index.htm

Imhoff ML, Bounoua L, Ricketts T, Loucks C, Harriss R, Lawrence WT (2004) Global patterns in human consumption of net primary production. Nature 4(29):870–873

Inglesby AE, Beatty DA, Fisher AC (2012) Rhodopseudomonas palustris purple bacteria fed Arthrospira maxima cyanobacteria: demonstration of application in microbial fuel cells. RSC. Adv. 2, 4829–4838
Jia J, Tang Y, Liu B, Wu D, Ren N, Xing D (2013) Electricity generation from food wastes and microbial community structure in microbial fuel cells. Biores Technol 144:94–99
Kalathil S, Pant D (2016) Nanotechnology to rescue the bacterial bidirectional extracellular electron transfer in bioelectrochemical systems. RSC Advances 6:30582–30597
Kalathil S, Patil SA, Pant D (2016) Microbial Fuel Cells: Electrode Materials. Encyclopedia of Interfacial Chemistry: Surface Science and Electrochemistry 1–10
Kim GT, Webster G, Wimpenny JWT, Kim BH, Kim HJ, Weightman AJ (2006) Bacterial community structure, compartmentalization and activity in a microbial fuel cell. J Appl Microbiol 101(3):698–710
Kim JR, Cheng S, Oh SE, Logan BE (2007) Power generation using different cation, anion, and ultrafiltration membranes in microbial fuel cells. Environ Sci Technol 4(3):1004–1009
Kokabian B, Gude VG (2015) Role of membranes in bioelectrochemical systems. Membr Water Treat 6(1):53–75. https://doi.org/10.12989/mwt.2015.6.1.000
Li XM, Cheng KY, Selvam A, Wong JWC (2013) Bioelectricity production from acidic food waste leachate using microbial fuel cells: effect of microbial inocula. Process Biochem 48: 283–288
Linden D, Reddy TB (2002) Handbook of batteries. McGraw–Hill, New York
Logan BE (2008) Microbial fuel cells. Wiley, Hoboken
Logan BE (2009) Exoelectrogenic bacteria that power microbial fuel cells. Nat Rev Microbiol 7:375–381
Logan BE (2010) Scaling up microbial fuel cell and other bioelectrochemical system. Appl Microbiol Biotechnol 85:1665–1671
Logan BE, Regan JM (2006) Microbial fuel cells-challenges and applications. Environ Sci Technol 40:5172–5180
Logan BE, Hamelers B, Rozendal R, Schröder U, Keller J, Freguia S, Aelterman P, Verstraete W, Rabaey K (2006) Microbial fuel cells: methodology and technology. Environ Sci Technol 40 (17):5181–5192
Logan B, Cheng S, Watson V, Estadt G (2007) Graphite fiber brush anodes for increased power production in air-cathode microbial fuel cells. Environ Sci Technol 41:3341–3346
Lovely DR (2006) Microbial energizers: fuel cells that keep on going. Microbe 1(7):323–329
Lovley DR (2008) The microbe electric: conversion of organic matter to electricity. Curr Opin Biotechnol 19:564–571
Lund H (2007) Renewable energy strategies for sustainable development. Energy 32:912–919
Ma Z, Song H, Wang H, Xu P (2017) Improving the performance of microbial fuel cells by reducing the inherent resistivity of carbon fiber brush anodes. J Power Sources 348:193–200
Mah TC, O'Toole GA (2001) Mechanisms of biofilm resistance to antimicrobial agents. Trends Microbiol 1:34–39
Manohar AK, Mansfeld F (2009) The internal resistance of a microbial fuel cell and its dependence on cell design and operating conditions. Electrochim Acta 54:1664–1670
Mashkour M, Rahimnejad M (2015) Effect of various carbon—based cathode electrodes on the performance of microbial fuel cell. Biofuel Res J 8:296–300
Mathuriya AS, Sharma VN (2009) J Biochem Technol 2(1):133–137
Mathuriya AS (2014) Inoculum selection to enhance performance of a microbial fuel cell for electicity generation during wastewater treatment. Environmental Technology: Sustain Technol Bioener Biofuel Biowaste Biomass 34:1957–1964
Mathuriya AS, Yakhmi J (2014) Microbial fuel cells—applications for generation of electrical power and beyond. Crit Rev Microbio 11–17
Membranes International Inc., CMI-7000 Cation Exchange Membranes. Technical Specifications Document URL: http://www.membranesinternational.com/tech-cmi.htm
Mohan Y, Muthukumar M, Das D (2008) Electricity generation using microbial fuel cells. Int J Hydrogen Energ 33(1):423–426

Mohan SV, Raghavulu V, Peri D, Sarma PN (2009) Integrated function of microbial fuel cell (MFC) as bio-electrochemical treatment system associated with bioelectricity generation under higher substrate load. Biosens Bioelectron 24:2021–2027

Mohan SV, Velvizhi G, Modestra JA, Srikanth S (2014) Microbial fuel cell: Critical factors regulating bio-catalyzed electrochemical process and recent advancements. Renew Sustain Energy Rev 40:779–797

Ng LY, Mohammad AW, Ng CY (2013) A review of nanofiltration membrane fabrication and modification using polyelectrolytes: effective ways to develop membrane selective barriers and rejection capabilities. Adv Colloid Interface Sci 197–198:85

Osman MH, Shah AA, Walsh FC (2010) Recent progress and continuing challenges in bio-fuel cells. Part II: microbial. Biosens Bioelectron 26:953–963

Pandey P, Shinde VN, Deopurkar RL, Kale SP, Patil SA, Pant D (2016) Recent advances in the use of different substrates in microbial fuel cells toward wastewater treatment and simultaneous energy recovery. Appl Energy 168:706–723

Pant D, Bogaert GV, Diels L, Vanbroekhoven K (2010) A review of the substrates used in microbial fuel cells (MFCs) for sustainable energy production. Biores Technol 101(6): 1533–1543

Pant D, Singh A, Van Bogaerat G, Gallego YA, Diels L, Vanbroekhoven K (2011) An introduction to the life cycle assessment LCA) of bioelectrochemical systems (BES) for sustainable energy and product generation: Relevance and key aspects. Renew Sustain Energy Rev 15:1305–1313

Pant D, Singh A, Bogaert GV, Olsen SI, Nigam PS, Diels L, Vanbroekhoven K (2012) Bioelectrochemical systems (BES) for sustainable energy production and product recovery from organic wastes and industrial wastewaters. RSC Advances 2:1248–1263

Passos VF, Neto SA, Andrade AR, Reginatto V (2015) Energy generation in a Microbial Fuel Cell using anaerobic sludge from a wastewater treatment plant. Scientia Agricola 73:424–428

Patil SA, Gildemyn S, Pant D, Zengler K, Logan BE, Rabaey K (2015) A logical data representation framework for electricity-driven bioproduction processes. Biotechnology Advances 33(6):736–744

Pham TH, Aelterman P, Verstraete W (2009) Bioanode performance in bioelectrochemical systems: recent improvements and prospects. Trends Biotechnol 27:168–178

Picioreanu C, Head IM, Katuri KP, van Loosdrecht MCM, Scott K (2007) Water Res 41 (13):2921–2940

Potter MC (1911) Electrical effects accompanying the decomposition of organic compounds. Proc R Soc Lond B 84:260–276

Qin M, Hynes EA, Abu-Reesh IM, He Z (2017) Ammonium removal from synthetic wastewater promoted by current generation and water flux in an osmotic microbial fuel cell. J Clean Prod 149:856–862

Rabaey K, Verstraete W (2005) Microbial fuel cells: novel biotechnology for energy generation. TRENDS Biotechnol 23(6):291–298

Rabaey K, Angenent L, Schroder U, Keller J (2009) Bioelectrochemical systems: from extracellular electron transfer to biotechnological application, 1st edn. IWA Publishing, Lodon

Read ST, Dutta P, Bond PL, Keller J, Rabaey K (2010) Initial development and structures of biofilm on microbial fuel cell anodes. BMC Microbiol 10:98

Reguera R, McCarthy KD, Mehta T, Nicoll JS, Tuominen MT, Lovley DR (2005) Extracellular electron transfer via microbial nanowires. Nature 435:1098–1101

Rosenbaum M et al (2005) In situ electrooxidation of photobiological hydrogen in a photobioelectrochemical fuel cell based on *Rhodobacter sphaeroides*. Environ Sci Technol 39:6328–6333

Rozendal RA, Hamelers HVM, Buisman CJN (2006) Effects of membrane cation transport on pH and microbial fuel cell performance. Environ Sci Technol 40:5206–5211

Salgado CA (2009) Microbial fuel cells powered by *Geobacter sulfurreducens*. Basic Biotechnol J 5:96–101

Sevda S, Sreekrishnan TR (2012) Effect of salt concentration and mediators in salt bridge microbial fuel cell for electricity generation from synthetic wastewater. J Environ Sci Health, Part A 47:878–886

Sevda S, Dominguez-Benetton X, Vanbroekhoven K, De Wever H, Sreekrishnan TR, Pant D (2013a) High strength wastewater treatment accompanied by power generation using air cathode microbial fuel cell. Appl Energy 105:194–206

Sevda S, Dominguez-Benetton X, Vanbroekhoven K, Sreekrishnan TR, Pant D (2013b) Characterization and comparison of the performance of two different separator types in air–cathode microbial fuel cell treating synthetic wastewater. Chem Eng J 228:1–11

Sevda S, Dominguez-Benetton X, Graichen FHM, Vanbroekhoven K, De Wever H, Sreekrishnan TR, Pant D (2014) Evaluation and enhanced operational performance of microbial fuel cells under alternating anodic open circuit and closed circuit modes with different substrates. Biochem Eng J 90:294–300

Sevda S, Chayambuka K, Sreekrishnan TR, Pant D, Dominguez-Benetton X (2015) A comprehensive impedance journey to continuous microbial fuel cells. Bioelectrochemistry 106:159–166

Sevda S, Dominguez-Benetton X, Graichen FHM, Vanbroekhoven K, De Wever H, Sreekrishnan TR, Pant D (2016) Shift to continuous operation of an air-cathode microbial fuel cell long-running in fed-batch mode boosts power generation. Int J Green Energy 13(1): 71–79

Shukla AK, Suresh P, Berchmans S, Rahjendran A (2004) Biological fuel cells and their applications. Curr Sci 87:455–468

Sonawane JM, Yadav A, Ghosh PC, Adeloju SB (2017) Recent advances in the development and utilization of modern anode materials for high performance microbial fuel cells. Biosens Bioelectron 90:558–756

Trapero JR, Horcajada L, Linares JJ, Lobato J (2017) Is microbial fuel cell technology ready? An economic answer towards industrial commercialization. Appl Energy 185:698–707

Wang X, Cheng SA, Feng YJ, Merrill MD, Saito T, Logan BE (2009) Use of carbon mesh anodes and the effect of different pretreatment methods on power production in microbial fuel cells. Environ Sci Technol 43:6870–6874

Xiao B, Yang F, Liu J (2011) Enhancing simultaneous electricity production and reduction of sewage sludge in two-chamber MFC by aerobic sludge digestion and sludge pretreatments. J Hazard Mater 1(899):444–449

Yang F, Ren L, Pu Y, Logan BE (2013) Electricity generation from fermented primary sludge using single-chamber air-cathode microbial fuel cells. Biores Technol 128:784–787

Yusoff MZM, Hu A, Feng C, Maeda T, Shirai Y, Hassan MA, Yu CP (2013) Influence of pretreated activated sludge for electricity generation in microbial fuel cell application. Biores Technol 145:90–96

Zhao F, Rahunen N, Varcoe JR, Chandra A, Avignone-Rossa C, Thumser AE, Slade RCT (2008) Activated carbon cloth as anode for sulfate removal in amicrobial fuel cell. Environ Sci Technol 42:4971–4976

Zuo Y, Cheng SA, Call D, Logan BE (2007) Tubular membrane cathodes for scalable power generation in microbial fuel cells. Environ Sci Technol 41:3347–3353

Chapter 12
Economics of Solid Waste Management

Karthik Rajendran, V. Sudharsan Varma, Durga Madhab Mahapatra and Dhamodharan Kondusamy

Abstract The concept of gazing trash or waste as a worthwhile resource is augmenting exponentially. Despite its worthiness, there are distinct factors which affect the profitability of a plant which processes waste to worthwhile products such as fuels, fertilizers, energy, or chemicals. There is the adequate literature of different processing methods of wastes; however, the majority of the studies do not consider economic perspective. This chapter focusses on the economics of handling wastes including collection and transportation and processing it into value-added products such as compost, electricity, or fuel. Different countries have different legislation for waste handling, and this chapter addresses the waste handling costs in different countries. Furthermore, the profitability of different processing methods was discussed. The critical factors affecting the profitability of waste treatment and handling were identified, and workable solutions or directions were provided to the scientific community to address the problems with an industrial outlook including profitability.

Keywords Economics · Solid waste management · Composting
Anaerobic digestion · Collection and transportation · Landfilling

---

K. Rajendran (✉)
MaREI Centre, Environmental Research Institute, University College Cork, Cork, Ireland
e-mail: karthik.1988@gmail.com; k.rajendran@ucc.ie

K. Rajendran
School of Engineering, University College Cork, Cork, Ireland

V. Sudharsan Varma
Agricultural Research Organization, Institute of Soil, Water, and Environmental Sciences, Newe Ya'ar Research Center, 30095 Ramat Yishay, Israel

D. M. Mahapatra
Department of Biological and Ecological Engineering, Oregon State University, Corvallis, Oregon 97331, USA

D. Kondusamy
Department of Civil Engineering, Indian Institute of Technology Guwahati, Guwahati, Assam, India

R. R. Singhania et al. (eds.), *Waste to Wealth*, Energy, Environment, and Sustainability, https://doi.org/10.1007/978-981-10-7431-8_12

12.1 Introduction

The municipal solid waste generation had exceeded more than 1.5 billion tons per year, and it is expected to double over the next decade. This is an alarming issue that needs immediate attention on waste minimization or waste treatment methodologies. Waste treatments have a lot of benefits including cleaning and greening the environment, reduction of noxious gases and global warming, saving carbon footprint (Daniel and Perinaz 2012; Hoornweg and Bhada-Tata 2012). Moreover, wastes have a possibility of recovering energy from them before it is dumped or landfilled. There are several products which could be obtained from the wastes including energy/fuels, electricity, fertilizers, and heat. To treat the waste, there are costs associated with it including collecting and transportation costs, capital expenses in building a treatment facility, operational expenses, and finally the cost for upgrading/or purifying the product produced. These costs will vary depending on the waste characteristics, type of treatment facility, location, and labor costs (Tchobanoglous et al. 1993).

In most developing countries, landfilling or dumping seems to be common options as most waste treatments are not profitable. However, it is crucial for any process to have its economic viability to promote and develop. Several waste treatment options are available including composting, anaerobic digestion, landfilling with gas collections, thermal processing technologies such as combustion, incineration, and gasification (Tchobanoglous et al. 1993; Tchobanoglous and Kreith 1999). Different treatments come at different costs, and it is chosen based on the policies developed in each country. For instance, in Japan, more than 70% of its waste are incinerated which could be due to the lack of footprint available (Taherzadeh and Rajendran 2014).

Different treatments results in different products including fertilizers from composting; fertilizer, fuel, and energy from anaerobic digestion; fuel and electricity from thermochemical processes; and methane collection from landfills. This chapter focuses on the economics of the wastes collection and transportation, and economics of different treatment methods including composting, anaerobic digestion, landfilling, and thermochemical processes. A deep insight is attempted to make understand fellow researchers and academicians the importance of economic viability and diverting the research focus with an emphasis on the economics of the current research that is deployed. As most technologies that are available today are not economically feasible which is why a waste management has not been successful in many countries.

12.2 Waste Collection and Transportation

MSW collection and transportation are considered as the most difficult operations faced by local authorities in many cities across the globe. The collection and transportation services are highly costly and require huge financial support from local taxpayers and/or state governments to operate. The World Bank estimated that in developing countries, municipalities spend between 20 and 50% of their annual budget on solid waste management. Of the allotted budget, up to 60% is spent only for collection and transportation services, which shows the determining factor of current day solid waste management. One of the crucial factors of this high cost is the wages to the employees which is higher and most works are still manual. Therefore, it is essential to plan effectively, the services and budgets for an optimum waste management strategy (Daniel and Perinaz 2012; Hoornweg and Bhada-Tata 2012).

In many developing countries, the improper management system and lack of technical support to the government bodies make the system a complex and ineffective. Moreover, factors such as political interference, legal, cultural, and environmental factors add to this complexity (Kum et al. 2005). Currently, the world cities are spending about $205.4 billion on effective solid waste management practices. This cost is expected to double by 2025, which affects the low- and lower middle-income countries such as Nigeria, Indonesia, and Vietnam. About 25% of the spending is expected to happen from Asia, meaning that a strong infrastructure development on waste management practices is necessary. In developing countries, the majority of the solid waste collected is either landfilled or dumped due to the costs associated with treating it. In Japan, most of the solid waste is incinerated, for which the country spends ten times the collection cost (Taherzadeh and Rajendran 2014).

The cost of municipal solid waste collection varies on different cities of the world due to their location, labor costs, equipment costs, etc. In the USA, the curbside collection and transportation to the landfill costs was $3.5 per ton mile, for Thailand the costs vary between $2.9 and $10.4 per ton, in Canada the tipping fee is between $80 and $120 per ton, while Florida spends about $16 per ton of solid waste, whereas Kuwait spends $24 per ton, and in Alaska each household pay $12 per month for curbside collection (Pollock 1987; Young 1991; Chung and Poon 1997; Koushki et al. 1997; Koushki et al. 2004; Affifi Consultants 2000; Rubenstein-Montano and Zandi 2000; Danteravanich and Siriwong 1998). Table 12.1 shows the cost each country group based on their income spent on solid waste management in 2010 and the cost to be expected in 2025 (Daniel and Perinaz 2012).

12.2.1 Collection and Logistics

Municipalities or corporations are generally responsible for the collection of MSW. Sometimes, the municipalities authorize private companies for the collection of MSW, which usually end up as a monopoly. Waste management services consume a huge amount of money, which is spent for collection, transportation, and disposal/ treatment services. Results indicate that that costs for collection are usually higher when there is more than one collector or service provider due to logistic issues.

Table 12.1 Estimated solid waste management costs 2010 and 2025 adapted from (Daniel and Perinaz 2012)

S. No	Country income group	2010 cost	2025 cost
1	Low income	1.5 billion $	7.7 billion $
2	Lower middle income	20.1 billion $	84.1 billion $
3	Upper middle income	24.5 billion $	63.5 billion $
4	Higher income	159.3 billion $	220.2 billion $

Waste collection and transportation, treating the waste are two different entities, which are subsequently followed one after another. Table 12.2 provides the estimation of solid waste management costs at disposal (Daniel and Perinaz 2012).

Let us take an example case study on the taxes paid by the local authorities of Catalonia, Spain, for the collected MSW. Table 12.3 indicates the charges paid by the local authorities for landfills and incineration depending on whether the waste contains bio-waste or not. This charges mentioned in the table do not include the gate fee, which needs to be paid on top of it. According to Catalan Law 8/2008 (modified in 2011) on financing infrastructure and waste management taxes, the authorities have to pay 10 €/T for landfilling and 5 €/T for incineration. However, from 2010, the charges increased as 20 €/T for landfilling and 15 €/T for incineration when they initiated separate collection schemes for bio-waste.

World Bank alarmed that due to increasing populations and rapid urbanization, managing trash would become an uphill task for many municipalities and it needs a

Table 12.2 Estimated solid waste management costs by disposal method (Daniel and Perinaz 2012)

Category	Low-income countries	Lower middle-income countries	Upper middle-income countries	High-income countries
Income (GNI/capita)	<$876	<$876–3465	$3465–10725	$10725
Waste generation (tons/capita/year)	0.22	0.29	0.42	0.78
Collection efficiency (percent collected)	43%	68%	85%	98%
Cost of collection and disposal (US $/tonne)				
Collection	20–50	30–75	40–90	85–250
Sanitary landfill	10–30	15–40	25–65	40–100
Open dumping	2–8	3–10	NA	NA
Composting	5–30	10–40	20–75	35–90
Waste to energy Incineration	NA	40–100	60–150	70–200
Anaerobic digestion	NA	20–80	50–100	65–150

Table 12.3 Taxes paid by the local authorities of Catalonia, Spain (R4R 2014)

Taxable item	2004–2008 €/t	2009–2010 €/t	2011 €/t	2012 €/t	2014 €/t
Landfill	10.0	10.0	12.0	12.4	15.8
Incineration	–	5.0	5.5	5.7	7.4
Landfill (no bio-waste collection)	–	20.0	21.0	21.3	25.4
Incineration (no bio-waste collection)	–	15.0	16.0	16.5	18.6

dedicated attention to overcome. It should also be noted that proper solid waste management is not only building a sustainable and clean city, but also linked to the climate change, health, and other social considerations. Furthermore, World Bank has supported 329 solid waste projects around the globe by spending $4.5 billion. Table 12.4 provides the overview of various solid waste management services provided by World Bank across the world.

Table 12.4 Services provided by World Bank for solid waste management (Daniel and Perinaz 2012)

S. No.	Region	Location	Services provided
1	Africa	Monrovia	The funds were focused on strengthening the financial management, procurement, contracts management, and cost recovery. One sanitary landfill and two waste transfer stations were built, which improved the waste collection from 13 to 50%
2	East Asia and the Pacific	China and Ningbo	Introduced household waste separation and public–private partnership to finance construction of a treatment plant to process kitchen waste
3	Europe and Central Asia	Azerbaijan	Increasing the use of recyclables and reducing the raw materials usage. Regional-based solid waste management services are implemented
4	Latin America and the Caribbean	Three cities in Argentina—Mar del Plata, Rosario, and Salta	A Huge amount of food waste produced in these cities are likely landfilled, instead, strategies are developed to encourage food donation, source separation, partnerships with the food industry and producing high-quality compost
5	The Middle East and North Africa	West Bank, Gaza Strip, Morocco, and other locations	Development of new landfills and recycling services benefitting the conditions of waste pickers. Also, initiation of waste-to-resource,

(continued)

Table 12.4 (continued)

S. No.	Region	Location	Services provided
			energy from waste, and employment opportunities for the informal recyclers
6	South Asia	Nepal and India	Projects are focused on bridging the gaps between the costs of solid waste management services and the revenues collected. Further, projects focusing the complete chain of waste management system from collection to disposal are introduced

Logistics management of MSW is necessary for the uncontrolled production of the waste stream and to reuse the non-renewable resources. Recycling could be the most suitable option for the effective waste management, which lowers consumption of energy sources and economic cost and also the environmental impact. The concept of circular economy and market for recycling needs to be promoted to establish recycling as a potential method of waste reduction activities. MSW management practices can be categorized into three levels, i.e., strategic, tactical, and operational. The major factors that influence the collection and transportation costs are time spent for collection, overall distances, and fuel consumption. The time spent for the collection and time spent for transportation to the landfill should be effectively planned. The distances and the route plan for the transportation to the landfill should be planned effectively, which reduces the fuel costs and saves time. Hence, the internal routing of the truck is always suggested to be studied in advance not to cross the same road twice. The fuel consumption and costs depend on the truck type and operation time. It is recommended that to determine the major costs and savings potential, collection costs and transportation costs need to calculated separately. The favorable cost management for collection and transportation services is mainly followed by low energy consumption and manpower service.

12.3 Economics on Waste Treatment Methods

12.3.1 Composting

Aerobic composting of organic fraction is one of the best methods for converting waste to worthy products. The solid waste in the presence of oxygen will lose a fraction of its weight in the form of carbon dioxide, while a major fraction is turned into compost, which could be potentially used as an organic fertilizer. The quality of the compost is an important factor in deciding the price, which is why it is essential to operate the composting plant under optimum conditions. Many composting methods are followed in different countries according to their economic and environmental conditions such as windrow composting, pile composting, in-vessel composting, and vermicomposting. Most of the Western countries such as USA, Canada, Israel, and European Union countries follow pile composting because of its low investment and operational cost compared to other methods. The cost of disposing or landfilling of organic materials and the problem of leachate collection had helped the composting. Moreover, it can earn revenue, where landfill and leachate do not have that option (El-Fadel et al. 2002; Pelaez et al. 2009).

In the USA, the most suitable options for composting are Ag-Bag and in-vessel technologies as it provides a suitable environment, including odor and vector management; however, it's expensive. In the case of turned windrows, it is cheapest nonetheless, potential odor and vector problem exist. Covered aerated static piles and vermicomposting are the least options as it is capital intensive (Pisarek 2012). Vermicomposting technology provides very good-quality compost; however, the associated costs are expensive making it not feasible for large-scale applications. Composting organic fraction of MSW in low- and middle-income countries is a difficult task due to the unavailability of segregated waste, moreover, financing is not that efficient, and law policies are also not supporting appropriately for its installation and operation.

Composting plant installed large scale in some Asian countries such as China and India faced problems due to the unsegregated waste. The market for MSW composting plants in developing countries is not good enough as the cost of chemical fertilizers is economically cheaper and high productivity reporting. However, using compost has many advantages such as land restoration, moisture maintenance, and increase soil fertility for the long term. Restoring the carbon lost to the atmosphere happens with the use of compost, unlike chemical fertilizers. The estimated cost analysis for the different composting process is provided in Table 12.5. The capital cost covers the land requirement, equipment's purchase, and others, whereas the operational cost covers running expenses such as fuel and lubrication, equipment's maintenance, electricity, material transfer, water, bulking material, and labor wages (van Haaren 2009).

Decentralizing a composting plant to a community scale decreases the cost of investment. Apartments or colony, commercial places, i.e., markets and public gathering places including schools, colleges, and others where the organic waste is generating about 5–50 tons per day can do compost in nearby barren lands or any corner of parks or gardens. The centralized way of composting is better for more than 10–500 tons of organic waste per day. Market considerations are only one piece of a complex puzzle, apart from that lot of other essential success factors such as financial management and the environmental policy of the particular locality. On the financial management side, experience has shown that while grants and loans are necessary for the initial buildout of the facility, operating costs must be financially self-sufficient. In India, a minimum net profit of ₹100 ($1.5) per MT of compost is achieved with higher margins received on compost traded by Karnataka Compost Development Corporation from other producers, and it showed the better way for other developing countries (van Haaren 2009).

Table 12.5 Estimated cost analysis for different composting technologies (Kaza et al. 2016; Chen 2016; Composting Council of Canada 2017; van Haaren 2009; Pisarek 2012)

Composting method	Capital investment ($/ton)	Operational cost ($/ton)	Process includes	Advantages and drawbacks
Windrow or pile	40–60	40–50	Material shredding, screening, and composting	Low investment but slow process with poor-quality compost
Aerated static pile	100–150	40–45	Material shredding, screening, aeration and composting	Medium investment, fast process, and good-quality compost
In-vessel	300–500	100–150	Material shredding, screening, rotation or aeration, and composting	High investment, fast process, less area requirement, and good-quality compost
Vermicomposting	200–400	<200	Material shredding, screening, seeding, Vermiculturing, and compost screening	High investment, slow process, labor intensive, and very good-quality compost

12.3.2 Anaerobic Digestion

Converting methane from organic fractions of the solid wastes by anaerobic digestion has multiple benefits over other waste to energy processes. It is best adaptable to most types of high-moisture organic wastes and is operated at mesophilic or thermophilic temperature range with relatively high-energy recovery. The anaerobic digestion technology for methane conversion is simple and well established, and research on the utilization of biogas for thermal, electrical, and mechanical power generation is progressive. In addition, the residues of anaerobic digestion have positive environmental value to utilize as a nutrient for plants, irrigation water, and animal/fish feed (Rajendran et al. 2012). Noteworthy research has been done in recent years to improve the performance and efficiency of anaerobic digestion, particularly when treating solid wastes, characterized by a high degree of particulate material.

Kitchen waste originating from door-to-door collection contains low levels of inert materials (plastic, glass, stones, etc.) and is a good feed for anaerobic digestion (AD), aiding biogas production in the range 160–180 m^3/ton of raw waste treated (Bolzonella et al. 2006).

Different digester types are available for the treatment of organic solid waste; few of them are covered anaerobic lagoons, landfill bioreactors, plug flow digesters, completed mixed reactors, two-stage anaerobic digesters, and more. Each reactor has its own techno-economic advantages and drawbacks. Covered anaerobic lagoons are the cheapest option in cost-wise for the treatment of MSW compared to others. The anaerobic lagoon is the simple technique of anaerobic treatment, which is based on a natural ecosystem. The relation between the cost of organic removal and capital or operation cost for various sewage treatment systems at various annual interest rates revealed that, for the Indian context, up-flow anaerobic sludge blanket reactor was the more appropriate in terms of expenses and treatment efficiency (Sato et al. 2007).

The capital expense of an organic fraction of a municipal waste treatment facility in Sweden processing 150 MT/day cost between $35 and $40 million USD, whereas the operational expenses varied between 5 and 10% of the capital expenses (Rajendran et al. 2014). Similarly, when forest residues were used as a feedstock, the CAPEX was ranged between $55 and $60 million USD which had a processing capacity of 20,000 MT/year (Kabir et al. 2015). Household digesters are another kind which is completely different from the industrial where the cost of a 1 m^3 digester could range between $50 and $100 (Rajendran et al. 2013). The profitability of an anaerobic digestion facility depends on the type of the product it is produced such as vehicle fuel, electricity, or CHP as there are costs associated with upgrading it. The upgrading costs depend on the type of the upgrading method employed, for instance, water scrubbing which is the most common upgrading method had an energy consumption of 0.20–0.30 kWh/m^3, whereas amine scrubbers require only half of it, making it economically feasible. However, for amine scrubbing, high throughput is necessary, otherwise, the economic feasibility is questioned (Petersson and WeLLInGer 2009). Table 12.6 shows the cost of anaerobic digestion facility in different economies around the world.

Table 12.6 Estimated cost analysis of an anaerobic digestion plant for different countries (Faulhaber and Raman 2011; Kaza et al. 2016; Moriarty 2013; McCrea et al. 2009; Patterson et al. 2011)

Anaerobic digestion plant	Capital cost ($/ton)	Operational cost ($/ton)	Advantages and drawbacks
High-income countries	300–500	40–80	Segregated waste available, high biogas production, successful running of plant but higher labor wages and another operational cost
Upper middle-income countries	400–500	30–60	Segregated waste available, high biogas production, successful running of plant but higher operational cost
Lower middle-income countries	200–300	20–35	No segregated waste available, average biogas production, hard running of plant but skilled labor available for lower wages
Low-income countries	NA	NA	No waste collection or no segregated waste in many countries

12.3.3 Incineration/Combustion

Incineration or combustion is a thermal process in which the refuse after biological treatment or dry waste is fed to the boiler, where steam is generated. The generated steam is converted to electricity or heat depending on the geographical conditions. For the incineration or combustion to process, it is important that the waste is dry or low moisture. When the moisture content of the solid waste is higher, a lot of energy is consumed to expel the moisture, resulting in lower net energy production. Due to this sometimes, a high-moisture waste requires more fuel than it is processed during combustion or incineration.

The capital cost of an incineration facility processing 100,000 MT/year costs between 35 and 50 Million Euro, and doubling the capacity costs between 50 and 80 Million Euro due to economies of scale. (Yassin et al. 2009). The OPEX of a combustion facility which has a treatment capacity of 50,000–400,000 MT/year ranges between 49 and 77 Euro/ton. Similarly, the OPEX for gasification or pyrolysis of plant capacity 30,000–360,000 MT/year was 28–77 Euro/ton (McLanaghan 2002). The operating costs for running different types of equipment was 5.8–9.2 Euro/MWh for a gas engine, 4.6–5.4 Euro/MWh for combined cycle gas turbine, and for steam turbine, it was 1.5–2.3 Euro/MWh. Comparing gas engine, combined cycle gas turbine, and steam turbine, gas engine was economical and yielded an IRR between 11.1 and 11.3% which could be attributed to its lower capital costs. However, as reported earlier the operating costs are also higher for the gas engines. The thermal energy obtained from the waste could be between 29 and 59 MW (Yassin et al. 2009; Martinez-Sanchez et al. 2016).

12.3.4 Landfills

Evaluation of the economics through the cost-benefit analysis in the life cycle of the landfill facility is crucial to assess the feasibility. Apart from the capital and operational costs, any incremental use of processes or technologies will affect the net expenditures and total turnover costs. Efficient and fast waste stabilization demands high energy and advanced infrastructure in terms of equipment and support facilities. There are also possibilities of additional costs for monitoring and surveillance of the high-throughput systems and for the maintenance of this equipment. However, these practices are beneficial in term of environmental gains and revenue saving (leachate disposal and energy production) that are most valued in the present context. Costs associated with substituting or adding one more unit process and the economics of risks, potential failures on catastrophes involving environmental liabilities are difficult to predict (Naveen et al. 2017). The various costs associated with the landfill facilities are illustrated in Fig. 12.1.

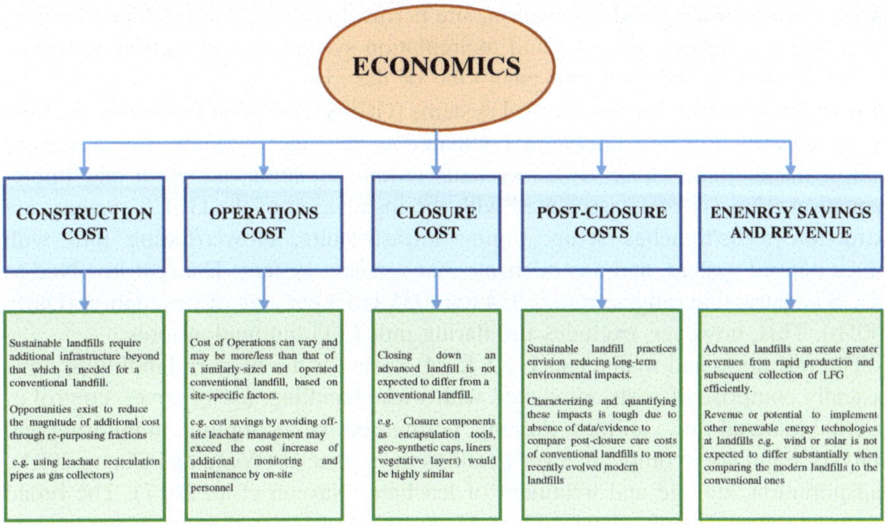

Fig. 12.1 Various cost components deciding the economy of the landfill facilities

Foremost significant and crucial part of the solid waste disposal through landfilling is the estimation of land acquisition, infrastructure, operation and maintenance costs and forestalled privileges and benefits in terms of revenues. The short-term and long-term costs and benefits impact the decisions on investments for the initial setup/construction and operation costs and are advisable to be dealt critically with comparisons with a base case. This aids in better decision making and predicting the risks and liabilities associated with adopting futuristic technologies and investment accordingly. Such kinds of economic projections for feasibility assessments before heading on to the construction phase often benefit the stakeholders. The initial cost prior to implementation basically comprises of the up-front land development costs before any construction ranges from 0.75 to 1 Million USD (Duffy 2005a). During the development phase, there is a need for land leveling and appropriation with the construction of access roads, installations of connections as electricity, water supply, and materials resources with temporary support structures for storage and future land explorations. The various cost elements in this process include site evaluation, planning, design, permitting, borrow source, land acquisition, site fencing and access control, site buildings/support structures, weigh scales, site utilities, access roads, landscaping, and financial assurance.

The cost components associated with construction phase comprise of earthwork for site preparations, implanting either a clay or geotextile liner, geomembrane liners, leachate collection, and containment systems. The cost associated with these constructional activities ranges from 0.15 to 0.45 M USD per the acre of the land constructed excluding the costs incurred for earthwork (ground improvement or grade preparation). Broadly, the various cost components during the construction phase are site clearing and excavation, site berms, liner systems, leachate pumping and storage, leachate collection and recirculation system, gas extraction and monitoring wells (groundwater and gas). One of the major capital-intensive installations is the gas collection and control systems (GCCS) that often requires more time for its construction and operation compared to the liner systems. For advanced landfill installations, these GCCs systems require an initial adoption and implementation. Various costs associated with the installation of the GCCS systems are extraction wells/trenches setup, piping infrastructure, blower/flaring unit with robust control system, and a condensate management system. The cost involved in GCCS construction ranges from 0.024 to 0.035 USD per acre of installation (Duffy 2005b). This, however, excludes the flaring unit LFG diminution tools.

The operation and maintenance (O & M) costs for the modern landfilling units basically comprise of costs associated with waste handling, use of cover, control of external material flow, utilities, training, permission fees, sampling, monitoring and analysis, regulatory compliance verifications through tests (Weng and Chang 2001), transportation, storage and treatment of leachate (Naveen et al. 2017). The broad cost components involved with O & M are (a) costs related to daily operations, (b) daily cover costs, and (c) monitoring cost. The O & M costs contribute to the bulk of expenditures incurred in the management of the landfill facilities mostly consisting of >50% of the landfills overall cost. The O & M cost for the GCCS is reported to be 0.0041 M USD per acre per annum (USEPA 2016) and varies depending upon a total number of gas collection units/wells, systems automation, and configuration, sampling and monitoring frequency. As a crucial activity, regular monitoring, surveillance and compliance check are required for landfill gas emissions during the entire lifetime and post-closure. Finally, after the target filling spatial extent is through, subsequent to completion of operational periods of the landfill facilities, the landfill closure happens. This involves a prominent cost required for the capping (final cover over the underlying fillings) that ranges from 0.15 to 0.3 M USD per acre (Duffy 2005a).

The post-closure various collection, monitoring, and management practices need to be continued for ensuring sound environmental compliance with regular checking of leachate removal and treatment, gas collections, maintenance of the capping/cover area. The cost incurred for the yearly post-closure surveillance and management (comprising of site security, environmental monitoring cap maintenance) ranges from 0.002 to 0.003 M USD per acre of the land maintained. The cost components associated with the final closure, post-closure, and future site utilities are (a) cost incurred for final cover, (b) cost incurred for post-closure care, and (c) final site utility with possibilities of urban structures. The owners of the operational landfill units derive the basic revenue as the fees charged to dispose of the waste in the landfill (tipping fee). This fee is mostly based on the quantity of the waste disposed of, observed by weighing machines near the entrance to the sites. This tipping fee is used for remitting the infrastructure and operational expenditures during the landfilling process with the costs associated with the closure and post-closure assignments. The average tipping fee reported ranges from 15 to 96 USD/ton of waste collected (Van Haaren et al. 2010). The other sources of revenue generation are through the auction of electricity and/or the LFG. The prices of the electricity developed and the gas produced after necessary treatments are on par with the prices from other sources and are often gain incentives as renewable power. Moreover, the opportunity cost of the landfill location as power hubs for installation of solar/the wind as renewable sources of energy or land reclamations through mining is attractive and can become a major material and energy recovery option.

As discussed in the previous section, there can be cost for technological interventions for resource recovery and possibilities of energy generation. A substantial cost is required for laying out the infrastructure (equipment, controls, and automation) and operation and management for (a) liquid injection, (b) leachate recirculation, (c) air injection, (d) sampling and monitoring systems, (e) side-slope control during the various constructional phases of GCCS. The technological interventions aids in enhancing the life of a landfill site with efficient airspace utilization, reduced treatment of leachate and consequent savings on disposal costs, land reuse and opportunities of large quantities of gas production that aids in generating additional revenue through energy production (Berge et al. 2009). The breakeven of the major cost components is depicted in Fig. 12.2 with assumptions for cost calculations based on (Berge et al. 2009).

In the above cases, the capital costs for the construction of infrastructure for both the conventional and modern landfill installations are almost comparable. However, the greatest difference appears to be the cost of the leachate treatment that is phenomenally high compared to modern landfills use to efficient conversion of leachate and reduction in the final volumes. The cost involved in infrastructure required for liquid addition is crucial is one of the aforementioned decisive steps in the adoption of the landfill practice. For example, the choice of selection of surface technique over subsurface technique in based on cost criteria. The subsurface

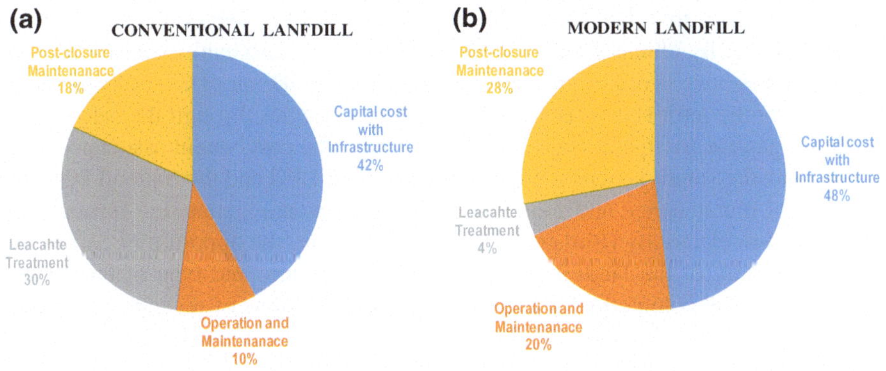

Fig. 12.2 Comparative account of component costs for **a** Conventional landfill and **b** Modern landfill adapted from Berge et al. (2009)

systems are costlier and require more resources. Reduced leachate volumes and fractional treatment result in significant cost savings. A higher leachate recirculation frequency ensures higher systems efficacy in terms of leachate volume generated and treatment and stabilization requirements. These results have been tested in earlier studies (Berge et al. 2009). In terms of gas management and associated costs, the option of air addition results in high quantities of LFG generation that can thus have many cumulative impacts on the economy of the process. Important drivers that decide the benefits and costs associated in LFG are rapid gas production during operations and difference in gas production after the closure of the site. Early stage gas collections are beneficial for the economy. However, inefficient collection in the active landfilling stages may lead to reduced benefits of LFG production and enhances the risks of GHG emissions with a foul odor. Many factors that influence the revenue generation and thus project economics from LFG are collection rates and efficiency, proximity to nearby industrial plants and their energy requirement, electricity prices, prices of natural gas. The LFG to electricity infrastructure cost ranges from 1400 to 5500 USD/kW. However, the O & M costs range from 130 to 380 per unit based on the scale of the project. The airspace recovery otherwise known for the potential creation of additional disposal capacity is very essential and decides the major chunk of benefit in terms of landfill economics. Large airspace is created with quicker decomposition rates and thus benefits in creating additional space for more decomposition. This has been demonstrated by studies conducted on aerobic vs. anaerobic systems that revealed high airspace volumes with short operations time aids in faster aerobic degradation (Powell and Townsend 2004).

One of the most important and among the critical overlooked costs is the social costs (Dijkgraaf and Vollebergh 2004). This mostly comprised of (a) amenities requirement and land use and land cover implications, (b) emission of pollutants, and (c) deterioration of the environment through potential GHG emissions (Sasao 2004). Such costs are very difficult to be quantifying and mostly require a stochastic approach for modeling and prediction. The loss in aesthetics and amenities includes noise, odor, reduced opportunity costs of the areas in the vicinity of the landfill site. Studies conducted on the damages caused due to a loss of aesthetics and amenities values were close to 1 USD/ton (Méry and Bayer 2005). GHG emissions can be greatly reduced by efficient gas capture and processing mechanisms (Amini et al. 2012). Studies have revealed costs of damages as 21 USD/ton of CO_2 emitted (16). This cost can be retrieved by potential marketing strategies through the sale/supply of landfill gas/electricity (Jamasb and Nepal 2010). The cost benefit associated with the post-closure stages is dependent on the amount of LFG and the tradeoff between the costs of maintenance and monitoring with the revenue generation through the sale of LFG or electricity (Barlaz et al. 2002). Moreover, the opportunity cost of the site after closure of the landfill is dependent on the nature and type of land cover intended after the completion of the landfill project. Options of revenue generation by the installation of solar/wind firms can greatly benefit the environment and potential hubs for renewable energy generation. Besides this material mining in landfills can also potentially generate substantial revenues that vary from 4 to 8 USD/m^3 (Innovative waste consulting services LLC 2009).

12.4 Conclusion

Economics is the most important aspects for industrializing the waste management practices. There are several factors which affect the economics of solid waste management including political, regional, geographical, habitual, policy level. The cost of running a waste management facility from the point of generation to the final treatment, i.e., landfilling was discussed in this chapter. It is essential to check and analyze several research methodologies to a feasibility analysis to reduce the research cycle and develop products for the waste management industry. The different technologies such as composting, anaerobic digestion, incineration, and landfilling and the cost of building and operating such facilities were also discussed in this chapter.

References

Affifi Consultants (2000) High variations in the costs of municipal solid waste management, vol 6

Amini HR, Reinhart DR, Mackie KR (2012) Determination of first-order landfill gas modeling parameters and uncertainties. Waste Manage 32(2):305–316

Barlaz MA, Rooker AP, Kjeldsen P, Gabr MA, Borden RC (2002) Critical evaluation of factors required to terminate the postclosure monitoring period at solid waste landfills. ACS Publications

Berge ND, Reinhart DR, Batarseh ES (2009) An assessment of bioreactor landfill costs and benefits. Waste Manage 29(5):1558–1567

Bolzonella D, Pavan P, Mace S, Cecchi F (2006) Dry anaerobic digestion of differently sorted organic municipal solid waste: a full-scale experience. Water Sci Technol 53(8):23–32

Chen Y-T (2016) A Cost analysis of food waste composting in taiwan. Sustainability 8(11):1210

Chung S-S, Poon C-S (1997) Quantifying externalities in solid waste management in Hong Kong. J Environ Eng 123(3):282–289

Composting Council of Canada (2017) Composting processing technologies. Composting Council of Canada, Canada

Daniel H, Perinaz B-T (2012) What a waste: a global review of solid waste management. World Bank

Danteravanich S, Siriwong C (1998) Solid waste management in Southern Thailand. J Solid Waste Technol Manage 25(1):21–26

Dijkgraaf E, Vollebergh HR (2004) Burn or bury? A social cost comparison of final waste disposal methods. Ecol Econ 50(3):233–247

Duffy D (2005a) Landfill economics part I: siting. MSW Manage

Duffy DP (2005b) Landfill economics part III: closing up shop. MSW Manage

El-Fadel M, Bou-Zeid E, Chahine W, Alayli B (2002) Temporal variation of leachate quality from pre-sorted and baled municipal solid waste with high organic and moisture content. Waste Manage 22(3):269–282

Faulhaber C, Raman DR (2011) Techno-economic analysis of farm scale plug-flow anaerobic digestion. Anim Ind Rep 657(1):82

Hoornweg D, Bhada-Tata P (2012) What a waste: a global review of solid waste management. Urban Dev Ser Knowl Pap 15:1–98

Innovative waste consulting services LLC (2009) Landfill reclamation demonstration project. Florida department of environmental protection, Florida, USA

Jamasb T, Nepal R (2010) Issues and options in waste management: a social cost–benefit analysis of waste-to-energy in the UK. Resour Conserv Recycl 54(12):1341–1352

Kabir MM, Rajendran K, Taherzadeh MJ, Horváth IS (2015) Experimental and economical evaluation of bioconversion of forest residues to biogas using organosolv pretreatment. Bioresour Technol 178:201–208

Kaza S, Yao L, Stowell A (2016) Sustainable financing and policy models for municipal composting

Koushki P, Al-Duaij U, Al-Ghimlas W (2004) Collection and transportation cost of household solid waste in Kuwait. Waste Manage 24(9):957–964

Koushki P, Hulsey J, Bashaw E (1997) Household solid waste: traits and disposal site selection. J Urban Plann Dev 123(1):1–9

Kum V, Sharp A, Harnpornchai N (2005) Improving the solid waste management in Phnom Penh city: a strategic approach. Waste Manage 25(1):101–109

Martinez-Sanchez V, Hulgaard T, Hindsgaul C, Riber C, Kamuk B, Astrup TF (2016) Estimation of marginal costs at existing waste treatment facilities. Waste Manage 50:364–375

McCrea M, Tan TK, Ting H-H, Zuo X (2009) A cost-benefit analysis of different waste-to-energy technologies for the management of municipal solid waste in Singapore

McLanaghan S (2002) Delivering the landfill directive: the role of new and emerging technologies. Rep Strategy Unit 8:2002

Méry J, Bayer S (2005) Comparison of external costs between dry tomb and bioreactor landfills: taking intergenerational effects seriously. Waste Manage Res 23(6):514–526

Moriarty K (2013) Feasibility study of anaerobic digestion of food waste in St. NREL, Bernard, Louisiana

Naveen B, Mahapatra DM, Sitharam T, Sivapullaiah P, Ramachandra T (2017) Physico-chemical and biological characterization of urban municipal landfill leachate. Environ Pollut 220:1–12

Patterson T, Esteves S, Dinsdale R, Guwy A (2011) An evaluation of the policy and techno-economic factors affecting the potential for biogas upgrading for transport fuel use in the UK. Energ Policy 39(3):1806–1816

Pelaez AI, Sanchez J, Almendros G (2009) Bioreactor treatment of municipal solid waste landfill leachates: characterization of organic fractions. Waste Manage 29(1):70–77

Petersson A, WeLLInGer A (2009) Biogas upgrading technologies–developments and innovations. IEA Bioenergy 20

Pisarek N (2012) Large-scale composting options for YVR: cost analysis

Pollock C (1987) Mining urban wastes: the potential for recycling. Worldwatch paper (USA) no 76

Powell J, Townsend T (2004) An economic analysis of aerobic versus anaerobic bioreactor landfills. In: 19th international conference on solid waste technology and management, Philadelphia, PA

R4R (2014) Good practice Catalonia: waste disposal tax. Regions for Recycling

Rajendran K, Aslanzadeh S, Johansson F, Taherzadeh MJ (2013) Experimental and economical evaluation of a novel biogas digester. Energy Convers Manage 74:183–191

Rajendran K, Aslanzadeh S, Taherzadeh MJ (2012) Household biogas digesters—a review. Energies 5(8):2911–2942

Rajendran K, Kankanala HR, Martinsson R, Taherzadeh MJ (2014) Uncertainty over techno-economic potentials of biogas from municipal solid waste (MSW): a case study on an industrial process. Appl Energy 125:84–92

Rubenstein-Montano B, Zandi I (2000) An evaluative tool for solid waste management. J Urban Plann Dev 126(3):119–135

Sasao T (2004) An estimation of the social costs of landfill siting using a choice experiment. Waste Manage 24(8):753–762

Sato N, Okubo T, Onodera T, Agrawal LK, Ohashi A, Harada H (2007) Economic evaluation of sewage treatment processes in India. J Environ Manage 84(4):447–460

Taherzadeh MJ, Rajendran K (2014) Factors affecting development of waste management. In: Waste management and sustainable consumption: reflections on consumer waste. Routledge, pp 67–87
Tchobanoglous G, Kreith F (1999) Handbook of solid waste management, 2nd edn. Mcgraw-Hill, USA
Tchobanoglous G, Theisen H, Vigil S (1993) Integrated solid waste management: engineering principles and management issues. McGraw-Hill, USA
USEPA (2016) LFG energy project development handbook. Landfill methane outreach program, Environmental Protection Agency, USA
van Haaren R (2009) Large scale aerobic composting of sourceseparated organic wastes: a comparative study of environmental impacts, costs, and contextual effects. Columbia University
Van Haaren R, Themelis N, Goldstein N (2010) The state of garbage in America: 17th nationwide survey of MSW management in the US. Joint study by BioCycle and the Earth Engineering Center of Columbia University
Weng Y, Chang N-B (2001) The development of sanitary landfills in Taiwan: status and cost structure analysis. Resour Conserv Recycl 33(3):181–201
Yassin L, Lettieri P, Simons SJ, Germanà A (2009) Techno-economic performance of energy-from-waste fluidized bed combustion and gasification processes in the UK context. Chem Eng J 146(3):315–327
Young JE (1991) Discarding the Throwaway society. Worldwatch Paper 101. ERIC, Washington DC, USA

Chapter 13
Biodiesel from Microalgae

Rozita Madadi, Meisam Tabatabaei, Mortaza Aghbashlo, Mohammad Ali Zahed and Ahmad Ali Pourbabaee

Abstract Widespread application of non-renewable energy resources such as fossil fuels is limited mainly due to their adverse environmental impacts by increasing the amount of greenhouse gas (GHG) emissions. A solution to limit fossil-fuel pollution is the use of renewable energy resources. In the recent years, microalgae have received considerable attention as a suitable feedstock for biofuel production. Microalgae can grow in various aquatic wastewater media and are able to produce biomass, lipids, and hydrocarbons. Using different types of wastewaters as media for algae cultivation could not only reduce their freshwater footprint but also the costs associated with algae cultivation and biofuel production. This chapter presents an overview on various algal cultivation systems as well as on optimization of algal

R. Madadi
Department of Agricultural Biotechnology, University College of Agriculture and Natural Resources, University of Tehran, Karaj, Iran

M. Tabatabaei (✉)
Biofuel Research Team (BRTeam), Karaj, Iran
e-mail: meisam_tab@yahoo.com; meisam_tabatabaei@abrii.ac.ir

M. Tabatabaei
Department of Microbial Biotechnology, Agricultural Biotechnology Research, Institute of Iran (ABRII), Agricultural Research, Education and Extension Organization (AREEO), Karaj, Iran

M. Aghbashlo (✉)
Department of Mechanical Engineering of Agricultural Machinery, Faculty of Agricultural Engineering and Technology, College of Agriculture and Natural Resources, University of Tehran, Karaj, Iran
e-mail: maghbashlo@ut.ac.ir

M. A. Zahed
Faculty of Biological Sciences, Kharazmi University, Tehran, Iran

A. A. Pourbabaee
Department of Soil Science, Faculty of Agricultural Engineering and Technology, University College of Agriculture and Natural Resources, University of Tehran, Karaj, Iran

R. R. Singhania et al. (eds.), *Waste to Wealth*, Energy, Environment, and Sustainability, https://doi.org/10.1007/978-981-10-7431-8_13

cultivations, while downstream processes including harvesting and drying of microalgae and lipid extraction systems are also reviewed and discussed. Subsequently, different microalgae biofuel production pathways are presented. Finally, the applications of microalgae in integrated systems, i.e., in wastewater treatment and biodiesel production systems and biofixation of carbon, are scrutinized.

Keywords Biodiesel · Algae · Waste treatment · CO_2 fixation

13.1 Introduction

Energy resources are divided into non-renewables and renewables, with the latter having a minor contribution to the global energy market at the present time. However, due to the limitations on non-renewable energy resources and increasing greenhouse gas (GHG) emissions as a result of using these fuels, the share of renewable energy resources such as biofuels, hydro, wind, solar, and geothermal energies is bound to increase (Bwatanglang et al. 2015). Among the above-mentioned renewables, only biofuels are being used globally in the transportation sector, the main GHG emitter. Based on their feedstocks, biofuels are classified into first-generation biofuels (FGBs) produced from sugar, starch, animal fats, and vegetable oils; second-generation biofuels (SGBs) produced from non-food crops, agro-forest residues, and wastes; and third-generation biofuels (TGBs) produced from microalgae (Demirbas and Demirbas 2011; Laghari et al. 2015).

Microalgae, photosynthetic microorganisms, could be grown on non-arable land and have an acceptable growth rate (20–30 times faster than other conventional energy crops) and high photosynthetic conversion efficiency (Ullah et al. 2014). Cultivation of microalgae consumes less water than land crops, and unlike corn, soybean, and palm as main sources of biofuel production, algae is not used as a primary food source for human being, affirming that they can be used distinctively as fuel while having less impact on food security. Moreover, due to their ability to withstand high CO_2 contents in gas stream, microalgae have high efficiency for CO_2 mitigation as well (Demirbas and Demirbas 2010; Mata et al. 2010; Wang et al. 2008; Zhang 2015).

Microalgae are classified into four main taxonomic groups: diatoms (*Bacillariophyceae*), green algae (*Chlorophyceae*), cyanobacteria or blue-green algae (*Cyanophyceae*), and golden algae (*Chrysophyceae*). These microorganisms contain high lipid, high protein, and low carbohydrate content. Nowadays, the main interest is in cultivating microalgae to produce lipid as feedstock for biodiesel (Markou and Nerantzis 2013), while other types of biofuels are also of some interest. The lipid content, lipid productivity, and different types of biofuel reportedly produced from different microalgae are shown in Table 13.1. Moreover, a comparison among different biodiesel feedstocks in terms of their oil properties is also presented in Table 13.2.

Table 13.1 Lipid content, lipid productivity and different types of biofuel production from different microalgae species[a]

Microalgae	Algae type	Lipid content (% dry wt. biomass)	Lipid productivity (mg L^{-1} d^{-1})	Volumetric productivity of biomass (g L^{-1} d^{-1})	Areal productivity of biomass (g m^{-2} d^{-1})	Major biofuel type
Arthrospira maxima	Blue-green	20.34	–	–	–	Hydrogen, biodiesel
Chlorella emersonii	Green	25.0–63.0	10.3–50.0	0.036–0.041	0.91–0.97	–
Chlorella protothecoides	Green	14.6–57.8	1214	2.00–7.70	–	Biodiesel
Chlorella sorokiniana	Green	19.0–22.0	44.7	0.23–1.47	–	–
Chlorella vulgaris	Green	5.0–58.0	11.2–40.0	0.02–0.20	0.57–0.95	Ethanol
Chlorella	Green	18.0–57.0	18.7	–	3.50–13.90	Biodiesel
Chlorococum sp.	Blue-green	19.3	53.7	0.28	–	Biodiesel, ethanol
Dunaliella salina	Green	6.0–25.0	116.0	0.22–0.34	1.6–3.5/20–38	–
Dunaliella sp.	Green	17.5–67	33.5	–	–	Ethanol
Haematococcus pluvialis	Red	25.0	–	0.5–0.6	10.2–36.4	Biodiesel
Nannochloris sp.	Green	20.0–56.0	60.9–76.5	0.17–0.51	–	–
Nannochloropsis oculata	Green	22.7–29.7	84.0–142.0	0.37–0.48	–	–
Nannochloropsis sp.	Eustigmatophytes	12.0–53.0	37.6–90.0	0.17–1.43	1.9–5.3	–

(continued)

Table 13.1 (continued)

Microalgae	Algae type	Lipid content (% dry wt. biomass)	Lipid productivity (mg L^{-1} d^{-1})	Volumetric productivity of biomass (g L^{-1} d^{-1})	Areal productivity of biomass (g m^{-2} d^{-1})	Major biofuel type
Neochloris oleabundans	Green	29.0–65.0	90.0–134.0	–	–	Biodiesel
Scenedesmus obliquus	Green	11.0–55.0	–	0.004–0.74	–	Methanol, hydrogen
Spirulina platensis	Green	4.0–16.6	–	0.06–4.3	1.5–14.5/24–51	Hydrogen

Mata et al. (2010), Maity et al. (2014)

Table 13.2 Comparison of different biodiesel feedstocks[a]

Conventional feedstocks	Oils or fats	Oil content (% oil by wt. biomass)	Physicochemical properties of biodiesel feedstocks							Water footprint (m^3 GJ^{-1})	Land use (m^2 GJ^{-1})	Biodiesel yield (L ha^{-1} a^{-1})
			Density (kg/m^3)	Kinematic viscosity at 40 °C (mm^2 s^{-1})	Cetane no. (°C)	High heating value (MJ kg^{-1})	Flash point (°C)	Saponification value	Iodine value			
Edible	Canola	41	911.5	34.72	37.6	39.7	246	189.80	–	383	258	1190
	Soybean	18–22	913.8	28.87	37.9	39.6	254	195.30	128–143	383	689	446
	Sunflower	25–35	916.1	35.84	37.1	39.6	274	193.14	125–140	61	323	951
	Palm	30–60	918.0	44.79	42.0		267	208.63	48–58	75	52	5906
	Peanut	42–52	902.6	39.60	41.8	39.8	271	191.50	84–100	58	220	1396
	Corn	48	909.5	30.75	37.6	39.5	277	183.06	103–128	–	–	152 (kg biodiesel ha^{-1} $year^{-1}$)
	Coconut	63–65	918	27.26	–	–	–	267.56	7.5–10.5	49	128	2399
	Cottonseed	18–25	914.8	33.50	–	39.4	234	198.50	103–115	135	945	325
Non-edible	Jatropha	30–40	940	33.90	–	38.65	225	200.80	82–98	383	258	1190
	Castor	48	955	251.20	42.3	37.4	–	191.08	83–86	–	–	1156 (kg/ha year)
	Microalgae *Chaetoceros* sp.	70	1.305 gm/ml	6.2	–	–	–	173.56 mg/mg of oil	–	<379	2–13	24355–136886
	Spirogyra	14.82	884	4.4	–	–	–	–	–	–	–	–
	Cladophora	11.76	892	3.8	–	–	–	–	–	–	–	–
	Tolypothrix	12.78	857	4.1	–	–	–	–	–	–	–	–
	Mahua	35	960	24.50	–	36.0	232	190.5	58–70	–	–	–
	Neem	30	918.5	50.30	–	–	–	209.66	65–80	–	–	–

[a]*Sources* Demirbas (2009), Demirbas and Demirbas (2010), Karmakar et al. (2010), Mata et al. (2010), Kumar et al. (2011), Moser and Vaughn (2012), Ananadhi Padmanabhan and Stanley (2012), Atabani and Silva César da (2014) and Singh et al. (2016)

Microalgae can grow in various aquatic environments, such as freshwater or marine water (Zhou et al. 2011a, b), industrial wastewaters (Wang et al. 2010, Zhou et al. 2013), municipal wastewaters (Kong et al. 2010), animal wastewaters (Wang et al. 2010; Zhou et al. 2012; Hu et al. 2012), and agricultural wastewaters. Accordingly, many studies have strived to promote biofuel production using wastewater resources as a means of improving the economic aspects of algal fuels production (Pittman et al. 2011; Wu et al. 2012). However, in a rather recent critical review, Chisti (2013) pointed out the constraints to microalgal biofuel commercialization. Among those was the calculations concerning the inadequacy of wastewater as a source of nitrogen and phosphorus for microalgal cultivation and that the algal biofuels produced using the wastewater generated in a metropolitan area such as New York city could only be sufficient to replace 1–3% of the petroleum demands of the city. As shown in Table 13.3, presenting the potential of algal fuel production from wastewater in major cities in the world, this is absolutely true and such a scenario would be totally inefficient if one places the main focus on biofuel production using wastewater.

To the contrary and by highlighting algal-based wastewater treatment instead of biofuels production, i.e., by looking at this scenario other way around, different conclusions could be made. Accordingly, biofuel production using wastewater would come second as a strategy to further justify, or economize, the algal-based treatment process of various types of wastewater. Table 13.4 tabulates the pros and cons of microalgal-based wastewater treatment systems and compares them with the conventional wastewater treatment procedures.

The present chapter aims to review the developments made and success stories reported in different aspects of algal cultivation and harvesting/extraction within the framework of integrated biofuel production/wastewater treatment systems. Furthermore, the application of microalgae in integrated systems, i.e., microalgae-driven wastewater treatment and algal-based carbon biofixation with simultaneous biofuels production, has been brought into attention.

13.2 Algae Cultivation Systems

13.2.1 Suspended Culture

The most common large-scale algae production systems are based on suspended culture. In these cultures, including open ponds and closed reactors, single cells and small groups of cells are maintained in liquid medium. This medium requires agitation and gas exchange.

Table 13.3 Potentials for application of wastewater generated in major cities around the world as a source of nutrients for algal cultivation

City	Population	Petroleum Consumption ($m^3\ yr^{-1}$)	Wastewater generation ($m^3 d^{-1}$)	Annual algal oil production potential from wastewater (m^3)	Max. potential of wastewater-driven algal oil to replace petroleum demand of city (%)	References
A large US city as model	10,000,000	35,770,000	3,780,000	425,000	1.07	Chisti (2013)
Toronto	5,132,794	19,015,718	1,940,196	218143.745	1.03	Present study[a]
Tehran	8,293,140	10,473,406	3,134,806	352458.45	1.67	Present study
New York	8,405,837	28,840,426	3,177,406	357248.0725	1.11	Present study
Beijing	21,150,000	87,233,175	7,994,700	898,875	9.29	Present study
Paris	2,273,305	3,592,844	859,309	96615.4625	2.42	Present study
Sydney	4,840,628	12,774,175	1,829,757	205726.26	1.45	Present study
Moscow	11,500,000	14,817,175	4,347,000	488m750	2.97	Present study
Tokyo	13,350,000	26,995,035	5,064,300	569398.8	1.97	Present study
Berlin	3,562,166	6,019,882	1,346,498	151392.05	2.26	Present study

[a]These values were calculated according to Chisti (2013) and are reported for the first time in the present work

Table 13.4 Pros and cons of microalgal-based wastewater treatment systems in comparison with the conventional wastewater treatment procedures

Wastewater treatment	Type	Advantages	Disadvantages	References
Algae-based systems	HRAP[a]	– Simple and cost effective	– Algal biomass harvesting is difficult – Risk of contamination is high – Control on algal species is low – Unapplicable water footprint	Park et al. (2011)
	Immobilized	– Algae harvesting is facilitated and economical – Possibility of nutrient removal as well as other pollutants such as heavy metals and industrial pollutants	– Phosphate-removal efficiency is dependent on elevated pH of the wastewater – It is always accompanied with enhance removal of nutrients	de-Bashan and Bashan (2010)
	Attached algal system	– Biomass harvesting is facilitated – Improved water quality	– There is no consensus on the best method of growing and harvesting algal biofilms	Christenson and Sims (2011)
Conventional methods	Chemical	– Low energy requirement	– Cost of treatment is higher than those of the other methods (physical and biological)	Gupta et al. (2012)
	Physical	– Possibility of volatile and semi-volatile organic compounds removal – Possibility of removal of coarse solids	– Energy requirements are high	https://www.teicrete.gr
	Biological	– Cost effective – Possibility of controlling the amount of aeration to avoid excessive dissolved oxygen – Improve efficiency of aeration system	– BOD removal by biological treatment requires higher energy than BOD removal by primary treatment	Mittal (2011)

[a]HRAP: High rate algal pond

13.2.1.1 Open Ponds

Open ponds can be categorized into natural waters (lakes, lagoons, and ponds) and artificial ponds. The most commonly used systems for algae cultivation include circular ponds and raceway ponds. Algae cultivation in open system has some disadvantages, such as the difficulties in controlling contamination, culture environment conditions, poor light utilization by the cells, and requirement for large areas of land, while biomass harvesting is costly as well (Carvalho et al. 2006). Circular ponds are generally round, simple, and mixed with a rotating circular arm fixed in the pond center (Lee and Lee 2001). Raceway ponds are shallow ponds and are used for commercial microalgal production, usually lined with plastic, with a 15–20 cm depth in which water and nutrients circulate around a racetrack with a rotating paddle wheel (Brennan and Owende 2010). High-rae algal ponds (HRAP)s are raceway-type ponds and have 0.2–1 m depth, paddle wheel-mixed, and provide improved wastewater treatment. They are efficient and cost-effective upgrades for treating municipal, industrial, and agricultural wastewater (Park et al. 2011; Craggs et al. 2012). These ponds are in fact a combination of algal reactor and amplified oxidation ponds.

13.2.1.2 Closed Reactors

Closed reactors are expensive to build. However, compared with open ponds, they are much easier to control contamination and environmental conditions. Closed reactors require chemical sterilizers to effectively sterilize. In such reactors, cost of harvesting is less than open ponds and the obtained biomass concentration is higher than open ponds (Lee and Lee 2001; Scott et al. 2010). There are four key requirements for algal growth in reactors. The photosynthetic activity of microalgae depends on light; therefore, light is one of the restrictive factors in the algae culture; if light is too low, growth of microalgae will be slow and their photosynthesis will decline. Conversely, if it is too high, photoinhibition and oxidative damage would occur (Kumar et al. 2010a, b, c). Another key parameter is temperature, and too low and too high values would result in slow growth and cell death, respectively. Fluctuations in temperature can lead to significant decreases in productivity, while the optimal growth temperature for microalgae is often in the range of 20–30 °C (Chisti 2008). Mixing is also an important parameter in microalgal cultivation that improves gas exchange, keeps cells in suspension, distributes the nutrients, and decreases photoinhibition on the surface (Ugwu et al. 2008). Mixing in photobioreactors (PBR) is provided by pumping or aeration through a variety of gas transferring systems. Finally, nutrients are also instrumental in achieving am efficient cultivation system. Low nutrient availability leads to growth inhibition, while high concentration may exert toxic effects. Essential elements for algal growth include nitrogen (N), phosphorus (P), and, in some cases, silicon (Chisti 2007).

Closed reactors can be categorized into flat-plate reactors and tubular reactors. Flat-plate reactors are vertical reactors made up of narrow panels with 10-mm glass plates that are pasted together. Tubular reactors are another type of closed reactors that can be categorized into: horizontal tubular, vertical airlift, and helical tubular. The only type of closed systems used on large scales is tubular reactors (Chisti 2007). The control of temperature and pH in tubular photobioreactors is better than that in open ponds. In comparison with open ponds, tubular photobioreactors can generally provide a better protection against culture contamination, less evaporative loss, better mixing, and higher cell densities (Mata et al. 2010). Horizontal tubular systems are composed of thin-diameter tubing lying or stacked horizontally, while vertical tubular systems are composed of vertical tubes that can be easily erased and kept sterile. This type of reactor is suitably set and manufactured at low cost (Ugwu et al. 2008). Helical tubular systems are constructed of tubing coiled around a circular framework, and hence, angle to sunlight is reduced; subsequently, the

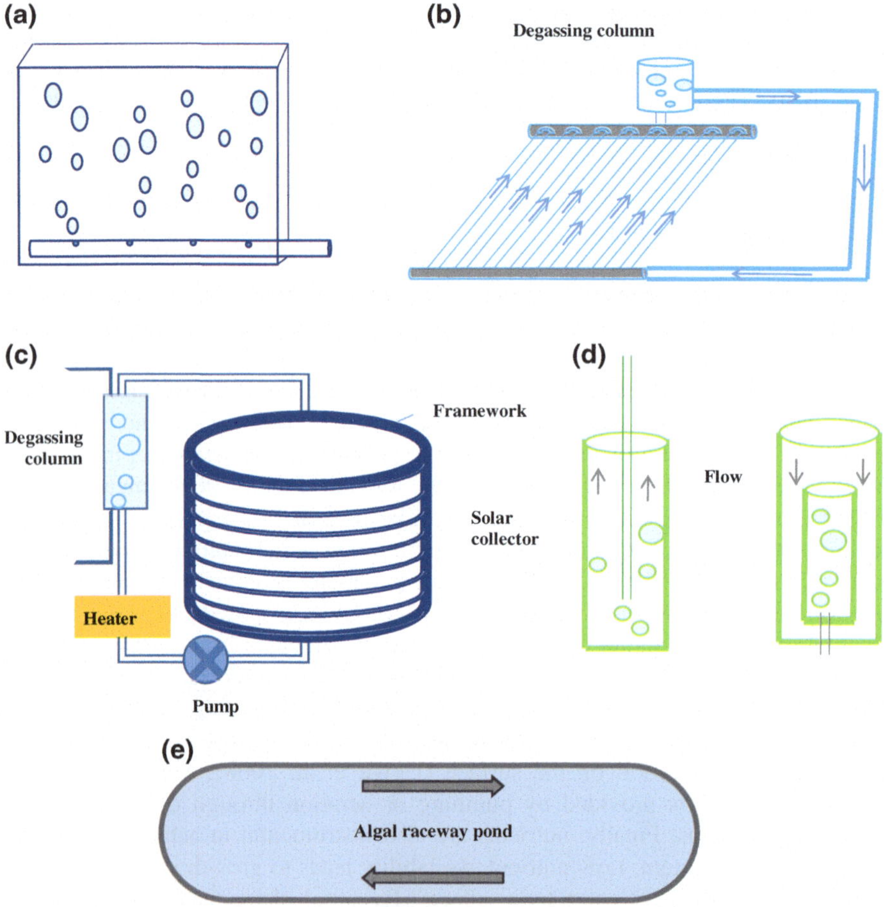

Fig. 13.1 Schematic diagrams for closed reactors: **a** flat plate, **b** horizontal tubular, **c** helical tubular, **d** vertical airlift, and **e** algal raceway pond (Chisti 2007; Xu et al. 2009; Mata et al. 2010; Park et al. 2012)

Table 13.5 Comparison of open and closed culture systems for microalgae[a]

Culture systems for microalgae	Open systems (Ponds)	Closed systems (PBRs)	
		Tubular photobioreactor	Flat-plate photobioreactor
Contamination control	Difficult	Easy	Easy
Species control	Difficult	Easy	Easy
Weather dependence	High light intensity, temperature, rainfall	Medium light intensity, cooling required	Medium light intensity, cooling required
Biomass productivity	Poor biomass productivity	Good biomass productivities	High biomass productivities
Sterility	None	Easy to sterilize	Easy to sterilize
Mixing	Poor mixing	Good mixing	Good mixing
Space required	Large area of land required	Requires large land space	Requires large surface area
Operation costs	Open systems $\ll$ closed systems	Expensive compared to open ponds	Expensive compared to open ponds
Illumination surface area	Light only effectively penetrates 2′–3″ in ponds	Large illumination surface area	Large illumination surface area
Temperature control	Difficult temperature control	More uniform temperature	Difficult temperature control
Evaporation of growth medium	High	Low	Low
Scalability	High	Medium	Difficult
Gas transfer control	Low	High	High
O_2 inhibition	Usually low enough because of continuous	High (O_2 must be removed to prevent photosynthesis inhibition)	High (O_2 must be removed to prevent photosynthesis inhibition)
Maintenance	Easy	Hard	Hard

[a]*Sources* Mata et al. (2010), Brennan and Owende (2010), Christenson and Sims (2011) and Singh et al. (2016)

required land area is reduced (Morita et al. 2001). A comparison of open and closed culture systems for microalgae is shown in Table 13.5 (Pluz 2001; Brennan and Owende 2010; Ugwu et al. 2008). An efficient hybrid system of photobioreactor and open ponds was also suggested by Narala and co-authors (2016). Schematic diagrams for closed reactors and open pond are shown in Fig. 13.1.

13.2.2 Immobilized Algal Culture

Harvesting microalgae is a major challenge in suspended culture at large-scale algae production system. Using immobilized cultures (attached algal processes) could play a major role in overcoming this major challenge to production, i.e., harvesting of microalgae (Hoffmann 1998). In addition to that, immobilization offers a number of other advantages over free-cell systems as well, including less space requirements, easier handling, higher resistance to unfavorable environmental conditions, and the possibility of using higher cell densities in the process as well as reusing the biomass for product generation (Mallick 2002; de-Bashan and Bashan 2010; Christenson and Sims 2011; Eroglu et al. 2015). It should also be mentioned that immobilized cells have been reported to possess higher biosorption capacity and bioactivity (Mallick 2002). These collectively mark immobilized algal cultivation systems as cost-effective processes for scale-up processing. Among the various immobilization processes, the most common ones are discussed herein.

13.2.2.1 Matrix-Immobilized Microalgae

In this method, microalgal cells are immobilized or entrapped in a 3D matrix made of natural (such as agar, cellulose, alginate) or synthetic (such as polyacrylamide, polyurethane, polyvinyl) polymers. de-Bashan and Bashan (2010) argued that the latter is comparatively more stable in wastewater samples, while natural polymers such as alginate are advantageous in terms of their higher nutrient/product diffusion rates and their eco-friendly features (de-Bashan and Bashan 2010). In spite of the promising aspects of matrix immobilization of algal cells, this method is still limited to laboratory scale for the cost of the immobilization matrix is yet to be further decreased in order to be economically justified (Chevalier et al. 2000).

13.2.2.2 Algal Biofilms

The main advantage of algal cultivation systems designed based on algal biofilm is the facilitated harvesting of algal cells by scraping. As mentioned earlier, expensive harvesting systems used in suspended cultures, e.g., flocculation and centrifugation, generally jeopardize the economic viability of these systems, and therefore, algal biofilms when become economically available could assist with overcoming this shortcoming.

13.3 Algal Cultivation Optimization

In order to achieve an economically viable biodiesel production system from microalgal biomass, optimization of algal cultivation in terms of algal biomass and lipid content is of prominent importance. Numerous attempts of different nature have been made in order to achieve the above-mentioned goals. For instance, Bohutskyi et al. (2014) introduced an innovative, mixed trophic state process based on *Auxenochlorella protothecoides* grown phototrophically, to obtain high lipid content for generating algal biodiesel. They argued that simultaneous nitrogen deprivation and glucose supplementation during the heterotrophic stage could boost total lipid content by over threefolds. They also proposed to couple biodiesel production with anaerobic digestion in order to produce biogas from the remaining biomass after oil extraction and stated that the overall energy output of the coupled process could be increased by up to 40% (Bohutskyi et al. 2014).

In a different study, various nutritional modes, including glucose supplementation, were investigated with an aim to enhance biomass and lipid productivity in different microalgal strains. They reported that lipid productivity ranged from 2 to 13% under photoautotrophic conditions, 1.7–32% under mixotrophic conditions, and 0.9–20% under heterotrophic conditions. While under heterotrophic conditions where glucose supplementation was practiced, polyunsaturated fatty acids (PUFA) fraction of the oil was decreased by around 2–4-folds depending on the microalgae strain under investigation. On the other hand, saturated fatty acid (SFA) fraction was also negatively impacted by glucose supplementation. Oils rich in SFA containing low PUFA are ideal feedstock for achieving high oxidative stability in biodiesel, and therefore, glucose supplementation could be serve this purpose well (Ratha et al. 2013).

Another strategy proposed by Duong et al. (2015) was to target both algal lipid and protein simultaneously to improve the economic viability of algal biodiesel production. More specifically, they tried to isolate algal strains meeting three criteria of fast growth, high lipid content, and protein-rich biomass, while that last could be used for animal feed (Duong et al. 2015). Converti et al. (2009) explored the effects of temperature concentration on lipid content in *Nannochloropsis oculata* and *Chlorella vulgaris*. They argued that variations in the investigated factor strongly impacted lipid content. For instance, a temperature boost from 20 to 25 °C increased lipid content by 100%, while an opposite was observed for *C. vulgaris* when the temperature was increased from 25 to 30 °C (Converti et al. 2009).

Nitrogen concentration in the cultivation media is also an important parameter. It is well documented that nitrogen deprivation could result in increased lipid content but could also negatively affect algal growth. Therefore, a trade-off should be observed to achieve the highest lipid productivity. In-depth understanding of the relationships between cell nitrogen content, growth, and cell composition is essential in order to be able to identify an optimal nitrogen content required for most favorable lipid productivity in batch or continuous cultivation modes (Griffiths et al. 2014). It should be highlighted that nitrogen deprivation could also improve the fatty acid profile of algal oils leading to more favorable biodiesel properties (Griffiths et al. 2014).

Nitrogen-to-phosphorus (N/P) ratio could also have a crucial effect on the biomass growth (Alketife et al. 2017). Xin et al. (2010) showed that this ratio is significantly effective on biomass yield and lipid accumulation of a freshwater microalga *Scenedesmus* sp. LX1. They claimed that under nitrogen (2.5 mg L^{-1}) or phosphorus (0.1 mg L^{-1}) limitations, the microalgae under investigation could accumulate lipids to as high as 30 and 53% of its algal biomass, respectively and lipid productivity was not enhanced reportedly. Similar observations were made by Kalla and Khan (2016) who also studied the effect of decreasing nitrogen and phosphorus concentrations on growth, biomass, and lipid content of *C. vulgaris*. They argued that significant decreases were recorded in growth and by decreasing nitrogen and phosphorus concentrations in the medium from (1.5–0.0 g/l) and (0.04–0.0 g/l), respectively. On the contrary, lipid accumulation was enhanced under the phosphorus and nitrogen limitations.

Different ions could also impact algal growth and lipid production significantly. For instance, Huang et al. (2014) investigated the effects of ferric ion concentrations on three species of microalgae (*Tetraselmis subcordiformis*, *Nannochloropsis oculata,* and *Pavlova viridis*). They concluded that growth, lipid content, as well as the fatty acid profiles of the studied microalgae varied in response to changes in ferric ion concentrations and that an optimum ferric ion concentration can improve the properties of respective algal biodiesels.

In a different study performed in high-glycerol content media, the effect of calcium and magnesium ions supplementation was studied using two fast-growing algal strains of *Aurantiochytrium* sp. DBTIOC-18 and *Schizochytrium* sp. DBTIOC-1 for biomass and lipid production (Singh et al. 2016). It was revealed that increasing both calcium and magnesium ions' concentration promoted glycerol utilization and resulted in a significant boost in biomass and lipid production. Such findings highlight the importance of calcium and magnesium ions' concentrations in preventing substrate inhibition under high nutrient concentrations, especially carbon sources to achieve high biomass and lipid yields (Singh et al. 2016).

Sulfate ions are also effective on growth of microalgae. In a recent study, Lv et al. (2017) strived to look into the responses of the self-flocculating microalga *Chlorococcum* sp. GD to different sulfate concentrations in a synthetic municipal wastewater. Their results showed that the microalgal cells grew better in the synthetic municipal wastewaters containing 18, 45, 77, 136, and 271 mg/L SO_4^{2-} than in the control wastewater without SO_4^{2-}. They argued that sulfate deprivation led to significant decreases in antioxidative enzymes and photosynthetic activities and that these in turn significantly weakened the growth and self-flocculation properties of the algal cells (Lv et al. 2017).

pH is also important for the microalgal growth and the accumulation of intracellular lipids. This was confirmed by the findings of Sakarika and Kornaros (2016) who investigated the impacts of various pH values on *C. vulgaris* cultivation. They also argued that the fatty acid composition of the algal cultures was not impacted by pH variations. Illumination, i.e., length of photoperiod and light intensity, could also result in changes in algal growth and lipid content and have to be optimized (Wahidin et al. 2013). Overall, producing high amounts of lipids while maintaining a high algal growth rate is critical for an economic algal biodiesel production simply because high algal biomass productivity would lead to high yield per harvest volume and high lipid content would decrease the cost of extraction per unit product (Tan and Lee 2016). On such basis and since high lipid content and high biomass growth rate basically contradict each other, efforts have been being made to construct algal strains capable of producing high amounts of lipids without sacrificing growth through genetic and metabolic engineering (Talebi et al. 2015).

13.4 Harvesting and Drying of Microalgae

Harvesting in general constitutes a major fraction (20–30%) of the costs associated with microalgal production (Ndikubwimana et al. 2016). Two-step separation, i.e., thickening followed by dewatering, is usually practiced to decrease the cost of the final product. The concentration of the algal cells is increased to approx. 2–7 and 15–25% (TSS basis) through the two stages, respectively. There are several methods for harvesting algae including (Christenson and Sims 2011): (1) filtration—algae can be filtered out by passing through membranes; in this method, recovery rate is high and lower energy inputs are involved, but dewatering might be required; (2) centrifugation—a mechanical method for harvesting microalgae which does not involve contamination with chemicals and, like filtration, the rate of recovery is high; (3) flocculation, a method for separating algae using chemicals that lead to

Table 13.6 List of some microorganisms used for bioflocculation of microalgae[a]

Microorganism	Type	Bioflocculated microalgae
Bacillus licheniformis	*Bacteria*	*Desmodesmus* sp.
Pseudomonas stutzeri and Bacillus cereus		*Pleurochrysis carterae*
Paenibacillus sp.		*Chlorella vulgaris*
Paenibacillus polymyxa		*Scenedesmus* sp.
Bacillus subtilis		*Chlorella vulgaris*
Bacillus sp.		*Nannochloropsis oceanica* sp.
Ankistrodesmus falcatus	*Fungi*	*Chlorella vulgaris*
Scenedesmus obliquus		*Chlorella vulgaris*
Tetraselmis suecica		*Nannochloropsis oleabundans*
Skeletonema sp.		*Nannochloropsis* sp.
Tetraselmis suecica	*Microalgae*	*Chlorella* sp. *and Nannochloropsis* sp.

[a]Al Hattab et al. 2015, Powell and Hill (2013), and Kawaroe et al. (2016)

aggregation of algal cells. In this method, destruction of algal cells is less than in centrifugation and low energy is required; (4) floatation is a separation method in which algae are floated into the surface using bubbling, often used in combination with flocculation for wastewater treatment. No disturbance is made to the cells, and low-energy requirement is also considered as an advantage of this system; (5) ultrasonic separation in which sound waves cause the cells to agglomerate.

Choice of harvesting methods depends on the characteristics of microalgal strain/consortium, while the type and value of the end product are also of importance (Barros et al. 2015). Among different methods, bioflocculation, i.e., the use of microorganisms for the recovery of microalgae biomass, has been most widely used as it is accompanied with significantly less dewatering cost which is economically critical for their full-scale application (Ndikubwimana et al. 2016). A list of microorganisms used in bioflocculation is tabulated in Table 13.6 (Al Hattab et al. 2015; Kawaroe et al. 2016). These microorganisms when added to an algal culture lead to the settlement of the algal cells by adhering and consequent weight increase (Al Hattab et al. 2015). For instance, Ndikubwimana et al. (2014) claimed 98% removal efficiency when they use *Bacillus licheniformis* as bioflocculant for harvesting *Desmodesmus* sp. culture. In a different study, Zhang and Hu (2012) employed a co-culture of *Chlorella vulgaris* and filamentous fungi and successfully extracted the oil for biodiesel production.

In general, both algal oil extraction and its conversion into biodiesel are strongly negatively affected by the presence of water and, therefore, algal biomass should be effectively dried prior to the transesterification reaction (Kumar et al. 2010a, b, c). As a result, different drying methods are usually employed after secondary dewatering (Richmond 2008). Solar drying is the most economically viable drying method especially in places where abundant sunlight is available throughout the year (Sharma et al. 2013). On the contrary, drying methods which are dependent on fossil-oriented energy carriers for their operation, e.g., spray drying and drum drying, are economically and environmentally justified for microalgae biodiesel production (Zhang et al. 2014).

13.5 Lipid Extraction

Microalgal lipids are divided into nonpolar (hydrocarbons, waxes, eicosanoids, fatty acids, and acylglycerols) and polar (phospholipids and glycolipids). There are several methods for cell disruption and extracting microalgal lipids such as mechanical (expeller press), physical (decompression, microwave, freeze-drying, and thermolysis), chemical (organic solvent, chelating agent, supercritical CO_2, detergent, and antibiotics), and enzymatic (lytic, autolysis, and phage) (Kumar et al. 2015).

Mechanical extraction methods offer a number of advantages over the other methods including less dependency on the type of microalgae species to be processed and no contamination of the extracted lipid (Ramesh 2013). Nevertheless, higher energy requirements are considered as a drawback of mechanical extraction methods. This is ascribed to the fact that heat is generated during mechanical extraction of lipids and, in order to prevent damages to the lipids, cooling needs to be performed whose energy and equipment costs negatively impact the overall economics of the process (Lee et al. 2012). Moreover, for a successful implementation of mechanical extraction methods, low-moisture-content algal biomass is required and, therefore, a drying stage needs to be included which could also considerably increase the overall extraction costs. It should also be noted that the amount of pressure employed during mechanical extraction is of critical importance. More specifically, increasing pressure to an optimal level could improve the extraction efficiency, while above-optimal pressure values could negatively affect the process leading to decreased lipid recovery and increased heat generation (Ramesh 2013). Expeller press is one of the simplest mechanical techniques for extracting various oil feedstocks including algae. Nevertheless, its major technical drawback is the presence of pigments along with oil. This method also requires huge amounts of energy, and its efficiency rate is low to moderate.

Among the physical extraction methods, microwave-assisted extraction has attracted a great deal of attention due to its effectiveness in disrupting algal cell walls, being non-toxic, and the possibility of reusing the media after extraction (Lee et al. 2010; Halim et al. 2012; Hattab and Ghaly 2015). Nevertheless, the high costs associated with its maintenance still limit its large-scale application. Freeze-drying and autoclave techniques are also classified among physical extraction methods. However, both these methods suffer from drawbacks such as high costs and long processing times (Hattab and Ghaly 2015).

Cell disruption and consequently extraction of lipids can be also achieved by using a large variety of chemical compounds including antibiotics, chelating agents, chaotropes, detergents, solvents, hypochlorites, acids, and alkali, through different mechanisms though (Günerken et al. 2015). For instance, basic compounds disrupt the cell membranes through saponification of the membrane lipids, while acidic compounds exert their disruptive properties through poration of the cell membrane/wall (Halim et al. 2012; Günerken et al. 2015). In general, lipid extraction from algal biomass is currently carried out using organic solvents such as chloroform, methanol, water, chloroform/methanol (1:2 v/v), chloroform/methanol/water (1:2:0.8 v/v/v), hexane, isopropanol, hexane/isopropanol (3:2 v/v), and ethanol (Zhang et al. 2014). It should be mentioned that organic solvent-based extraction is time- and labor-demanding and, more importantly, it is most efficient for lipid extraction from some algal strains, while it is not reportedly applicable for all algal strains (Ranjith Kumar et al. 2015).

Extracting oil from algal cells is generally limited due to the presence of algal cell wall (Johnson and Wen 2009). Therefore, the use of enzymes such as cellulase, neutral protease, alkaline protease, papain, and lysozyme has been practiced to facilitate cell disruption (Taher et al. 2014; Hattab and Ghaly 2015). Compared to mechanical and chemical methods, enzymatic extraction of algal lipids is very efficient and rapid while causing no corrosion as is the case when chemical

Table 13.7 Different processes used for converting algal biomass to various types of biofuels[a]

Conversion process			Final product	Advantages and limitations
Biochemical conversion		Photobiological hydrogen production	Hydrogen	–
		Fermentation	Bioethanol, acetone, bioethanol	Co-products can be utilized, conversion of sugar to bioethanol possible, long processing time required, biomass has to be preprocessed to be converted to sugars
		Anaerobic digestion	Methane, hydrogen	
Thermochemical conversion	Dry feedstock	Gasification	Syngas	–
		Pyrolysis	Bio-oil–charcoal–syngas	High bio-oil yields possible(up to 57.5% *w/w* for fast and flash pyrolysis, high-energy content required to dry feedstock
	Wet feedstock	Liquefaction	Bio-oil	Algal wet slurry can be used, energy (and cost) reduction, high yields possible (up to 60% *w/w*), reactors are complex and expensive
		Direct combustion	Power generation	–
Chemical reaction		Transesterification	Biodiesel	Enhanced physical properties of renewable fuels, biodiesel has a current market that simplifies commercialization, limited to conversion of lipids and does not utilize carbohydrate and protein fractions of feedstock

[a]Tsukahara and Sawayama (2005) and Vardon et al. (2012)

extraction methods are used. However, the application of enzyme-based methods is limited owing to the high cost of enzymes.

13.6 Microalgae Biofuel Production Pathways

After oil extraction from microalgae for biodiesel production, the remaining biomass can be converted into different types of biofuels, i.e., biohydrogen (Fedorov et al. 2005; Kapdan and Kargi 2006), biomethane (Sialve et al. 2009), and bioethanol (Dexter and Fu 2009) (Table 13.7).

13.6.1 Biochemical Conversion of Algal Biomass

Technologies for biochemical conversion of algal biomass include anaerobic digestion (or biomethanation) and fermentation. More specifically, in biochemical conversion, carbohydrates are digested into sugars using bacteria, microorganisms, and enzymes, which are then transformed into gaseous or liquid fuels, such as biogas (biomethane and biohydrogen) and bioethanol (Zamalloa et al. 2012). For instance, Batista et al. (2015) converted the biomass of an algal consortium (*Chlorella vulgaris, Scenedesmus obliquus*) grown on wastewater into biohydrogen through dark fermentation by an *Enterobacter aerogenes* strain. The highest biohydrogen production yield achieved was 56.8 mL H_2/gVS.

13.6.2 Thermochemical Conversion of Algal Biomass

Thermochemical conversion involves the use of heat to convert algal biomass into gaseous or liquid fuels. Thermochemical conversion can be classified according to the primary desired product (solid, liquid, gas) and the water content of the feedstock (dry or wet).

13.6.2.1 Biocrude Oil Production by Hydrothermal Liquefaction (HTL) of Wet Algal Biomass

The thermochemical conversion of wet algal biomass (75–98% moisture) into biocrude oil in the presence of a solvent at 200–350 °C temperatures and 5–25 MPa pressure to maintain water in the liquid state is called hydrothermal liquefaction (HTL) (Biller et al. 2011). In HTL, biomass is broken down into shorter carbon chains that have a higher energy density (Brennan and Owende 2010). Oxygen, sulfur, and water contents are very low in crude HTL oil. HTL oil recovers more than 70% of the feedstock carbon content. The product is a heavy oil or tarry material, which is called biocrude oil (Biller et al. 2011). The size of biomass particles, residence time, solvent media type, and hydrogen donor solvents are effective for the bio-oil yield and the product quality (Akhtar and Amin 2011). The basic reaction mechanisms involve: (a) depolymerization of the biomass, (b) decomposition of biomass monomers, and (c) recombination of reactive fragments (Toor et al. 2011).

13.6.2.2 Biofuel Production by Pyrolysis of Algal Biomass

Pyrolysis is one of the subclasses of thermochemical conversion in which dry algal biomass is decomposed in the absence of oxygen (or any halogen) and converted into biofuels such as bio-oil–charcoal–syngas. This conversion occurs in the temperature range of 401.85–701.85 °C and 0.1–0.5 MPa pressure (Demirbas 2006). On the basis of operation conditions, pyrolysis process is classified into: (1) slow pyrolysis with operation temperature of 286.75–676.85 °C (Bridgwater 2003), (2) fast pyrolysis with operation temperature of 577–977 °C under inert atmospheric conditions (Mohan et al. 2006), and (3) flash pyrolysis with operation temperature of 777–1027 °C (Balat et al. 2009).

13.6.2.3 Syngas *Production* Through Gasification of Microalgal Biomass

Syngas (a combination of hydrogen, carbon monoxide, and carbon dioxide) is usually produced through the gasification of different carbonous materials including algal biomass (Brown et al. 2010). Gasification process is in fact a partial oxidation process that converts dry algal biomass for instance into a mixture of gases. Gasification is classified into low temperature gasification (700–1000 °C) and high temperature gasification (1200–1600 °C) (McKendry 2002). Yield of syngas depends on various factors including microalgal biomass quality, the equipment (gasifier) used, as well as process parameters (e.g., temperature and catalysis used). In a study, Raheem et al. (2015a, b) reported that syngas yield increases from 28 to 57% by increasing temperature from 552 to 952 °C. The generated syngas could eventually be used for hydrogen production, liquid biofuels production, synthetic natural gas (SNG) production, etc. (Mondal et al. 2011).

13.6.3 Chemical Reaction

13.6.3.1 Biodiesel Production by Transesterification of Algal Oil

Biodiesel, also known as methyl or ethyl esters of long-chain fatty acids, is an alternative to mineral diesel fuel produced from vegetable oils, animal fats, and algal oil mainly through the transesterification reaction with an alcohol (methanol and/or ethanol) and in the presence of a catalyst (mostly NaOH or KOH). The main advantages of biodiesel as fuel include widespread availability, renewability, clean-burning features compared with mineral diesel, and lower sulfur and aromatic contents (Demirbas 2007). These are numerous reports confirming that biodiesel lowers exhaust emissions from diesel engines (Hayyan et al. 2010), i.e., particulate matter (PM) (Kolesárová et al. 2011), unburned hydrocarbons (HC), and carbon monoxide (CO). On the contrary, there is no consensus on the impact of biodiesel on nitrogen oxide (NO_x) emission as there are reports claiming increases in NO_x due to the oxygen content of biodiesel (Sharma et al. 2008). There are four methods for biodiesel production and utilization, direct use and raw oils blending, microemulsions, pyrolysis, and transesterification. As mentioned earlier, the last procedure is most commonly used (Demirbas 2003). Through transesterification, biodiesel and its co-product, i.e., glycerin, is produced in several stages. Afterward, the excess methanol is recovered from the methyl esters through evaporation, and the final biodiesel is eventually washed with water, neutralized, and dried (Xu et al. 2006). Since fossil oil is derived from spores and planktonic algae that were under high pressure and temperature over millions of years, the chemical properties of microalgal lipids and the consequent biodiesel are also very similar to those of mineral diesel (Demirbas and Demirbas 2011).

Transesterification reaction can be acid/base/enzyme catalyzed. Alkaline catalysts include NaOH, NaO^-, KOH, and KO^{-1}, while acid catalysts include HCL and H_2SO_4. Enzymatic catalysts such as lipases that are able to catalyze the transesterification of triglycerides effectively in either aqueous or nonaqueous systems are more environmentally friendly than the other two groups as they result in no wastewater and the produced glycerin needs minimal purification (Fukuda et al. 2001). In another word, the weak points of transesterification reaction by alkaline catalysts are difficult recovery of glycerol, the need for alkaline wastewater treatment, free fatty acid and water interference with the reaction, energy intensity, and the necessity of removing the catalyst from the product (Meher et al. 2006). Some properties of diesel, biodiesel from various oil feedstocks, and microalgae biodiesel are shown in Table 13.8 (Kiss et al. 2007; Huang et al. 2010; Veillette et al. 2012).

13.7 Applications of Microalgae in Integrated Systems

Integration of algal biodiesel production with other activities such as wastewater treatment with an aim to enhance the economic viability of the whole process could be regarded as an efficient strategy to overcome most of the challenges faced. For instance, and as mentioned earlier, wastewater resources are rich in nutrients, such as nitrogen and phosphorus, and could be served microalgal growth as cultivation medium. In fact, using wastewater for microalgal biofuel production not only can reduce freshwater footprint and the cost of these fuels (Clarens et al. 2010) but also could offer new algal-based wastewater treatment systems (Table 13.5). It is worth mentioning that non-fuel products such as fertilizers, chemicals, pharmaceutics, dyes, paints, and animal feeds could also be obtained from microalgaes grown on wastewater (Bhatt et al. 2014).

Table 13.8 Properties of diesel and biodiesel produced from various oil feedstocks*

Property	Diesel	Biodiesel							
		Average biodiesel	Soybean	Jatropha	Rapeseed	Crambe	Corn	Microalgae	*Nannochloropsis oculata*
Fuel composition	C10–C21 HC[a]	C12–C22 FAME[b]	–	–	–	–		–	–
Degree of unsaturation (DU)	–	–	143.70	120.17	123.20	43.60	87	–	53.20
Saponification value (SV)	–	–	202.26	198.94	196.73	72.60	165.26	–	186.67
Iodine value (IV)	–	–	136.84	108.78	111.17	44.00	78.67	–	50.79
Oxidation stability (OS) (h)	–	–	4.56	5.52	6.52	11.52	184.2	–	93.31
Heating value (MJ L^{-1})	36–38	32–36	39.6	39.04[c]	39.3[c]	11.54[c]	28.7[c]	35.40	34[c]
Kinetic viscosity, $mm^2\ s^{-1}$ (at 40 °C)	1.9–3.8	2.8–5.7	1.29	1.31	1.33	0.5	0.84	3.87–5.2	1.12
Density (kg L^{-1})	0.838	0.84–0.90	0.89	0.87	0.87	0.32	0.65	0.864	0.76
Cetane number	40–55	45–70	42.5	49.26	49.3	111.58	61.63	39–54	64.11
Specific gravity, (15.5 °C)	0.81–0.86	0.86–0.89	–	–	–	–	–	0.864	–
Boiling point (°C)	188–343	182–338	–	–	–	–	–	–	–
Flash point (°C)	60–80	100–170	–	–	–	–	–	115	–
Cold filter plugging point (°C)	−3(max −6.7)	Summer max. 0	−1.3	−2.43	−11.29	−0.11	−10	−11	−4.79
Pour point (°C)	−35 to −15	−15 to 10	−5.79	−4.87	−10.81	−11.10	−7.67	–	6.15
Sulfur (wt%)	0.01–0.04	0.0000–0.0024	0.004	0.001	–	–	–	0.0069	–
Stoichiometric air/fuel ratio (AFR)	15	13.8	–	–	–	–	–	–	–
Acid value (mg KOH g^{-1})	Max 0.5	Max 0.5	0.21	0.21	0.98	0.33	–	0.374	–
Lubricity (25 °C)	0.509–0.283	0.114	–	–	–	–	–	–	–
H/C ratio	1.81	–	–	–	–	–	–	1.81	–

[a]HC—hydrocarbons
[b]FAME—fatty acid methyl esters
[c]MJ/Kg
**Sources* Kiss et al. (2007), Huang et al. (2010), Veillette et al. (2012), Uthman and Saka (2013), Oliveira and Silva (2013), Islam et al. (2013), http://www.chempro.in, and http://www.brteam.ir/biodieselanalyzer

Beside biofuel production, biofixation of carbon could also be the secondary objective of algal-based biofuel production systems. This is increasingly important given the criticality of climate change and the very recent international call for immediate action to address this crisis to the level that even the leader of the Catholic Church Pope Francis raised the issue during his visit to the USA in September 2015 (The Gurdian 2015).

The following sections summarize the efforts made during the last several years in order for integrating algal biofuel production systems with wastewater treatment and carbon biofixation.

13.7.1 Algal Biofuel Production and Wastewater Treatment

A major requirement of an efficient wastewater treatment is obviously the need to remove high concentrations of nutrients, in particular N and P. As mentioned earlier, microalgae are capable of uptaking such nutrients as well as heavy metals and organic pollutants from wastewater and producing biomass. Thus, it offers great promises for the treatment of various municipal, agricultural, and industrial wastewaters (Feng et al. 2011; Zhu et al. 2013). However, there are many reports indicating that most of the microalgal species with high lipid contents do not adapt well to grow in wastewater (Xin et al. 2010). Contrary to these reports, there are also a number of success stories through which efficient integration of algal biofuels production and wastewater treatment has been accomplished (Zhou et al. 2011a, b, 2013; de Alva et al. 2013; Hena et al. 2015). For instance, Zhou et al. (2011a, b) claimed that five species of microalgae isolated from Minnesota wastewaters including *Chlorella* sp., *Heynigia* sp., *Hindakia* sp., *Micractinium* sp., and *Scenedesmus* sp. showed high growth rate (0.455–0.498 d^{-1}) and lipid productivities (74.5–77.8 mg L^{-1} d^{-1}) on municipal wastewater.

In a more recent investigation, de Alva et al. (2013) also cultivated *Scenedesmus acutus* in pretreated municipal wastewater with a dual focus on biomass productivity and lipid accumulation. They argued that *S. acutus* could successfully remove nutrients from the wastewater and that they achieved 249.4 mg L^{-1} biodiesel from the referred algal oil. It should be pointed out that in a series of experimental surveys, Chinnasamy et al. (2010a, b), Kong et al. (2010), and Zhou et al. (2011a, b) revealed that municipal wastewater was a better option compared with industrial wastewater for algal biomass and lipid production. A comparison of biomass productivity, lipid content, and lipid productivity of different microalgae species grown on various wastewater is tabulated in Table 13.9. The following sections present the integration of biofuels production with algal-based treatment of different types of wastewaters, namely high N/P content wastewater, high heavy metal-content wastewater, and high organic-content wastewater (PAHs aromatic hydrocarbons and polychlorinated biphenyls (PCBs)).

Zhu et al. (2013) proposed *Chlorella zofingiensis* cultivation on piggery wastewater with a dual purpose of wastewater treatment and biodiesel production.

Table 13.9 Comparison of biomass and lipid productivities in microalgae grown in various wastewater conditions

Microalgae species	Wastewater	Biomass (DW*) productivity (mg L^{-1} d^{-1})	Lipid content (% DW)	Lipid productivity (mg L^{-1} d^{-1})	References
Auxenochlorella protothecoides	Municipal centrate	268.80	28.90	77.7	Zhou et al. (2011a, b)
B. braunii	Industrial (carpet mill)	34.00	13.20	4.50	Chinnasamy et al. (2010a, b)
Chlamydomonas reinhardtii (biocoil-grown)	Municipal centrate	2000	25.25	500	Kong et al. (2010)
Chlamydomonas mexicana	Piggery wastewater	Not available	33[a]	0.31[a]	Abou-Shanab et al. (2013)
Chlorella sp.	Agricultural (dairy)	2.6 gm^{-2} d^{-1}	9	230 mg m^{-2} d^{-1}	Johnson and Wen (2010)
Chlorella sp.	Municipal centrate	231.40	33.53	77.50	Zhou et al. (2011a, b)
Chlorella sp.	Municipal centrate	241.70	30.91	74.70	Zhou et al. (2011a, b)
Chlorella saccharophila	Industrial (carpet mill)	23.00	18.10	4.20	Chinnasamy et al. (2010a, b)
Chlorella sp.	Agricultural (digested dairy manure, 20 × dilution)	81.4[b]	13.6[c]	11[c]	Wang et al. (2010)
C. pyrenoidosa	Piggery wastewater	–	–	6.3	Wang et al. (2012)
Mix of *Chlorella* sp., *Micractinium* sp., *Actinastrum* sp.	Agricultural (dairy wastewater, 25% dilution)	59[d]	29	17	Woertz et al. (2009)
Mix of *Chlorella* sp., *Micractinium* sp., *Actinastrum* sp.	Municipal (primary treated + CO_2)	270[e]	9	24.4	Woertz et al. (2009)

(continued)

Table 13.9 (continued)

Microalgae species	Wastewater	Biomass (DW*) productivity ($mg\ L^{-1}\ d^{-1}$)	Lipid content (% DW)	Lipid productivity ($mg\ L^{-1}\ d^{-1}$)	References
Dunaliella tertiolecta	Industrial (carpet mill)	28	15.20	4.30	Chinnasamy et al. (2010a, b)
Pleurochrysis carterae	Industrial (carpet mill)	33	12	4	Chinnasamy et al. (2010a, b)
Scenedesmus obliquus	Municipal sewage	26[f]	31.4[g]	8[g]	Krishna et al. (2012)
Scenedesmus sp.	Municipal centrate	247.50	30.90	74.50	Zhou et al. (2011a, b)
Scenedesmus acutus	Municipal	79.9	–	280 $mg\ L^{-1}$	de Alva et al. (2013)
Botryococcus braunii	Agricultural	700	–	69	Krishna et al. (2012)
Hindakia sp.	Municipal centrate	275	28.30	77.80	Zhou et al. (2011a, b)

*DW—dry weight
[a]Estimated from biomass value of 1000 $mg\ L^{-1}$ after 40 d
[b]Estimated from biomass value of 1.71 $g\ L^{-1}$ after 21 d
[c]Fatty acid content and productivity determined rather than total lipid
[d]Estimated from lipid productivity and lipid content value
[e]Estimated from biomass value of 812 $mg\ L^{-1}$ after 3 d
[f]Estimated from biomass value of 197 $mg\ L^{-1}$ after 31 d
[g]Fatty acid content and productivity determined rather than total lipid

In a different investigation, Maity et al. (2014) investigated the integration of biofuel and bioelectricity production with wastewater treatment using one species of microalgae, i.e., *Leptolyngbya* sp. JPMTW1 (KF977831). They argued that only after 7 d of cultivation, biomass production, rate of biomass production, lipid production, and rate of lipid production stood at 3300 mg L^{-1}, 471.42 mg L^{-1} day^{-1}, 1068.383 mg g^{-1} dry wt. biomass, 152.62 mg g^{-1} dry biomass/day, respectively. The also reported that over the same period, electrical conductivity (EC), chemical oxygen demand (COD), and total dissolved solid (TDS) decreased from 982 to 854 (mS/cm), 255 to 112 mg L^{-1}, and 490–427 mg L^{-1}, respectively. Overall, their findings were indicative of the possibility of the production of biofuel, bioelectricity, and wastewater treatment by *Leptolyngbya* sp. JPMTW1. In another study, Chen et al. (2014) produced biocrude

Table 13.10 Nutrient removal efficiency of microalgal species

Microalgal species	Wastewater type	Nitrogen (%)	Phosphate (%)	COD removal (%)	References
Chlorella Mexicana	Piggery	62	28%	–	Abou-Shanab et al. (2013)
Chlorella vulgaris	Textile	44.4–45.1	33.1–33.3	38.3–62.3	Lim et al. (2010)
Chlorella vulgaris	Municipal	55–88	12–100	–	Khan and Yoshida (2008), Ruiz-Marin et al. (2010)
Chlorella kessleri	Artificial medium	8–19[a]	8–20[b]	–	Cai et al (2013)
Chlorella sp.	Municipal centrate	89.1	80.9	90.8	Li et al. (2011)
Chlorella sp.	Dairy manure	75.7–82.5	62.5–74.7	27.4–38.4	Wang et al. (2010)
Chlorella pyrenoidosa	Industrial	87–89	70	–	Hongyang et al. (2011)
Chlorella minutissima	Primary- and tertiary-treated	70–80	60–70	–	Malla et al. (2015)
Chlamydomonas reinhardtii	Artificial medium	12–83	13–14	–	Kong et al. (2010)
Chlamydomonas polypyrenoideum	Dairy	74–90	70	–	Kothari et al. (2012)
Scenedesmus obliquus	Municipal	79–100[a]	47–98	–	Cai et al. (2013)
Scenedesmus acutus	Municipal	66	94	–	de Alva et al. (2013)
Euglena sp.	Sewage treatment plant	93	66	–	Mahapatra et al. (2013)

[a]Nitrate, nitrite
[b]Total orthophosphates

oils from a mixed-culture algal biomass harvested from a functioning wastewater treatment system as well.

13.7.1.1 High N/P Content Wastewater

Nitrogen is a critical nutrient required for algal growth, and the application of nitrogen starvation for enhancing algal cell lipid content is well documented (Brennan and Owende 2010). Likewise, another key factor in algal energy metabolism is phosphorus which is found in a variety of biological substances, such as nucleic acids, lipid, proteins, and intermediates of carbohydrate metabolism. All eukaryotic algae require inorganic nitrogen, while some algal species are capable of using both inorganic and organic phosphorus (Liang 2013). In recent years, investigations into the ability of microalgae to simultaneously grow on wastewater streams and remove nutrients have revealed many microalgae species with high protectional for N and P removal from wastewaters (Table 13.10). For instance, Cai et al. (2013) achieved an N removal efficiency of 79–100% by *S. obliquus* from municipal wastewater. Earlier in the year 2010, Lim et al. made an attempt to treat textile wastewater medium using *C. vulgaris* and reportedly managed to remove N and P by 45 and 33%, respectively. A wide range of N (55–88%) and P (12–100%) removal has been reported when municipal wastewater was used as the waste stream (Khan and Yoshida 2008; Ruiz-Marin et al. 2010; Li et al. 2011). Mixed municipal and industrial wastewater was used by Gentili (2014) to produce *Selenastrum minutum* algal biomass and lipid, while effective wastewater treatment was also targeted. Their results showed that ammonium and phosphate contents were decreased from 96 to 99% and 91 to 99%, respectively, while the highest biomass and lipids yields (dry matter basis) reaching 37%.

Lu et al. (2015) used meat processing wastewater for the cultivation of the microalgae *Chlorella* sp. (UM6151) aiming at simultaneous biomass production, wastewater treatment, and nutrient removal. They implemented an innovative cultivation approach based on wastewater mixing to supply nutrients and improve biomass yield at economic rates. They claimed that algal biomass yield (0.675–1.538 g/L) achieved using mixed wastewater was much higher than those obtained using individual wastewater and synthetic medium. Moreover, they achieved improved ammonia nitrogen removal efficiencies (68.75–90.38%) and total nitrogen removal efficiencies (30.06–50.94%). Interestingly, by using wastewater mixing, algal protein content was also enhanced reaching as high as 60.87–68.65%.

In an effort, Abou-Shanab et al. (2013) strived to integrate biofuel production and the treatment of piggery wastewater (TN: 56 $\pm$ 2 and TP:13.5 $\pm$ 0.6 mg/L). They reported that six microalgal species including *Ourococcus multisporus*, *Nitzschia cf. pusilla*, *Chlamydomonas mexicana*, *S. obliquus*, *Chlorella vulgaris*, and *Micractinium reisseri* were capable of efficiently treating wastewater and producing high oil content for biodiesel production. Among the studied species, *C. Mexicana* was proven to have the highest removal rates, i.e., N (62%), phosphorus (28%), and inorganic carbon (29%). Hence and due to the higher lipid productivity and lipid content (0.31 ± 0.03 g/L and 33 ± 3%, respectively), compared with the other species, the authors suggested that *C. mexicana* could be a suitable candidate for integrated biodiesel production and wastewater treatment.

In a study, Min et al. (2014) suggested an efficient method, i.e., a pilot-scale stacked-tray bioreactor to increase nutrient removal rate from piggery wastewater coupled with biofuel production. Through their proposed cultivation system, algal biomass productivity (based on TSS) was enhanced from 19.15 to 23.19 g m^{-2} day^{-1} and they achieved lipid contents ranging between 1.77 and 3.55%. Wang et al. (2012) looked into the impact of dilution on algal biodiesel production and nutrient removal from high N/P content wastewater. In their study, primary piggery wastewater was used as the waste stream and was treated by mixotrophic cultivation of *Chlorella pyrenoidosa*. They stated that there was a positive linear correlation between algal biomass productivity and the initial COD values ranging from 250 to 1000 mg L^{-1}. The maximal lipid productivity 6.3 mg L^{-1} day^{-1} was recorded with an initial COD of 1000 mg L^{-1}, while nutrients such as ammonium were removed efficiently at rates as high as >90% in all diluted samples. Can et al. (2015) also explored the potential of microalgae *Spirulina platensis* for biofuel and biochemical production coupled with domestic wastewater treatment. Similar to Wang et al. (2012), in their experimental approach, wastewater was also diluted with distilled water to achieve different concentrations of 100, 75, 50, and 25%. Their findings were in line with those of Wang et al. (2012), revealing that the highest biomass yield was recorded when the wastewater without dilution (100%) was used. In terms of lipid production, however, the maximal value was measured in 25% wastewater. Therefore, a trade-off should be observed in order to maximize lipid productivity.

In a different study, Malla et al. (2015) also studied the potential of *Chlorella minutissima* for biodiesel production coupled with wastewater treatment. Their results indicated that after 12 d of the experiment, *C. minutissima* removed about 90–98% TDS, 70–80% N, 60–70% P, and 45–50% K from the high N/P content wastewater. They also converted the algal lipid extracted to biodiesel as part of the integrated system. Hena et al. (2015) investigated the potentials of a consortium of native microalgae species grown on a dairy farm treated wastewater for biodiesel production. The claimed that biomass production and lipid content of the consortium were 153.54 t ha^{-1} $year^{-1}$ and 16.89%, respectively, and that 72.70% of the algal lipid obtained could be converted into biodiesel.

13.7.1.2 High Heavy Metal Content Wastewaters

Heavy metals mainly include transition metals, metalloids, lanthanides, and actinides. These metals have a highly specified gravity and are toxic to a level that even at low concentrations represents a significant environmental concern (Bhargava et al. 2012). Various methods have been investigated for heavy metal removal, which are tabulated in Table 13.11 (Fu et al. 2011). Among these methods, algae have been proposed as ideal candidates for heavy metal removal from various

Table 13.11 Heavy metal wastewater treatment techniques (Fu and Wang 2011)

Technique	Conventional processes	Material used in the process	Removed ions
Chemical precipitation	Hydroxide precipitation	$Ca(OH)_2$, NaOH	Zn^{2+}, Cr^{3+}, Pb^{2+}, Hg^{2+}, Cu^{2+}
	Sulfide precipitation	Iron sulfide (FeS)	Pb^{2+}, Cu^{2+}, Cd^{2+}
	Heavy metal chelating precipitation	1,3-benzenediamidoethanethiol, hexahydrotriazine dithiocarbamate (HTDC), ethyl xanthate	Hg^{2+}, Cu^{2+}
Ion exchange	–	Clinoptilolite	Pb^{2+}, Ni^{2+}, Zn^{2+}
Adsorption	Activated carbon adsorbents	–	Pb^{2+}, Cu^{2+}
	Low-cost adsorbents	Chemically modified plant wastes, agricultural waste material, industrial by-products such as lignin, natural substances	Pb^{2+}, Ni^{2+}, Cd^{2+}
	Carbon nanotube absorbents	(1) Single-walled CNTs (SWCNTs)	Pb^{2+}, Ni^{2+}, Cd^{2+}, Cu^{2+}
		(2) Multi-walled CNTs (MWCNTs)	
	Bioadsorbents	Non-living such as potato peels, sawdust, coffee husks as well as living such as algal biomass and microbial biomass	Pb^{2+}, Cd^{2+}, Cu^{2+}
Membrane filtration	Ultrafiltration	Micellar-enhanced ultrafiltration (MEUF) and polymer-enhanced ultrafiltration (PEUF)	Pb^{2+}, ASO^{4-}, Cd^{2+}, Zn^{2+}, Cr(III), Cr(VI), Cu^{2+}, Cr^{3+}, Ni^{2+}
	Reverse osmosis	–	Zn^{2+}, As, Cu^{2+}, Ni^{2+}
	Nanofiltration	NF90 andN30F	Cr(VI), Cu^{2+}
	Electrodialysis	–	Pb^{2+}, Cr(III)
Coagulation and flocculation	–	Polyferric sulfate (PFS), polyacrylamide (PAM)	Ni^{2+}, Cu^{2+}, Pb^{2+}, Zn^{2+}
Electrochemical treatment	–	–	Zn^{2+}, Ag^{+}, Cu^{2+}, Ni^{2+}
Flotation	–	–	Cd^{2+}, Pb^{2+}, Cu^{2+}

wastewaters through either uptake or accumulation of Hg, Cd, Zn, Au, Ag, Co, Mn, Cs, Ni, Fe, Cu, and Cr from their environment (Chekroun and Baghour 2013). In fact, algae produce polypeptides called chelating agents capable of binding to heavy metals. Apart from that, large surface area of algal cells is also effective in removing heavy metals (Kumar et al. 2015). More specifically, metal absorption by microalgae occurs at two stages: first, at the surface of algal cells through very quick physical adsorption or ion exchange. The second stage, also called

Table 13.12 Heavy metal removal by microalgae from different wastewater source

Microalgae species	Wastewater type	Metal studied	Removal efficiency or accumulation	References
Scenedesmus sp.		Cu, Ni		Kumar et al. (2015)
Chlorella vulgaris	Synthetic wastewater	Cr	43.3 mg g^{-1} biomass	Xie et al. (2014)
Spirulina maxima and Chlorella vulgaris	Secondary effluent	Cu	81.7%	Chen et al. (2014)
Pavlova lutheri, Tetraselmis chuii, Nannochloropsis, and *Chaetoceros muelleri*	Municipal wastewater	Leachate		Richards and Mullins (2013)
Scenedesmus quadricauda	Synthetic wastewater	Pb	82%	Mirghaffari et al. (2015)
Phaeodactylum tricornutum	Seawater enriched	Hg	2229 mg g^{-1} biomass	Deng and Lu (2013)
Chlorella vulgaris, Spirulina maxima and *Synechocystis sp.*	Wastewater treatment plant discharge	Cu, Zn		Chan et al. (2013)
Scenedesmus bijuga, Oscillatoria quadripunctulata	Sewage wastewater and petrochemical effluents	Cu, Co, Zn, Pb		Ajayan et al. (2011)
Dictyosphaerium chlorelloides	Leather tanning, tincture wood preservatives, and the electroplating industry wastewater	Cr (III)		Pereira et al. (2010)

chemisorption, takes place at a slower rate intracellularly and is driven by metabolic processes involving active binding groups (Zhou et al. 2012).

Richards and Mullins (2013) studied algal-based bioremediation of municipal leachate by using a consortium of four marine microalgae species, i.e., *Pavlova lutheri, Tetraselmis chuii, Nannochloropsis*, and *Chaetoceros muelleri* while also targeting enhanced lipid production. Their results revealed that algal-based bioremediation was a feasible method for simultaneous treatment of waste streams and lipid production. Yang et al. (2015) proposed an integration of heavy metal wastewater utilization and biofuel production as an alternative solution to address energy shortage and environmental concerns. They claimed that *Chlorella minutissima* UTEX 2341 had strong resistance to cadmium, copper, manganese, and zinc ions under heterotrophic culture condition and could efficiently remove these heavy metals through intracellular accumulation and extracellular immobilization. Moreover, lipid accumulation was not negatively affected by heavy metals. Heavy metal removal by some species of microalgae from various wastewater sources is depicted in Table 13.12.

Table 13.13 Degradation of organic pollutants by algal species

Microalgae species	Organic pollutant	References
Monoraphidium braunii	Bisphenol	Gattullo et al. (2012)
Chlamydomonas reinhardtii	Herbicide (fluroxypyr)	Zhang and Hu (2012)
Pediastrum tetras *Ankistrodesmus fusiformis* *Amphora coffeaeformis*	Herbicide (mesotrione)	Valiente Moro et al. (2012)
Scenedesmus quadricauda	Herbicide (isoproturon)	Dosnon-Olette et al. (2010)
Scenedesmus obliquus GH2	Crude oil degradation	Tang et al. (2011)
Scenedesmus obliquus	Nonylphenol, octylphenol	Zhu et al. (2013)
Skeletonema costatum	Phenanthrene, fluoranthene	Hong et al. (2008)
Selenastrum capricornutum	Benzene, toluene, chlorobenzene, 1,2-dichlorobenzene, nitrobenzene, naphthalene, 2,6-dinitrotoluene, phenanthrene, di-n-butylphthalate, pyrene	Lei et al. (2007), Gavrilescu (2010)
Nitzschia sp.	Phenanthrene, fluoranthene	Hong et al. (2008)
Chlorella sp. *Scenedesmus obliquus* *Stichococcus sp.*	Phenol	Zhang and Hu (2012)
Chlorella vulgaris	Atrazine	Dosnon-Olette et al. (2010)
Chlorella fusca var. *vacuolata*	2,4-Dichlorophenol	Zhang and Hu (2012)
Chattonella subsalsa *Chattonella marina* var. *marina* *Chattonella marina* var. *ovata*	PCB (Aroclor 1242)	Niestroy et al. (2014)

13.7.1.3 High Organic-Content Wastewater

Organic pollutants are chemical substances that persist in an environment through industrial discharges and agricultural usages. They are also resistant to environmental degradation through chemical, biological, and photolytic processes and have harmful effects on human health. Among these organic pollutants, polycyclic aromatic hydrocarbons (PAHs) and polychlorinated biphenyls (PCBs) are highly persistent compounds and, if introduced into the food chain, they have been proven to be carcinogenic (Gilden et al. 2010). Microalgae are capable of decomposing different kinds of organic pollutants including phenolics, pesticides, as well as PAHs and PCBs. *Ankistrodesmus braunii, Scenedesmus quadricauda, Ochromonas danica,* and *Monoraphidium braunni* are examples of microalgae species that can biodegrade phenolic and biophenolic compounds (Mukherjee et al. 2013; Gattullo et al. 2012). Ali et al. (2012) introduced microalgae such as *chlorella vulgaris* as a low-cost adsorbent for removing organic pollutants from wastewaters. Attempts for degradation of organic pollutants by some species of microalgae are summarized in Table 13.13.

- *PAHs aromatic hydrocarbons*

  PAHs and polyaromatic hydrocarbons are ubiquitous environmental pollutants which are found in petroleum and fossil fuels, or are formed during the incomplete combustion of these energy carriers (Chekroun et al. 2014). These are neutral and nonpolar hydrocarbons that are composed of two or more benzene rings or pentacyclic molecules. Certain types of PAHs including benzo [a] anthracene, chrysene, benzo [b] fluoranthene, benzo [a] pyrene, and benzo [ghi] perylene are potentially carcinogenic for human beings and, due to their carcinogenic and mutagenic characteristics, are dangerous air pollutants (Gariazzo et al. 2015). Some types of the PAHs such as fluoranthene (Fla), pyrene (Pyr), benz[a]anthracene (BaA), chrysene (Chr), benzo[a]pyrene (BaP), benzo[k]fluoranthene (BkF), and dibenzo[a,h] anthracene (DA)] have half-lives of about 1000–3000 h in aquatic environments (Luo et al. 2014). Absorption, chemical degradation, photolysis, and volatilization and microbial degradation are significant methods for PAH removal. Nevertheless, the major process of removing PAH contamination in the environment is microbial degradation and algae are no exception (Ukiwe et al. 2013).

  Microalgae release biosurfactants that could further enhance phenanthrene degradation. Moreover, microalgae are able to produce the O_2 required by acclimatized bacteria to biodegrade hazardous pollutants such as polycyclic aromatic hydrocarbons, phenolics, and organic solvents (Chekroun et al. 2014). For example, some kinds of marine algae such as cyanobacteria, *Oscillatoria,* and *Agmenellum* spp. are known to degrade naphthalene through pathways that are similar to fungus (Haritash and Kaushik 2009; Barrios et al. 2011). The capability of *S. obliquus* and *Nitzschia linearis* in removing n-alkanes and PAHs has also been reported (Subashchandrabose et al. 2013).
- *Polychlorinated biphenyls (PCBs)*

  PCBs; organic chemical compounds of chlorine attached to 'biphenyl, are a class of the worst persistent organic pollutants (POPs) (Gauthier et al. 2014). Due to their characteristics such as high toxicity, carcinogenicity, and slow biodegradation, exposure to PCBs can cause neurological disorders, reproductive toxicity, endocrine disruption, cancer, and even at extremely low concentrations (Pandelova et al. 2010). There are several technologies for PCB remediation, including biological treatment (phytoremediation, aerobic biodegradation, anaerobic dechlorination), physical methods, thermal treatment, and chemical treatment (Gomes et al. 2013). Bioaccumulation of PCBs by algae has attracted a great deal of attention (Chekroun et al. 2014), while its integration with biodiesel production is also of interest (Usher et al. 2014). The efficiency of algal-based remediation of PCBs could be influenced by water quality, chlorination, phytoplankton composition, the structure of the PCBs, and the algal cell wall (Zhao et al. 2014).

13.7.2 Biofixation of Carbon and Biofuel Production Systems

Microalgae use inorganic carbon for growth, while they can also fix CO_2 from industrial exhaust gases (Shilton et al. 2008). Utilization of microalgae for biofixation of carbon has numerous advantages as follows: (1) Microalgae have much higher CO_2 fixation abilities compared with other crops, since they have a higher growth rate (Chisti 2007; Li et al. 2008), and (2) microalgae are able to convert CO_2 into chemical energy through photosynthesis, which can then be converted into biofuels (Demirbas et al. 2004). Therefore, combination of wastewater treatment, biofuel production, and biofixation of CO_2 and GHG may provide a very promising alternative to climate change mitigation strategies.

For instance, CO_2 fixation rate (g/m^3/h) by *Chlorella vulgaris* has been reported at 80–260 (Cheng et al. 2006). Yoo et al. (2010) studied three species of microalgae, *Botryococcus braunii, Chlorella vulgaris*, and *Scenedesmus* sp., cultivated with ambient air containing 10% CO_2 and flue gas. Their results showed that the biomass and lipid productivity in flue gas condition rose by 1.9-fold (39.44 mg L^{-1} d^{-1}) and 3.7-folds (20.65 mg L^{-1} d^{-1}), for *Scenedesmus* sp and *B. braunii,* respectively. Moreover, they suggested that *B. braunii* was suitable for biodiesel production, due to its high lipid content, whereas *Scenedesmus* sp. was suitable for mitigating CO_2 as a result of high biomass productivity. In another study by Tang et al. (2011), two species of microalgae, *S. obliquus* and *Chlorella pyrenoidosa,* were explored as suitable species for mitigating CO_2 in the flue gases and biodiesel production.

CO_2 removal efficiency (%) by *Euglena gracilis, Porphyridium* sp., *S. platensis* has also been recorded at 3.1, 3–18, 38.3–60, respectively (Chae et al. 2006; Shibata et al. 2004; Kumar et al. 2010a, b, c). Nayak et al. (2013) also demonstrated biomass productivity and CO_2 biofixation of three strains of *Scenedesmus* sp. in the presence of different NaOH concentrations in algae cultivation media. They stated that under their experimental conditions, the algal lipids were mainly composed of C16/C18 fatty acids and were favorable for biodiesel production.

Exogenous CO_2 concentration could also impact algal biomass yield, nutrients removal rate, as well as biodiesel production potentials. For instance, in a study, Li et al. (2011) looked into the effects of environmental factors including exogenous CO_2 concentration on wastewater nutrient removal and biodiesel production using 14 strains of microalgae belonging to the genus of *Chlorella, Haematococcus, Scenedesmus, Chlamydomonas,* and *Chloroccum* cultivated. The results of this study proved that the environmental factors had effects on the yields of algal biomass and lipid accumulation which could consequently result in significantly different biodiesel production potentials. Among the algal strains investigated, *Chlorella kessleri* and *Chlorella protothecoides* represented the highest biomass accumulation of 2.01, 1.31 g/L, respectively. Overall, biomass accumulation, biodiesel production rate, and the removal rates of nitrogen and COD were increased by higher light intensity and exogenous CO_2 concentration as well as longer lighting period, while higher phosphorus removal rates were achieved in lower exogenous CO_2 concentrations.

13.8 Conclusions and Future Prospects

Widespread utilization of fossil fuels is among the major causes of GHG emissions and the resultant tragic environmental consequences such as global warming. Biofuels such as biodiesel produced from algae could be regarded as a promising solution to turn this scenario around. However and in spite of these attractive features of algal fuels, current technologies are yet to be further improved to lead to economically justified production of these alternative fuels. Accordingly, it seems that the integration of algal fuels production with wastewater treatment and/or carbon biofixation could potentially serve as cost-effective and eco-friendly platform to achieve the above-mentioned goals.

References

Abou-Shanab RA, Ji MK, Kim HC, Paeng KJ, Jeon BH (2013) Microalgal species growing on piggery wastewater as a valuable candidate for nutrient removal and biodiesel production. J Environ Manage 115:257–264

Ajayan K V, Selvaraju M, Thirugnanamoorthy K (2011) Growth and Heavy Metals Accumulation Potential of Microalgae Grown in Sewage Wastewater and Petrochemical Effluents. Pak J Bio Sci 14(16):805–811

Akhtar J, Amin NAS (2011) A review on process conditions for optimum bio-oil yield in hydrothermal liquefaction of biomass. Renew Sustain Energy Rev 15(3):1615–1624

Al Hattab M, Ghaly A, Hammoud A (2015) Microalgae harvesting methods for industrial production of biodiesel: critical review and comparative analysis. J Fundam Renew Energy Appl 5:1000154

Ali I, Asim M, Khan TA (2012) Low cost adsorbents for the removal of organic pollutants from wastewater. J Environ Manage 113:170–183

Alketife AM, Judd S, Znad H (2017) Synergistic effects and optimization of nitrogen and phosphorus concentrations on the growth and nutrient uptake of a freshwater Chlorella vulgaris. Environ Technol 38(1):94–102

Ananadhi Padmanabhan MR, Stanley SA (2012) Microalgae as an oil producer for biofuel applications. Res J Recent Sci 1(3):57–62

Atabani AE, da Silva César A (2014) Calophyllum inophyllum L.-A prospective non-edible biodiesel feedstock. Study of biodiesel production, properties, fatty acid composition, blending and engine performance. Renew Sustain Energy Rev 37:644–655

Balat M, Balat M, KÄrtay E, Balat H (2009) Main routes for the thermo-conversion of biomass into fuels and chemicals. Part 1: pyrolysis systems. Energy Convers Manage 50(12):3147–3157

Barros AI, Gonçalves AL, Simões M, Pires JC (2015) Harvesting techniques applied to microalgae: a review. Renew Sustain Energy Rev 41:1489–1500

Batista AP, Ambrosano L, Graça S, Sousa C, Marques PA, Ribeiro B … Gouveia L (2015) Combining urban wastewater treatment with biohydrogen production–an integrated microalgae-based approach. Bioresour Technol 184: 230–235

Bhargava A, Carmona FF, Bhargava M, Srivastava S (2012) Approaches for enhanced phytoextraction of heavy metals. J Environ Manage 105:103–120

Bhatt NC, Panwar A, Bisht TS, Tamta S (2014) Coupling of algal biofuel production with wastewater. Sci World J Article ID 210504

Biller P, Riley R, Ross A (2011) Catalytic hydrothermal processing of microalgae: decomposition and upgrading of lipids. Biores Technol 102(7):4841–4848

Bohutskyi P, Kula T, Kessler BA, Hong Y, Bouwer EJ, Betenbaugh MJ, Allnutt FT (2014) Mixed trophic state production process for microalgal biomass with high lipid content for generating biodiesel and biogas. BioEnergy Res 7(4):1174–1185

Brennan L, Owende P (2010) Biofuels from microalgaeâ-a review of technologies for production, processing, and extractions of biofuels and co-products. Renew Sustain Energy Rev 14 (2):557–577

Bridgwater AV (2003) Renewable fuels and chemicals by thermal processing of biomass. Chem Eng J 91(2–3):87–102

Brown TM, Duan P, Savage PE (2010) Hydrothermal liquefaction and gasification of Nannochloropsis sp. Energy Fuels 24(6):3639–3646

Bwatanglang IB, Faruq M, Gupta AK, Yusof NA (2015) Algae-derived biomass for sustainable and renewable biofuel production. In: Agricultural biomass based potential materials. Springer International Publishing, Berlin, pp 341–373

Cai T, Park SY, Li Y (2013) Nutrient recovery from wastewater streams by microalgae: status and prospects. Renew Sustain Energy Rev 19:360–369

Carvalho P, Goder V, Rapoport TA (2006) Distinct ubiquitin-ligase complexes define convergent pathways for the degradation of ER proteins. Cell 126(2):361–373

Chae S, Hwang E, Shin H (2006) Single cell protein production of Euglena gracilis and carbon dioxide fixation in an innovative photo-bioreactor. Biores Technol 97(2):322–329

Chan A, Salsali H, McBean E (2013) Heavy metal removal (copper and zinc) in secondary effluent from wastewater treatment plants by microalgae. ACS Sustain Chem Eng 2(2):130–137

Chekroun KB, Baghour M (2013) The role of algae in phytoremediation of heavy metals: a review. J Mater Environ Sci 4(6):873–880

Chekroun KB, Sánchez E, Baghour M (2014) The role of algae in bioremediation of organic pollutants. Int Res J Public Environ Health 1(2):19–32

Chen WT, Zhang Y, Zhang J, Yu G, Schideman LC, Zhang P, Minarick M (2014) Hydrothermal liquefaction of mixed-culture algal biomass from wastewater treatment system into bio-crude oil. Biores Technol 152:130–139

Cheng L, Zhang L, Chen H, Gao C (2006) Carbon dioxide removal from air by microalgae cultured in a membrane-photobioreactor. Sep Purif Technol 50(3):324–329

Chevalier P, Proulx D, Lessard P, Vincent W, De la NoÃŒe J (2000) Nitrogen and phosphorus removal by high latitude mat-forming cyanobacteria for potential use in tertiary wastewater treatment. J Appl Phycol 12(2):105–112

Chinnasamy S, Bhatnagar A, Claxton R, Das K (2010a) Biomass and bioenergy production potential of microalgae consortium in open and closed bioreactors using untreated carpet industry effluent as growth medium. Biores Technol 101(17):6751–6760

Chinnasamy S, Bhatnagar A, Hunt RW, Das K (2010b) Microalgae cultivation in a wastewater dominated by carpet mill effluents for biofuel applications. Biores Technol 101(9):3097–3105

Chisti Y (2007) Biodiesel from microalgae. Biotechnol Adv 25(3):294–306

Chisti Y (2008) Biodiesel from microalgae beats bioethanol. Trends Biotechnol 26(3):126–131

Chisti Y (2013) Constraints to commercialization of algal fuels. J Biotechnol 167(3):201–214

Christenson L, Sims R (2011) Production and harvesting of microalgae for wastewater treatment, biofuels, and bioproducts. Biotechnol Adv 29(6):686–702

Clarens AF, Resurreccion EP, White MA, Colosi LM (2010) Environmental life cycle comparison of algae to other bioenergy feedstocks. Environ Sci Technol 44(5):1813–1819

Converti A, Casazza AA, Ortiz EY, Perego P, Del Borghi M (2009) Effect of temperature and nitrogen concentration on the growth and lipid content of Nannochloropsis oculata and Chlorella vulgaris for biodiesel production. Chem Eng Process 48(6):1146–1151

Craggs R, Sutherland D, Campbell H (2012) Hectare-scale demonstration of high rate algal ponds for enhanced wastewater treatment and biofuel production. J Appl Phycol 24(3):329–337

de Alva MS, Luna-Pabello VM, Cadena E, Ortiz E (2013) Green microalga Scenedesmus acutus grown on municipal wastewater to couple nutrient removal with lipid accumulation for biodiesel production. Bioresour Technol 146:744–748

de-Bashan LE, Bashan Y (2010) Immobilized microalgae for removing pollutants: review of practical aspects. Biores Technol 101(6):1611–1627

Demirbaş A (2003) Biodiesel fuels from vegetable oils via catalytic and non-catalytic supercritical alcohol transesterifications and other methods: a survey. Energy Convers Manag 44(13):2093–2109

Demirbas A (2007) Converting biomass derived synthetic gas to fuels via fisher-tropsch synthesis. Energy Sour Part A 29(16):1507–1512

Demirbas A (2009) Biofuels securing the planet's future energy needs. Energy Convers Manag 50 (9):2239–2249

Demirbas A, Demirbas MF (2010) Algae energy: algae as a new source of biodiesel. Springer Science & Business Media, Berlin

Demirbas A, Demirbas MF (2011) Importance of algae oil as a source of biodiesel. Energy Convers Manag 52(1):163–170

Demirbas E, Kobya M, Senturk E, Ozkan T (2004) Adsorption kinetics for the removal of chromium (VI) from aqueous solutions on the activated carbons prepared from agricultural wastes. Water Sa 30(4):533–539

Deng DJ, Lu ZM (2013) Differentiation and adaptation epigenetic networks: translational research in gastric carcinogenesis. Chin Sci Bull 58(1):1–6

Dosnon-Olette R, Trotel-Aziz P, Couderchet M, Eullaffroy P (2010) Fungicides and herbicide removal in Scenedesmus cell suspensions. Chemosphere 79(2):117–123

Dexter J, Fu P (2009) Metabolic engineering of cyanobacteria for ethanol production. Energy Environ Sci 2(8):857–864

Duong VT, Ahmed F, Thomas-Hall SR, Quigley S, Nowak E, Schenk PM (2015) High protein-and high lipid-producing microalgae from northern australia as potential feedstock for animal feed and biodiesel. Front Bioeng Biotechnol 3:53

Eroglu E, Smith SM, Raston CL (2015) Application of various immobilization techniques for algal bioprocesses. In: Biomass and biofuels from microalgae. Springer International Publishing, Berlin, pp 19–44

Fedorov AS, Kosourov S, Ghirardi ML, Seibert M (2005) Continuous hydrogen photoproduction by Chlamydomonas reinhardtii. Paper presented at the twenty-sixth symposium on biotechnology for fuels and chemicals

Feng Y, Li C, Zhang D (2011) Lipid production of Chlorella vulgaris cultured in artificial wastewater medium. Biores Technol 102(1):101–105

Fu F, Wang Q (2011) Removal of heavy metal ions from wastewaters: a review. J Environ Manage 92(3):407–418

Fukuda H, Kondo A, Noda H (2001) Biodiesel fuel production by transesterification of oils. J Biosci Bioeng 92(5):405–416

Gariazzo C, Lamberti M, Hänninen O, Silibello C, Pelliccioni A, Porta D … Forastiere F (2015) Assessment of population exposure to Polycyclic Aromatic Hydrocarbons (PAHs) using integrated models and evaluation of uncertainties. Atmos Environ 101:235–245

Gattullo CE, Bährs H, Steinberg CE, Loffredo E (2012) Removal of bisphenol a by the freshwater green alga Monoraphidium braunii and the role of natural organic matter. Sci Total Environ 416:501–506

Gavrilescu M (2010) Environmental biotechnology: achievements, opportunities and challenges. Dyn Biochem Process Biotech Mol 4(1):1–36

Gentili FG (2014) Microalgal biomass and lipid production in mixed municipal, dairy, pulp and paper wastewater together with added flue gases. Biores Technol 169:27–32

Gilden RC, Huffling K, Sattler B (2010) Pesticides and health risks. J Obstet Gynecol Neonatal Nurs 39(1):103–110

Gomes HI, Dias-Ferreira C, Ribeiro AB (2013) Overview of in situ and ex situ remediation technologies for PCB-contaminated soils and sediments and obstacles for full-scale application. Sci Total Environ 445:237–260

Griffiths MJ, van Hille RP, Harrison ST (2014) The effect of nitrogen limitation on lipid productivity and cell composition in Chlorella vulgaris. Appl Microbiol Biotechnol 98 (5):2345–2356

Günerken E, d'Hondt E, Eppink MHM, Garcia-Gonzalez L, Elst K, Wijffels RH (2015) Cell disruption for microalgae biorefineries. Biotechnol Adv 33(2):243–260

Gupta VK, Ali I, Saleh TA, Nayak A, Agarwal S (2012) Chemical treatment technologies for waste-water recycling—an overview. Rsc Adv 2(16):6380–6388

Halim R, Harun R, Danquah MK, Webley PA (2012) Microalgal cell disruption for biofuel development. Appl Energy 91(1):116–121

Haritash A, Kaushik C (2009) Biodegradation aspects of polycyclic aromatic hydrocarbons (PAHs): a review. J Hazard Mater 169(1):1–15

Hattab MA, Ghaly A (2015) Microalgae oil extraction pretreatment methods: critical review and comparative analysis. J Fundam Renew Energy Appl 5:172

Hayyan A, Alam MZ, Mirghani ME, Kabbashi NA, Hakimi NINM, Siran YM, Tahiruddin S (2010) Sludge palm oil as a renewable raw material for biodiesel production by two-step processes. Biores Technol 101(20):7804–7811

Hena S, Fatimah S, Tabassum S (2015) Cultivation of algae consortium in a dairy farm wastewater for biodiesel production. Water Resour Ind 10:1–14

Hoffmann JP (1998) Wastewater treatment with suspended and nonsuspended algae. J Phycol 34 (5):757–763

Hong YW, Yuan DX, Lin QM, Yang TL (2008) Accumulation and biodegradation of phenanthrene and fluoranthene by the algae enriched from a mangrove aquatic ecosystem. Marine Poll Bull 56(8):1400–1405

Hongyang S, Yalei Z, Chunmin Z, Xuefei Z, Jinpeng L (2011) Cultivation of Chlorella pyrenoidosa in soybean processing wastewater. Bioresour Technol 102(21):9884–9890

Hu B, Min M, Zhou W, Du Z, Mohr M, Chen P … Ruan R (2012) Enhanced mixotrophic growth of microalga Chlorella sp. on pretreated swine manure for simultaneous biofuel feedstock production and nutrient removal. Bioresour Technol 126:71–79

Huang G, Chen F, Wei D, Zhang X, Chen G (2010) Biodiesel production by microalgal biotechnology. Appl Energy 87(1):38–46

Huang X, Wei L, Huang Z, Yan J (2014) Effect of high ferric ion concentrations on total lipids and lipid characteristics of Tetraselmis subcordiformis, Nannochloropsis oculata and Pavlova viridis. J Appl Phycol 26(1):105–114

Islam MA, Magnusson M, Brown RJ, Ayoko GA, Nabi MN, Heimann K (2013) Microalgal species selection for biodiesel production based on fuel properties derived from fatty acid profiles. Energies 6(11):5676–5702

Johnson MB, Wen Z (2009) Production of biodiesel fuel from the microalga Schizochytrium limacinum by direct transesterification of algal biomass. Energy Fuels 23(10):5179–5183

Johnson MB, Wen Z (2010) Development of an attached microalgal growth system for biofuel production. Appl Microbiol Biotechnol 85(3):525–534

Kalla N, Khan S (2016) Effect of nitrogen, phosphorus concentrations, pH and salinity ranges on growth, biomass and lipid accumulation of Chlorella vulgaris. Int J Pharm Sci Res 7(1):397

Kapdan IK, Kargi F (2006) Bio-hydrogen production from waste materials. Enzyme Microb Technol 38(5):569–582

Karmakar A, Karmakar S, Mukherjee S (2010) Properties of various plants and animals feedstocks for biodiesel production. Biores Technol 101(19):7201–7210

Kawaroe M, Prartono T, Sunuddin A, Saputra D (2016) Marine microalgae Tetraselmis suecica as flocculant agent of bio-flocculation method. HAYATI J Biosci 23(2):62–66

Khan M, Yoshida N (2008) Effect of L-glutamic acid on the growth and ammonium removal from ammonium solution and natural wastewater by Chlorella vulgaris NTM06. Biores Technol 99 (3):575–582

Kiss AA, Dimian AC, Rothenberg G (2007) Biodiesel by catalytic reactive distillation powered by metal oxides. Energy Fuels 22(1):598–604

Kolesárová N, Hutňan M, Bodík I, Špalková V (2011) Utilization of biodiesel by-products for biogas production. BioMed Res Int

Kong Q-X, Li L, Martinez B, Chen P, Ruan R (2010) Culture of microalgae Chlamydomonas reinhardtii in wastewater for biomass feedstock production. Appl Biochem Biotechnol 160 (1):9–18

Kothari R, Pathak VV, Kumar V, Singh DP (2012) Experimental study for growth potential of unicellular alga Chlorella pyrenoidosa on dairy waste water: an integrated approach for treatment and biofuel production. Bioresour Technol 116:466–470

Krishna AR, Dev L, Thankamani V (2012) An integrated process for Industrial effluent treatment and Biodiesel production using Microalgae. Res Biotechnol 3(1)

Kumar A, Ergas S, Yuan X, Sahu A, Zhang Q, Dewulf J et al (2010a) Enhanced CO_2 fixation and biofuel production via microalgae: recent developments and future directions. Trends Biotechnol 28:371–380. https://doi.org/10.1016/j.tibtech.2010.04.004

Kumar A, Ergas S, Yuan X, Sahu A, Zhang Q, Dewulf J, Van Langenhove H (2010b) Enhanced CO_2 fixation and biofuel production via microalgae: recent developments and future directions. Trends Biotechnol 28(7):371–380

Kumar A, Yuan X, Sahu AK, Dewulf J, Ergas SJ, Van Langenhove H (2010c) A hollow fiber membrane photoâ-bioreactor for CO_2 sequestration from combustion gas coupled with wastewater treatment: a process engineering approach. J Chem Technol Biotechnol 85(3):387–394

Kumar KS, Dahms H-U, Won E-J, Lee J-S, Shin K-H (2015) Microalgaeâ-A promising tool for heavy metal remediation. Ecotoxicol Environ Saf 113:329–352

Kumar P, Suseela MR, Toppo K (2011) Physico-chemical characterization of algal oil: a potential biofuel. Asian J Exp Biol Sci 2(3):493–497

Laghari SM, Isa MH, Abdullah AB, Laghari AJ, Saleem H (2015) Microwave individual and combined pre-treatments on lignocellulosic biomasses. IOSR J Comput Eng 4(2):14–28

Lee AK, Lewis DM, Ashman PJ (2012) Disruption of microalgal cells for the extraction of lipids for biofuels: processes and specific energy requirements. Biomass Bioenerg 46:89–101

Lee HJ, Lee SY (2001) Heat transfer correlation for boiling flows in small rectangular horizontal channels with low aspect ratios. Int J Multiph Flow 27(12):2043–2062

Lee JY, Yoo C, Jun SY, Ahn CY, Oh HM (2010) Comparison of several methods for effective lipid extraction from microalgae. Biores Technol 101(1):S75–S77

Lei AP, Hu ZL, Wong YS, Tam NFY (2007) Removal of fluoranthene and pyrene by different microalgal species. Bioresour Technol 98(2):273–280

Li Y, Horsman M, Wu N, Lan CQ, Dubois-Calero N (2008) Biofuels from microalgae. Biotechnol Prog 24(4):815–820

Li Y, Zhou W, Hu B, Min M, Chen P, Ruan RR (2011) Integration of algae cultivation as biodiesel production feedstock with municipal wastewater treatment: strains screening and significance evaluation of environmental factors. Biores Technol 102(23):10861–10867

Liang Y (2013) Producing liquid transportation fuels from heterotrophic microalgae. Appl Energy 104:860–868

Lim SL, Chu WL, Phang SM (2010) Use of Chlorella vulgaris for bioremediation of textile wastewater. Bioresour Technol 101(19):7314–7322

Lu Q, Zhou W, Min M, Ma X, Chandra C, Doan YT … Chen P (2015) Growing Chlorella sp. on meat processing wastewater for nutrient removal and biomass production. Bioresour Technol 198:189–197

Luo L, Wang P, Lin L, Luan T, Ke L, Tam NFY (2014) Removal and transformation of high molecular weight polycyclic aromatic hydrocarbons in water by live and dead microalgae. Process Biochem 49(10):1723–1732

Lv J, Guo J, Feng J, Liu Q, Xie S (2017) Effect of sulfate ions on growth and pollutants removal of self-flocculating microalga Chlorococcum sp. GD in synthetic municipal wastewater. Biores Technol 234:289–296

Mahapatra DM, Chanakya HN, Ramachandra TV (2013) Euglena sp. as a suitable source of lipids for potential use as biofuel and sustainable wastewater treatment. J Appl Phycol 25(3):855–865

Maity JP, Bundschuh J, Chen CY, Bhattacharya P (2014) Microalgae for third generation biofuel production, mitigation of greenhouse gas emissions and wastewater treatment: Present and future perspectives–a mini review. Energy 78:104–113

Malla FA, Khan SA, Sharma GK, Gupta N, Abraham G (2015) Phycoremediation potential of Chlorella minutissima on primary and tertiary treated wastewater for nutrient removal and biodiesel production. Ecol Eng 75:343–349

Mallick N (2002) Biotechnological potential of immobilized algae for wastewater N, P and metal removal: a review. Biometals 15(4):377–390

Markou G, Nerantzis E (2013) Microalgae for high-value compounds and biofuels production: a review with focus on cultivation under stress conditions. Biotechnol Adv 31(8):1532–1542

Mata TM, Martins AA, Caetano NS (2010) Microalgae for biodiesel production and other applications: a review. Renew Sustain Energy Rev 14(1):217–232

McKendry P (2002) Energy production from biomass (part 2): conversion technologies. Biores Technol 83(1):47–54

Meher L, Sagar DV, Naik S (2006) Technical aspects of biodiesel production by transesterificationâ-a review. Renew Sustain Energy Rev 10(3):248–268

Mirghaffari N, Moeini E, Farhadian O (2015) Biosorption of Cd and Pb ions from aqueous solutions by biomass of the green microalga, Scenedesmus quadricauda. J Appl Phycol 27 (1):311–320

Mittal A (2011) Biological wastewater treatment. Water Today 1:32–44

Mohan D, Pittman CU, Steele PH (2006) Pyrolysis of wood/biomass for bio-oil: a critical review. Energy Fuels 20(3):848–889

Mondal P, Dang GS, Garg MO (2011) Syngas production through gasification and cleanup for downstream applications—recent developments. Fuel Process Technol 92(8):1395–1410

Morita M, Watanabe Y, Okawa T, Saiki H (2001) Photosynthetic productivity of conical helical tubular photobioreactors incorporating Chlorella sp. under various culture medium flow conditions. Biotechnol Bioeng 74(2):136–144

Moser BR, Vaughn SF (2012) Efficacy of fatty acid profile as a tool for screening feedstocks for biodiesel production. Biomass Bioenerg 37:31–41

Mukherjee K, Saha R, Ghosh A, Saha B (2013) Chromium removal technologies. Res Chem Intermed 39(6):2267–2286

Narala RR, Garg S, Sharma KK, Thomas-Hall SR, Deme M, Li Y, Schenk PM (2016) comparison of Microalgae cultivation in Photobioreactor, open raceway pond, and a two-stage hybrid system. Front Energy Res 4:29

Nayak M, Thirunavoukkarasu M, Mishra BK (2013) Maximizing biomass productivity and CO_2 biofixation of microalga, Scenedesmus sp. by using sodium hydroxide. J Microbiol Biotechnol 23(9):1260–1268

Niestroy J, Martínez AB, Band-Schmidt CJ (2014) Analysis of concentration-dependent effects of copper and PCB on different Chattonella spp. microalgae (raphidophyceae) cultivated in artificial seawater medium. EXCLI J 13:197

Ndikubwimana T, Zeng X, Liu Y, Chang JS, Lu Y (2014) Harvesting of microalgae Desmodesmus sp. F51 by bioflocculation with bacterial bioflocculant. Algal Res 6:186–193

Ndikubwimana T, Zeng X, Murwanashyaka T, Manirafasha E, He N, Shao W, Lu Y (2016) Harvesting of freshwater microalgae with microbial bioflocculant: a pilot-scale study. Biotechnol Biofuels 9(1):47

Pandelova M, Piccinelli R, Kasham S, Henkelmann B, Leclercq C, Schramm K-W (2010) Assessment of dietary exposure to PCDD/F and dioxin-like PCB in infant formulae available on the EU market. Chemosphere 81(8):1018–1021

Park J, Craggs R, Shilton A (2011) Wastewater treatment high rate algal ponds for biofuel production. Biores Technol 102(1):35–42

Park J, Seo J, Kwon EE (2012) Microalgae production using wastewater: effect of light-emitting diode wavelength on microalgal growth. Environ Eng Sci 29(11):995–1001

Pereira FV, Gurgel LVA, Gil LF (2010) Removal of Zn^{2+} from aqueous single metal solutions and electroplating wastewater with wood sawdust and sugarcane bagasse modified with EDTA dianhydride (EDTAD). J Hazard Mater 176(1):856–863

Pittman JK, Dean AP, Osundeko O (2011) The potential of sustainable algal biofuel production using wastewater resources. Biores Technol 102(1):17–25

Powell RJ, Hill RT (2013) Rapid aggregation of biofuel-producing algae by the bacterium Bacillus sp. strain RP1137. Appl Environ Microbiol 79(19):6093–6101

Raheem A, Azlina WW, Yap YT, Danquah MK, Harun R (2015a) Thermochemical conversion of microalgal biomass for biofuel production. Renew Sustain Energy Rev 49:990–999

Raheem A, WAKG WA, Yap YT, Danquah MK, Harun R (2015b) Optimization of the microalgae Chlorella vulgaris for syngas production using central composite design. RSC Adv 5 (88):71805–71815

Ramesh D (2013) Lipid identification and extraction techniques. In: Biotechnological applications of microalgae: biodiesel and value-added products, pp 89–97

Ranjith Kumar R, Hanumantha Rao P, Arumugam M (2015) Lipid extraction methods from microalgae: a comprehensive review. Front Energy Res 2:61

Ratha SK, Babu S, Renuka N, Prasanna R, Prasad RBN, Saxena AK (2013) Exploring nutritional modes of cultivation for enhancing lipid accumulation in microalgae. J Basic Microbiol 53 (5):440–450

Richards RG, Mullins BJ (2013) Using microalgae for combined lipid production and heavy metal removal from leachate. Ecol Model 249:59–67

Richmond A (ed) (2008) Handbook of microalgal culture: biotechnology and applied phycology. Wiley, New Jersey

Ruiz-Marin A, Mendoza-Espinosa LG, Stephenson T (2010) Growth and nutrient removal in free and immobilized green algae in batch and semi-continuous cultures treating real wastewater. Biores Technol 101(1):58–64

Sakarika M, Kornaros M (2016) Effect of pH on growth and lipid accumulation kinetics of the microalga Chlorella vulgaris grown heterotrophically under sulfur limitation. Biores Technol 219:694–701

San Martín YB (2011) Bioremediation: a tool for the management of oil pollution in marine ecosystems. Biotecnol Apl 28(2):69–76

Scott SA, Davey MP, Dennis JS, Horst I, Howe CJ, Lea-Smith DJ, Smith AG (2010) Biodiesel from algae: challenges and prospects. Curr Opin Biotechnol 21(3):277–286

Sharma KK, Garg S, Li Y, Malekizadeh A, Schenk PM (2013) Critical analysis of current microalgae dewatering techniques. Biofuels 4:397–407

Sharma Y, Singh B, Upadhyay S (2008) Advancements in development and characterization of biodiesel: a review. Fuel 87(12):2355–2373

Shibata T, Kawaguchi S, Hama Y, Inagaki M, Yamaguchi K, Nakamura T (2004) Local and chemical distribution of phlorotannins in brown algae. J Appl Phycol 16(4):291–296

Shilton A, Mara D, Craggs R, Powell N (2008) Solar-powered aeration and disinfection, anaerobic co-digestion, biological CO_2 scrubbing and biofuel production: the energy and carbon management opportunities of waste stabilisation ponds. Water Sci Technol 58(1):253

Sialve B, Bernet N, Bernard O (2009) Anaerobic digestion of microalgae as a necessary step to make microalgal biodiesel sustainable. Biotechnol Adv 27(4):409–416

Singh D, Barrow CJ, Puri M, Tuli DK, Mathur AS (2016) Combination of calcium and magnesium ions prevents substrate inhibition and promotes biomass and lipid production in thraustochytrids under higher glycerol concentration. Algal Res 15:202–209

Subashchandrabose SR, Ramakrishnan B, Megharaj M, Venkateswarlu K, Naidu R (2013) Mixotrophic cyanobacteria and microalgae as distinctive biological agents for organic pollutant degradation. Environ Int 51:59–72

Taher H, Al-Zuhair S, Al-Marzouqi AH, Haik Y, Farid M (2014) Effective extraction of microalgae lipids from wet biomass for biodiesel production. Biomass Bioenerg 66:159–167

Talebi AF, Tohidfar M, Mousavi Derazmahalleh SM, Sulaiman A, Baharuddin AS, Tabatabaei M (2015) Biochemical modulation of lipid pathway in microalgae Dunaliella sp. for biodiesel production. BioMed Res Int

Tan KWM, Lee YK (2016) The dilemma for lipid productivity in green microalgae: importance of substrate provision in improving oil yield without sacrificing growth. Biotechnol Biofuels 9 (1):255

Tang D, Han W, Li P, Miao X, Zhong J (2011) CO_2 biofixation and fatty acid composition of Scenedesmus obliquus and Chlorella pyrenoidosa in response to different CO_2 levels. Biores Technol 102(3):3071–3076

The Guardian (2015) Pope Francis issues final call for unity at end of historic US visit. Available at https://www.theguardian.com/world/2015/sep/28/pope-francis-leaves-us-end-visit. Accessed on 19 June 2017

Toor SS, Rosendahl L, Rudolf A (2011) Hydrothermal liquefaction of biomass: a review of subcritical water technologies. Energy 36(5):2328–2342

Tsukahara K, Sawayama S (2005) Liquid fuel production using microalgae. J Jpn Pet Inst 48 (5):251

Ugwu C, Aoyagi H, Uchiyama H (2008) Photobioreactors for mass cultivation of algae. Biores Technol 99(10):4021–4028

Ukiwe LN, Egereonu UU, Njoku PC, Nwoko CI, Allinor JI (2013) Polycyclic aromatic hydrocarbons degradation techniques: a review. Int J Chem 5(4):43

Ullah K, Ahmad M, Sharma VK, Lu P, Harvey A, Zafar M … Anyanwu CN (2014) Algal biomass as a global source of transport fuels: overview and development perspectives. Prog Nat Sci: Mater Int 24(4):329–339

Usher PK, Ross AB, Camargo-Valero MA, Tomlin AS, Gale WF (2014) An overview of the potential environmental impacts of large-scale microalgae cultivation. Biofuels 5(3):331–349

Uthman H, Saka AA (2013) Comparatives study of production biodiesel from soybean oil and Jatropha Curcas seeds oil. Distrib Gener Altern Energy J 28(2):31–42

Valiente Moro C, Bricheux G, Portelli C, Bohatier J (2012) Comparative effects of the herbicides chlortoluron and mesotrione on freshwater microalgae. Environ Toxicol Chem 31(4):778–786

Vardon DR, Sharma BK, Blazina GV, Rajagopalan K, Strathmann TJ (2012) Thermochemical conversion of raw and defatted algal biomass via hydrothermal liquefaction and slow pyrolysis. Biores Technol 109:178–187

Veillette M, Nikiema J, Heitz M, Chamoumi M, Faucheux N (2012) Production of biodiesel from microalgae. INTECH Open Access Publisher, London

Wahidin S, Idris A, Shaleh SRM (2013) The influence of light intensity and photoperiod on the growth and lipid content of microalgae Nannochloropsis sp. Biores Technol 129:7–11

Wang B, Li Y, Wu N, Lan CQ (2008) CO_2 bio-mitigation using microalgae. Appl Microbiol Biotechnol 79(5):707–718

Wang H, Xiong H, Hui Z, Zeng X (2012) Mixotrophic cultivation of Chlorella pyrenoidosa with diluted primary piggery wastewater to produce lipids. Biores Technol 104:215–220

Wang L, Min M, Li Y, Chen P, Chen Y, Liu Y, Ruan R (2010) Cultivation of green algae Chlorella sp. in different wastewaters from municipal wastewater treatment plant. Appl Biochem Biotechnol 162(4):1174–1186

Woertz I, Feffer A, Lundquist T, Nelson Y (2009) Algae grown on dairy and municipal wastewater for simultaneous nutrient removal and lipid production for biofuel feedstock. J Environ Eng 135(11):1115–1122

Wu X, Ruan R, Du Z, Liu Y (2012) Current status and prospects of biodiesel production from microalgae. Energies 5(8):2667–2682

Xie Y, Li H, Wang X, Ng I S, Lu Y, Jing K (2014) Kinetic simulating of Cr (VI) removal by the waste Chlorella vulgaris biomass. J Taiwan Inst Chem Eng 45(4):1773–1782

Xin L, Hong-ying H, Ke G, Ying-xue S (2010) Effects of different nitrogen and phosphorus concentrations on the growth, nutrient uptake, and lipid accumulation of a freshwater microalga Scenedesmus sp. Biores Technol 101(14):5494–5500

Xu H, Miao X, Wu Q (2006) High quality biodiesel production from a microalga Chlorella protothecoides by heterotrophic growth in fermenters. J Biotechnol 126(4):499–507

Yang J, Cao J, Xing G, Yuan H (2015) Lipid production combined with biosorption and bioaccumulation of cadmium, copper, manganese and zinc by oleaginous microalgae Chlorella minutissima UTEX2341. Biores Technol 175:537–544

Yoo C, Jun S-Y, Lee J-Y, Ahn C-Y, Oh H-M (2010) Selection of microalgae for lipid production under high levels carbon dioxide. Biores Technol 101(1):S71–S74

Zamalloa C, Boon N, Verstraete W (2012) Anaerobic digestibility of Scenedesmus obliquus and Phaeodactylum tricornutum under mesophilic and thermophilic conditions. Appl Energy 92:733–738

Zhang J, Hu B (2012) A novel method to harvest microalgae via co-culture of filamentous fungi to form cell pellets. Biores Technol 114:529–535

Zhang X (2015) Microalgae removal of CO_2 from flue gas. IEA Clean Coal Centre, UK

Zhang X, Rong J, Chen H, He C, Wang Q (2014) Current status and outlook in the application of microalgae in biodiesel production and environmental protection. Front Energy Res 2:32

Zhao Z, Jiang Y, Xia L, Mi T, Yan W, Gao Y, Hussain J (2014) Application of canonical correspondence analysis to determine the ecological contribution of phytoplankton to PCBs bioaccumulation in Qinhuai River, Nanjing, China. Environ Sci Pollut Res 21(4):3091–3103

Zhou N, Zhang Y, Wu X, Gong X, Wang Q (2011a) Hydrolysis of Chlorella biomass for fermentable sugars in the presence of HCl and $MgCl_2$. Biores Technol 102(21):10158–10161

Zhou W, Hu B, Li Y, Min M, Mohr M, Du Z, Ruan R (2012) Mass cultivation of microalgae on animal wastewater: a sequential two-stage cultivation process for energy crop and omega-3-rich animal feed production. Appl Biochem Biotechnol 168(2):348–363

Zhou W, Li Y, Min M, Hu B, Chen P, Ruan R (2011b) Local bioprospecting for high-lipid producing microalgal strains to be grown on concentrated municipal wastewater for biofuel production. Biores Technol 102(13):6909–6919

Zhou Y, Schideman L, Yu G, Zhang Y (2013) A synergistic combination of algal wastewater treatment and hydrothermal biofuel production maximized by nutrient and carbon recycling. Energy Environ Sci 6(12):3765–3779

Zhu L, Wang Z, Takala J, Hiltunen E, Qin L, Xu Z, Yuan Z (2013) Scale-up potential of cultivating Chlorella zofingiensis in piggery wastewater for biodiesel production. Biores Technol 137:318–325

Chapter 14
Food Waste Valorization by Microalgae

R. Yukesh Kannah, Chinnathambi Velu, J. Rajesh Banu, Kirsten Heimann and Obulisamy Parthiba Karthikeyan

Abstract In early twenty-first century, both developed and developing countries aim to avoid burning of fossil fuel in an effort to reduce the greenhouse gas emissions and impacts on global warming. Microalgae are potential key players for tackling greenhouse gas emissions and for providing feedstock for renewable energy production. Microalgae utilize freely available solar radiation as an energy source to extract protons and electrons from water to ultimately convert atmospheric carbon dioxide into organic carbon manifested in the growth rates and biomass concentrations. The microalgal biomass consists of biopolymers (protein and carbohydrate), lipid and pigments, which provides a platform for producing value-added products or for utilization as renewable energy resources. However, carbon and nutrient requirements for their cultivation are major bottlenecks adding to the overall production costs. Alternatively, food waste could be used for cultivation of microalgae after suitable pretreatment to solubilize organic carbon polymers. In an integrated bio-refinery approach, harvested microalgal biomass, value-added products are extracted sequentially, with the leftover components (those that do not have a significant market value) to be used in energy generation through anaerobic digestion/fermentation processes. This chapter will provide an overview on food waste valorisation by and most suitable species of microalgae, a brief discussion on adopting various pretreatment techniques for solubilization of carbon from food waste for easy valorisation by microalgae.

Keywords Food waste · Pretreatment · Bio-refinery · Microalgae
Value products

R. Y. Kannah · J. Rajesh Banu (✉)
Department of Civil Engineering, Regional Centre of Anna University, Tirunelveli, India
e-mail: rajeshces@gmail.com

C. Velu · K. Heimann · O. P. Karthikeyan
College of Science and Engineering, James Cook University, Townsville, Australia

O. P. Karthikeyan
Department of Biology, Hong Kong Baptist University,
Hong Kong SAR, People's Republic of China

R. R. Singhania et al. (eds.), *Waste to Wealth*, Energy, Environment, and Sustainability, https://doi.org/10.1007/978-981-10-7431-8_14

14.1 Introduction

Food waste is characterized with the high moisture content and biodegradable organic fractions such as 5–10% of protein, 20–45% carbohydrate and 10–40% of lipid (Karthikeyan et al. 2016). Improper disposal of food waste creates environmental issues like malodour, breading of diseases causing files, emission of toxic gases and groundwater contamination via biological degradation (Chen et al. 2013; Menon et al. 2016). According to FAO (2011), nearly 1.3 billion tonnes of food were accumulated as waste throughout the worldwide and its project to be 2.2 billion by 2025. Predominantly, the developing countries are reported to have lack of proper food waste management systems and policies, which create stress to discover suitable management options (Karthikeyan et al. 2017; Pleissner et al. 2017). On seeing the benefits of food waste and typical composition, the food waste can be a raw material for value product production through bio-valorisation approaches or resource for bioenergy recovery (i.e. biomethane, biohydrogen and biohythane) through anaerobic digestion/fermentation (Kannah et al. 2017; Ghimire et al. 2017; Akinbomi and Taherzadeh 2015) as shown in Fig. 14.1.

Food waste also holds high valuable calorific nutrient source such as phosphate, fatty acid, amino acid and sugar monomers (Pleissner et al. 2014b). Recovery of nutrients and calories from food waste will be possible only if the integrated bio-refinery approach was established and practised. The integrated bio-refinery approach aims to produce more than one value produce from the food waste under lower energy/operational costs (Karthikeyan et al. 2017). However, food waste

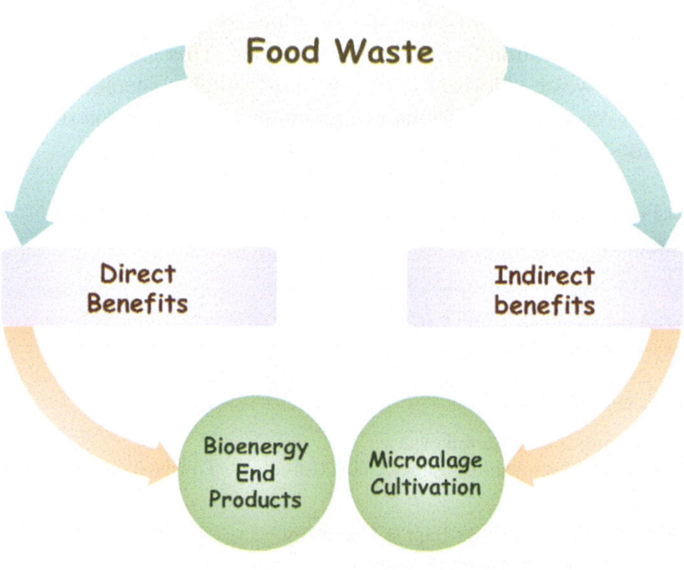

Fig. 14.1 Food waste value additions through direct and indirect approaches

required certain degree of pretreatment to solubilize some sugars and essential nutrients in liquid phase for bio-valorisation into value products using bacteria, yeast or microalgae. The microalgae-based food waste valorisation approaches are more superior to yeast or bacteria that combined the greenhouse gas emission reduction and produce number of high-value products (e.g. pigments, biolipids, biodiesel precursor molecules and vitamins). The residual solids from pretreatment process could be digested/fermented for bioenergy recovery. Therefore, the pre-treatment process is the key for successful development of integrated bio-refinery process required better understanding of overall food waste composition and value addition processes with respect to the choice of specific microalgal species. In this chapter, we have provided a detailed review of different pretreatment methods to process the food waste for bio-valorisation and bioenergy recovery processes under the "integrated bio-refinery" concept.

14.2 Food Waste Generation, Composition and Characteristics

In global scale, the term food waste can be referred as food ingredients that are wasted between field and fork. In other words, food waste means the food, which is unfit for human intake or rejected after its get rotten or uneaten part. It can be thrown as waste, the food, which has no longer value or in a state of decomposition. Food waste and food loss around the world were surveyed and published in a report by Food and Agricultural Organization, United Nations (FAO 2011) (Table 14.1). South and Southeast Asia contribute the major share of 50% with the rest of the world. According FAO, the maximum food waste is occurring at two stages (i) consumer waste and (ii) retailer waste. Whereas, food loss is defined as food ingredients that wasted during various stages of food flow. The food loss happen during the following five stages of food flow such as (i) production, (ii) handling & storage, (iii) processing, (iv) distribution and marketing and (v) consumption.

The food waste is heterogeneous in composition at macroscale that is influenced by the season, culture, income and policies of particular country/region. In microscale level, food waste from residential (kitchen waste), commercial (cooked, uncooked and bakery waste), institutional (canteen waste), industrial (processed and packed waste) and agricultural field (harvested waste) is varied widely. However, a typically elemental composition (i.e. carbon, nitrogen, hydrogen and sulphur) of food waste found to be similar.

The carbon accounts for 45–55%, which essential element for building block for fat, protein and carbohydrates for many microbial processes. Similarly, other essential elements such as nitrogen 1.5–4%, oxygen 25–40%, hydrogen 6–8%, sulphur 0.01–0.2% and remaining other elementals. The remaining other elements are mainly trace elements and dissolvable inorganic salts. Usually, trace element accounts for very less amount in food waste and they are iron (Fe), Copper (Cu) and

Table 14.1 Food waste and food loss around world

S. no	Region	Food waste		Food loss				
		Consumer waste	Retailer waste	Production	Handling and storage	Processing	Distribution and market	Consumption
		(kg/capita/year)		(%)				
1.	South and Southeast Asia	110	15	32	37	4	15	13
2.	Sub-Sahara Africa	150	10	39	37	7	13	5
3.	Industrialized Asia	155	85	17	23	2	11	46
4.	North America and Oceania	175	120	17	6	9	7	61
5.	North Africa, West and Central Asia	175	35	23	21	4	18	34
6.	Europe	180	100	23	11	5	9	52
7.	Latin America	200	25	28	22	6	17	28

Table 14.2 Macronutrient composition of food waste

Type of food waste	Starch	Protein	Carbohydrate	Lipid	References
	(mg/g)				
Canteen waste	–	120 ± 5	580 ± 65	269 ± 25	Pleissner et al. (2017)
Cake waste	458 ± 30	141 ± 8	643 ± 12	161 ± 8	Kiran and Liu (2015)
Mixed waste	603 ± 38	86 ± 4	768 ± 52	146 ± 31	
Bakery waste	316.7	98.2	654.5	265.8	Pleissner et al. (2014a)
Food waste A	612.3	57.9	738.4	73.8	
Food waste B	361.5	99.3	470.3	373.7	
Cake waste	458 ± 30	141 ± 8	168 ± 5	161 ± 7.5	Kiran et al. (2015)
Mixed food waste	461 ± 32	111 ± 18	82 ± 7	153 ± 21	
Food waste	–	103.7 ± 13.7	332.6 ± 4.7	150.3 ± 11.8	Pleissner et al. (2013a)
Bakery waste	–	42.5 ± 3.5	620.1 ± 87.0	190.2 ± 33.4	
Bread	59.8	8.9	46.8	0.9	Han et al. (2015a)
Pastry	44.6	7.1	33.5	35.2	
Cake	12.9	17.0	62.0	19.0	
Restaurant	36.2–61.2	5.8–9.9	47.0–73.8	7.4–37.4	
Food waste	40.6 ± 0.6	10.5 ± 0.5	42.7 ± 0.8	6.2 ± 0.7	Han et al. (2015b)

Zinc (Zn). On seeing dissolvable salts, they are categorized into five: sodium (Na), magnesium (Mg), chlorine (Cl), potassium (K) and calcium (Ca) (Menon et al. 2016). If not, the food waste composition is characterized based on macronutrients such as starch, protein, carbohydrate and lipids (as shown in Table 14.2).

14.3 Food Waste Management and Bio-Valorisation

Food waste management is a global issue that required better understanding of quality and quantity of food waste generation from a particular region. According to the food waste management hierarchy, reduction/prevention is the most preferred options and landfilling/incineration is the least preferred options (Fig. 14.2). In addition, it is also suggested to feed hungry people or animals where the cultural and social factors are the critical barriers, if not able to reduce/prevent the food waste generation. There are conventional technologies such as composting and anaerobic digestion which are feasible option to treated food waste. The main advantages of these processes are (i) produce bio-fertilizer and bioenergy; (ii) help to circulate the food waste carbon back into the fossil economy; and (iii) reduce the greenhouse gas emissions and support carbon credits (Karthikeyan et al. 2017). Alternatively, a number of value-added industrial chemicals/feed stocks/fuels/

Fig. 14.2 Food waste management hierarchy

bio-products could also be produced from food waste by integration of different process through bio-refinery approaches. The bio-refinery concepts are not new; however, it required better understanding for food waste valorization into commodity chemicals.

The focus of the original bio-refinery concept is to produce transportation biofuels that can be mixed or substituted with gasoline, kerosene, diesel or other fuel oils. The volume and price should be competitive enough, while the quality should be equally good as petrochemical-based fuels are the key for bio-refinery industry developments in next few years (de Jong and Jungmeier 2015). Food bio-valorisation into oil and biofuels will be attractive option under bio-refinery concepts, while the selection of appropriate microbial agent is quite challenging. The microalgae are reported to produce two magnitudes of higher oil accumulation capacities (up to 58,700 L oil per hectare cultivation) than energy crops (Chisti 2008). Microalgae have several advantages over oil crops cultivation for biofuel such as (a) they can grow in nutrient-rich wastewater and do not compete for freshwater; (b) do not require arable land; they have high rapid growth rate; (c) high oil content ranging from 20–50% dry weight of biomass; (d) utilizes atmospheric carbon dioxide (1.83 kg of carbon dioxide can be fixed by 1 kg of algal biomass (DW)) which mitigates global warming; (e) microalgae do not require herbicides and pesticides; and (f) microalgae produce several co-products (Brennan and Owende 2010). Microalgal biodiesel are better than petroleum diesel, as it is renewable, non-toxic, contains reduced amounts of particulates, carbon monoxide, hydrocarbons and SO_x (Brennan and Owende 2010). Thus, the microalgae-based food waste bio-volarization will be more attractive and cost-effective approaches to produce high-quality transportation fuels. However, food waste required certain degree of pretreatment before bio-valorisation using microalgae.

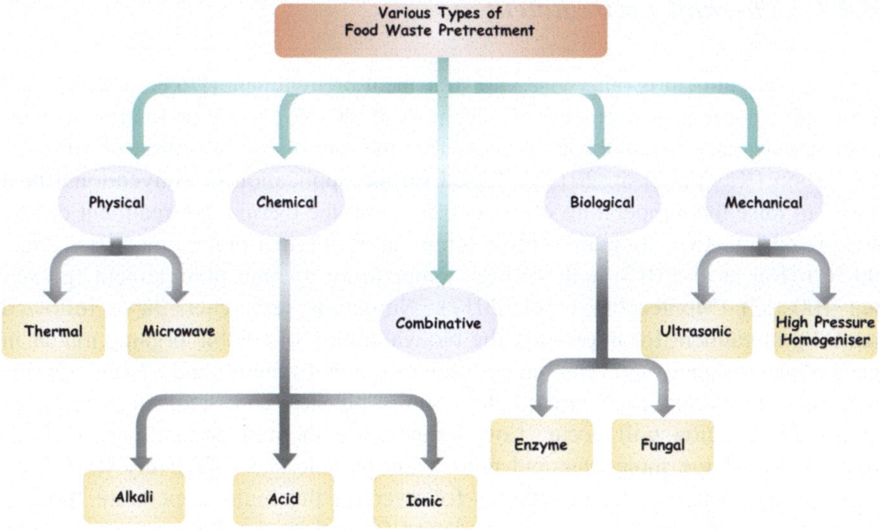

Fig. 14.3 Food waste pretreatment options

14.4 Various Types of Food Waste Pretreatment

Various pretreatment methods adopted for food waste to efficiently solubilize the nutrients and carbon for microalgae valorisation. Figure 14.3 represents the various types of food waste pretreatment. Many researcher have adopted various pretreatments on organic waste biomass such as thermal (Hao et al. 2015), microwave (Zhang et al. 2016) alkaline (Banu et al. 2012), acid (Razaghi et al. 2016), thermochemical (Banu et al. 2011), ultrasonication (Guo et al. 2014), high pressure homogenizes (Ma et al. 2011), fungal (Yang et al. 2015) and enzyme (Kavitha et al. 2013) pretreatment methods in literature. Only very few considered the pretreatment followed by microalgae valorisation.

14.4.1 Physical Pretreatment

Physical pretreatment is effective in disintegration of organic fraction by applying external pressure or mechanical means. Most probably, the physical pretreatment is combined with chemical pretreatment to improve the synergetic effect in disintegration of organic fraction (Do et al. 2012).

14.4.2 Thermal Pretreatment

Thermal pretreatment is otherwise known as heat pretreatment (Do et al. 2009) and sterilisation pretreatment process (Razaghi et al. 2016). In this pretreatment, temperature plays an extensive role to accelerate the rate of solubilization of nutrients and carbon (Kavitha et al. 2015b). Based on the application of conventional heat energy to raise the temperature of organic fraction, the thermal pretreatment can be categorized into two; they are (i) low-temperature thermal pretreatment (less than 100 °C) (Raj et al. 2013) and (ii) high-temperature thermal pretreatment (greater than 100 °C) (Montecchio et al. 2017). Numerous researchers have followed thermal pretreatment for improving the bioavailability of soluble organic matter in liquid phase and their optimum temperature ranges between 60 and 270 °C. On the other hand, in thermal pretreatment alteration of physical and chemical structure of the organic fraction will occur. Low temperature thermal pretreatment (LTTP) might destroy large molecules and reduce the particle size effectively (Raj et al. 2013). Even though same effect was followed in the high-temperature thermal pretreatment, it required longer duration and formation of refractory compounds, while also consumes more energy (Li and Jin 2015; Lagerkvist and Morgan-Sagastume 2012). But, the high temperature pretreatments are not/less reported for food waste valorisation.

14.4.3 Microwave Pretreatment

Microwave pretreatment is also known as alternative to conventional method of heat pretreatment (Rani et al. 2013). It is a promising technique to increase the bioavailability of supernatant for effective conversion of bioenergy (Ebenezer et al. 2015b). Many researchers have suggested that microwave pretreatment results in higher solubilization of organic matter from various organic substrates including food waste with minimal input energy consumption, when compared to conventional method of heat pretreatment (Kavitha et al. 2016b). In microwave, a rapid electromagnetic radiation was induced to heat the organic matter would effectively vaporises the water molecules. Once the vaporization process complete, internal pressure created on organic matter which accelerates the disintegration process simultaneously (Eswari et al. 2016). On the other hand, it accounts for higher operational cost than conventional method (Ebenezer et al. 2015a). The application of microwave pretreatment was not yet reported in pilot scale yet. The merits of microwave pretreatment method over conventional heat treatment are (i) effectively reduce heat loss and (ii) environmental conservation. Microwave pretreatment has following demerits they are (i) water loss (ii) denaturation of protein molecules through excessive heat (iii) maillard reaction (degradation of sugar molecules).

14.4.4 Chemical Pretreatment

Chemical pretreatment is cost-effective and easily solubilize the organic matter which can improve the bioenergy production. On the other hand, chemical pretreatment combined with physical and mechanical will result in reduction of operational cost and energy (Gayathri et al. 2015). This creates positive impact on chemical pretreatment, and a number of researches report that it is the best option. The chemical pretreatment is further categorized into (i) alkaline pretreatment, (ii) acid pretreatment, (iii) ionic liquid pretreatment.

14.4.4.1 Alkaline Pretreatment

Alkaline pretreatment is a well-known method for effective in liquefaction organic waste (Kavitha et al. 2016b). The pretreatment was carried out use of strong and weak alkaline-based chemicals. In general, three types of alkaline are frequently used by many researcher such as sodium hydroxide (NaOH) (Menon et al. 2016), potassium hydroxide (KOH) (Kim and Shin 2008) and calcium hydroxide ($Ca(OH)_2$) (Junoh et al. 2016). There have been numerous reports that suggest sodium hydroxide as strong alkaline shows best result in improving the release of bioactive compounds into the liquid steam than other alkaline. Some other alkaline such as sodium chloride (Kavitha et al. 2015a) and calcium chloride (Kavitha et al. 2015c) are used for alkaline pretreatment of organic waste. Alkaline pretreatment is cost-effective than other pretreatment methods (Rani et al. 2012b). During alkaline pretreatment the hydroxyl radical (OH^-) will cause salvation and saponification which results in effective liquefaction of organic matter. During this process, the complex food substance surface area gets enlargement, which accelerates the biodegradability. As a result, the rate of liquefaction of food waste will drastically increase bioactive substance in liquid stream, which makes best environment for the effective growth of microalgae.

14.4.4.2 Acid Pretreatment

Acid pretreatment was cost-effective method and increasing solubilization of food waste (Kim and Shin 2008). The pretreatment was carried out by the use of strong and weak acid-based chemicals. The following acid can be utilized for acid pretreatment, for example hydrochloric acid (Kim et al. 2014), sulphuric acid (Del Campo et al. 2006) and citric acid (Gayathri et al. 2015). In general, hydrochloric acid was frequently followed by many researchers. On the other hand, sulphuric acid has limited in usage due to formation of hydroxyl-furfural compound (Razaghi et al. 2016). According to Taherzadeh and Karimi (2007) during acid pretreatment on organic substrate leads to formation different kinds of recalcitrant compounds. The following inhibitors by-products such as furans, carboxylic acids and phenolic were achieved due to low pH on substrate.

14.4.4.3 Ionic Liquid Pretreatment

Ionic liquid pretreatment is an efficient technique used to solubilize the lignocellulosic biomass (Groff et al. 2013) particularly as well as food waste (Allison et al. 2016). In ionic liquid pretreatment, salts play a major role; at ambient temperature, they are in the liquid form. Ionic liquids are otherwise known as green solvents. The liquidized salts have very low vapour pressure at room temperature, and they are stable over 300 K. During pretreatment, liquidized salts target solubilization of predominant components such as cellulose, hemicellulose and lignin as well as interaction depends on combination particular substances cation and anion (Singh and Simmons 2013). The advantages over this pretreatment techniques enhance the enzyme accessibility for effective biofuel production and highly reduce the lignin concentration. On the other hand, recovery of ionic liquid from pretreated biomass is a challenging task and the capital investment cost of this pretreatment is high.

14.5 Biological Pretreatment

Biological pretreatment is most attractive because eco-friendly in nature and it does not cause any undesirable impact on environment (Kavitha et al. 2017a). In addition, (i) on considering economic aspect—the process requires low initial investment cost and (ii) on considering energy consumption aspect—the process requires minimum input energy to yield maximum output energy. Biological pretreated food waste supernatant was commonly used to for microalgae cultivation. Because it holds soluble macronutrient and which is essential carbon source for microalgae growth, remaining solid portion has capable of bioenergy production. It can be of two categories: (i) enzymatic pretreatment and (ii) fungal pretreatment.

14.5.1 Enzymatic Pretreatment

Enzyme pretreatment on organic food waste is cost-effective, and it holds dual benefits. Commercial available protease and amylase secreting enzymes are most probably used for enzymatic pretreatment (Kavitha et al. 2014a, b). The protease and amylase secreting enzymes can able to degrade soluble protein into amino acid and soluble carbohydrate into simpler sugar monomers (Kavitha et al. 2017b). It holds leading advantages, there is no formation inhibitory by-products during enzyme pretreatment.

14.5.2 Fungal Pretreatment

The fungi species such as *Aspergillus, Rhizopus and Monascus* are most commonly used to pretreat the food waste. The above-stated fungi species are capable of degrade the soluble protein into amino acid and soluble carbohydrate into simpler sugar monomers (Lam et al. 2015). Several researchers have reported that the sub-species of aspergillus such as *Aspergillus awamori* and *Aspergillus oryzae* are predominately used for food waste fungal pretreatment (Han et al. 2016). *A. awamori* are capable of secreting insignificant amount of citric acid. Simultaneously, it has capable converting complex starch substance into simple sugar monomers. Similarly, *A. oryzae* are capable of secreting insignificant amount of amylase. Then it was the best example for industrialized fungal highly secreting amylase like α-amylase and glucoamylase. Recycling of liquid steam to microalgae growth will highly impact on the cultivation cost. Then remaining solid residues holds organic matter which will be helpful for bioenergy generation.

14.6 Mechanical Pretreatment

Mechanical pretreatment is well known for reducing the particulate size and effective in release of organic fraction into the liquid phase (Kavitha et al. 2016a). Typically, mechanical pretreatment was combined with physical or chemical techniques to minimize the negative impact of investment and operational cost (Tamilarasan et al. 2017). Many research articles account for mechanical pre-treatment, which have been creates a wider platform for maximum bioenergy yield. The mechanical pretreatment was primarily explained about two types; they are (i) ultrasonication pretreatment, (ii) high pressure homogenizer pretreatment.

14.6.1 Ultrasonication Pretreatment

Ultrasonic pretreatment is well known for mechanical method of solubilizing the complex organic substrate. The word ultrasonic refers to propagation sound wave with higher frequency (>20 kHz). The following pretreatment on complex organic substrate could reduce the particle size and improve the substrate biodegradability (Packyam et al. 2015). During ultrasonic pretreatment, complex organic matter was broken down into small fragments by micro bubble induced cavitation effect. This action improves the transformation of bioactive substance into liquid portion

(Kavitha et al. 2016c). However, an additional cavitation effect created would reduce the surface tension (Ushani et al. 2017). According to Guo et al.(2013), ultrasonication pretreatment has no negative impact on environment but it has high positive impact on disintegration of complex organic substrate which was demonstrated.

14.6.2 *High-Pressure Homogenizer Pretreatment*

High-pressure homogenizer pretreatment is very effective in increasing the rate of solubilization and emulsification properties (Rani et al. 2012b). Conversely, HPH is the supreme method to break down larger particles size to smaller. It holds high-positive impact in size reduction (Kavitha et al. 2016d) and it was successful method which was widely practise in pilot scale. The HPH is mainly focused on food processing industry, especially for homogenization process. During HPH pretreatment size of organic particles get reduced this in turns increase in surface area for subsequent liquefaction of organic matter. In addition, during this pretreatment the viscosity of substrate was also reduced with the help of high pressure induced. Macromolecules were breakdown into small soluble substances, which drastically increase the content range in the supernatant (Tamilarasan et al. 2017).

14.7 Combinative Pretreatment

The term combinative pretreatment is referring to combination of two or more pretreatment techniques to achieve higher organic matter solubilization. The main advantage of combined pretreatment is to increase liquefaction of waste biomass with minimal energy consumption. Thermochemical pretreatment is the combinations of both thermal and chemical which are frequently reported as best in solubilization of sugars and sterilization of materials (Rani et al. 2012a). Another effective combined pretreatment of food waste is hydrothermal pretreatment. The hydrothermal refers water and thermal is heat pretreatment. According to Jia et al. (2017) at 90 °C for 30 min, SCOD release of 132 g/L was achieved. Similarly, for Ding et al. (2017) 140 °C for 20 min, SCOD release of 70,800 mg/L was achieved. There are many combinative pretreatments are adopted organic waste to attain effective solubilization.

Followed by pretreatment, the food waste slurry that contains micro- and macro-nutrients is diverted for microalgae cultivation. The remaining food waste residual solid could be digested for bioenergy production. On economic point of view, biological pretreatment is more cost-effective method than others in the bio-refinery concept. It helps to selectively extract the nutrients from food waste for microalgae cultivation and provide a better medium. However, different pretreatment methods will have subsequent effect on microalgae growth due to toxic accumulation. Similarly, the choice of microalgae and their cultivation conditions are also playing a crucial role in the valorisation of food waste nutrients from soluble medium.

14.8 Microalgae Cultivation for Value Products and Biofuels

Microalgae are photosynthetic microorganisms which include eukaryotic photoautotrophic microorganisms and prokaryotic blue-green algae which are also called as cyanobacteria (Hunter-Cevera and Belt 1996). These microorganisms are responsible for 50% of global carbon dioxide fixation. They are rapidly growing microbes. As they are unicellular or simple multicellular structure, they could live in harsh conditions. Though more than 50,000 of species exist, however, only around 30,000 species only studied and analysed (Richmond 2008). Microalgal genera classified into five phyla based on their diversity and commercial benefits (Heimann and Huerlimann 2015).

1. Chlorophyta
2. Rhodophyta
3. Haptophyta
4. Stramenopiles
   - 4.1 Eustigmatophyceae
   - 4.2 Basillariophyceae
   - 4.3 Labyrinthulomycetes
5. Dinophyta.

There are number of microalgae have been widely used for commercial and industrial production of other value products such as human nutrition, pigment and fatty acids productions for cosmetics, food colouring and pharmaceutical applications as antioxidants, toxins and isotopes, animal feed and industry and biofuels such as bioethanol, acetone, bioethanol and biomethane. Microalgae also used as biofertilizer and environmental bioremeidation applications such as nutrient and heavy metal removal from wastewater. The most commonly reported industrial microalgae such as *Chlorella* sp., *Dunaliella* sp., *Spirulina* sp., *Scenedesmus* sp., *Nannochloropsis* sp., *Haematococcus pluvialis*, *Crypthecodinium* sp., *Porphyridium* sp. and *Rhodella* sp. and their commercial values are given in Table 14.3. Thus, the microalgae show great promises for both developed and developing countries in term of supporting green industry developments, new jobs and product markets. But, biofuel production from microalgae is known for centuries; however, the most feasible method is yet to be discovered in terms of cost-, energy- and water-smart efficiency.

Table 14.3 Commercial product and market values of microalgae

Product	Microalgae	Use	Market price (US$)	Reference
Health foods	*Chlorella* *Spirulina*	Dietary supplement Gastric ulcers	US$ 40 million in 2005	Borowitzka (2006) Singh et al. (2005)
Carotenoids (β-carotene astaxanthin)	*Dunaliella* *Haematococcus*	Food colours Cosmetics Supplements for human Animal feed	US$ 300–3000 kg^{-1}	Singh et al. (2005) Borowitzka (2006)
Phycobiliproteins (phycocyanin and phycoerythrin)	*Porphyridium* *Spirulina*	Food colours Cosmetics Pharmaceuticals	US$ 3–25 mg^{-1}	Spolaore et al. (2006) Becker (1994)
Fatty Acids (PUFA) (Omega -3, γ-linolenic acid) (Docosahexaenoic acid) (Eicosapentaenoic acid)	*Crypthecodinium cohnii* *Spirulina* *Ulkenia* *Phaeodactylum tricornutum* *Porphyridium* *Nannochloropsis*	Infant formulas for full-term infants Nutritional supplements Heart diseases Brain functioning	US $ 300 million per annum (Marktek) US$4602 kg^{-1}	Spolaore et al. (2006)
Special products	Many microalgal species	Stable Isotopic biochemicals Toxins	US$ 5900 g^{-1}	Spolaore et al. (2006

14.8.1 Cultivation System for Microalgae

The microalgae are classified based on their cultivation conditions and carbon requirements. Current cultivation system of algae is based on natural conditions, which is also called as photoautotrophic cultivation systems. Since algae absorb sunlight and sequestrate CO_2 from atmosphere and other nutrients from the aquatic system, we need to provide an enhanced optimal natural conditions to cultivate algae (Brennan and Owende 2010). Photoautotrophic is the only economically viable method for large-scale cultivation of microalgae. They are open pond and closed photobioreactor technologies. Each of the technologies has their own pros and cons with their specifications based on climate conditions, cost of land and water (Borowitzka 1997, 1999).

Open pond system is cheaper than closed photobioreactors for large-scale algal cultivations; it requires low energy input (Rodolfi et al. 2009). It does not compete with land which is required for agriculture crops (Chisti 2008). *Dunaliella salina* is the most cultivated strain in 2008 in open pond system and costed € 2.55 kg^{-1} of dry biomass. Since open pond system is very prone to other algae and protozoa contamination, it requires very selective environment. Contaminations can be avoided by providing extreme environmental conditions such as high salinity, nutrient rich and high alkali for respective algal strains to maintain monoculture cultivation. (*Chlorella*—high nutrients, *D.salina*—high salinity and *Spirulina*—high alkalinity). However, bacterial and other biological contaminations will occur during long-term productions (Lee 2001). In terms of biomass productivity, open pond system produces less biomass due to several factors such as evaporation losses, temperature fluctuation, carbon dioxide deficiency and insufficient mixing. These factors reduce 37 tonnes ha^{-1} biomass productivity compared to closed photobioreactors (Chisti 2008).

Closed systems such as tubular, flat plate, and column photobioreactors are avoiding risk of contamination and maintaining single microalgal culture for long duration; these closed systems are more suitable for sensitive algal strains. Though harvesting cost is cheaper than open pond systems, the cost of constructing the closed systems is higher (Carvalho et al. 2006). The photobioreactors are mainly made up of an array of straight glass or plastic tubes. The tubular array could be placed horizontally or vertically as a helix and designed to overserve the sunlight (Brennan and Owende 2010). Mechanical pump or airlift systems are being used in closed systems for re-circulating the algal culture and mixing of CO_2 (Eriksen 2008). The advantages and disadvantages of tubular, flat plate and column photobioreactor are given in Table 14.3. Flat plate photobioreactors are more appropriate for mass cultivation of algae, as low dissolved oxygen accumulations. Tubular photobioreactors are more suitable for large-scale outdoor cultivation since they have large area surface to absorb sunlight (Brennan and Owende 2010). Column photobioreactors provide best-controlled growth conditions and offer efficient mixing for culture. These systems are cheaper and easy to handle and operate. Closed photobioreactors have more attention for microalgal research in recent years. These closed systems are promising high-biomass productivity compared to open raceway ponds. Therefore, closed photobioreactors are potentially more suitable for biofuel and bioproduct development from food waste bio-valorisation.

14.8.2 Bio-Valorisation of Food Waste by Microalgae

Though the microalgae are autophotrophic, and they can utilize sunlight and carbon dioxide to make their carbon source, heterotrophic, mixotrophic or phototrophic cultivations using organic carbon enhances more biomass productivities and high lipid content. Hence, utilizing the nutrients recovered from food waste will be an economically feasible way to cultivate microalgae (Lau et al. 2014; Li et al. 2007). The species such as *C. vulgaris, B. braunii, Dunaliella tertiolecta, Chlorella prothothecoides, C. prothothecoides and Microcystis aeruginosa* are reported for biofuel (biodiesel) productions at large scale. In addition, the fatty acid methyl ester (FAME) distributions are the key factor to consider for the selection of microalgae for bio-valorisation studies. The C18 and C16 of FAME determine the final fuel quality and cetane number (Table 14.4). The above-mentioned microalgal species are containing highest amount of polyunsaturated FAME. These polyunsaturated FAMES are being used for biodiesel production with the low (42.47–50.52) cetane number (CN), the high (101.33–136.97) iodine values (IV) and the low oxidation stability. The high-quality biodiesel should be having lowest cetane number (CN), the highest iodine value (IV) and lowest oxidation capacity. The higher levels of saturated FAME in the oils of *Chlamydomonas* sp., *Chlorella*, *Dunaliella* sp. and *Scenedesmus obliquus* indicated them as source of biodiesel with higher oxidation stability, higher CN (63.63–64.94) and lower IV (27.34–35.28). Hence, these microalgal species are promising candidates for biodiesel production. (Nascimento et al. 2013).

Pleissner et al. (2013a) investigated growth performance of two heterotrophic microalgae *Schizochytrium mangrovei* and *Chlorella pyrenoidosa* in the fungal hydrolyzed food waste hydrolysate to develop cost-efficient production of food, feed and biofuels (Pleissner et al. 2013b). In this study, food waste (rice, noodles, meat and vegetables) hydrolyzed using fungus *A. awamori* and *A. oryzae* and food waste hydrolysate contained ~60% carbohydrate, ~20% proteins and ~10% lipids (w/w). Additionally, phosphate added in this study.

The growth rate of *S. mangrovei* and *C.pyrenoidos* were twofold higher in food waste hydrolysate than in the conventional growth media. There was no considerable depletion of phosphate in both media; however, during exponential growth most of the FAN was utilized until glucose is available in both medium. The protein concentration of *S. mangrovei* biomass grown in food waste hydrolysate was threefold higher than grown in conventional media, however, protein, lipids and fatty acids content similar in both media. For *C. pyrenoidos*, protein content was twofold higher when grown in food waste hydrolysate than grown in conventional

Table 14.4 Fatty acids content of commercial microalgae (mg g^{-1} dry biomass)

FAME	[3]*Chlorella* sp.	[1]*Scenedemus* sp.	[1]*Nannochlopsis* sp.	[1]*Phaeodactylum* sp.	[2]*Haematococcus Sp.*	[4]*Dunaliella* sp.	[5]*Isochrysis* sp.
C14:0	2.5	0.4	15.3	6.1	0.7	NA	28.2
C15:0	3.2	0.3	1.4	0.8	0.3	NA	NA
C16:0	19.1	13.1	85.4	43.8	12.7	28.1	12.9
C16:1 (7)	8.1	1.3	78.4	89.3	0.7	0.0	6.7
C16:1 (9)	0.0	3.0	1.0	0.0	0.0	0.0	NA
C16:2 (7,10)	13.9	1.7	0.0	2.6	0.0	2.8	1.7
C16:2 (9,12)	0.0	0.7	0.0	0.0	0.0	0.0	NA
C16:3(cis 6,9,12)	0.0	0.5	0.0	9.5	0.0	1.4	NA
C16:3(7,10,13)	0.0	1.3	0.0	0.0	0.0	0.0	NA
C16:4 (4,7,10,13)	0.0	12.9	0.0	0.0	0.0	NA	NA
C17:0	3.4	0.3	1.0	0.0	0.2	NA	NA
C18:0	3.6	0.5	2.6	1.5	4.79	0.6	0.6
C18:1 (9)	6.8	6.3	53.3	6.7	11.1	19.3	8.4
C18:2	22.4	10.6	3.4	0.0	13.0	14.7	5.7
C18:3 all cis 6,9,12	12.1	0.8	0.0	0.0	2.8	3.2	0.7
C18:3 (9,12,15)	0.0	20.7	0.0	0.0	1.8	44.1	8.4
C18:4	0.0	2.9	0.0	0.0	0.0	NA	13.5
C20:0	2.4	0.0	0.0	0.0	0.4	NA	NA
C20:4	0.0	0.0	0.0	0.0	1.8	NA	NA
C20:5	0.0	0.0	21.9	22.5	0.9	NA	0.6
C22:6	0.0	0.0	0.0	0.0	0.0	NA	6.6
SFA %	37.2	18.9	0.5	28.1	NA	28.7	41.7
SMUFA %	14.9	17.4	50.0	54.1	NA	NA	16.8
SPUFA %	48.4	63.7	9.5	17.8	NA	71.3	37.7

[1]Islam et al. (2013); [2]Lei et al. (2012); [3]Li et al. (2011); [4]Li et al. (2011); [5]Renaud et al. (2002)

media whereas carbohydrate is lower when grown in food waste (Pleissner et al. 2013b).

Lau et al. (2014) cultivated *Chlorella vulgaris* heterotrophically using different concentrated food waste hydrolysates to investigate *C. vulgaris*'s carbohydrate, protein, lipid and fatty acids contents, in order to evaluate the microalgal sustainability as a feedstock for food, feed and fuel productions (Li et al. 2007). Lau et al. (2014) utilized bakery (cake and pastry) and food (noodles, rice, meat and vegetables) for this study. *C. vulgaris* grew well in all different concentrated food waste hydrolysate (2.5, 5, 10, 20, 30, 40, 50 and 20% + nitrate) and utilized 2.5–50% (v/v) of supplied nutrients. *C. vulgaris* grew two times faster in 20% (v/v) concentrated food waste hydrolysate than modified basal media. Since food waste rich in trace elements, it is unnecessary to provide trace elements such as Co^{2+}, Fe^{2+} and Mn^{2+} to grow *C. vulgaris.* Additionally, bench-top scale fermentation was carried out, in order to produce sufficient amount of *C. vulgaris* in 20% food waste hydrolysate and growth rate plus carbohydrate, protein and lipid contents were estimated from the biomass. Nutrients levels were measured from media. *C. vulgaris* was efficient to convert glucose into the biomass yield of 0.9 g g^{-1} glucose. Carbohydrate contents were increased in the biomass from 200 mg g^{-1} to 400 mg g^{-1} from day 4 to day 6, but protein and lipids content were decreased from 300 mg g^{-1} to 200 mg g^{-1} from day 4 to day 6; however, lipid content again increased to 300 mg g^{-1} on day 7. The depletion of carbon, nitrogen and phosphorus sources in the media correlated with the changes in carbohydrate, protein and lipid concentration in the media (Lau et al. 2014).

Sloth et al. (2017) investigated the growth performance and phycocyanin synthesis of heterotrophic microalgae *Galdieria sulphuraria* grown in food waste obtained from restaurants and bakeries (Sloth et al. 2017). The restaurant waste (noodles, potatoes, vegetables, rice, meat and sauce) contained ~35% of starch, ~14.8. % of protein, ~12.9% of fat and 8.5% of free sugars. After hydrolysis, the restaurant waste contained 103 g L^{-1} of glucose, 6 g L^{-1} of sucrose, 13 g L^{-1} of fructose and 14 g L^{-1} xylose. And also contained 1 g L^{-1} of lactic acid. The concentration of free amino acid nitrogen (FAN) was increased during the hydrolysis which was from 0.23 to 0.45 g L^{-1}. The bakery waste (wasted bread) contained ~61% of starch and 8.3% or protein, whereas after hydrolysis 128 g L^{-1} of glucose, 72 g L^{-1} of sucrose, 6 g L^{-1} of fructose and 7 g L^{-1} of xylose were present. The FAN concentration remained unchanged during hydrolysis (Sloth et al. 2017).

G. sulphuraria was able to grow in restaurant, and bakery was hydrolysate without any ammonium or inorganic nutrients. Since there was no ammonium in restaurant hydrolysate, *G. sulphuraria* utilized all organic nitrogen sources. However, adding ammonium increased the specific growth rate of the algae (Sloth et al. 2017). Whereas, in bakery waste hydrolysate, *G. sulphuraria* did not grow well since there is no enough any organic bound nitrogen. However, algae showed maximum growth rate when all inorganic substances supplemented in both hydrolysate. Based on glucose, yield of biomass observed high in restaurant waste hydrolysate than in the actual algal medium. The culture utilized all other nutrients presented in restaurant food waste hydrolysate, whereas, biomass grown well in bakery waste hydrolysate only after adding ammonium. Another interesting fact is that no phycocyanin content was observed in the biomass grown in both hydrolysates without inorganic nutrient supplement. The specific phycocyanin content of 5.5 and 4.4 mg g^{-1} biomass was observed when all other inorganic nutrients plus ammonium supplemented with restaurant and bakery waste hydrolysate, respectively (Sloth et al. 2017).

Another study was carried by Pleissner et al. (2017) to investigate saturated and unsaturated fatty acids production of heterotrophic microalgae *Chlorella pyrenoidosa* grown in food waste. Food waste such as noodles, rice, meat, eggs, bread, cake, vegetables was together enzymatically hydrolyzed. They used two enzymes, namely amylolytic and proteolytic for hydrolyzing the food waste. Glucose was rich and free amino nitrogen (FAN) was poor in amylolytic digestion (122.3 g L^{-1} glucose, 0.4 g L^{-1} fructose, 0.24 g L^{-1} FAN and 0.17 g L^{-1} phosphate), whereas glucose was less and FAN was rich in proteolytic digestion (30.7 g L^{-1} glucose, 0.7 g L^{-1} fructose, 0.99 g L^{-1} FAN and 0.29 g L^{-1} phosphate). They investigated both batch and continues culture system and in continues culture system they have varied dilution rate *C. pyrenoidosa* increasing biomass concentration was observed from 0.5 to 14.1 g L^{-1} within 2.8 days at a specific growth rate of 1.4 d^{-1}, when 30 g of glucose L^{-1} provided. *C. pyrenoidosa* consumed fructose as an additional carbon source after glucose concentration was decreased below 5 g L^{-1}. *C. pyrenoidosa* contained 103.8 mg lipids g^{-1} biomass, when sufficient nutrients were supplemented, whereas three fold higher lipid content (317 mg lipid g^{-1} biomass) were observed during nutrient-limited conditions in continues culture (Sloth et al. 2017).

14.9 Summary and Conclusion

Food waste bio-valorisation by microalgal to produce food, feed and biofuels would be effective approach under heterotrophic or mixotrophic cultivation conditions in closed type photobioreactors. Effective pretreatment should be considered to maximize the bio-valorisation efficiency, which will rely on the choice of microalgae used. However, the lipid and FAME quantities are to be taken into consideration during the selection of microalgae for food waste valorisation, if not the other value products. Among the different types of food waste, the restaurant food waste shows more promising opportunity than bakery waste to produce high-quality biodiesel precursors using microalgae. Although, microalgae have huge inevitable potential commercial applications, cultivating them in large scale and bioprocessing them for bio-products are still have several challenges such as nutrients requirement, cost of closed cultivation system, contaminations in open outdoor cultivation systems and dewatering cost. Therefore, better analysis of different components, i.e. cost, technology and carbon foot print, of food waste bio-valorisation using microalgae is required and should be considered as future research direction.

References

Akinbomi J, Taherzadeh MJ (2015) Evaluation of fermentative hydrogen production from single and mixed fruit wastes. Energies 8(5):4253–4272

Allison BJ, Cádiz JC, Karuna N, Jeoh T, Simmons CW (2016) The effect of ionic liquid pretreatment on the bioconversion of tomato processing waste to fermentable sugars and biogas. Appl Biochem Biotechnol 179(7):1227–1247

Banu JR, Kaliappan S, Kumar A, Yeom IT, Uan DK (2011) Effect of low temperature thermochemical pretreatment on sludge reduction potential of membrane bioreactor treating primary treated dairy wastewater. Water Qual Res J 46(4):312–320

Banu JR, Do Khac U, Kumar SA, Ick-Tae Y, Kaliappan S (2012) A novel method of sludge pretreatment using the combination of alkalis. J Environ Biol 33(2):249

Becker EW (1994) Microalgae: biotechnology and microbiology, vol 10. Cambridge University Press, Cambridge

Borowitzka MA (1997) Microalgae for aquaculture: opportunities and constraints. J Appl Phyco 9(5):393–401

Borowitzka MA (1999) Commercial production of microalgae: ponds, tanks, tubes and fermenters. J Biotechnol 70(1):313–321

Borowitzka M (2006) Biotechnological & environmental applications of Microalgae

Brennan L, Owende P (2010) Biofuels from microalgae—a review of technologies for production, processing, and extractions of biofuels and co-products. Renew Sust Energ Rev 14(2):557–577

Carvalho AP, Meireles LA, Malcata FX (2006) Microalgal reactors: a review of enclosed system designs and performances. Biotechnol Prog 22(6):1490–1506

Chen Y, Luo J, Yan Y, Feng L (2013) Enhanced production of short-chain fatty acid by co-fermentation of waste activated sludge and kitchen waste under alkaline conditions and its application to microbial fuel cells. Appl Energy 102:1197–1204

Chisti Y (2008) Biodiesel from microalgae beats bioethanol. Trends Biotechnol 26(3):126–131

de Jong E, Jungmeier G (2015) Biorefinery concepts in comparison to petrochemical refineries. Industrial Biorefineries & White Biotechnology. Elsevier, Netherlands

Del Campo I, Alegría I, Zazpe M, Echeverría M, Echeverría I (2006) Diluted acid hydrolysis pretreatment of agri-food wastes for bioethanol production. Ind Crops Prod 24(3):214–221

Ding L, Cheng J, Qiao D, Yue L, Li Y-Y, Zhou J, Cen K (2017) Investigating hydrothermal pretreatment of food waste for two-stage fermentative hydrogen and methane co-production. Bioresour Technol 241:491–499

Do KU, Banu JR, Chung IJ, Yeom IT (2009) Effect of thermochemical sludge pretreatment on sludge reduction and on performances of anoxic-aerobic membrane bioreactor treating low strength domestic wastewater. J Chem Technol Biot 84(9):1350–1355

Do K-U, Banu RJ, Son D-H, Yeom I-T (2012) Influence of ferrous sulfate on thermochemical sludge disintegration and on performances of wastewater treatment in a new process: anoxic–oxic membrane bioreactor coupled with sludge disintegration step. Biochem Eng J 66:20–26

Ebenezer AV, Arulazhagan P, Kumar SA, Yeom I-T, Banu JR (2015a) Effect of deflocculation on the efficiency of low-energy microwave pretreatment and anaerobic biodegradation of waste activated sludge. Appl Energy 145:104–110

Ebenezer AV, Kaliappan S, Kumar SA, Yeom I-T, Banu JR (2015b) Influence of deflocculation on microwave disintegration and anaerobic biodegradability of waste activated sludge. Bioresour Technol 185:194–201

Eriksen NT (2008) The technology of microalgal culturing. Biotechnol Lett 30(9):1525–1536

Eswari P, Kavitha S, Kaliappan S, Yeom I-T, Banu JR (2016) Enhancement of sludge anaerobic biodegradability by combined microwave-H2. Environ Sci Pollut Res 23(13):13467–13479

FAO (2011) Global food losses and food waste: extent, causes and prevention. Food and Agriculture Organization of the United Nations, Rome

Gayathri T, Kavitha S, Kumar SA, Kaliappan S, Yeom IT, Banu JR (2015) Effect of citric acid induced deflocculation on the ultrasonic pretreatment efficiency of dairy waste activated sludge. Ultrason Sonochem 22:333–340

Ghimire A, Kumar G, Sivagurunathan P, Shobana S, Saratale GD, Kim HW, Luongo V, Esposito G, Munoz R (2017) Bio-hythane production from microalgae biomass: key challenges and potential opportunities for algal bio-refineries. Bioresour Technol 241:525–536

Groff D, George A, Sun N, Sathitsuksanoh N, Bokinsky G, Simmons BA, Keasling JD (2013) Acid enhanced ionic liquid pretreatment of biomass. Green Chem 15(5):1264–1267

Guo W-Q, Yang S-S, Pang J-W, Ding J, Zhou X-J, Feng X-C, Zheng H-S, Ren N-Q (2013) Application of low frequency ultrasound to stimulate the bio-activity of activated sludge for use as an inoculum in enhanced hydrogen production. RSC Advances 3(44):21848–21855

Guo W, Wu Q, Yang S, Luo H, Peng S, Ren N (2014) Optimization of ultrasonic pretreatment and substrate/inoculum ratio to enhance hydrolysis and volatile fatty acid production from food waste. RSC Advances 4(95):53321–53326

Han W, Lam WC, Melikoglu M, Wong MT, Leung HT, Ng CL, Yan P, Yeung SY, Lin CSK (2015a) Kinetic analysis of a crude enzyme extract produced via solid state fermentation of bakery waste. ACS Sustain Chem Eng 3(9):2043–2048

Han W, Ye M, Zhu AJ, Zhao HT, Li YF (2015b) Batch dark fermentation from enzymatic hydrolyzed food waste for hydrogen production. Bioresour Technol 191:24–29

Han W, Huang J, Zhao H, Li Y (2016) Continuous biohydrogen production from waste bread by anaerobic sludge. Bioresour Technol 212:1–5

Hao H-TN, Karthikeyan OP, Heimann K (2015) Bio-Refining of Carbohydrate-Rich Food Waste for Biofuels. Energies 8(7):6350–6364

Heimann K, Huerlimann R (2015) Microalgal classification: major classes and genera of commercial microalgal species. Academic Press, Cambridge

Hunter-Cevera JC, Belt A (1996) Maintaining cultures for biotechnology and industry. Academic Press, Cambridge

Islam MA, Magnusson M, Brown RJ, Ayoko GA, Nabi MN, Heimann K (2013) Microalgal species selection for biodiesel production based on fuel properties derived from fatty acid profiles. Energies 6(11):5676–5702

Jia X, Xi B, Li M, Xia T, Hao Y, Liu D, Hou J (2017) Evaluation of biogasification and energy consumption from food waste using short-term hydrothermal pretreatment coupled with different anaerobic digestion processes. J Clean Prod 152:364–368

Junoh H, Yip C, Kumaran P (2016) Effect on Ca $(OH)_2$ pretreatment to enhance biogas production of organic food waste. In: IOP conference series: earth and environmental science, vol 1. IOP Publishing, p 012013

Kannah RY, Kavitha S, Banu JR, Yeom IT, Johnson M (2017) Synergetic effect of combined pretreatment for energy efficient biogas generation. Bioresour Technol 232:235–246

Karthikeyan OP, Sanjeet M, Wong JWC (2017) Food waste bio-refinery for fuel and energy. In: ICEER 2017, Porto, Portugal, 17–20 July 2017

Karthikeyan OP, Selvam A, Wong JW (2016) Hydrolysis–acidogenesis of food waste in solid–liquid-separating continuous stirred tank reactor (SLS-CSTR) for volatile organic acid production. Bioresour Technol 200:366–373

Kavitha S, Kumar SA, Yogalakshmi K, Kaliappan S, Banu JR (2013) Effect of enzyme secreting bacterial pretreatment on enhancement of aerobic digestion potential of waste activated sludge interceded through EDTA. Bioresour Technol 150:210–219

Kavitha S, Jayashree C, Kumar SA, Yeom IT, Banu JR (2014a) The enhancement of anaerobic biodegradability of waste activated sludge by surfactant mediated biological pretreatment. Bioresour Technol 168:159–166

Kavitha S, Kumar SA, Kaliappan S, Yeom IT, Banu JR (2014b) Improving the amenability of municipal waste activated sludge for biological pretreatment by phase-separated sludge disintegration method. Bioresour Technol 169:700–706

Kavitha S, Kaliappan S, Kumar SA, Yeom IT, Banu JR (2015a) Effect of NaCl induced floc disruption on biological disintegration of sludge for enhanced biogas production. Bioresour Technol 192:807–811

Kavitha S, Kannah RY, Yeom IT, Do K-U, Banu JR (2015b) Combined thermo-chemo-sonic disintegration of waste activated sludge for biogas production. Bioresour Technol 197:383–392

Kavitha S, Saranya T, Kaliappan S, Kumar SA, Yeom IT, Banu JR (2015c) Accelerating the sludge disintegration potential of a novel bacterial strain Planococcus jake 01 by $CaCl_2$ induced deflocculation. Bioresour Technol 175:396–405

Kavitha S, Banu JR, IvinShaju C, Kaliappan S, Yeom IT (2016a) Fenton mediated ultrasonic disintegration of sludge biomass: biodegradability studies, energetic assessment, and its economic viability. Bioresour Technol 221:1–8

Kavitha S, Banu JR, Kumar JV, Rajkumar M (2016b) Improving the biogas production performance of municipal waste activated sludge via disperser induced microwave disintegration. Bioresour Technol 217:21–27

Kavitha S, Brindha GJ, Gloriana AS, Rajashankar K, Yeom IT, Banu JR (2016c) Enhancement of aerobic biodegradability potential of municipal waste activated sludge by ultrasonic aided bacterial disintegration. Bioresour Technol 200:161–169

Kavitha S, Pray SS, Yogalakshmi K, Kumar SA, Yeom I-T (2016d) Effect of chemo-mechanical disintegration on sludge anaerobic digestion for enhanced biogas production. Environ Sci Poll Res 23(3):2402–2414

Kavitha S, Preethi J, Banu JR, Yeom IT (2017a) Low temperature thermochemical mediated energy and economically efficient biological disintegration of sludge: simulation and prediction studies for anaerobic biodegradation. Chem Engin J 317:481–492

Kavitha S, Subbulakshmi P, Banu JR, Gobi M, Yeom IT (2017b) Enhancement of biogas production from microalgal biomass through cellulolytic bacterial pretreatment. Bioresour Technol 233:34–43

Kim S-H, Shin H-S (2008) Effects of base-pretreatment on continuous enriched culture for hydrogen production from food waste. Int J Hydrogen Energy 33(19):5266–5274

Kim D-H, Jang S, Yun Y-M, Lee M-K, Moon C, Kang W-S, Kwak S-S, Kim M-S (2014) Effect of acid-pretreatment on hydrogen fermentation of food waste: microbial community analysis by next generation sequencing. Int J Hydrogen Energy 39(29):16302–16309

Kiran EU, Liu Y (2015) Bioethanol production from mixed food waste by an effective enzymatic pretreatment. Fuel 159:463–469

Kiran EU, Trzcinski AP, Liu Y (2015) Enhancing the hydrolysis and methane production potential of mixed food waste by an effective enzymatic pretreatment. Bioresour Technol 183:47–52

Lagerkvist A, Morgan-Sagastume F (2012) The effects of substrate pretreatment on anaerobic digestion systems: a review. Waste Manag 32(9):1634–1650

Lam WC, Kwan TH, Lin CSK (2015) Enzymes in valorization of food and beverage wastes. Enzymes in food and beverage processing, p 479

Lau KY, Pleissner D, Lin CSK (2014) Recycling of food waste as nutrients in *Chlorella vulgaris* cultivation. Bioresour Technol 170:144–151

Lee Y-K (2001) Microalgal mass culture systems and methods: their limitation and potential. J Appl Phyco 13(4):307–315

Lei A, Chen H, Shen G, Hu Z, Chen L, Wang J (2012) Expression of fatty acid synthesis genes and fatty acid accumulation in *Haematococcus pluvialis* under different stressors. Biotechnol Biofuels 5(1):18

Li Y, Jin Y (2015) Effects of thermal pretreatment on acidification phase during two-phase batch anaerobic digestion of kitchen waste. Renew Energy 77:550–557

Li X, Xu H, Wu Q (2007) Large-scale biodiesel production from microalga *Chlorella* protothecoides through heterotrophic cultivation in bioreactors. Biotechnol Bioeng 98 (4):764–771

Li Y, Chen Y-F, Chen P, Min M, Zhou W, Martinez B, Zhu J, Ruan R (2011) Characterization of a microalga *Chlorella* sp. well adapted to highly concentrated municipal wastewater for nutrient removal and biodiesel production. Bioresour Technol 102(8):5138–5144

Ma J, Duong TH, Smits M, Verstraete W, Carballa M (2011) Enhanced biomethanation of kitchen waste by different pretreatments. Bioresour Technol 102(2):592–599

Menon A, Ren F, Wang J-Y, Giannis A (2016) Effect of pretreatment techniques on food waste solubilization and biogas production during thermophilic batch anaerobic digestion. J Mater Cycles and Waste 18(2):222–230

Montecchio D, Gallipoli A, Gianico A, Mininni G, Pagliaccia P, Braguglia C (2017) Biomethane potential of food waste: modeling the effects of mild thermal pretreatment and digestion temperature. Environ Technol 38(11):1452–1464

Nascimento IA, Marques SSI, Cabanelas ITD, Pereira SA, Druzian JI, de Souza CO, Vich DV, de Carvalho GC, Nascimento MA (2013) Screening microalgae strains for biodiesel production: lipid productivity and estimation of fuel quality based on fatty acids profiles as selective criteria. Bioenergy Res 6(1):1–13

Packyam GS, Kavitha S, Kumar SA, Kaliappan S, Yeom IT, Banu JR (2015) Effect of sonically induced deflocculation on the efficiency of ozone mediated partial sludge disintegration for improved production of biogas. Ultrason Sonochem 26:241–248

Pleissner D, Lam WC, Sun Z, Lin CSK (2013a) Food waste as nutrient source in heterotrophic microalgae cultivation. Bioresour Technol 137:139–146

Pleissner D, Lam WC, Sun Z, Lin CSK (2013b) Food waste as nutrient source in heterotrophic microalgae cultivation. Biores Technol 137:139–146

Pleissner D, Kwan TH, Lin CSK (2014a) Fungal hydrolysis in submerged fermentation for food waste treatment and fermentation feedstock preparation. Bioresour Technol 158:48–54

Pleissner D, Kwan TH, Lin CSK (2014b) Fungal hydrolysis in submerged fermentation for food waste treatment and fermentation feedstock preparation. Bioresour Technol 158:48–54

Pleissner D, Demichelis F, Mariano S, Fiore S, Gutiérrez IMN, Schneider R, Venus J (2017) Direct production of lactic acid based on simultaneous saccharification and fermentation of mixed restaurant food waste. J Clean Prod 143:615–623

Raj SE, Banu JR, Kaliappan S, Yeom I-T, Kumar SA (2013) Effects of side-stream, low temperature phosphorus recovery on the performance of anaerobic/anoxic/oxic systems integrated with sludge pretreatment. Bioresour Technol 140:376–384

Rani RU, Kaliappan S, Kumar SA, Banu JR (2012a) Combined treatment of alkaline and disperser for improving solubilization and anaerobic biodegradability of dairy waste activated sludge. Bioresour Technol 126:107–116

Rani RU, Kumar SA, Kaliappan S, Yeom I-T, Banu JR (2012b) Low temperature thermo-chemical pretreatment of dairy waste activated sludge for anaerobic digestion process. Bioresour Technol 103(1):415–424

Rani RU, Kumar SA, Kaliappan S, Yeom I, Banu JR (2013) Impacts of microwave pretreatments on the semi-continuous anaerobic digestion of dairy waste activated sludge. Waste Manag 33 (5):1119–1127

Razaghi A, Karthikeyan O, Hao HN, Heimann K (2016) Hydrolysis treatments of fruit and vegetable waste for production of biofuel precursors. Bioresour Technol 217:100–103

Renaud SM, Thinh L-V, Lambrinidis G, Parry DL (2002) Effect of temperature on growth, chemical composition and fatty acid composition of tropical Australian microalgae grown in batch cultures. Aquaculture 211(1):195–214

Richmond A (2008) Handbook of microalgal culture: biotechnology and applied phycology. John Wiley & Sons

Rodolfi L, Chini Zittelli G, Bassi N, Padovani G, Biondi N, Bonini G, Tredici MR (2009) Microalgae for oil: Strain selection, induction of lipid synthesis and outdoor mass cultivation in a low-cost photobioreactor. Biotechnol Bioeng 102(1):100–112

Singh S, Simmons BA (2013) Ionic liquid pretreatment: mechanism, performance, and challenges. Aqueous pretreatment of plant biomass for biological and chemical conversion to fuels and chemicals pp 223–238

Singh S, Kate BN, Banerjee U (2005) Bioactive compounds from cyanobacteria and microalgae: an overview. Crit Rev Biotechnol 25(3):73–95

Sloth JK, Jensen HC, Pleissner D, Eriksen NT (2017) Growth and phycocyanin synthesis in the heterotrophic microalga *Galdieria sulphuraria* on substrates made of food waste from restaurants and bakeries. Bioresour Technol 238:296–305

Spolaore P, Joannis-Cassan C, Duran E, Isambert A (2006) Commercial applications of microalgae. J Biosci Bioeng 101(2):87–96

Taherzadeh MJ, Karimi K (2007) Acid-based hydrolysis processes for ethanol from lignocellulosic materials: a review. BioResour 2(3):472–499

Tamilarasan K, Kavitha S, Banu JR, Arulazhagan P, Yeom IT (2017) Energy-efficient methane production from macroalgal biomass through chemo disperser liquefaction. Bioresour Technol 228:156–163

Ushani U, Banu JR, Tamilarasan K, Kavitha S, Yeom IT (2017) Surfactant coupled sonic pretreatment of waste activated sludge for energetically positive biogas generation. Bioresour Technol 241:710–719

Yang X, Zhu M, Huang X, Lin CSK, Wang J, Li S (2015) Valorisation of mixed bakery waste in non-sterilized fermentation for L-lactic acid production by an evolved *Thermoanaerobacterium* sp. Strain. Bioresour Technol 198:47–54

Zhang J, Lv C, Tong J, Liu J, Liu J, Yu D, Wang Y, Chen M, Wei Y (2016) Optimization and microbial community analysis of anaerobic co-digestion of food waste and sewage sludge based on microwave pretreatment. Bioresour Technol 200:253–261

Chapter 15
High-Value Coproducts from Algae—An Innovational Way to Deal with Advance Algal Industry

Preeti Mehta, Dilip Singh, Rohit Saxena, Rekha Rani, Ravi Prakash Gupta, Suresh Kumar Puri and Anshu Shankar Mathur

Abstract Expanded worldwide energy consumption and usage of fossil fuel cause its exhaustion and create energy crises, fuel security, global warming that have prompted a development of energy from alternative biomass that is renewable, economical, and eco-friendly. First- and second-generation biomass types, nonetheless, are frequently reprimanded because of displacement of food and the amount of crops it takes to deliver a gallon of oil. Algae to biodiesel (third-generation biofuel) have gained attention by many researchers, experts from petroleum industry as inexhaustible reliable and secure source of energy. Department of Energy, Govt. of USA, has investigated that algae grow much faster than terrestrial plants which give 30 times more energy yield per acre than land crops such as soybeans. Algae are a renewable bioresource that use sunlight, mitigate CO_2 emissions, reduced nutrients (N, P, and K) from waste streams and water, and produce biomass in the form of sugars, proteins, and oils that can be processed into both biofuels and valuable coproducts. In light of utilization, worldwide algal products are separated into nutraceuticals, nourishment and bolster supplements, pharmaceuticals, paints, colorants, etc. Algal-derived coproducts such as carotenoids, β-carotene, omega 3 polyunsaturated fatty acids (docosahexaenoic acid and eicosahexaenoic), astaxanthin, squalene, phycobiliproteins have increased popularity from the neutraceuticals and pharmaceutical industry and are relied upon to give high income to the algae producing companies around the world. A few algal strains with a high wholesome esteem and vitality content are developed industrially as aquaculture feed and are also a potential source of lipids, ethanol, and hydrogen. In this chapter, we attempt to elucidate the primary existing and potential high-value coproducts and its commercial significance, algal species used and market sizes, trends, and future prospects.

P. Mehta · D. Singh · R. Saxena · R. Rani · R. P. Gupta · S. K. Puri · A. S. Mathur (✉)
DBT-IOC Centre for Advanced Bio-Energy Research, Research & Development Centre, Indian Oil Corporation Limited, Sector-13, Faridabad 121007, India
e-mail: mathuras@indianoil.in

R. R. Singhania et al. (eds.), *Waste to Wealth*, Energy, Environment, and Sustainability, https://doi.org/10.1007/978-981-10-7431-8_15

Keywords Algae · High-value coproducts · Protein · Fatty acids
Market size

15.1 Introduction

Growing world population and expanding energy demand create an enormous competition for the scarce resources of the planet. The yearly world essential energy consumption was evaluated at 11,295 million tonnes of oil proportional (Worldometers 2015). Among various non-renewable major fuels, oil represented the highest energy consumption about 35% share, followed by coal (29%) and natural gas (24%), while 5% by nuclear energy and hydroelectricity denotes 6% of the total primary energy consumption (Center for Biological Diversity 2012). It has turned out to be progressively clear that continued reliance on fossil fuel energy resources is unsustainable, contribute to lessen world reserves and causes the green house gas emissions. With the expansion in anthropogenic GHG emissions, climatic change projections could have significant outcomes for nature as well as human frameworks (Brennan and Owend 2009). Thus, there is need for enhancement of worldwide strategies for energy security and mitigation of CO_2 emissions by using sustainable energy sources and production processes. Therefore, there are vigorous research activities started that aimed to developing alternative energy resources like first-generation and second-generation biofuels. However, first-generation biofuels rely on edible crops such as sugarcane, sugar beet, maize, and rapeseed which put a tremendous effect on world food markets. Second-generation biofuels derived from lignocellulosic agriculture/forest wastes and from non-food crop feed stocks overwhelm some of the problem arose by first-generation biofuels. However, there is concern over competing land use or required land use changes.

In the middle of the world energy crisis and major problems associated with first- and second-generation biofuels, researchers have diverted their research on third-generation biofuels (derived from microbes) which is considered as sustainable alternative solution toward a bio-based circular economy (Borowitzka 2013a, b). The significance of new options offered by algae cultivation is motivated by the fact that algae are very efficient at converting light, water and carbon dioxide (CO_2) into biomass and does not require arable land. Algae produce varying amount of lipids (high–low) content, carbohydrates based on different type of species and its growth conditions that are converted into biofuels including biodiesel or bioethanol. Algal oils for biofuels are a low-regard, high-volume thing, and the huge obstacle to the commercialization of algae-inferred biofuels is the high cost of creation. The evaluated expenses of a barrel of algae-based fuel using current innovations are US $300–2600, compared with $40–80 for petroleum (Hannon et al. 2010). Monetary evaluations of algal biodiesel given in Table 15.1 depending on production system vary from USD/L 0.42 to 7.50 in open ponds and 1.25–72 USD/L in closed PBRs (Louw et al. 2016).

Table 15.1 Cost appraisals of algal biodiesel generation based on cultivation systems (Louw et al. 2016)

Cultivation method/scale	Biomass productivity	Lipid content (%)	USD/L algal lipid	References
PBR/500 ha	40–90 g/m^2/day	40	25–33	Amer et al. (2011)
Ponds/500 ha	24 g/m^2/day	40	4–7.50	Amer et al. (2011)
Ponds/1950.58 ha	25 g/m^2/day	25	2.25	Davis et al. (2011)
PBR/1950.58 ha	1.25 kg/m$^{3/}$day	25	4.78	Davis et al. (2011)
Ponds/333.3 ha	20–30 g/m^2/day	20–50	1.3–2.53a	Delrue et al. (2012)
Ponds/400 ha	30 g/ m^2/day	50	0.42–0.97	Nagarajan et al. (2013)
PBR/100,000 t year^{-1}	80–120 t/ha/ year	20–40	1.25–2.5a	Brownbridge et al. (2014)

In this manner, it is significant to investigate ways for cost reduction of microbial biofuel processes, by using cost-effective raw materials and/or producing high value-added coproducts. With increasing health awareness among consumers, research has been focused to develop novel products with functional ingredients from algae as commercial sources of high-value compounds which have good health effects as these microorganisms deliver polyunsaturated fatty acids, polysaccharides, pigments, minerals, vitamins, enzymes, and bioactive peptides. This chapter briefly reviews the potential and at present accessible high-value Co-products which can be obtained from algae, their wide range of applications commercially available in the market.

15.2 Algae Bioproducts and Biorefineries Approach

To accomplish a practical algal-inferred biofuel, there is need of distinguishing high-esteem essential and bioproducts by expanding the inherent value of the algal biomass for different conversions. The algae can biosynthesize, process, aggregate, and discharge an awesome assorted variety of essential and auxiliary metabolites more than 15,000 biomaterials, huge numbers of which are important substances with potential applications in the sustenance, pharmaceutical, and beautifying agents enterprises (Fig. 15.1).

In spite of the fact that the various research activities based on micro-algae derived nutrients are very encouraging, the products at present on the market are still limited. A large number of potential products are identified which break out bioproducts by their approximate concentration in algal biomass and their anticipated market estimate.

There are two main classes of food market products acquired from micro-algae. The first category is dried algae (i.e., *Chlorella* and *Spirulina*) which is high source of protein and carbohydrate content along with vitamin B_{12}, C, and D_2 and directly

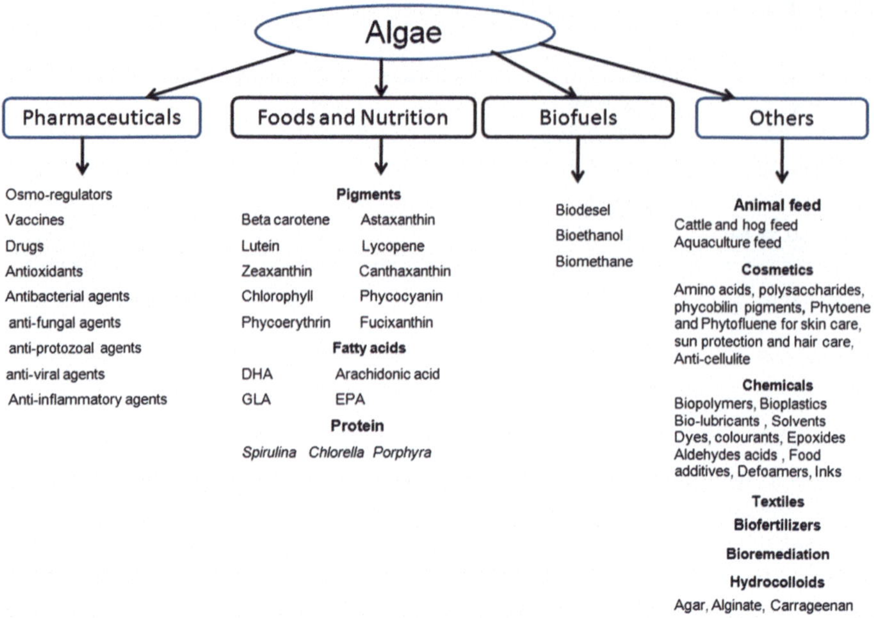

Fig. 15.1 Prominent algal-derived potential products

sold as health dietary supplements. The second type is secondary metabolites isolated and extracted from the micro-algae, i.e., astaxanthin, ß-carotene, phycocyanin and omega-3, docosahexaenoic acid, and eicosapentaenoic acid that can be used as food additive to boost their nutritional value. Their composition in oil likewise made them a conceivable source for the production of environmentally friendly biofuel. The huge amount of work on algae as a source of single-cell protein during the 1970s was started (Soeder and Pabst 1970; Venkataraman et al. 1977).

Among various species of algae, *Chlorella* and *Spirulina* were first commercialized as a health food in Mexico, Japan, and Taiwan (Sánchez et al. 2003; Borowitzka 2013a, b). By the 1980s, expansive scale algal growth generation offices were set up in Asia, India, the USA, Israel, and Australia (Enzing et al. 2014). After single-cell protein, scientist focused on extraction of β-carotene from *Dunaliella salina* which was commercialized in the 1980s (Borowitzka and Borowitzka 1989) followed by astaxanthin extraction from *Haematococcus pluvialis* (Lorenz and Cysewski 2000) and docosahexaenoic acid from *Crypthecodinium cohnii* (Kyle 2005). These items might be exhausted and used in a genuine biorefinery process in various enterprises, for example, sustenance, pharmaceutical, nutraceutical, cosmetics, and synthetic. Such approach will exploit the different items produced by the algae or its biomass, in this way expanding the value derived from the entire process, with a coveted insignificant natural affect. Thusly, the economics of the process may be fundamentally moved forward, by the coproduction of the high value-added products for sustainable and feasible biofuel production.

15.3 Current and Emerging Potential for High-Value Coproducts

In the market, a few existing algal-based items are accessible, because of various clusters for the creation of significant worth included items from algae; new items are likely going to be produced in the following decade. In contrast with items got from conventional yields, the aggregate market size and generation volumes of miniaturized scale algal-based sustenance and bolster items are as yet littler in estimate. In this way, products extracted from algae should be built up in business sectors in such a route, thus to the point that it can supplant the effectively commanded items like petrochemical encourage stocks and rival-entrenched supply chains for power, plastics, etc. (Da Silva et al. 2014). Be that as it may, some item bunches got from algal growth, for example, hydrocolloids or nourish for fish hatcheries, have useful points of interest over the current advancements. This coproduction is becoming an important option to make the process economic viable. Market and economic factors play a significant role in enhancing the production and commercialization of algal-derived high-value coproducts; be that as it may, need in writing about the monetary feasibility of these items and their business sectors is rare which make it hard to know their modern potential. Be that as it may, algal-based projects are driven by public and private dares to set up show offices through forefront restrictive innovation yet they do not uncover the certifiable circumstance of market (Table 15.2). Some researchers provide in their publications the global estimates production of high-value coproducts (Borowitzka 2013a, b; Milledge 2012; Vigani et al. 2015), but evidences on the development of this sector are not available at EU level. The present chapter summarizes the market figure of micro-algal-based products in Table 15.3 provided by the various available literatures. The estimated global algae market is projected to reach US$1143.0 Mn by 2024 from US$608.0 Mn in 2015. The algal market is relied upon to achieve 27,552.11 tons by 2024, extending at a CAGR of 5.32% in the vicinity of 2016 and 2024 regarding volume (http://www.credenceresearch.com/industry/agriculture-market). This section provides insights into the various current and emerging products of algae, its applications, and market potential.

15.4 Algae for Human Consumption

15.4.1 Algae as Protein Source

As aforementioned, micro-algae like *Chlorella vulgaris*, *Haematococcus pluvialis*, *Dunaliella salina,* and *Spirulina maxima* are broadly marketed and utilized as dietary supplements for people and as creature encourage added substances in different nations. These algal sources are rich in carbohydrates, protein, vitamins and minerals like vitamin A, C, B_1, B_2, B_6, niacin, iodine, potassium, iron,

Table 15.2 Worldwide distribution of private companies producing commercial algae-derived products

Country name	Name of companies	Products
Australia	Cognis Nutrition and Health, Muradel, Origin Oil Bio Fuels Pty Ltd	Dried algae, β-carotene, biofuels
Canada	Algae Can's, Pond Technologies, Algabloom International	Dried algae, astaxanthin, algal paste
China	Yunnan GinkoAsta Biotech Co., Ltd., Fuji Chemical Industry Co Ltd., AstaReal, Algatechnologies Ltd., AstaPure, Cyanotech Corporation, Nasdaq Capital Market, Jiangsu Tiankai Biotechnology Co., Ltd., Far East Algae Ind Co., Ltd	Dried algae, β-carotene, EPA/DHA
France	Aleor, Roquette	EPA/DHA and other dietary supplements
Germany	Breen Biotec GmbH, Phytolutions GmbH, Greenovation, Subitec GmbH, Astaxa, Salata GmbH, Novagreen GmbH, Algae growth, Blue Biotech GmbH	Dried algae, β-carotene EPA/DHA, astaxanthin aquaculture feed
India	N B Laboratories Pvt Ltd, Global Green Company Ltd. Parry Nutraceuticals, Energy algae, Jovialis, Sateera Nutria Biotech	Dried algae, β-carotene, astaxanthin
Israel	Algatech, Seambiotic, Nature Beta Technologies, TransAlgae	Dried algae, β-carotene, astaxanthin
Japan	Sunchlorella, Chlorella Industry Co., Yaeyama Shokusan Co. Ltd., Chlorella Industry Co., Ltd., Nihon Vitamin Chemical Co., Ltd., Nikken Sohonsha Company	Dried algae, β-carotene, astaxanthin, phycocyanin
Mexico	Recursos Renovables Alternativos	Dried algae
Myanmar	Myanmar Pharmaceutical Industries	Dried algae
Malaysia	Algaetech International	Dried algae
The Netherlands	Evonik, DSM	Dried algae, EPA/DHA, β-carotene
New Zealand	Aquaflow Bionomic Corporation, SeaDragon	EPA/DHA
Sweden	Asta real, BioReal	Astaxanthin
Taiwan	Taiwan Chlorella Manufacturing Co., Ltd.	Dried algae
USA	Algae Biosciences Algenol, Aurora Algae Inc, Cyanotech, Bodega Algae, Kent BioEnergy Corporation, TerraVia, Sapphire Energy, Solazyme, Inc., Solix BioSystems, Algae Systems, Algae to Omega Holdings, Inc., Algae Fuel, Algaewheel Algal Oil Diesel, Algenol, Algoil Energy, Algoil Industries, Inc., Applied Research Associates, Inc., Aquatic Energy, Aurora Algae, Diversified Energy Corporation, Global Green Solutions, Greener Bio-Energy	Dried algae, astaxanthin, EPA/DHA ethanol

Table 15.3 Market size of algal-derived products (NREL 2016)

Feedstock	Product	Market size (T)
Fatty acids	Hydrocarbon fuel products	5,000,000
Omega-3-fatty acids	Polyols, polyurethane, nutraceuticals	11,000,000, 11,000,000, 22,000
Hydroxy fatty acids	Surfactants, fuel additives	3,500,000
Branched chain fatty acids	Surfactants, fuel additives	3,500,000
Fatty alcohols	Surfactants, fuel additives	3,500,000
Sterols	Surfactants/emulsifiers Hydrocarbon fuel products Phytosterol nutra/pharmaceuticals	2,000,000 5,000,000 25,000
Phytol	Raw material for vitamin E, fragrance Surfactants, fuel additives	3,500,000
Polar lipids	Ethanolamine Phosphatidylcholine, phosphoinositol, and phosphatidyl ethanolamine(lecithin)	600,000 20,000–30,000
Glycerol	Di-acids for nylon production Feed, pharmaceuticals Polylactic acid (PLA) polymer	2,500,000 25,000 300,000
Fermentable sugars (glucose, mannose)	Di-acids Ethanol	2,500,000 68,000,000
Mannitol	Polyether polyols	2,300,000
Alginate	Alginate additives	12,000
Starch	Polysaccharide-derived bioplastics	2,000,000
Protein	Thermoplastics	5,000,000
Amino acids/peptides	Polyurethane	11,000,000
Amino acids/peptides	Biobutanol, mixed alcohol fuels	40,000,000

magnesium, and calcium. Due to balanced proportions of all the amino acids, algae proteins are gaining interest as alternative plant proteins. These are presently rivaling other plant and animal-based proteins, for example, soya, eggs, milk. Some of the key health benefits of using algal proteins are boosting the immune system, improve digestion, reduce fatigue, build endurance, cleanse the body, boost energy levels and appetite, and improve cardiovascular, liver, and kidney functions. It is therefore a sustainable solution to the problem of malnutrition as opposed to food fortification.

Athrospira, a single-cell algae (another name *Spirulina*) known for its high protein content, is robust algal species which has cell wall consists of polysaccharides that can be easily digestible by the human body (Priyadarshani and Rath 2012). Apart from this, *Spirulina* produces phycobiliproteins and a yellow-white protein extract that have several applications in food industries. The residual biomass after extraction could be processed for usage as biofuels or other products. The market size of *Spirulina* is about 10,000 metric tons per year, and its cost is US

$20/kg worldwide. As far as volume, the algae market is relied upon to reach 26,849.11 tons by 2022. The estimated market around the world was accounted for USD616.0 million in 2016 and is supposed to reach USD1128.0 million by 2022 (www.mordorintelligence.com/industry-reports/algae-protein-market).

Green algae, *Chlorella,* are the potential algae sold as human health, fish food, and nutritional supplements in stores in the form of tablets, capsules, and liquids. The composition of dried *chlorella* biomass is 45% protein, 20% carbohydrate, 20% fat, 10% minerals, 5% fiber, and vitamins. *Chlorella* provides various preventive measures against cancer, kidney damage, Crohn's disease, ulcers, colitis, fibromyalgia, asthma attacks, and diverticulosis and reduces side effects of radiation therapy, acts as restorative of the immune system and also helps in curing premenstrual syndrome. *Chlorella* boosts the digestion system by detoxification with the help of the presence of magnesium which ultimately improves the mental ability.

The production of *Chlorella* as a human health supplement accounted for about 2000 tons per year. The market value for *Chlorella* is US$44/kg (www.oilgae.com/non_fuel_products/chlorella). The major players in chlorella market are Yaeyama Shokusan Co. Ltd (Japan), Maypro Industries Inc., Taiwan *Chlorella* Manufacturing Co., Ltd (Taiwan), Roquette Klotze GmbH & Co. KG (Germany).

15.4.2 Algae in Nutraceuticals, Pharmaceuticals, and Cosmeceuticals

In spite of food, algae contribute to large number of applications that benefit the human beings in the form of health supplements, i.e., vitamins, medicines, vaccines, nutraceuticals, etc., that may be out of reach or too expensive to deliver using plants or animals. The manufacturing demand for health food or nutraceuticals by various biotechnology and pharmaceutical industries is increasing which prominently command the algae market. Neutraceuticals otherwise called practical/nourishment supplements are gotten from characteristic sources, and their utilization is probably going to profit human well-being by giving sustenance and pharmaceutical advantages to the body, for example, counteractive action and treatment of illnesses (Borowitzka 2013a, b). The developing utilization of algae biomass for nutraceutical purposes for existing is relied upon to give an alluring income stream for algae producers. The present market values and demands for the nutraceutical products are extremely higher but the production by algae is very small. Among industries, metabolites like carotenoids, phycobilins, DHA, EPA, polysaccharides, vitamins, sterols are currently being commercialized. The forthcoming segments will bring into center the utilization of algae as a potential source of pharmaceutical and nutraceutical ingredients.

Their capacity as antioxidants in the plant demonstrates intriguing parallels with their potential part as anti-oxidation agents in nourishments and people (Munir et al.

2013). Since algae are considered as an incredible wellspring of characteristic colorants and nutraceuticals, subsequently, it is normal that algal pigments generation will outperform synthetics and additionally other common sources because of their manageability and sustainable nature (Dufosse et al. 2005).

15.4.2.1 Pigments

Algae contain multiple combinations of pigments amalgamated with light irradiance depending on definite chemical composition which gives color to algal thallus. Apart from chlorophyll as the primary photosynthetic compound, they produce other pigments that enhance the light capturing ability with the generation of phycobiliproteins and protection against excessive solar radiation by the production of carotenoids. Their capacity as antioxidants in the plant demonstrates intriguing parallels with their potential part as anti-oxidation agents in nourishments and people (Munir et al. 2013). Since algae are considered as an incredible wellspring of characteristic colorants and nutraceuticals, subsequently, it is normal that algal pigments generation will outperform synthetics and additionally other common sources because of their manageability and sustainable nature (Dufosse et al. 2005).

Chlorophyll

Chlorophyll is the important pigment for photosynthesis and present in all plants, algae, and cyanobacteria. Among different chlorophyll pigments, algae contain chlorophyll abundantly. The *Spirulina platensis* is additionally an appealing option wellspring of the chlorophyll and has been investigated and marketed by cosmetics, pharmaceutical, and nourishment ventures. Chlorophyll content in algae is more than plants; for example, in Brazil, the chlorophyll obtained from spinach contains only 0.06 mg/g, whereas *Spirulina* sp. biomass contains 1.15 mg/g of chlorophyll (Chauhan and Pathak 2010). Large number of researches are also reporting health benefits from consumption of high chlorophyll diet. In light of its strong green color and consumers developing inclination for natural nourishments, their preferences are changing from artificial colorant in food to algal-based chlorophyll as natural colorant. The valuable health impact of chlorophyll is its anti-oxidation activity, relieves constipation and regulated proper digestion, recovery of harmed liver cells, and furthermore expands dissemination to all the organs by widening veins.

Carotenoids

Carotenoids from algae are well established, and its global market has been expanding because of its wide usage in human healthcare applications like in nutraceuticals, food industries, cosmetics. The worldwide carotenoids showcase is assessed to be esteemed at USD1.24 billion in 2016 and anticipated to achieve

USD1.53 billion by 2021, at a CAGR of 3.78% from 2016 to 2021 (www.marketsandmarkets.com/Market-Reports/carotenoid-market-158421566).
Carotenoids are organic pigments that are found in various plants and organisms. They act as antioxidants, secure the body against endless sicknesses, treatment of diabetes, cancer, and eye disorder, and prevent from cellular damage and aging effects. Carotenoids are commercially available in several forms including beta-carotene, lutein, lycopene, astaxanthin, zeaxanthin, annatto, and canthaxanthin.

The β-carotene section is anticipated to be the quickest developing business sector from 2016 to 2021 because of its high adequacy and therapeutic properties (Fig. 15.2). Beta-carotene is for the most part utilized as a colorant and as a wellspring of provitamin A for the stronghold of multivitamin juices, drinks for competitors, nourishment supplement arrangements, health foods, and so forth. The first alga to be commercialized as a source of a high-value β-carotene was *Dunaliella salina*, and depending upon its purity, it sells at the cost of around USD300–1500/kg in market (Borowitzka and Borowitzka 1988). Nature Beta Technology, Nutralite, Western Biotechnology Ltd, and Betatene Ltd were the four companies started producing and selling β-carotene in 1980 (Borowitzka 2013a, b). India and China have also started cultivating *Dunaliella* and commercialized β-carotene at small scale. High demand of synthetic β-carotene makes simpler and snappier way for the establishment of algal-based natural β-carotene. The β-carotene produced from *Dunaliella* was first sold in Japan where it was easiest to achieve approval for human use. The aggregate market for β-carotene is assessed at USD270 million (Costa Perez 2003; Borowitzka 2010).

Astaxanthin is another high-esteem carotenoid got from algae that is making business progress. Astaxanthin is available in nature, primarily biosynthesized in the algae, and is consumed by fish or zooplankton that accumulates the astaxanthin. The significant market for astaxanthin is as a pigmentation source in aquaculture. Aquafeed overwhelmed the business request and represented over 40% of the aggregate volume. Worldwide aquafeed showcase measure was worth over USD69

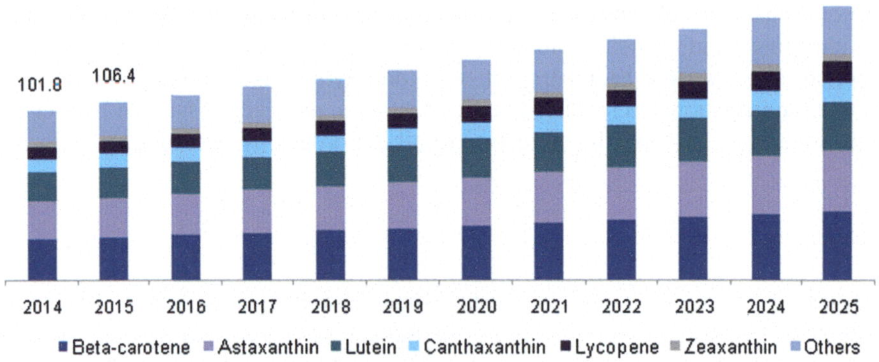

Fig. 15.2 Market size and demands of pigments

billion out of 2014 and is probably going to develop at a CAGR of over 11% from 2016 to 2023. The current worldwide market size of regular astaxanthin for the human market is assessed to be about $200 million. This is anticipated to hit $700 million by 2017. The average market price of astaxanthin is US$2500/kg (www.oilgae.com/ref/report/non-fuel-algae-products). Albeit 95% of this market expends artificially inferred astaxanthin; however, because of mindfulness among buyers about medical advantages, interest for common items makes the synthetic colors substantially less attractive. Among algae, *Haematococcus pluvialis* speaks to the wealthiest wellspring of the item by gathering more than 3% of astaxanthin and picked up acknowledgment in aquaculture and different markets as a "concentrated" type of normal astaxanthin. It is likewise affirmed by the US FDA for dietary supplements and furthermore endorsed in a few European countries for human utilization (Borowitzka et al. 2013a, b). Countries like USA, Israel, Japan and India are the main producer of astaxanthin in the world. Among them, USA is probably going to be the predominant provider over the conjecture time frame. DSM, ADM, Phasex Corporation, Cyanotech Corporation, Viva Labsa, and Otsuka Pharmaceutical are the real organizations which have put resources into numerous nations to improve market position. The benefits of astaxanthin are innumerable which promotes eye health, provides strength to muscle, protects premature aging of skin and inflammation, and helps in improving reproduction, immune function, and regeneration. It has additionally been shown that astaxanthin has a free radical battling limit worth 500 times that of vitamin E (Kim and Pangestuti 2011). Increment in healthful inadequacy combined with particular proteins request is probably going to drive astaxanthin request in nourishment and drinks applications and is additionally endorsed by FDA.

Among the emerging algae carotenoids compounds, Lutein is receiving increasing interest in Pharma industries because of its potential application in keeping the beginning of age-related macular degeneration (AMD) in the developing maturing population. The market estimation of lutein was around $233 million of every 2010 and is relied upon to reach $309 million by 2018 with an exacerbated yearly development rate of 3.6%. Lutein is also called as "the eye vitamin." Lutein helps to maintain macular pigment optical density which supports healthy vision. Healthy nutritional intake provides good eyesight. Rising eye diseases like glaucoma, cataracts, diabetic-related eye problems, and age-related macular degeneration may effect industry growth. Lutein is also used in ice cream, yogurt, fruit, beverages, and desserts. Apart from caring eyes, lutein is also used for curing many healthcare problems like it prevents several types of cancer which include breast cancer, type 2 diabetes and colon, skin disorders, and coronary heart disease. **Zeaxanthin**—zeaxanthin alone with lutein is used for eye health and to help prevent cataract. **Fucoxanthin**—fucoxanthin is used as preventive measures for obesity, tumor, diabetes, and inflammatory properties.

Phycobilins are the water soluble pigments composed of protein with bilin as prosthetic group and some cysteine residues. It is classified into phycocyanin, phycoerythrin, and allophycocyanin which are only present in *Spirulina,*

Porphyridium, *Rhodella,* and *Galdieria* algal species and not found in other taxa. The different phycobilins give different ranges of colors that have been used as natural colorants in different industries like food, cosmetics, and pharmaceuticals. Phycoerythrin is the main light-harvesting pigment that has wide applications which is used as fluorescent probe and analytical reagent in photodynamic therapy for cancer patients (Hu et al. 2008; Deng and Chow 2010). The cost of phycoerythrin is exceeding \$10,000/kg. In food and cosmetics industries, phycocyanin commercialized under the name as "Lina Blue" is used as colorants in chewing gums, dairy products, lipstick, eyeliners, etc., and is considered as safe by US Food and Drug Administration (FDA; GRAS 2012). Phycocyanin offers for between US \$500 and 100,000 kg^{-1} depending upon its purity. The present aggregate market an incentive for phycobiliprotein items is assessed to be more prominent than US\$60 million. Europa Bioproducts (Cambridge, UK) and Invitrogen (USA) are the major players in the Phycoerythrin Market. The use of phycobilins has been somehow limited because its purification process is quite complex and time taking where ion-exchange chromatography, gel filtration, and chromatography on hydroxylapatite are required for adequate amounts of the pure protein (Rito-Palomares et al. 2001).

15.4.2.2 Fatty Acids

Algal sources are explored for quite sometime as a source of long-chain polyunsaturated fatty acids (PUFAs) especially omega-3-fatty acids, i.e., ALA (alpha-linolenic acid), arachidonic acid, DHA (docosahexaenoic acid), and EPA (eicosapentaenoic acid) which have various applications in nutraceautical and pharmaceutical industries. Among them, DHA overwhelms the general market with more than income share in 2015, while the ALA and EPA combined together possessed one-fourth of the income share. ALA is the quickest developing sort of omega-3 with a CAGR of 15.8% from 2016 to 2022. The commercialized infant formula under the name DHASCO™ is the DHA rich oil from *Crypthecodinium cohnii* (Wynn et al. 2010) commercialized by DSM, USA.

The demand of PUFA has been increasing vigourously because of the health awareness among the people. Since the fish oil are the major source of PUFA or omega-3-fatty acid, but due to overfishing, oil processing cost, worldwide supply of fish oil becoming static or decreasing. The principle wellspring of EPA and DHA enrich fish oils (30% of total oil) is South American anchovy fishery which can supply around 300,000 t per year because of popularity of omega-3 unsaturated fats in market, and the anchovy supply alone will not have the capacity to take care of market demand. The estimated consumption of PUFAs is 123.8×10^3 metric tons worldwide which was cost around US\$2.3–2.5 billion in 2013 and 2014, respectively. By 2020, it is anticipated that demand for PUFAs globally would achieve $241X10^3$ metric tons with an estimation of US\$4.96 billion (Bermudez et al. 2010). Thus, alternative sources of omega-3 fatty acid from algae are on the rise to fill the demand gap. The selling price of algal omega-3 oil is US$140 kg^{-1} which is much higher as compared to the price of fish oils. Heterotrophic algae such as *Thraustochytrids, Crypthecodinium, Schizochytrium,* and *Ulkenia* are most widely commercialized strains for omega-3-fatty acid than phototrophic algae because of the less yield (Ratledge 2004; Barclay et al. 2010; Wynn et al. 2010).

The utilizations of algal omega-3 unsaturated fat are partitioned into various enterprises in view of dietary supplements, pharmaceuticals, newborn child equations, sustenance and drinks, pet nourishment, and aquaculture feed. Among different pharmaceutical and functional food applications, nutritional supplements are considered the largest component (around 59% of the global volume). North America is the biggest region with an income share more than 33% of the aggregate in 2015, trailed by Asia-Pacific and Europe. The main pharmaceutical application affirmed is for triglyceride decrease (e.g., Omacor/Lovaza delivered by Pronova BioPharma). Other EPA- and DHA-enriched TAGs (triacylglycerols) diminishment items and other pharmaceutical applications are a work in progress; however, it will take some time before these accomplish administrative endorsement and possible commercialization. Major players in the market of PUFAs are Cargill, Incorporated (US), Croda International Plc, Royal DSM, OLVEA Fish Oils, Omega Protein Corporation, GC Rieber Oils, Ltd., Pharma Marine AS, and Polaris.

Different research considers (Adarme-Vega et al. 2012) have underscored that these long chain ω-3 PUFA give huge medical advantages that are required for typical body development and mental health and furthermore keeps from heart assault, hypertension, sorrow, rheumatoid joint inflammation, asthma rheumatoid joint pain, Crohn's illness, ulcers. Moreover, in pregnant ladies, the sufficient and accurate ratio of EPA and DHA is critical for nourishment and development of the fetal mind.

15.4.2.3 Sterols

Phytosterols have developed in ubiquity because of their well-being advancing exercises in the course of recent decades. The primary modern wellsprings of sterols are vegetable oils, vegetables, nuts, seeds, entire grains, and dried organic products. Since the demand of phytosterols is increasing, algae provide as one of the best alternatives which could offer different types of phytosterols for functional food and pharmaceutical industries at a considerably higher effectiveness than terrestrial plants. Among various algal species, sterols are majorly found in *Isochrysis galbana, Nannochloropis gaditana, Nannochloropsis* sp., and *Phaeodactylum tricornutum* at a range of 7–34 g per kg (0.7–3.4%). Algae produce wide range of phytosterols depending on the taxonomic affiliation of the alga that includes brassicasterol, sitosterol, and stigmasterol (e.g., Patterson et al. 1994; Volkman 2003; Francavilla et al. 2010). The worldwide phytosterol showcase is at present about US$300 million and developing at around 7–9% per annum.

Because of the exhibited pharmacological capability of these marine sources and their high phytosterol content, they are obligated to be utilized as a part of non-pharmacological therapeutics and in mix with existing medications. Phytosterols have been accounted for to have numerous helpful well-being impacts in people, including immunomodulatory, mitigating, hostile to hyper cholesterolemic, and prevent cancer and anti-diabetic. Table 15.4 compresses the micro-algal phytosterols experiencing useful tests. As appeared, despite the fact that algae-inferred phytosterols are various, restricted investigations have tended to their well-being advancing activities.

LDL-C by dietary intercession can lessen the danger of CHD. Subsequently, items supplemented phytosterols from algae ought to be investigated keeping in mind the end goal to diminish the rate of these ailments and maybe ought to be devoured not just by individuals who need to bring down their blood cholesterol levels yet additionally by the individuals who need to decrease the symptoms of pharmacological treatment.

Table 15.4 Different types of phytosterols derived from algae (Luo et al. 2015)

Algae species	Major Phytosterols	Industrial importance	References
Chlorella vulgaris	Ergosterol, 7-dehydroporiferasterol, ergosterol peroxide, 7-dehydroporiferasterol peroxide	Anti-inflammatory	Yasukawa et al. (1996)
Chlorella vulgaris	Ergosterol peroxide	Anticancer	Yasukawa et al. (1996)
Nostoc commune	Lipid extract	Cholesterol-lowering activity	Rasmussen et al. (2008)
Isochrysis galbana	24-oxocholesterol acetate, ergost-5-en-3β-ol, cholest-5-en-24-1, 3-(acetyloxy)-3β-ol	Anti-tuberculosis	Prakash et al. (2010)
Dunaliella tertiolecta	Ergosterol, 7-dehydroporiferasterol	Immuno modulatory, Anti-inflammatory	Caroprese et al. (2012)
Dunaliella tertiolecta	Ergosterol, 7-dehydroporiferasterol	Neuro-modulatory	Francavilla et al. (2012)
Navicula incerta	Stigmasterol, 5β-hydroxysitostanol	Anticancer	Kim et al. (2014)
Schizochytrium sp.	Lathosterol, ergosterol, stigmastero	Cholesterol-lowering activity	Chen et al. (2014)
Schizochytrium aggregatum	Campesterol, 24-methylene cholesterol, 24-Methyl-cholest-7-en-3β-ol, ergosterol, stigmasterol	Antioxidant	Lv et al. (2015)

15.4.3 Algal Polysaccharides Bioproducts

Polysaccharides produced by algae *Porphyridium*, *Rhodella*, cyanobacteria have been explored for their nature and potential applications (Arad and Levy-Ontman 2010; De Philippis et al. 2011). Seaweed is known for the production of agar that has large applications range in food industry and pharmaceutical field. It is used in frozen foods, bakery products (like cake, pastries, candies, fruit juices), used as laxatives, anticoagulants (Cardoso et al. 2014; Cardozo et al. 2007). Carrageen is an another group of polysaccharides that are more known and widely used than agar as stabilizers, emulsifiers in various dairy products. Apart from this, carrageenan is also used in several pharmaceutical applications (like antitumor, antiviral, anticoagulant, and immune stimulators). Brown sea weed also produces alginate that has chelating ability and has the applications in the food, pharmaceutical, and textile industry worldwide (Muller and Alegre 2007).

15.4.4 Algae-Based Feedstocks for Commodity Bioproducts

Other than high value-added coproducts, algal biomass derived from cultivation can be refined to generate a wide array of bio-based products for different applications (e.g., paint, chemical building blocks, food and feed ingredients, and biofuels). The technology for production is still immature, but if developed, it is expected that more market combinations of commodities could be within reach. The production of chemicals from algae is an emerging field of biotechnology which is categorized as biopolymers, bioplastics, biolubricants, solvents, dyes and colorants, biofuels, agrochemicals, epoxides, aldehydes, acids, commodity chemicals, food additives, defoamers, inks. It is assessed that worldwide chemical deals and barring pharmaceuticals would reach $2.183 trillion by 2025, evaluating a 3–6% yearly development rate. The price of value-added products produced by algae such as lactic acid, polyhydroxyalkanoates, and butanol price varies from US$1300 to 7000 per tone. Solazyme, Blue Marble Biomaterials, Aquaflow, Solix, and BASF are the companies which diversified into algae chemicals business.

Macroalgae-derived fibers give a unique raw material to manufacture special clothing suitable for military applications such as protective fireproof clothing with. The market cost alginate fiber produced by algae ranges from $8000 to $10,000/ton. In textiles industries, algae have been used as colorant and its demand is increasing day-by-day as natural colors. The common fiber composite materials requests in showcase are relied upon to develop to US$531.3 million out of 2019 with 11% CAGR throughout the following five years. Similarly, for the production of biopolymers and bioplastics, algae are considered as excellent feedstock and its global market is expected to reach to $10bn by 2020 from $1bn in 2007. The companies like Dow, PetroSun, and Cereplast are active in the area of algae biopolymer research (www.oilgae.com/ref/report/non-fuel-algae-products).

15.4.5 Algae-Based Bioenergy Products

Apart from non-energy high-value products that have industrial values, algae are also known to produce biofuel and other bioenergy-related products (Fig. 15.3). Algal biomass has focused nowadays by researchers for biodiesel production since they can contain potentially over 80% total lipids under nutrient (especially nitrogen)-starved condition. The lipid concentration is lower (<40%) under normal growth conditions with higher biomass yield. Under nitrogen-starved condition, biomass production starts decreasing and thus enhances lipid content. The non-lipid part of the biomass produced after starvation can be further used as a source for coproducts, for example, *Botryococcus* sp., does not produce the higher lipids but it contains hydrocarbons of longer chain, which cannot be used as substrate for biodiesel production. Thus in this case, this higher carbon chain along with algal cell wall polysaccharides can be used as substrate for the bioethanol production like the process of cellulosic ethanol production (Hossain et al. 2015).

The advantage of ethanol production from algal feedstock has advantage over cellulosic biomass that it rarely contains lignin and thus can be easily degradable. Another fate of algal biomass is to convert organic material into biogas mainly composed of 60–70% biomethane by anaerobic digestion. CO_2 released during anaerobic digestion can be fed back to the algae system as carbon source and thus make the close-loop system for energy generation. Another advantage of this process is that we can use wet biomass as such thus reducing the drying cost. The residual or digested biomass containing nutrients can then be regained from the

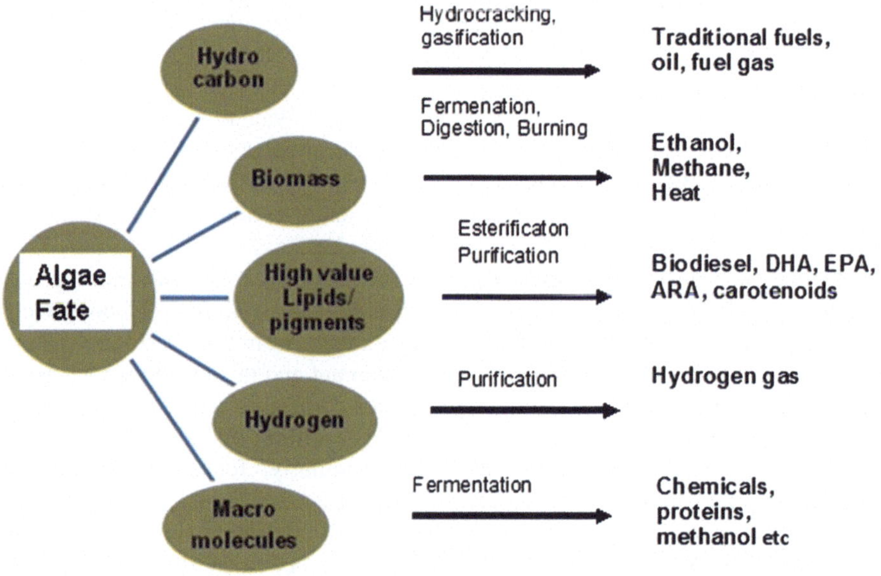

Fig. 15.3 Fate of algae into different potential products

liquid and solid phases for further use as animal feed. Because of the high cost of feedstock (algal biomass) at present, the chemical conversion into biogas production or other chemical processes is not viable till date, although it is presently one of the cheapest biofuel that can be recovered from algal biomass.

Recently, the biomass undergoes a chemical conversion depending on the water content under extreme conditions like high temperature and pressure due to which the biomass carbon changes into three different phases, i.e., raw gaseous, liquid, or solid phase which further can be customized for the use of biofuel. This process is energy intensive as energy input is higher as compared to biogas production. Various researchers are focusing on biohydrogen production from algal biomass, but the process of hydrogen production is in infancy stage and not cost effective as the yield of this process is very low. Algal cells use energy to form hydrogen due to which unable to produce much biomass; therefore, there is little potential for coproduction of other metabolites.

15.4.6 Algae as Fertilizer

After oil or carbohydrate extraction, some of the nutrients remain in the processed/residual biomass that can be used as biofertilizer for the growth of plants. Biofertilizers mean a promising other option to synthetic manures for endurable agri-based practices that meet present and future societal requirements for nourishment, healthful environments, and lives. Biofertilizers application to agricultural fields could expand the measure of amount of carbon that accumulate in these soils and contribute fundamentally to the decrease of greenhouse gas emissions by depleting the necessity of fossil-derived fuels through the recovery of N: P: K from wastewater streams. Biofertilizers hold alive or dormant microbes alone or in combination, which have the ability of fixing atmospheric nitrogen or solubilize the insoluble soil nutrients with the help of secretion of growth-promoting substances which ultimately enhance crop yield and supply stabilization of soil aggregates. In rice, blue green algae (BGA) under the brand name "Algalization" have been used as biofertlizer since past years that helps in creating an eco-friendly agro-ecosystem that makes sure of economic viability in paddy cultivation. This biofertilizer has been established to fix nitrogen under anaerobic conditions which bring about 25–30 kg N/hectare/season which enhances the crop yield by 10–15% (Paudel et al. 2012; Theil et al. 2014). Seaweed because of its mineral content and high water binding capacity of the soil has also been used as a fertilizer worldwide. When algal biofertlizers applied in agriculture, the nutrients are exudated slowly which promote germination, leaf or stem growth, flowering and also provide biological protection against plant diseases that directionally benefit plant growth and ultimately decrease the GHG emissions (Pulz and Gross 2004; Mulbry et al. 2008). The demand of algal-based fertilizers is increasing in the coming years which enhance its volume potential in market which reduces the usage of energy-intensive chemical fertilizers.

15.5 Conclusions

Biofuel is the most attractive product from algae; however, numerous algal fuel companies understand that it could take more amount of time to generate algal-based biofuels for commercialization. Thus, many companies are exploring venturing into algae-based value-added products business that can help to minimize the high cost of biodiesel production, which in turn make the process more feasible. Algae can synthesize, acquire, and secrete various types of metabolites and bioactive compounds, which have broad applications in the food, pharmaceutical, cosmetics, and agri-based industries. Evidence in this chapter suggests that the simultaneous extraction of valuable coproducts (viz., pigments, omega-3 and -6 fatty acids, vitamins and whole algae as human and animal food/feed items and for cosmetics, fertilizers, and hydrocolloids) with biofuel production has significant potential. Potential demand, market size, and acceptance by the consumers of algal-based coproducts and its competition with non-algae source may influence the development of algal-based coproducts. There exists a bridge between the well-recognized commercial potential of many algae species and the actual technical know-how of industrial process. The global algae biomass market is worth between US$5 and 7 billion. From this total, the health food sector accounts for US $2 billion and the fish meal applications account for US$0.7 billion. Endorsed omega-3-based pharmaceuticals on the planet together represent US$1.5 billion deals. *Spirulina* and *Chlorella* are potential single-cell protein sources and commercialized by many companies but their market value is not very high as the cost of *Spirulina* was US$20/kg in 2010, and *Chlorella* was sold at a price of US$44/kg in 2010. Algal-based coproducts need to break into set-up business sectors overwhelmed by other, frequently petrochemical feedstocks, and rival-entrenched supply chains (e.g., for fuels and plastics). Various research programs are in progress for developing new algal-based food and feed products with a huge possibilities of reaching the market; however, the commercial production of proteins, fatty acids and carbohydrates from micro-algae is still in its nascent stage, not yet able to significantly contribute to the reduction of the food-feed insecurity world-wide. Governments and privately owned businesses are as of now committing vast speculations to encourage the genetic modification using molecular approach on algae. Additionally, more recent research is expected to get more detailed financial and market information on generation volumes, turnovers, expenses, and input uses, both at organizations level and by nation. This would allow to get more definite data on the algal-based health/food sector division in the EU and to elucidate the real commitment that algal-based items can give to the advancement of the bio-economy in the EU.

Acknowledgements We acknowledge the Department of Biotechnology, Government of India and DBT-IOC Centre for Advanced Bio-Energy Research, Indian Oil Corporation Ltd., Research & Development Centre, Faridabad, India, for funding and support.

References

Adarme-Vega TC, Lim DKY, Timmins M, Vernen F, Li Y, Schenk PM (2012) Microalgal biofactories: a promising approach towards sustainable omega-3 fatty acid production. Microb Cell Factories 11:1–10

Amer L, Adhikari B, Pellegrino J (2011) Technoeconomic analysis of five algae-to-biofuels processes of varying complexity. Biores Technol 102:9350–9359

Arad S, Levy-Ontman O (2010) Red microalgal cell-wall polysaccharides: biotechnological aspects. Curr Opin Biotechnol 21:358–364

Barclay W, Weaver C, Metz J (2010) Development of docosahexaenoic acid production technology using Schizochytrium: historical perspective and update. In: Cohen Z, Ratledge C (eds) Single cell oils, microbial and algal oils. AOCS Press, Urbana, pp 75–96

Borowitzka MA (2010) Carotenoid production using microorganisms. In: Cohen Z, Ratledge C (eds) Single cell oils, microbial and algal oils. AOCS Press, Urbana, pp 225–240

Borowitzka MA (2013a) High-value products from algae—their development and commercialisation. J Appl Phycol 25:743–756

Borowitzka MA (2013b) Energy from algae: a short history. In: Borowitzka MA, Moheimani NR (eds) Algae for biofuels and energy. Spinger, Dordrecht, pp 1–15

Borowitzka MA, Borowitzka LJ (1988) Dunaliella. In: Borowitzka MA, Borowitzka LJ (eds) Micro-algal biotechnology. Cambridge University Press, Cambridge, pp 27–58

Borowitzka LJ, Borowitzka MA (1989) ß-Carotene (Provitamin A) production with algae. In: Vandamme EJ (ed) Biotechnology of vitamins, pigments and growth factors. Elsevier Applied Science, London, pp 15–26

Brennan L, Owende P (2009) Biofuels from algae—a review of technologies for production, processing, and extractions of biofuels and co-products. Renew Sustain Energy Rev. https://doi.org/10.1016/j.rser.2009.10.009

Brownbridge G, Azadi P, Smallbone A et al (2014) The future viability of algae-derived biodiesel under economic and technical uncertainties. Biores Technol 151:166–173

Cardoso SM, Carvalho LG, Silva PJ et al (2014) Bioproducts from seaweeds: a review with special focus on the Iberian Peninsula. Curr Org Chem 18(7):896–917

Cardozo KHM, Guaratini T, et al (2007) Metabolites from algae with economical impact. Comp Biochem Physiol Toxicol Pharmacol 146(1–2):60–78

Caroprese M, Albenzio M, Ciliberti MG et al (2012) A mixture of phytosterols from *Dunaliella tertiolecta* affects proliferation of peripheral blood mononuclear cells and cytokine production in sheep. Vet Immunol Immunopathol 150:27–35

Center for Biological Diversity (2012) http://www.biologicaldiversity.org/programs/climate_law_institute/energy_and_global_warming/index.html

Chauhan UK, Pathak N (2010) Effect of different conditions on the production of chlorophyll by *Spirulina platensis*. J Algal Biomass Utln 1(4):89–99

Chen J, Jiao R, Jiang Y et al (2014) Algal sterols are as effective as β-sitosterol in reducing plasma cholesterol concentration. J Agric 62:675–681

Costa Perez J (2003) Method of producing beta-carotene by means of mixed culture fermentation using (+) and (−) strains of *Blakeslea trispora*. European Patent Application 1,367,131

Cuellar-Bermudez SP, Aguilar-Hernandez I, Cardenas-Chavez DL et al (2010) Extraction and purification of high-value metabolites from algae: essential lipids, astaxanthin and phycobiliproteins. Microbial Biotechnolo 8(2):190–209

Davis R, Aden A, Pienkos PT et al (2011) Techno-economic analysis of autotrophic algae for fuel production. Appl Energy 88:3524–3531. https://doi.org/10.1016/j.apenergy.2011.04.018

De Philippis R, Colica G, Micheletti E (2011) Exopolysaccharide producing cyanobacteria in heavy metal removal from water: molecular basis and practical applicability of the biosorption process. Appl Microbiol Biotechnol 92:697–708

Delrue F, Setier P, Sahut C et al (2012) An economic, sustainability, and energetic model of biodiesel production from algae. Biores Technol 111:191–200

Deng R, Chow TJ (2010) Hyperlipidemic, antioxidant and antiinflammatory activities of algae *Spirulina*. Cardiovasc Ther 28:33–45

Dufosse L, Galaupa P, Yaron A (2005) Microorganisms and algae as sources of pigments for food use: a scientific oddity or an industrial reality? Trends Food Sci Technol 16:389–406

Enzing C, Ploeg M, Barbosa M (2014) Algae based products for the food and feed sector: an outlook for Europe. In: Vigani M, Parisi C, Rodíguez-Cerezo E (eds) JRC scientific and policy reports. https://ec.europa.eu/jrc/sites/default/files/final_version_online_ipts_jrc_85709.pdf

Francavilla M, Trotta P, Luque R (2010) Phytosterols from *Dunaliella tertiolecta* and *Dunaliella salina*: a potentially novel industrial application. Bioresour Technol 101:4144–4150

Francavilla M, Colaianna M, Zotti M, Morgese MG, Trotta P, Tucci P, Schiavone S, Cuomo V, Trabace L (2012) Extraction, characterization and in vivo neuromodulatory activity of phytosterols from microalga *Dunaliella tertiolecta*. Curr Med Chem 19:3058–3067

GRAS (2012) GRAS Notice 000424: notice to US Food and Drug Administration the use of CyaninPlus™ is generally recognised as safe. https://www.accessdata.fda.gov/scripts/fcn/gras_notices/GRN000424.pdf

Hannon M, Gimpel J, Tran M et al (2010) Biofuels from algae: challenges and potential. Biofuels 1(5):763–784

Hossain NB, Basu JK, Mamun M (2015) The production of ethanol from micro-algae *Spirulina*. Proc Eng 105:733–738

Hu L, Huang B, Zuo M, Wei H (2008) Preparation of the phycoerythrin subunit liposome in a photodynamic experiment on liver cancer cells. Acta Pharmacol Sin 29:1539–1546

Kim SK, Pangestuti R (2011) Biological properties of cosmeceuticals derived from marine algae. In: Kim S-K (ed) Marine cosmeceuticals. CRC Press, Boca Raton, pp 191–200

Kim YS, Li XF, Kang KH et al (2014) Stigmasterol isolated from marine algae *Navicula incerta* induces apoptosis in human hepatoma Hep G2 cells. BMB Rep 47:433

Kyle DJ (2005) The future development of single cell oils. In: Cohen Z, Ratledge C (eds) Single cell oils. AOCS, Urbana, pp 239–248

Lorenz RT, Cysewski GR (2000) Commercial potential for *Haematococcus* algae as a natural source of astaxanthin. Trends Biotechnol 18:160–167

Louw TM, Griffiths MJ, Jones SMJ et al (2016) Techno-economics of algal biodiesel algae biotechnol. https://doi.org/10.1007/978-3-319-12334-9_7

Lv J, Yang X, Ma H et al (2015) The oxidative stability of algae oil (*Schizochytrium aggregatum*) and its antioxidant ability after gastrointestinal digestion: relationship with constituents. Eur J Lipid Sci Technol. https://doi.org/10.1002/ejlt.201400588

Milledge JJ (2012) Algae: commercial potential for fuel, food and feed. Food Sci Technol 26 (1):28–31

Mulbry W, Kondrad S, Buyer J (2008) Treatment of diary and swine manure effluents using freshwater algae: fatty acid content and composition of algal biomass at different manure loading rates. J Appl Phycol 20:1079–1085

Muller JM, Alegre RM (2007) Alginate production by *Pseudomonas mendocina* in a stirred draft fermenter. World J Microbiol Biotechnol 23(5):691–695

Munir N, Sharif N, Naz S (2013) Algae: a potent antioxidant source. Sky J Microbiol Res 1(3): 22–31

Nagarajan S, Chou SK, Cao S et al (2013) An updated comprehensive techno-economic analysis of algae biodiesel. Biores Technol 145:150–156. https://doi.org/10.1016/j.biortech.2012.11.108

NREL U S Department of Energy (2016) National Algal Biofuels Technology Review. https://energy.gov/eere/bioenergy/downloads/2016-national-algal-biofuels-technology-review

Patterson GML, Larsen LK, Moore RE (1994) Bioactive natural products from blue-green algae. J Appl Phycol 6:151–157

Paudel YP, Pradhan S, Pant B (2012) Role of blue green algae in rice productivity. Agric Biol J N Am 3(8):332–335

Prakash S, Sasikala S, Aldous VHJ (2010) Isolation and identification of MDR-Mycobacterium tuberculosis and screening of partially characterised antimycobacterial compounds from chosen marine micro algae. Asian Pac J Trop Med 3:655–661

Priyadarshani I, Rath B (2012) Commercial and industrial applications of micro algae–A review. J Algal Biomass Utln 3(4):89–100

Pulz O, Gross W (2004) Valuable products from biotechnology of algae. Appl Microbiol Biotechnol 65(6):635–648

Rasmussen HE, Blobaum KR, Park YK et al (2008) Lipid extract of *Nostoc commune var. sphaeroides* Kützing. J Nutr 138:476–481

Ratledge C (2004) Fatty acid biosynthesis in microorganisms being used for single cell oil production. Biochimie 86:807–815

Rito-Palomares M, Nuñez L, Amador D (2001) Practical application of aqueous two-phase systems for the development of a prototype process for c-phycocyanin recovery from *Spirulina maxima*. J Chem Technol Biotechnol 76:1273–1280

Sánchez M, Bernal Castillo J, Rozo C (2003) *Spirulina* (*Arthrospira*): an edible microorganism: a review. Univ Sci 8:7–24

Soeder CJ, Pabst W (1970) Gesichtspunkte fur die Verwendung von Mikroalgen in der Ernahrung von Mensch und Tier. Ber Deutsch Bot Ges 83:607–625

Da Silva, TL, Gouveia L, Reis A (2014) Integrated microbial processes for biofuels and high value-added products: the way to improve the cost effectiveness of biofuel production. Appl Microbiol Biotechnol 98(3):1043–1053

Thiel T, Pratte BS, Zhong J et al (2014) Complete genome sequence of *Anabaena variabilis* ATCC 29413. Stand Genomic Sci 9(3):562–573

Venkataraman LV, Becker WE, Shamala TR (1977) Studies on the cultivation of the alga *Scenedesmus acutus* as a single cell protein. Life Sci 20:223–234

Vigani M, Parisi C, Rodrıguez-Cerezo E (2015) Food and feed products from algae: market opportunities and challenges for the EU. Trends Food Sci Technol 42:81–92

Volkman JV (2003) Sterols in microorganisms. Appl Microbiol Biotechnol 60:495–506

Worldometers (2015) http://www.worldometers.info/world-population

Wynn J, Behrens P, Sundararajan A (2010) Production of single cell oils from Dinoflagellates. In: Cohen Z, Ratledge C (eds) Single cell oils. Microbial and algal oils. AOCS Press, Urbana, pp 115–129

Yasukawa K, Akihisa T, Kanno H et al (1996) Inhibitory effects of sterols isolated from *Chlorella vulgaris* on 12-*O*-tetradecanoylphorbol-13-acetate-Induced inflammation and tumor promotion in mouse skin. Biol Pharm Bull 19:573–576

Chapter 16
Wastewater Algae to Value-Added Products

Durga Madhab Mahapatra, V. Sudharsan Varma, Shanmugaprakash Muthusamy and Karthik Rajendran

Abstract Globally, treatment and management of wastewater are a serious challenge. Voluminous wastewaters are generated on a day-to-day basis that is being either partly treated or untreated finds its ways into surface and groundwater thus enriching the systems with nutrients, pollutants and pathogens. In purview of the increasing water scarcity, rapid water deterioration, higher primary productivity in surface waters due to nutrient enrichment, towering wastewater production and complications related to its treatment, the understanding of water footprint, underlying mechanisms of wastewater treatment and transformation of terrestrial nutrients into value-added products and various downstream processes for their recovery needs to be understood. In this context, the algal treatment systems not only provides a simple and economical solution to wastewater treatment but also aids in the production of many valued bio-based products like lipids as feedstock for biofuels, single-cell proteins, Omega 3 fatty acids, carotenoids as astaxanthin and β-Carotene. The present chapter throws light on various mechanisms and strategies of wastewater transformations into value-added products while evaluating the techno-economics and feasibility of such systems for assessing its potential to be a bio-based industry. Various strategies for algal species selection targeting

D. M. Mahapatra (✉)
Department of Biological and Ecological Engineering, Oregon State University, Corvallis OR 97331,, Oregon, USA
e-mail: durgamadhab.mahapatra@gmail.com; mahapatd@oregonstate.edu

V. S. Varma
Agricultural Research Organization, Institute of Soil, Water, and Environmental Sciences, Newe Ya'ar Research Center, 30095 Ramat Yishay, Israel

S. Muthusamy
Downstream Processing Laboratory, Department of Biotechnology, Kumaraguru College of Technology, Coimbatore 641049, India

K. Rajendran
MaREI Centre, Environmental Research Institute, University College Cork, Cork, Ireland

K. Rajendran
School of Engineering, University College Cork, Cork, Ireland

R. R. Singhania et al. (eds.), *Waste to Wealth*, Energy, Environment, and Sustainability, https://doi.org/10.1007/978-981-10-7431-8_16

specific wastewater pollutants grown either as natural population or as engineered consortia with numerous wastewater treatment approaches until the production of valorized biomass is being discussed. Lastly, key techno-economics, environmental challenges and the scope of wastewater transformations into bio-based products are enumerated.

Keywords Wastewater · Algae · Bio products · Biofuel · Bioprocess
Techno-economic assessment

16.1 Introduction

With increased globalization, there has been an increasing stress on natural resources as land, water and air. The freshwater sources have been rampantly used either for various activities in a day-to-day fashion for domestic purpose's as drinking, bathing, etc. or for agricultural and industrial purposes. All of the water used in such cases loses its integrity due to the addition of organic and inorganic matter into it collated and termed as wastewater. A substantial quantum of wastewater is generated from urban conglomerates due to higher population densities and shortfalls in wastewater treatment and reuse (Mahapatra 2015). Whereas conventional mechanical-based technologies promise faster wastewater treatment with use of higher energy, they have been found unsuitable for the developing nations in terms of socioeconomic challenges requiring vast infrastructure, operation and maintenance cost and with absolutely no revenue or services to the local people (Ramachandra et al. 2014, 2015).

Algae have been highly studied over the past few decades due to its advantages of faster growth, its ability to assimilate, metabolize and fix essential elements (CHONPK) from wastewaters (Chanakya et al. 2012, 2013). Today most of the waters are nutrient enriched and are naturally infested with algae of various capacities and verities (Mahapatra et al. 2013b). Globally due to their innate ability to mobilize nutrients and proliferate in open waters, these organisms control the global nutrient cycles as the C, N and P pools. They are thus very essential for wastewater remediation and production of enormous biomass as primary productivity. Algal systems not only provide food for the higher trophic members but also produce oxygen that is considered as one of the greatest ecosystem services and is often valued in wastewater treatment (Mahapatra et al. 2017). Besides this, the algae have been also beneficial in terms of water detoxification and heavy metal removal/accumulation (Oswald 1988; Hoffman 1998; Ramachandra et al. 2014, 2015). Various cellular metabolites as phytochelatins and metallothioneins aid in the formation of complexes with heavy metals and trap them as an essential step for heavy metal removal (Suresh and Ravishankar 2004). Adsorption and absorption are the two processes that help in bioremediation of heavy metals (Lomax et al. 2011).

Multiple benefits in the algal route for treatment of wastewater promote algal bioprocesses for sustainable wastewater treatment and biomass production.

Although the natural algal-bacterial symbiotic processes for wastewater treatment have been documented from 1950s (Oswald and Gotaas 1957; Mogens 2008), the algal-based approaches have yet to find its own space and establish its niche in the commercial wastewater treatment systems (Mahapatra 2015). The algal treatment options besides removal of organic and inorganic nutrients, heavy metals and some xenobiotic compounds also provide a scope for generation of revenue by the production of value-added products that can be suitably produced and commercially sold. The advantages of their low-cost treatment together with the production of valorisable biomass makes them suitable candidates for turning 'waste to wealth'. These algal systems have been working effectively with concentrated wastewater streams from livestock (dairy, swine, poultry, etc.), agriculture, industrial and municipal wastewaters (Kaplan et al. 1988; Sivakumara et al. 2012; Mahapatra et al. 2014). Moreover, the algal biomass generated as a by-product of the treatment process has been found to be rich in lipids, proteins, carbohydrates, essential pigments, antioxidants, vitamins and minerals and other valued bio products of commercial interests. These major metabolic compounds are enriched with essential amino acids and fatty acids as Omega-3 fatty acids. Higher productivity of the essential metabolites of commercial importance can be obtained through imparting varying degrees of stress or suitable nutrient limitations. Wastewater algae have been reported to be a good source of lipids. In addition, after lipid extraction, the residual algal biomass that is left out often contains other high-value metabolites as proteins and carbohydrates (Mahapatra et al. 2016). This left out biomass can also be further processed either through biochemical or thermochemical processes for the generation of bioenergy through anaerobic digestion producing methane and fermentation producing ethanol. Through thermochemical treatment the spent biomass can be converted sub-critically, critically and super critically into biofuels or biocrude and syngas via hydrothermal liquefaction, pyrolysis and gasification (Mahapatra 2015).

This additional potential and scope for revenue generation gives them the edge over the conventional technologies. However, concentrating and harvesting of the algal biomass (due to their small size) have been one of the most important challenges in bioproduct development and processing. In order to have higher biomass harvesting efficiencies, auto-flocculation (Guo et al. 2013), bio flocculation (Salim et al. 2010) and other physicochemical methods using chitosan (Renault et al. 2009) for algal biomass separation based on cell wall characteristics, mucilaginous secretions, pH-/ORP-based agglomeration followed by settling have been recently studied and have shown promising outcomes (Harith et al. 2009). Besides these various approaches with co-cultivation strategies involving algal-bacterial consortia (Mahapatra 2015), cyanobacterial–algal consortia (Mahapatra et al. 2014), algal–fungal co-cultivation (Zhou et al. 2012) and protozoan/crustacean based co-culture techniques have been used for efficient algal harvest. Moreover, a fixed film-based algal biofilms have been recently studied for higher biomass retention compared to algal suspension-based treatment systems (Christenson and Sims 2012). Thus, avoiding loss of algal biomass as suspensions in the effluent after suitable wastewater treatment and better harvesting using algal turf scrubbers, etc. (Craggs et al. 1996).

Algal research has gained special impetus and scope in the present era. The exploitation of the algal whole cell biomass for extraction of value-added products (Ravindra 2000; Ramachandra et al. 2009) in tandem with wastewater treatment ensures its techno-economic feasibility with no environmental externalities. A major area of interest in algal biotechnology is the economic benefits associated with the mass culture and biomass use of algal species thriving in tropical wastewaters. Most of the wastewater generated in the developing world remains unattended and enriches the surface and the groundwaters with higher loads of nutrients rendering it unsuitable for any use (Mahapatra et al. 2011a, 2011b, 2011c, 2017). The algal species can be used as robust treatment systems to recover nutrients from these wastewater and municipal landfill leachate (Naveen et al. 2016) by harvesting whole cell algal biomass (Mahapatra and Ramachandra 2013). The algal biomass has shown potential to remove nutrients from C, N and P as uni-algal cultures or as microbial consortia comprising of *Chlorella, Chlorococcum, Monoraphidium, Ankistrodesmus, Scenedesmus, Chlamydomonas, Euglena, Lepocinclis, Phacus, Oscillatoria, Phomidium, Micractinium* and *Spirulina* (Mahapatra et al. 2014). These algal assemblages have been also recorded in wastewater treatment ponds (Mahapatra et al. 2013b). Laboratory-based experiments have shown higher production of valorisable algal biomass, with a significant quantity of lipids (Mahapatra and Ramachandra 2013; Mahapatra et al. 2013a, 2014; Ramachandra et al. 2015). Large-scale algae-based treatment systems have been also studied and have showed a higher techno-economic feasibility (Chanakya et al. 2012, 2013) for algal single-cell proteins (SCP) (Mahapatra et al. 2013b; 2016) with a lower environmental impact (Ramachandra and Mahapatra 2015; Ramachandra et al. 2015). The algal modules designed at the Indian Institute of Science (IISc), Bangalore, have been successfully employed for nutrient (C, N and P) capture and recovery at the Jakkur Lake in Bangalore city and show promising augmentation to the conventional treatment systems and providing purified water that can be recycled and reused (Ramachandra et al. 2013; Mahapatra 2015) and potential scope for biomass harvest.

16.2 Wastewater Characterization

The quality and quantity of any wastewater depends on its source and are not the same. The wastewater composition differs from residential, municipal, commercial, industrial or agricultural sectors. The physicochemical and biological characterization of wastewater significantly differ from each other and are dependent on many factors, which include lifestyle, social behaviour, a technical and juridical framework for households. While, industrial wastewaters have also the similar considerations, but depending on the type of raw material processed, the pollution characteristics vary. A huge quantity of wastewater is generated continuously around the globe and is estimated to be 330 km^3/year for only municipal wastewater. The nutritional value of the generating wastewater is theoretically sufficient to irrigate and fertilize millions of hectares of crops and energy production to supply millions of households. Several, municipal and industrial wastewater treatment plants are involved in treating the majority of the wastewater generated and in recycling operations. Globally, 60% of the produced municipal wastewater is treated. The treatment process and the maintenance of the plant are the most difficult and expensive operations. The treatment processes always result in the huge production of sludge, which is rich in organic matter and nutritional values (carbohydrates, proteins, lipids and micronutrients). Besides the traditional application of bio-solids and composting, recent biotechnological advancements have opened new dimensions for the recovery of value-added products from wastewater and sludge as raw material. These value-added products include biopesticides, bioherbicides, enzymes, bioplastics, biofloculants and biofuel, in which the value of the final product is much higher than the cost of processing. The production of value-added products and processing depends on the type of wastewater/sludge used as raw material. This section majorly focusses on the different types of wastewater and its characterization, followed by the algal treatment and production. The major constituents classifying any wastewater will fall into three categories, which include physical, chemical and biological characterization. The list of parameters that classify the physical, chemical and biological characterization of wastewater is provided in Table 16.1. Thus, characterization of wastewater is very important to design the treatment system and to monitor the pollution control before and after the treatment.

The wastewater characterization provides a wide range of information regarding the type and concentration of the specific pollutant of interest. The general characteristics of municipal wastewater are presented in Table 16.2. Industrial wastewater is considered as one of the important pollution sources next to the domestic wastewater. The pollution strength and characteristics of the industrial wastewater vary depending on the type of industry and contaminants. Each of the industries releases specific and combined pollutants of its own. The pollutants released from different industrial sectors are listed in Table 16.3.

The conventional sewage treatment system is designed to remove BOD, SS, nutrients, coliform bacteria and other toxic chemicals. The treatment processes have

Table 16.1 Physical, chemical and biological characterization of wastewater

S. No	Wastewater characterization	Constituents/parameters
1	Physical	Temperature, colour, odour and solids
2	Chemical	*Organic constituents*—carbohydrates, proteins, fats, oils and grease, surfactants, volatile organics, pesticides, phenols and others *Inorganic constituents*—pH, alkalinity, chlorides, nitrogen, phosphorous and heavy metals *Gases*—hydrogen sulphides, ammonia, methane and oxygen
3	Biological	Bacteria, viruses, parasites and worms eggs

Table 16.2 Typical characteristics of municipal wastewater

S. No	Parameter	Concentration (mg/l)
1	pH	7.0–8.0[a]
2	BOD	230–560
3	COD	500–1200
4	Total suspended solids	250–600
5	TKN	30–100
6	Ammonia (NH_3–N)	20–75
7	Total phosphorous	6–25
8	Oils, fats and grease	50–100
9	Total inorganic constituents (Na, Cl, Mg, S, Ca, K, Si, Fe)	100
10	Faecal coliforms	2–30 × 10^6/100 ml[a]

[a]Unit (mg/l) not applicable to pH and coliforms

Table 16.3 Wastewater characteristics of different industries

S. No	Industry	Pollutants
1	Chemicals	COD, surfactants, emulsifiers, petroleum hydrocarbons, phenols, organic chemicals, heavy metals, SS and cyanide
2	Iron and steel	BOD, COD, oil, acids, phenols, cyanides, sulphur compounds, dust, metal ions, ashes, slags and ore particles
3	Textiles and leather	BOD, COD, solids, sulphates, ammonia and chromium
4	Pulp and paper	BOD, solids, sulphates, chromium chloroform, dioxins, furans and phenols
5	Petrochemicals and oil refineries	BOD, COD, mineral oils, phenols and chromium
6	Mining	Acids, salts, metals and solids
7	Food processing	BOD, TS, salt, flavourings, colouring material, oils, fats and acids or alkali
8	Distillery, molasses or sugar factory	BOD, COD, solids, sugar, proteins and metals

several stages; the initial preliminary stage removes large solid particles by screening through 20–60 mm. Then, the primary treatment involves sedimentation of settleable solids, which can remove 40% of the BOD. The secondary treatment targets to reduce BOD by using biological process. The final tertiary treatment is to remove ammonium, nitrate and phosphate, in which algal cultures are very effective compared to other techniques (Abdel-Raouf et al. 2012). The industrial effluent wastewater also follows the similar primary, secondary and tertiary treatments, however, the final toxic chemicals and heavy metals require advanced treatment process. These additional stages include coagulation, filtration, activated carbon adsorption, electrodialysis, reverse osmosis, ozonation and advanced oxidation processes, etc. depending on the target pollutant removal. The cost of the process depends on the type of treatment adopted and is usually higher for these physicochemical processes compared to the algal treatment.

Even though conventional technologies are effective in removing aqueous pollutants, these biological methods require longer time in degrading persistent organic pollutants (POPs) such as polychlorinated biphenyls (PCBs), organochlorine pesticides, dichlorodiphenyltrichloroethane (DDT), dioxins and furans. The recalcitrant nature of these compounds makes the biological methods unsuitable due to biomass poisoning. However, the risk associated with these compounds could be tackled by introducing advanced oxidation processes (AOPs). These AOPs are highly reactive by releasing hydroxyl radicals and are non-selective, thereby very effective in oxidizing pollutants to CO_2 and inorganic ions. Few other methods such as flocculation, precipitation, adsorption, air stripping and reverse osmosis are also applicable; however, they need post-treatment stages to reduce the contamination levels (Danis et al. 1998).

The application of using algae for secondary and tertiary wastewater treatment started during the 1970s. The major purpose of introducing algae pond was to prevent eutrophication in the secondary effluent. The application of using algae to treat wastewater is highly beneficial economic wise (Singh and Olsen 2011). Algae can utilize the organic matter and micronutrients to form cellular constituents like carbohydrates and lipids. In addition, a wide range of biofuels including hydrogen, bioethanol, biodiesel, methane and dried mass could be produced (Pittman et al. 2011). The algal species offer effective removal of COD, nitrogen and phosphorus during the secondary and tertiary treatment. They are also effective in removing heavy metals and other toxic organic compounds. In addition, the release of oxygen into the water system during photosynthesis is very effective in disinfection. The algal cultures of *Chlorella, Scenedesmus, Chlamydomonas, Dunaliella, Golenkinia, Euglena, Oscillatoria* and *Micractinium* are widely used for the treatment process (Palmer 1974; Tam and Wong 1989; Wang et al. 2010; Chamberlin 2016). Thus, the mass production of algal biomass from the processes has the dual advantage of bioremediation and as well as for the biofuel production (Chamberlin 2016).

16.3 Algal Bioprocess Cultivation Systems

Algal cultivation systems have evolved with time. From algal growth in natural open water surfaces, selected algal species have been screened and tested for their efficacy in producing value-added products and have been grown in confined and controlled environments called photobioreactors. Apart from the type of cultivation, systems, i.e. open or closed, optimum growth and culture conditions are vital for the propagation of desired algal species. Primarily, four types of cultivation modes are in practise—autotrophic, heterotrophic, mixotrophic and photoheterotrophic or hybrid process. Table 16.4 represents the energy and C sources of the various modes of algal cultivation.

Table 16.4 Energy and carbon source for various cultivation and nutrition modes in algal bioprocesses

Name	Energy source	Carbon source	Algal species
Photoautotroph	Light	Inorganic	Cyanobacteria, e.g. *Spirulina*
Heterotroph	Organic	Organic	*Euglena*
Mixotroph	Light and organic	Inorganic and organic	*Chlorella* (Photoautotrophic heterotroph)
Photo-heterotroph hybrid	Light	Organic	Some green algae and purple and green photosynthetic bacteria, i.e. *Rhodospirillum*

16.3.1 Autotrophic Mode

This is the most commonly used mode for growth and propagation of algal communities for various applications. Autotrophy or photoautotrophic mode of nutrition uses light as a sole energy source, producing chemical energy through the light-dependent photosynthetic process. In case of such systems, the inputs are solar light, CO_2 as inorganic C source, inorganic nutrients and water as the medium for growth and development. In this case, photosynthesis is the principal C fixation mechanism and they build food materials as starch and C cytoskeleton for the algal cells without the need for external organic carbon in the system. The algal systems use either the ammonium nitrogen or the nitrate-nitrogen as a nitrogen source and carbon dioxide as its carbon source to biosynthesize biomass. The biosynthesis of algal biomass can be shown by the following stoichiometric relationships.

(a) C source as CO_2 and ammonia as the nitrogen source,

$$16NH_4^+ + 92\ CO_2 + HPO_4^{2-} + 92\ H_2O + 14\ HCO^{3-} \rightarrow C_{106}H_{263}O_{110}N_{16}P + 106\ O_2 \quad (1)$$

(b) C source as CO_2 and nitrate as the nitrogen source,

$$16NO_3^- + 124\ CO_2 + HPO_4^{2-} + 140\ H_2O \rightarrow C_{106}H_{263}O_{110}N_{16}P + 138\ O_2 + 18\ HCO^{3-} \quad (2)$$

where $C_{106}H_{263}O_{110}N_{16}P$ represents the stoichiometric formula for algae Both reactions are endothermic, i.e. require energy input, supplied by solar energy. Given a choice, ammonia is preferentially taken up by algal phytoplankton over nitrate. In general, the photoautotrophic metabolism is given by

$$H_2O + HCO_3^- \rightarrow C\ (algae) + 0.5\ O_2 + 3\ OH^- \quad (3)$$

16.3.2 Heterotrophic Mode

Heterotrophy is the utilization of sole carbon and energy source from organic compounds, and thus, the requirement for light is eliminated. Heterotrophy provides the ability to increase biomass concentration and thereby productivity when it functions independently and simultaneously with autotrophy. Such modes of nutrition can be implemented by the application of high-cell density techniques, frequently used for yeasts cultures, such as fed-batch, chemostat culture and membrane cells recycle systems (Chen 1996; Chen et al. 1996; Chen and Zhang 1997). Heterotrophic growth in aerobic processes was confirmed experimentally for the following strains of microalgae *Chlamydomonas reinhardtii*, *Chlorella pyrenoidosa, Chlorella regularis, Chlorella sorokiniana*, *Chlorella vulgaris*, and, *Scenedesmus* sp., *Scenedesmus obliquus*, *Spirulina* sp., *Haematococus* sp., *Nitzschia laevis* (Chen 1996; Neilson et al. 1973), *Synechocystis* sp., *Plectonema boryanum* and *Nostoc* sp. (Zhang et al. 1998a, 1998b)., where the organic compounds introduced were: glucose, peptone and acetate, although there is a range of other organic compounds that can be potentially used for growth in heterotrophic conditions. Based on stoichiometry, it has been also observed that the nature of the heterotrophic biomass is phenomenally different from those of photographically grown biomass (Mahapatra 2015). The production of heterotrophic biomass is provided in Eq. 4.

$$(1 + x)\ CH_2O + O_2 \rightarrow C\ (algae) + x\ CO_2 + (1 + x)\ H_2O \tag{4}$$

The biomass formula of a heterotrophically grown organism comprises of the low relative proportion of oxygen compared to the autotrophically grown biomass. This makes the biomass highly valorisable and is often attractive as biofuel feedstocks. Furthermore, it has also been reported that heterotrophically grown algal cells show high digestibility and biological value as feed for various animals and fish. The overall growth yield (0.46 g biomass/g glucose) was comparable to the yield values of well-known heterotrophs such as yeasts and other aerobic-heterotrophs. *Chlorella regularis* have been reported to grow in dark-heterotrophic continuous culture using acetate (Endo et al. 1977). Various advantages and disadvantages of heterotrophic algal cultures as different phases of the cultivation process are elucidated in Table 16.5.

16.3.3 Mixotrophic

Mixotrophy (photolithographic heterotrophy) is a metabolic process, in which photosynthesis is the main energy source, although both organic compounds and CO_2 are essential. Amphitrophy (a subtype of mixotrophy) refers either autotrophy or heterotrophy depending on the ratio of organic substrate concentration to light

Table 16.5 Pros and cons of algal heterotrophic cultures

S. No	Phases of bioprocess	Pros	Cons
1	Pre-treatment sterilization	Elimination of predatory organisms by sterilization	Potential contamination with bacteria
2	Illumination	No need to supply light	In some cases, inability to produce light-induced metabolites or high valuable substances that can be solved by shifting to mixotrophic growth conditions
3	Culturing	Easy maintenance of optimal condition for growth and production	High costs of growth medium, axenic cultures required
4	Biomass yields productivity	Higher biomass concentration, growth rate and productivity	Growth limitations when organic substrate concentration decreases. O_2 concentration should be controlled

intensity (Kaplan et al. 1986). According to another definition, a mixotroph is an organism able to assimilate organic compounds as carbon sources while using inorganic compounds as electron donors (Table 16.1). In mixotrophic culture, a simultaneous uptake of organic compounds and CO_2 takes place as carbon sources for cell synthesis and it is then expected that CO_2 released via respiration will be rapidly trapped and reused under sufficient light intensity (Martinez et al. 1997; Hata et al. 2000). Thus, mixotrophic cells acquire the energy by catabolizing organic compounds via respiration and converting light energy into chemical energy via photosynthesis (Hata et al. 2000).

$$\mathrm{y\ HCO_3 + z\ CH_2O \rightarrow (y + (z - x))\ C\ (algae) + 3\ OH^- + x\ CO_2} \quad (5)$$

The addition of organic substrate resulted in the increase in the growth rate, as well as in the final biomass concentration in mixotrophic growth modes. In mixotrophic culture, no photoinhibition was observed that typically occurred above 20 Klux in photoautotrophic culture due to the protective role of glucose and shift of light intensities that trigger photoinhibition (Ogawa and Terui 1972). Other effective carbon sources utilized in mixotrophic cultures in addition to glucose and peptone are arginine, aspartic acid, leucine, proline, TCA-cycle organic acids, acetic, butyric, tartaric and malic acids.

16.3.4 Photoheterotrophic Culture (Hybrid Systems)

Photoheterotrophic mode (photoorganotrophy, photoassimilation, photometabolism) designates metabolism in which light is required to use organic compounds as a carbon source. Earlier studies have stated photoheterotrophs as organisms able to use light as a source of energy and organic materials as a carbon source (Table 16.1). Photoheterotrophy or hybrid systems are also described as the use of sugar exclusively as a carbon source, ATP provided by electron transfer via photosystem I or respiration (Zhang et al. 1998). Inhibition of photosynthesis by DCMU [(3-(3,4-dichlorophenol)-1,1-dimethyl urea] causes inhibition of heterotrophy in any case with or without light supply. Photoheterotrophism occurs under the light natural condition of dim light that does not support photoautotrophic growth that was stimulated by inhibiting photosystem II with DCMU (Orus et al. 1991). In this category, *Synechocystis* sp. uses glucose with light supply but does not grow without both light and glucose (Kaplan et al. 1986; Zhang et al. 1998). The metabolic pathway for a photoheterotrophic algal system is elucidated in Fig. 16.1.

Photoheterotrophic organisms cannot grow solely on glucose nor solely on light but can grow only when light and glucose are present at the same time (scenario IV, Fig. 16.2). This means that glucose is used only as a building material, but not as an energy source, seems to be, there is no Krebs cycle actively working. Microalgae, which can perform mixotrophic growth, can easily shift between autotrophic and heterotrophic nutrition mode, driving scenarios I, II and III illustrated in Fig. 16.2, depending on environmental condition especially light and organic compound availability.

16.4 Intracellular Products from Wastewater Algae

Wastewater algae act as rapid biocatalysts for fixation of organic and inorganic nutrients into valued intracellular compounds as proteins, lipids, carbohydrates, pigments, vitamins and minerals. Out of these cell constituents, the lipids, carbohydrates and proteins are dominant. According to the mode of cultivation and stress type/nutrient limitations, algae can accumulate substantial proteins (12–60%) that is

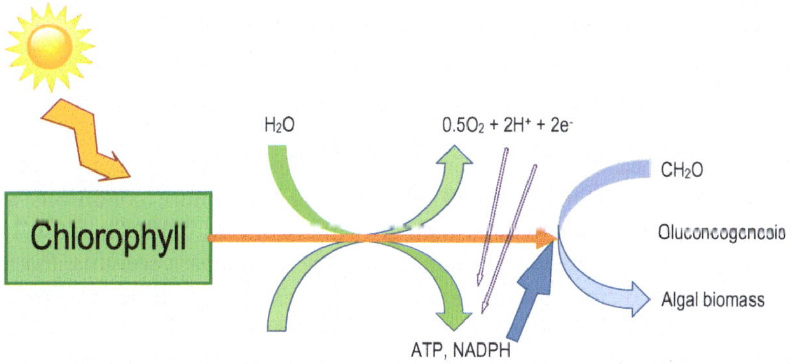

Fig. 16.1 Metabolic pathway: photoheterotrophic mode—light as the only source of energy

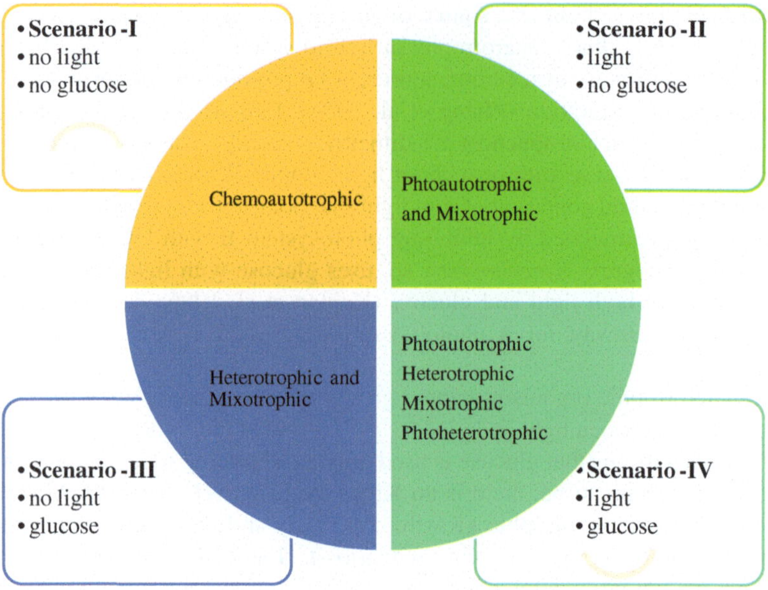

Fig. 16.2 Various trophic scenarios for the growth of algal species

rich in essential amino acids. Among the other constituents, lipids from algal sources are also suitable for biodiesel production through transesterification generating fatty acid methyl ester (FAME). Unsaturated fatty acids as alpha-linolenic acid (ALA, C18:3), eicosapentaenoic acid (EPA, C20:5) and docosahexaenoic acid (DHA, C22:6) are essential for metabolism and is directly used as food/feed ingredients. Starch, cellulose and certain others polysaccharides (marine sources) comprise the carbohydrates from algal sources. Algal species, growth conditions, stress and environmental factors govern the carbohydrate synthesis and its transformation to storage compounds. The various intracellular products have been discussed in the following section.

16.4.1 Lipids

The lipid content and composition vary with the microorganisms metabolic machinery, and can be manipulated by changing the culture conditions as pH, temperature, nutrients supplied and final growth conditions during harvest. In case of oleaginous microorganisms (oil contents >20% of biomass weight), the rate of lipid accumulation increases at N-limited conditions, where all the C directed for growth are routed towards lipids as triacylglycerides due to restricted cell division (Dong et al. 2016) typically increases during nitrogen limitation when cells still assimilate excess carbon, but cell division is inhibited funnelling carbon into triacylglycerides (TAGs) (Dong et al. 2016). High lipid productivity and rapid growth in microalgae such as *Chlorella* sp., *Nannochloropsis* sp. and *Scenedesmus* sp. make them attractive for biofuel production. Lipids manifest in number of forms in *Oleaginous* microorganisms as acylglycerides, phospholipids, glycolipids, sphingolipids, lipoprotein, free fatty acids, sterols. Such classes are characterized by viscosity, solubility, polarity (Guckert et al. 1988). Various lipids classes in cell biomass are listed in Table 16.6.

Apart from being a dense energy source in lipid bodies, they tend to accumulate differentially in various locations within a cell and serve myriads of functions. The cell cytoplasm houses the energy dense lipid bodies that mainly comprises of triglycerides and sterol esters surrounded by a phospholipid monolayer (Ryckebosch et al. 2014). Cell disruption techniques such as ultrasound, bead mills, enzymatic methods, high-pressure homogenizer, osmotic shock, microoven method and sub-critical water hydrolysis techniques used to extract the lipid present in the cells (Mahapatra et al. 2013a, 2013b; Ramachandra et al. 2013).

16.4.2 Hydrocarbons

Crude oil can be considered the precursor of the naturally existing hydrocarbons; however, these crude oil sources as oil shells have been reported to be evolved from algal biomass (Cane 1969). With increased oil prices and GHG emissions, algal hydrocarbons are gaining quick interest (Ladygina et al. 2006). Species of all algal phyla are capable to produce hydrocarbons. However, the overall hydrocarbon content in algal cells in generally below 2% with primary odd C members (C15, C17, or C21) (Qin 2010). Table 16.7 depicts the hydrocarbon variations in different algal members. Botryococcane is a unique algal derived hydrocarbon and is a member of a cyclic alkane from green alga *Botryococcus braunii* (Maxwell et al. 1968).

Botrycoccus braunii is a colonial unicellular green alga and can stock hydrocarbons up to 86% of the dry weight (Brown et al. 1969) and is commonly found in brackish and fresh waters across continents (Metzger et al. 1991). This alga is widespread in fresh and brackish waters of all continents. *Botrycoccus braunii* synthesizes and accumulates a variety of lipids including a variety of specific ether lipids with more than 30 specific chemical structures with terminal unsaturation.

16.4.3 Carbohydrates

Algae of marine origins show interesting characteristic polysaccharides comprising of structural, muco and storage polysaccharides (Murata and Nakazoe 2001). Polysaccharides comprise of simple sugars linked with glycosidic bonds are defined as polymers of simple sugars (monosaccharides) linked together by glycosidic

Table 16.6 Various lipid classes found in oleaginous microorganisms

Species	Cultivation	Neutral lipids		Free fatty acids	Phospholipid	Glycollipid	References
		Acyl glycerides	Sterol ester				
Nitzschia laevis	Heterotrophic, Lewin's marine diatom medium with glucose, 23 °C, pH: 8.5	79.2	–	–	11.6	8.1	(Chen et al. 2007)
Pavlova lutheri	Artificial seawater (ASW), 20 °C, 0.3% CO_2, pH: 8, light = 20 W/m^2	56.6	5	0.67	9.7	18.9	(Meireles et al. 2003)
Chlorella sorokiniana	Heterophic, Kuhl medium with glucose, 37 °C, pH: 7	78.9	2.7	11.2	7.1	–	(Zheng et al. 2013)
Schizochytrium limacinum	Heterotrophic, glycerol as carbon source (Seambiotic Let, Tel Aviv, Israel)	69	–	12.6	14	–	(Wang and Wang 2012)
Gymnodinium sp	GSe medium, 18.5 °C, 80 lmol/m^2/s	7.5–28.8	1.3–3.0	0.7–1.3	66.4–84.7	–	(Mansour et al. 2003)

Table 16.7 Hydrocarbon variation in various algae (modified from Nevenzel 1989)

Algal groups	Mean and range of hydrocarbon (% dry weight)	Number of species included
Blue-green algae	0.14 (0.02–0.44)	6
Red algae	0.02 (0.0003–0.073)	24
Dinoflagellates	1.34 (0.015–5.8)	9
Diatoms	0.56 (0.0041–0.66)	15
Brown algae	0.049 (0.00049–0.66)	45
Golden algae	0.76 (0.0034–0.0034)	4
Englenoids	0.51 (0.22–0.79)	3
Yellow algae	1.75 (1.4–2.1)	2
Green algae	0.68 (0.000044–17)	33

bonds, are processed and commercially used as emulsifiers, thickeners, stabilizers, food/feed, beverages (Tseng 2001). Highest polysaccharide content has been recorded in *Porphyra, Palmaria, Ulva* and *Ascophyllum* while the average ranges from 4 to 76% on a dry wt. basis.

As a basis of classification, the storage products are different in different marine algal members. As. green algae contain sulphated polysaccharides as sulphated galactan and xylans. The brown algae comprise of alginic acid, fucoidan (sulphated fucose), laminarin (β-1, 3 glucans) and sargassan. And, the red algae contain agars, carrageenans, water-soluble sulphated galactan, xylans, floridian starch (amylopectin like glucan) and porphyrin as mucopolysaccharides within the intercellular spaces. The three main polysaccharides obtained from marine algal sources are, (a) alginate, (b) carrageenan and (c) agar.

16.4.3.1 Brown Seaweeds and Alginates

Brown seaweeds have been the most commercially exploited marine algae for alginate production (~26 kT at US$318 annually). Although brown seaweeds have versatile temperature tolerance, these are predominant in cold waters (best up to 20 °C). *Laminaria Ascophyllum* and *Mycrocystis* are the major commercial sources for alginates with minor sources as *Lessonia, Turbinaria, Sargassum, Durvillea and Eklonia* (Bixler and Porse 2011). Alginates have been mostly used in cosmetic creams, pharmaceutical, processed foods and paper and cardboard making.

16.4.3.2 Red Seaweeds and Carrageenan's

A linear polysaccharide chain with sulphate half-esters attached to the sugar unit ascribes Carrageenans. Carrageenans readily dissolve in water, forming highly viscous solutions and are stable over a wide pH range. The kappa, lambda and iota are the various structural forms of carrageenans, with specific gelling properties. *Kappaphycus alvarezii and Eucheuma denticulatum are sources of kappa* and iota carrageenan, respectively. (Rasmussen and Morrissey 2007) for Carrageenan's (E 407) are used as food/feed and are major constituents in canned foods, desert mousses, salad dressings, bakery fillings, ice cream, instant desserts and canned pet foods and are also used in the brewing industry for clarifying beer, wines and honey.

16.4.3.3 Red Seaweeds and Agar

Agar comprising of agarose and agaropectin are structurally and functionally similar to carrageenans. *Gelidium* spp., and *Gracilaria* spp is the major sources of agar. (Rasmussen and Morrissey 2007). At higher temperatures, agar melts and gels as carrageenan and is used in pastry fillings and glazes that can be applied before the pastry is baked without melting in the pastry oven. Although in processed meats, carrageenan is the favoured water binder or texturing agent, however, agar rules the gelatin replacement market in canned meats and aspics. The agar texture in fruit jellies is preferred as compared to kappa carrageenan jellies in Japan and other parts of Asia. Various polysaccharides such as alginates, agars, carrageenans, fucoidan, mannitol, laminarin and ulvan are obtained from algal sources and applied in numerous fields.

16.4.4 Proteins

Single-cell proteins (SCP) are protein pull out from cultured biomass mainly used for protein supplementation. The term 'single-cell protein' was coined in 1968 at a meeting held at the Massachusetts Institute of Technology (MIT) as a substitute for the originally used terminologies referring to less aesthetic 'microbial protein' and 'petroprotein' (Tannenbaum and Mateles 1968; Tannenbaum and Wang 1975). Microbes as bacteria, algae and fungi are key sources of SCP (Mahapatra et al. 2016). *Spirulina* is most widely used alga for SCP, often utilized as a rich nutrient source during space explorations. Wastewater cultivated *Euglena* species have also shown potential use as SCP (Mahapatra et al. 2016).

The nutrition and amount of essential amino acids in primarily based on the nature and the type of SCP and its composition. Members of algal origin are naturally blessed with enormous proteins (40–60%) and essential vitamins as A and D besides other ingredients (Brock 1989) (Table 16.8).

Table 16.8 Algae as food in different regions of the world (modified from Ravindra 2000)

Algae	Uses
Alaria	Stipes called 'Saumen' are dried, salted and sold
Ascophyllum, Fucus, Laminaria	Feed for cattle, poultry and pigs
Laminaria	Stipes called 'Kombu' are dried, salted and sold
Laminaria, Ecklonia, Eisenia	Chopped and used for chickens and sheep
Rhodymenia palmata, Gelidium, Grateloupia, Fucus	Chopped and added to other dishes
Rhodymenia, Chlorella pyrenoidosa, Spirulina, Synechococcus	Oriental region regular portion of diet
Spirogyra, Oedogonium	Dried and used to make soup

16.4.5 Pigments

Natural pigments are essential in photosynthetic metabolism and pigmentation in algae. Because of its various biological properties such as antioxidant, anti-cancer, anti-inflammatory, anti-obesity, antiangiogenic and neuroprotective, it serves a vital role in human dietary. Based on its structure and action, it is classified as chlorophylls, carotenoids and phycobilins.

16.4.5.1 Phycobilins

Being readily soluble in water and easy to isolate and purify, phycobilins (or phycobiliproteins) are important pigment protein molecules. Phycobilins are accessory for collecting light during photosynthesis. For example, at four phycobilin classes are produced in red algae (Rhodophyta and Glaucophytes): allophycocyanin (green–bluish), phycocyanin (blue), phycoerythrin (purple) and phycoerythrocyanin (orange) (Cuellar-Bermudez et al. 2015). Phycobilins are widely used in industry and immunology laboratories, due to their high molar absorbance coefficients, high fluorescence quantum yield, large Stokes shift, high oligomer stability and high photostability absorption properties. They are often employed in molecular biology as fluorescent markers, used in immune assays and as fluorescent dyes for microscopy. In addition, phycobiliproteins are widely used in industry and clinical or research immunology laboratories. (Spolaore et al. 2006).

16.4.5.2 Chlorophylls

Chlorophylls are greenish and fat-soluble pigments with a porphyrin ring (Cuellar-Bermudez et al. 2015). Its main role is to convert solar energy into chemical energy during photosynthesis. Most microalgae have chlorophyll *a*, and some proportions of chlorophyll *b* and *c*. Generally, microalgae contain around 0.5–1.0% of chlorophyll per dry weight. Chlorophyllin a derivate of chlorophyll in which the magnesium is replaced by sodium or copper and phytol chains are lost. Chlorophyllins have been used to control body odour of geriatric patients, and dietary supplement, antimutagenic and anticarcinogenic.

16.4.5.3 Carotenoids

Carotenoids are the fat-soluble pigments, with colour ranging from brown, red, orange and yellow. The average carotenoid content in algae ranges from 0.1 to 0.2% and can go as high as 14% on dry weight basis. Fat solubility aids them to enter the blood circulation and gets attached to different lipoprotein. Inability to cannot synthesize these essential pigments, in human body, makes it a critical supplement in diets.

Based on the chemical structure, carotenoids are divided into two groups, (a) carotenes including beta-carotene and lycopene and (b) xanthophylls including astaxanthin, lutein and canthaxanthin. Carotenoids directly involved in photosynthesis (beta-carotene and lutein) and function as light-harvesting apparatus and photoprotection are primary in nature, and those that are not are secondary (astaxanthin and canthaxanthin) in nature. *Haematococcus pluvialis* is commercially well known as a major source of natural astaxanthin and produces more astaxanthin under nutrient limitation comprising 0.2–3% astaxanthin on dry weight basis (Batista et al. 2013). Beta-carotene has a very increasing demand as a natural pigment or nutritional supplementation, being a major precursor for vitamin A. *Dunaliella salina* has been the most industrially used strain for generation of beta-carotene (14%) (Spolaore et al. 2006).

16.5 Algal Bioprocess Conversion

Algae could be processed into various products such as biogas, ethanol, diesel, char, electricity and hydrogen. The conversion of algae could be processed via biochemical, chemical or thermochemical methods to convert it to value-added products.

16.5.1 Biodiesel

The predominant feedstock's for biodiesel were lipid-/fat-rich contents such as soybean oil, rapeseed oil, castor oil, canola oil; however, algae is also one of the other potential sources for biodiesel production. Processing algae to biodiesel are well known; however, the producing algae from wastewaters and using it for biodiesel production could be challenging. Microalgae are a promising feedstock for biodiesel production due to its biomass conversion efficiency, high growth rates, productivity compared with other conventional sources. It is possible to obtain more than 25 times oil yield potential than other substrates, and produced about 70% of its weight as oil in dried biomass (Scott et al. 2010; Arenas et al. 2016).

Though microalgae has several advantages in terms of biodiesel production, there are certain challenges as well which include lack of cost-effective technologies, efficient harvest and drying of biomass, protocols and methodologies for full-scale facilities, the downstream processing including lipid extraction and biodiesel purification (2). Transesterification is the important process in which lipid gets converted to biodiesel where oils/lipids get transformed to alkyl esters in the presence of an alcohol such as methanol or ethanol and a catalyst (sodium hydroxide, potassium hydroxide and sodium methoxide). *Chlorella* sp., in general, possess a higher lipid yield and efficiency ranging between 15 and 50% for lipid yield and 70–95% in terms of efficiency (Arenas et al. 2016).

Brazil has wastewater treatment facility, which produces algae. The growth of algae is controlled for efficient harvesting based on the different parameters such as nutrient limitation, retention time and biomass recirculation. Commonly, algae from wastewaters are cultivated at high-rate ponds, unlike photo bioreactors due to its energy requirements and operational expenses. The potential of algal oil in Brazil was 4.25×10^8 kg algal oil from 10^7 inhabitants, which correspond to 0.00035 L oil/L wastewater. The biodiesel from algae produced from municipal wastewaters could be 3 million m^3 which is equivalent to 21% of the current biodiesel production in Brazil. For the biodiesel production to be economically feasible, the energy consumption and operational costs need to be reduced through efficient and better technologies (Kligerman and Bouwer 2015).

16.5.2 Bioethanol

Ethanol, one of the potential liquid fuels could be produced from algae and cyanobacteria, which is third-generation feedstocks. Ethanol could be produced from algae or cyanobacteria through three pathways including (a) traditional hydrolysis and fermentation with yeast, (b) dark fermentation and (c) photo fermentation. The traditional conversion of hydrolysis is more efficient route than others due to their well-known enzyme efficiency and yeast interactions. Nonetheless, this route has several steps involved in which makes it energy intensive and expensive. The common carbohydrates present in algal biomass include starch, glycogen and cellulose of which degradation of cellulose or starch is well understood. Acid hydrolysis at mild conditions (120 °C and 20 min) resulted in between 70 and 95% of the carbohydrates released (de Farias Silva and Bertucco 2016).

Dark fermentation is more common to produce hydrogen; however, most of the intermediates are less carbon chain organic acids and alcohols, which could be converted to ethanol. Ethanol is produced through the accumulation of carbohydrates in the algal cells during photosynthesis which is forced to produce ethanol during fermentative metabolism under dark conditions. This route is not as efficient as other routes mentioned above (Farias Silva and Bertucco 2016). Finally, the photo fermentation is gaining adequate interest in the recent times because of its ability to produce fermentation products through efficient metabolic pathways (Farias Silva and Bertucco 2016). Photanol just not only produces ethanol but also other naturally occurring products through glycolysis fermentation pathways. Important species which have been explored for photanol production includes *Synechocystis* sp. PCC 6803, *Synechococcus elongatus* sp. PCC 7992, *Synechococcus* sp., PCC 7002 and *Anabaena* sp. PCC 7120 (Hellingwerf and Mattos 2009). The production of ethanol after the fixation of inorganic carbon through Calvin cycle, followed by phosphoglycerate production which is further converted to pyruvate by pyruvate decarboxylase and alcohol dehydrogenase and finally it is converted to ethanol (Farias Silva and Bertucco 2016). Ho et al. reported a 92% theoretical yield of ethanol using *Chlorella vulgaris* using sonication of biomass as a pre-treatment method (Ho et al. 2013). Many other studies reported a similar yield between 70 and 95% theoretical yields of bioethanol based on their carbohydrates content (Simas-Rodrigues et al. 2015).

16.5.3 Biogas

Biogas is one of other products which could be produced from algae and has been studied extensively. Algae correspond to the third-generation feedstocks for renewable energy production, which has the advantage of no competition with food crops, low lignin and high growth rate. On the other side, improvements are necessary for algae to biogas processes on reducing the water content, seasonal chemical composition and avoiding inhibitory phenomenon during biogas production. The C/N ratio in algae is around 10:1, where the nitrogen content is high for the anaerobic digestion, which is why co-digestion is essential when algae are used as feedstock. Between 100 and 200 ml/g VS (volatile solids), methane yield could be obtained using algae as a feedstock depending on the processing conditions including temperature, retention time and loading rate (Montingelli et al. 2015). However, this yield could be increased by 30 and 70% when co-digested with other carbon-rich substrates. The common inhibitions for biogas production using algae as a feedstock include ammonia inhibition, which is due to the high nitrogen content, and volatile fatty acids accumulation due to the processing parameters. Mild pre-treatments should be sufficient for efficient release of carbohydrates from algae, unlike lignocellulose pre-treatments, which require harsh pre-treatments due to the presence of lignin (8). Dewatering the algal biomass is crucial for economic feasibility or alternative reactors such as UASB should be employed, which could produce biogas at a faster rate. The challenge with the UASB reactors is that it needs rich volatile fatty acids to process it to methane at a faster rate; however, the concentration of volatile fatty acids is lower which makes it inefficient to use UASB or a pre-treatment step is required before it could be processed (Ward et al. 2014).

16.5.4 Biochar

Biochar could be produced from algae through slow pyrolysis operated between 250 and 400 °C (Bird et al. 2012). Biochar helps is carbon sequestration that restores minerals and nutrients in the soil, improves soil quality, thereby reducing the need for other chemical fertilizers. Plant growth rate increases between 15 and 30 times when biochar was added compared with plants grown without biochar/fertilizer (Bird et al. 2011). Algal biochar possessed similar physicochemical properties in comparison with biochar produced from several other feedstocks such as lignocelluloses, and poultry litter (Bird et al. 2011). However, the carbon in the algal biomass is lower compared with lignocelluloses which make the amount of carbon sequestered lesser. The loss of volatiles increases with the increase in the temperature during pyrolysis (Bird et al. 2011).

16.5.5 Biohydrogen

As discussed in ethanol section, through dark fermentation and photo fermentation hydrogen could be produced from microalgae. One of the important concerns over the hydrogen production is its negative energy balances during dark fermentation, meaning that it requires more energy than energy produced from hydrogen production. For economic viability, positive energy balance is required and adequate research is essential in integrating hydrogen production and other value-added products from algae. Major studies reported a biodegradability of algal biomass for hydrogen production between 10 and 30%, which correspond to more than 70% of the biomass which is not utilized. More research is required to increase the yield of hydrogen from algae and integrative approach to producing other value-added products besides hydrogen (Sambusiti et al. 2015; Eroglu and Melis 2016).

16.5.6 Bioelectricity

Electricity could be produced from algae on wastewaters using microbial fuel cells where algae could act as a cathode. The feasibility of this integration is based on the algal growth and cathode reactions in MFCs. For efficient activity in MFC, algae should grow efficiently which when obtained could reduce more than 90% of COD from wastewaters. The light source is crucial for this complete synergy to be optimal (Xu et al. 2016).

16.6 Techno-Economics, Environmental Feasibility as Bio-Based Industry and Sustainability of Wastewater Bioprocess

Algal biomass popularly known as the third-generation feedstock for bioenergy is promising with a huge potential for bio-based industries, simultaneously offering a range of valued bio products. Apart from GHG mitigation, the algal bioprocess offers phytoremediation and other ecosystem services. However, the biggest constraints in algal bioprocesses are the production of enormous algal biomass. For sustainable algal industries to be a reality, there is a huge requirement of millions of tons of algal biomass with myriads of beneficial products extracted during the downstream processing. One of the major resource confinements is the availability of large quantum of water for their cultivation. Although various alternatives to freshwater resources as marine water, brackish water, industrial effluents, livestock wastewater, agricultural effluents and primarily municipal wastewaters are being studied recently for the growth and development of algal cultures. The techno-economic assessment of algal bioprocesses reveals lacks of economic processes and routes for algal biomass harvest and concentration. A cost-based analysis, however, shows higher biomass costs for the algae cultivated in closed photo bioreactors compared to open ponds. However, the open pond systems suffer from very frequent contaminations and grazing issues that bring down the overall biomass productivity. Multistage algal bioprocess with multiple product developments with a complete bio refinery approach is essential for techno-economic feasibility with a lower environmental footprint. Therefore, process integration of algal systems with wastewater treatment can lead of sustainable algal bioprocesses for 4 F's as fuel/food/feed/fertilizer productions.

The environmental sustainability of algal bioprocess depends on the use and reuse of natural and renewable resources. Essentially, the algal growth is powered by photosynthesis that requires solar light and CO_2 from the environment that is abundant. Biofuels and bio products from algal sources are eco-friendly as they release low levels of NOx and SOx after combustion. Especially the algal biodiesel are efficient sources of bio-based energies and are compatible with existing combustion engines that require no further modifications. Moreover, algal biodiesel has similar fuel properties as petroleum-derived diesel in terms of heating values, viscosity, density, cold flow properties, flash point. Furthermore, large-scale production of algal biomass leads to sequestration of atmospheric CO_2. This aids in restricting GHG emissions and devising key strategies for climate change mitigation. Notably, ~1.8 tons of CO_2 are utilized to produce 1 ton of algal biomass. Thus, this provides ample scope for pushing gaseous pollutants as flue gas, etc. into algal cultures for purification of contaminants. The same strategy can be also implemented for absorbing CO_2 from biomethanation units as essential strategies for upgradation of biogas for bioenergy production. The C footprint of algal bio products is much lower than petroleum and other synthetic industrial products. Algal growth on wastewater also avoids usage of fresh water and brings down the usable water footprint. The algal use for nutrient recovery also maintains the ecosystems integrity by closing the nutrient cycles and ensures subsequent use of nutrient upon mineralization. Such ecosystem services and options for revenue generation can dramatically bring down the cost of production of nutrient required for synthesis of agricultural chemical fertilizers.

The economic sustainability of algae-based products is based on return on investments in algal production. A comparative account of the cost of production of 1 kg of algal biomass in open ponds and closed photo bioreactors shows 7.32 and 1.54 USD, respectively. While the cost of 1 kg of algal-based oils grown on open ponds and photo bioreactors are 7.64 and 24.6 USD, respectively. Compared to the petroleum-derived fuels (PDF) (60 USD/barrel), the costs of algal biofuels (biodiesel) are strikingly high (~2000 USD/barrel) and need suitable process optimization and technological interventions for its economic acceptability as renewable and green fuel. In context to US transportation, the cost of replacement of 1% of annual PDF is ~30 million tons of algal biomass with an average 40% oil content (Chsiti 2007). To bring down the cost of production of algal biofuels and make it more economically sustainable, more research and emphasis have to be laid on cost-effective cultivation, harvesting, extraction through optimal downstream processing. Integration of both biochemical and thermochemical approaches with innovative technologies and optimization of output parameters by means of suitable control systems can pave the path for a sustainable algal bio-based economy that fosters clean development. Schematic representation of an algal cultivation system with waste gases and wastewater feed with various bio products generation is depicted in Fig. 16.3.

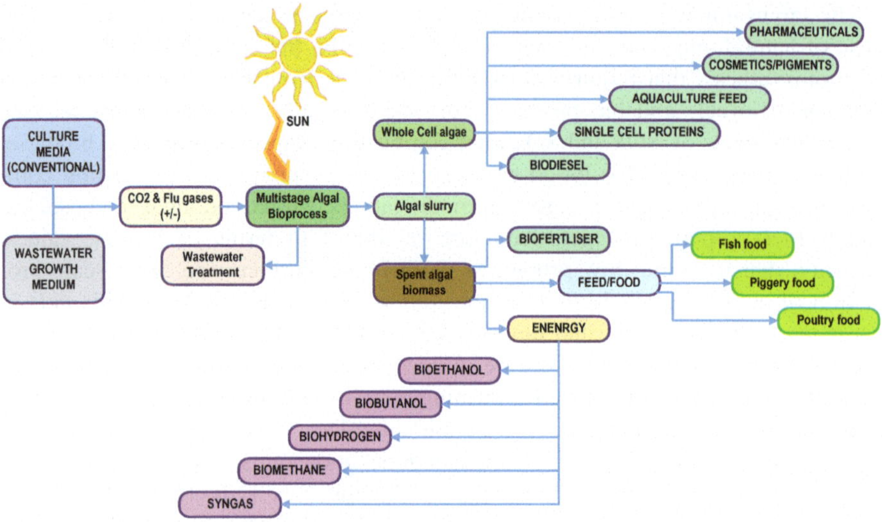

Fig. 16.3 A schematic representation of algal bioprocess and the range of valued products from the alga industry

16.7 Conclusions

In spite of the availability of advanced bioreactor designs, bioprocessing technologies and high-throughput screening techniques there is a large gap for algal bio products commercialization. This is owing to heterogeneity in production capacities and lack of standardization in the ongoing studies. Investigations tested within a narrow set of environmental parameters and species limit the scope of algal applications. Therefore, a sound understanding and identification of the basic drivers of algal biomass production and its subsequent conversion into valuable bio products are of utmost importance. In addition, the present-day research on wastewater remediation using various algal cultivation strategies needs to incorporate variability in growth conditions during outdoor applications.

Successful integration of algal cultivation systems with wastewater treatment processes on a larger scale taking into account that the real conditions with uncertainties in nutrient regimes and seasonal dynamics are essential. A thorough understanding of the linkages between algal growth with wastewater characteristics from different sources with various growth platforms, algal assemblages, at various microclimatic variabilities, nutrient uptake and mobilization trends, and biochemical composition is vital for a complete utilization of biomass for diverse applications following the biorefinery approach. More emphasis on mass and energy balance and systems stability by optimization of flow properties, materials for cell attachment and improved post-processing through catalytic processes is necessary. Studies on natural algal assemblages with heterotrophic–autotrophic interactions with the community characterization in static and dynamic environments will eventually lead to the design of innovative bioprocesses. An integration of various feasible approaches ensuring the technical feasibility, economic viability and environmental sustainability would lead algal bioprocess technologies and bio-based products a reality.

Acknowledgements The corresponding author would like to acknowledge Science and Engineering Research Board (SERB), Department of Science and Technology (DST), Government of India for Funding and Support.

References

Abdel-Raouf N, Al-Homaidan AA, Ibraheem IBM (2012) Microalgae and wastewater treatment. Saudi J Biol Sci 19:257–275

Arenas EG, Palacio R, Juantorena AU, Fernando SEL, Sebastian PJ (2016) Microalgae as a potential source for biodiesel production: techniques, methods, and other challenges. Int J Ene Res (inpress)

Batista AP, Gouveia L, Bandarra NM, Franco JM, Raymundo A (2013) Comparison of microalgal biomass profiles as novel functional ingredient for food products. Algal Res 2(2):164–173

Bird MI, Wurster CM, de Paula Silva PH, Bass AM, De Nys R (2011) Algal biochar–production and properties. Bioresour Technol 102(2):1886–1891

Bird MI, Wurster CM, de Paula Silva PH, Paul NA, De Nys R (2012) Algal biochar: effects and applications. GCB Bioene 4(1):61–69

Bixler HJ, Porse H (2011) A decade of change in the seaweed hydrocolloids industry. J Appl Phycol 23(3):321–335

Brock T (1989) A textbook of industrial microbiology. Sinauer Associates Inc, Sunderland, MA, pp 306–316

Brown A, Knights B, Conway E (1969) Hydrocarbon content and its relationship to physiological state in the green alga *Botryococcus braunii*. Phytochem 8(3):543–547

Cane R (1969) Coorongite and genesis of oil shale. Geochim Cosmochim Acta 33(2):257

Chamberlin JF (2016) Algal wastewater treatment and biofuel production: an assessment of measurement methods, and impact of nutrient availability and species composition. Theses and Dissertations. 1511

Chanakya HN, Mahapatra DM, Sarada R, Abitha R (2013) Algal biofuel production and mitigation potential in India. Miti Adapt Strat Global Change 18:113–136

Chanakya HN, Mahapatra DM, Sarada R, Chauhan VS, Abitha R (2012) Sustainability of large-scale algal biofuel production in India. J IISc 92(1):63–98

Chen F (1996) High cell density culture of microalgae in heterotrophic growth. Tibtech 14: 421–426

Chen F, Zhang Y (1997) High cell density mixotrophic culture of *Spirulina platensis* on glucose for phycocyanin production using a fed-batch system. Enzy Microbial Technol 20:221–224

Chen F, Zhang Y, Guo S (1996) Growth and phycocyanin formation of *Spirulina platensis* in photoheterotrophic culture. Biotechnol Lett 18:603–608

Chen GQ, Jiang Y, Chen F (2007) Fatty acid and lipid class composition of the eicosapentaenoic acid-producing microalga, *Nitzschia laevis*. Food chem 104(4):1580–1585

Chisti Y (2007) Biodiesel from microalgae. Biotechnol Adv 25:294–306

Christenson LB, Sims RC (2012) Rotating algal biofilm reactor and spool harvester for wastewater treatment with biofuel by-products. Biotechnol Bioeng 109(7):1674–1684

Craggs RJ, Adey WH, Jenson KR, St. John MS, Green GF, Oswald WJ (1996) Phosphorous removal from wastewater using and algal turf scrubber. Water Sci Technol 33:191–198

Cuellar-Bermudez SP, Aguilar-Hernandez I, Cardenas-Chavez DL, Ornelas-Soto N, Romero-Ogawa MA, Parra-Saldivar R (2015) Extraction and purification of high-value metabolites from microalgae: essential lipids, astaxanthin and phycobiliproteins. Microbial biotechnol 8(2):190–209

Danis TG, Albanis TA, Petrakis DE, Pomonis PJ (1998) Removal of chlorinated phenols from aqueous solutions by adsorption on alumina pillared clays and mesoporous alumina aluminum phosphates. Water Res 32:295–302

de Farias Silva CE, Bertucco A (2016) Bioethanol from microalgae and cyanobacteria: a review and technological outlook. Process Biochem 51(11):1833–1842

Dong T, Knoshaug EP, Pienkos PT, Laurens LM (2016) Lipid recovery from wet oleaginous microbial biomass for biofuel production: a critical review. Appl Ene 177:879–895

Endo H, Hosoya H, Koibuchi T (1977) Growth yields of *Chorella regularis* in dark-heterotrophic continuous culture using acetate. J Ferment Technol 55:369–379

Eroglu E, Melis A (2016) Microalgal hydrogen production research. Int J Hyd Energy 41 (30):12772–12798

Guckert JB, Cooksey KE, Jackson LL (1988) Lipid sovent systems are not equivalent for analysis of lipid classes in the microeukaryotic green alga, *Chlorella*. J Microbiol Methods 8(3): 139–149

Guo SL, Zhao XQ, Wan C, Huang ZY, Yang YL, Alam MA, Ho SH, Bai FW, Chang JS (2013) Characterization of flocculating agent from the self-flocculating microalga *Scenedesmus obliquus* AS-6-1 for efficient biomass harvest. Bioresour Technol 145:285–289

Harith Z, Yusoff F, Mohamed M, Shariff M, Din M, Ariff A (2009) Effect of different flocculants on the flocculation performance of flocculation performance of microalgae, *Chaetoceros calcitrans*, cells. Afr J Biotechnol 8. doi:https://doi.org/10.4314/AJB.V8I21.66083

Hata J, Hua Q, Yang C, Shimizu KM, Taya M (2000) Characterization of energy conversion based on metabolic flux analysis in mixotrophic liverwort cells *Marchantia polymorpha*. Biochem Engin J 6:65–74

Hellingwerf KJ, De Mattos MT (2009) Alternative routes to biofuels: light-driven biofuel formation from CO_2 and water based on the 'photanol' approach. J Biotechnol 142(1):87–90

Ho SH, Huang SW, Chen CY, Hasunuma T, Kondo A, Chang JS (2013) Bioethanol production using carbohydrate-rich microalgae biomass as feedstock. Bioresour Technol 135:191–198

Hoffman JP (1998) Wastewater treatment with suspended and nonsuspended algae: minireview. J Phycol 34:757–763

Kaplan D, Christiaen D, Arad S (1988) Binding of heavy metals by algal polysaccharides. In: Stadler T, Mollion J, Verdus MC, Karamanos Y, Morvan H, Christiaen D (eds) Algal biotechnology. Elsevier Applied Science, London, pp 179–187

Kaplan D, Richmond AE, Dubinsky Z, Aaronson S (1986) Algal nutrition. In: Richmond A (ed) CRC handbook of microalgal mass culture CRC press Boca Raton FL

Kligerman DC, Bouwer EJ (2015) Prospects for biodiesel production from algae-based wastewater treatment in Brazil: a review. Ren Sust Ene Rev 52:1834–1846

Ladygina N, Dedyukhina E, Vainshtein M (2006) A review on microbial synthesis of hydrocarbons. Process Biochem 41(5):1001–1014

Lomax C, Liu WJ, Wu L, Xue K, Xiong J, Zhou J, McGrath SP, Meharg AA, Miller AJ, Zhao FJ (2011) Methylated arsenic species in plants originate from soil microorganisms. New Phytol 193:665–672

Mahapatra DM (2015) Algal bioprocess development for sustainable wastewater treatment and biofuel production. Ph. D. Thesis. Indian Institute of Science, Bangalore, India

Mahapatra DM, Chanakya HN, Ramachandra TV (2011a) Role of *Macrophytes* in urban sewage fed lakes. I Inte Omics Appl Biotech 2(7):1–9

Mahapatra DM, Chanakya HN, Ramachandra TV (2011b) C: N ratio of sediments in a sewage fed Urban Lake. Inte J Geol 5(3):86–92

Mahapatra DM, Chanakya HN, Ramachandra TV (2011c) Assessment of treatment capabilities of Varthur Lake, Bangalore. Inte J Env Tech Mgmt 14(1–4):84–102

Mahapatra DM, Chanakya HN, Ramachandra TV (2013a) *Euglena* sp. as a suitable source of lipids for potential use as biofuel and sustainable wastewater treatment. J Appl Phycol 25: 855–865
Mahapatra DM, Chanakya HN, Ramachandra TV (2013b) Treatment efficacy of algae based sewage treatment plants. Env Mon Assess 185:7145–7164
Mahapatra DM, Chanakya HN, Ramachandra TV (2014) Bioremediation and lipid synthesis of mixotrophic algal consortia in municipal wastewater. Bioresour Technol 168:142–150
Mahapatra DM, Chanakya HN, Ramachandra TV (2016) Algae derived single-cell proteins: economic cost analysis and future prospects. Protein byproducts: transformation from environmental burden into value-added products, 1st edn. Elsevier, pp 275–301
Mahapatra DM, Joshi NV, Ramachandra TV (2017) Insights to bioprocess and treatment competence of Urban wetlands. J Env Manage (accepted)
Mahapatra DM, Ramachandra TV (2013) Algal biofuel: bountiful lipid from *Chlorococcum* sp. proliferating in municipal wastewater. Curr Sci 105:47–55
Mansour MP, Volkman JK, Blackburn SI (2003) The effect of growth phase on the lipid class, fatty acid and sterol composition in the marine dinoflagellate, Gymnodinium sp. in batch culture. Phytochem 63(2):145–153
Martinez E, Camacho F, Jimenez JM, Espinola JB (1997) Influence of light intensity on the kinetic and yield parameters of *Chlorella pyrenoidosa* mixotrophic growth. Process Biochem 32(2): 93–98
Maxwell J, Douglas A, Eglinton G, McCormick A (1968) The Botryococcenes—hydrocarbons of novel structure from the alga *Botryococcus braunii,* Kützing. Phytochem 7(12):2157–2171
Meireles LA, Guedes AC, Malcata FX (2003) Lipid class composition of the microalga Pavlova lutheri: eicosapentaenoic and docosahexaenoic acids. J Agri Food Chem 51(8):2237–2241
Metzger P, Villarreal-Rosales E, Casadevall E (1991) Methyl-branched fatty aldehydes and fatty acids in *Botryococcus braunii.* Phytochem 30(1):185–191
Mogens H (2008) Biological wastewater treatment: principles modelling and design. In: Henze M, van Loosdrecht MCM, Ekama GA, Brdjanovic D (eds) ISBN: 9781843391883. Published by IWA Publishing, London, UK
Montingelli ME, Tedesco S, Olabi AG (2015) Biogas production from algal biomass: a review. Ren Sust Ene Rev 43:961–972
Murata M, Nakazoe J (2001) Production and Use of marine aigae in Japan. JARQ 35(4):281–290
Naveen BP, Mahapatra DM, Sitharam TG, Sivapullaiah PV, Ramachandra TV (2016) Physico-chemical and biological characterization of urban municipal landfill leachate. Environ Poll. 220, Part A:2–12. http://dx.doi.org/10.1016/j.envpol.2016.09.002
Neilson AH, Blankley WF, Lewin RA (1973) Growth with organic carbon and energy sources. In: Stein JRD (ed) Handbook of phycological methods.Cambridge University Press Cambridge, pp 275–285
Nevenzel J (1989) Biogenic hydrocarbons of marine organisms. Marine Biog Lipids, Fats Oils 1:3–72
Ogawa T, Terui G (1972) Growth kinetics of spirulina platensis in autotrophic and mixotrophic cultures. Proceedings of IV IFS: fermentation technology today, pp 543–549
Orus MI, Marco E, Martinez F (1991) Suitability of chlorella vulgaris UAM 101 for heterotrophic biomass production. Bioresour Technol 38:179–184
Oswald W, Gotaas H (1957) Photosynthesis in sewage treatment. Trans Amer Soc Civil Eng, United States, p 122
Oswald WJ (1988) Large-scale algal culture systems (engineering aspects). Micro-algal biotechnology. Cambridge University Press, Cambridge, pp 357–394
Palmer CM (1974) Algae in American sewage stabilization's ponds. Rev Microbiol (S-Paulo) 5:75–80
Pittman JK, Dean AP, Osundeko O (2011) The potential of sustainable algal biofuel production using wastewater resources. Bioresour Technol 102:17–25. https://doi.org/10.1016/j.biortech.2010.06.035

Qin J (2010) Hydrocarbons from algae. In: Handbook of hydrocarbon and lipid microbiology, Springer, pp 2817–2826

Ramachandra TV, Mahapatra DM, Asulabha KS, Sincy V, Bhat S (2014) Integrated wetlands ecosystem: Sustainable model to mitigate water crisis in Bangalore (2014) ENVIS technical report-76. Environmental Information System, CES, IISc, Bangalore, p 560012

Ramachandra TV, Mahapatra DM (2015) Science of carbon footprint analysis. In: The handbook of carbon footprint assessment CRC press, Taylor & Francis Group., pp 1–44

Ramachandra TV, Mahapatra DM, Bhat S, Joshi NV (2015) Biofuel production along with remediation of sewage water through algae. In: Algae and environmental sustainability, vol 7. Developments in Applied Phycology, Springer series, pp 33–51

Ramachandra TV, Mahapatra DM, Karthick B, Gordon R (2009) Milking diatoms for sustainable energy: biochemical engineering vs. gasoline secreting diatom solar panels. Ind Eng Chem 48 (19):8769–8788

Ramachandra TV, Mahapatra DM, Samantray S, Joshi NV (2013) Biofuel from urban wastewater: scope and challenges. Ren Sus Ene Rev 21:767–777

Rasmussen RS, Morrissey MT (2007) Marine biotechnology for production of food ingredients. Adv Food Nut Res 52:237–292

Ravindra P (2000) Value-added food: single cell protein. Biotechnol Adv 18(6):459–479

Renault F, Sancey B, Badot P, Crini G (2009) Chitosan for coagulation/flocculation processes—an eco-friendly approach. Eur Polym J 45:1337–1348. https://doi.org/10.1016/j.eurpolymj.2008.12.027

Ryckebosch E, Bermúdez SPC, Termote-Verhalle R, Bruneel C, Muylaert K, Parra-Saldivar R, Foubert I (2014) Influence of extraction solvent system on the extractability of lipid components from the biomass of *Nannochloropsis gaditana*. J Appl Phycol 26(3):1501–1510

Salim S, Bosma R, Vermue MH, Wijffels RH (2010) Harvesting of microalgae by bioflocculation. J Appl Phycol 23:849–855

Sambusiti C, Bellucci M, Zabaniotou, Beneduce L, Monlau F (2015) Algae as promising feedstocks for fermentative biohydrogen production according to a biorefinery approach: A comprehensive review. Ren Sust Ene Rev 44:20–36

Scott SA, Davey MP, Dennis JS, Horst I, Howe CJ, Lea-Smith DJ, Smith AG (2010) Biodiesel from algae: challenges and prospects. Current Opi Biotecnol 21(3):277–286

Simas-Rodrigues C, Villela HD, Martins AP, Marques LG, Colepicolo P, Tonon AP (2015) Microalgae for economic applications: advantages and perspectives for bioethanol. J Expt Botany, 130

Singh A, Olsen SI (2011) A critical review of biochemical conversion, sustainability and life cycle assessment of algal biofuels. Appl Ene 88(10):3548–3555

Sivakumara G, Xu J, Thompson RW, Yang Y, Smith PR, Weathers PJ (2012) A review on integrated green algal technology for bioremediation and biofuel. Bioresour Technol 107:1–9

Spolaore P, Joannis-Cassan C, Duran E, Isambert A (2006) Commercial applications of microalgae. J Biosci Bioeng 101(2):87–96

Suresh B, Ravishankar GA (2004) Phytoremediation: a novel and promising approach for environmental cleanup. Crit Rev Biotechnol 24:97–124

Tam NFY, Wong YS (1989) Wastewater nutrient removal by *Chlorella pyrenoidosa* and *Scenedesmus* sp. Environ Poll 58:19–34

Tannenbaum SR, Mateles RI (1968) Single cell protein. Science 4(5):87

Tannenbaum SR, Wang DI (1975) Single-cell protein. MIT Press Cambridge, MA

Tseng C (2001) Algal biotechnology industries and research activities in China. J Appl Phycol 13 (4):375–380

Wang G, Wang T (2012) Characterization of lipid components in two microalgae for biofuel application. J Amer Oil Chem Soc 89(1):135–143

Wang L, Min M, Li Y, Chen P, Chen Y, Liu Y, Wang Y, Ruan R (2010) Cultivation of green algae *Chlorella* sp. in different wastewaters from municipal wastewater treatment plant. Appl Biochem Biotechnol 162(4):1174–1186

Ward AJ, Lewis DM, Green FB (2014) Anaerobic digestion of algae biomass: a review. Algal Res 5:204–214
Xu L, Zhao Y, Doherty L, Hu Y, Hao X (2016) The integrated processes for wastewater treatment based on the principle of microbial fuel cells: a review. Crit Rev Env Sci Technol 46(1):60–91
Zhang Ch-C, Jeanjean R, Joset F (1998) Obligate phototrophy in cyanobacteria: more than a lack of sugar transport. FEMS Microbiol Rev 161:285–292
Zhang XW, Zhang YM Chen F (1998) Kinetic models for phycocyanin production by high cell density mixotrophic culture of the microalga *Spirulina platensis*. J Ind Microbiol 21:283–288
Zheng Y, Li T, Yu X, Bates PD, Dong T, Chen S (2013) High-density fed-batch culture of a thermos tolerant microalga *Chlorella sorokiniana* for biofuel production. Appl ene 108: 281–287
Zhou WG, Cheng Y, Li Y, Wan Y, Liu Y, Lin X, Ruan R (2012) Novel fungal palletization assisted technology for algae harvesting and wastewater treatment. Appl Biochem Biotechnol 167(2):214–228

Chapter 17
The Pretreatment Technologies for Deconstruction of Lignocellulosic Biomass

Manali Kapoor, Surbhi Semwal, Ruchi Gaur, Ravindra Kumar, Ravi P. Gupta and Suresh K. Puri

Abstract Owing to the finite supply of fossil fuels, greenhouse gasses emission, global warming, increasing price, and unexpected fluctuations, there is a need to pay attention for alternative energy resources and thus interest in ethanol which is renewable, environmentally sustainable, and economically viable fuel has been strengthened. Due to economic and environmental concerns cropped up with the use of the first-generation ethanol processes, second-generation ethanol processes which comprise the use of waste biomass, viz., agricultural crop residues, municipal solid waste, sludge, livestock manure, etc., has been contemplated to be the hot emerging field. However, due to many technological issues, development of an effective technology is still a challenge. This chapter, therefore, provides insight into the pretreatment technologies involved in the production of free sugars which can be fermented to ethanol along with discussion on the merits and demerits of each of the technologies and their future prospects. This chapter also deals with various biomass-related issues and the updated technology status along with commercial aspects.

Keywords Pretreatment · Biomass · Dilute acid · Steam explosion
Fermentable sugars · Ethanol

17.1 Introduction

Consumption of energy resources has increased tremendously and anticipated to increase by 0.7–1.2% yearly for the next 20 years. About 64% of the increase in demand is contributed by transportation fuel. Owing to many issues like: (i) greenhouse gasses (GHG) emission; (ii) limited supply; (iii) augmenting price with unanticipated fluctuations which are linked with the use of fossil fuels, interest

M. Kapoor · S. Semwal · R. Gaur · R. Kumar (✉) · R. P. Gupta · S. K. Puri
Research & Development Centre, Indian Oil Corporation Limited, Sector-13, DBT-IOC Centre for Advance Bioenergy Research, Faridabad 121007, India
e-mail: kumarr3@indianoil.in

R. R. Singhania et al. (eds.), *Waste to Wealth*, Energy, Environment, and Sustainability, https://doi.org/10.1007/978-981-10-7431-8_17

in alternative, renewable, and sustainable biofuels such as bioethanol has been aroused (Mood et al. 2013; Silveira et al. 2015). Ethanol being a bio-based and oxygenated (35% oxygen) compound enables better hydrocarbon oxidation compared to gasoline resulting in reduction of GHG emissions (Balat 2007). To be employed in internal combustion engines as a fuel, it may either be applied as such or mixed with gasoline in varying amounts. High octane number makes it appropriate to use in a normal gasoline engine as can be easily mixed with gasoline (Li et al. 2005). In Brazil, ethanol obtained from sugarcane is used as such or mixed with gasoline in a 24% ratio to obtain gasohol, whereas in different areas of USA, 10% ethanol is added to gasoline to obtain E10. In several other countries like China, Columbia, Peru, Paraguay, Australia, and Thailand, blend of ethanol, i.e., E10, is prevalent while in India it is E5 (Dias De Oliveira et al. 2005; Malça and Freire 2006).

Ethanol which is commonly available in the market is produced from starchy or sugary crops via fermentation and is commonly mentioned as the first-generation ethanol (Naik et al. 2010). Its main application is in the transport sector where it is progressively used throughout the world as a substitute to fossil fuels. The main disadvantage is that the feedstock used for its generation is used as human food, and thus increase in the production of these fuels can result in the rising food prices, thereby leading to ethical discussions on food versus fuel. There is also a considerable debate about the sustainability of it as this can influence environment and carbon balances. Therefore, it is highly required to exploit the inedible materials for huge scale supplies of ethanol in the future (Laursen 2006; Gray and Zhao 2006).

In this respect, second-generation ethanol which involves the use of lignocellulosic biomass (LCB), i.e., woody crops, agricultural residues are interesting alternatives as these are cheap, renewable, and readily available (Mood et al. 2013). There have been large case studies which state that ethanol derived from LCB can decrease GHG emissions by more than 80% (Soam et al. 2016). Since it is a waste material, it has no competition with the food crop and has socioeconomic benefits and hence very relevant in the current context, however, ethanol production from LCB is very complicated unlike that from first-generation ethanol, e.g., from corn starch (which involves simple hydrolysis followed by fermentation) or sugarcane juice (direct fermentation). Owing to complex structure of biomass, it needs to be deconstructed to make the cellulose easily available to enzymes so that it can be easily hydrolyzed to fermentable sugars. There are many technological issues associated with the pretreatment which has limited the commercialization of biomass to ethanol refinery. In spite of the commissioning of 100 pilot plants and about 10 demonstration and commercial plants globally over the last decade, cellulosic ethanol production technologies have struggled universally to be economically and technologically feasible.

This book chapter, therefore, provides insight into the pretreatment technologies involved in the production of free sugars which can be fermented to ethanol along with discussion on the merits and demerits of each of the technologies and their future prospects. This chapter also deals with various biomass-related issues and the updated technology status along with commercial aspects.

17.2 Lignocellulosic Biomass

Lignocellulosic biomass (LCB) comprises agricultural waste material, i.e., wheat straw, rice straw, corn stover, and sugarcane bagasse; forest crops which include softwood and hardwood and energy crops, i.e., salix, switchgrass (Kumar et al. 2009). The LCB has a very complex architecture consisting of three predominant structural units which are cellulose, hemicellulose, and lignin and also contains slight quantities of extractives, ash, and pectin. The characteristic features of main components of lignocellulosic biomass are given in Fig. 17.1. Cellulose, a polymer of D-glucose subunits is connected via β-(1,4)-glycosidic bonds, whereas hemicelluloses, polymer of xylose, arabinose, and mannose, etc., consist of small branches at the backbone which are connected via β-(1,4)-glycosidic bonds and periodically β-(1,3)-glycosidic bonds (Kumar et al. 2009; Brodeur et al. 2011). The cellulose chains are connected jointly by bonds such as Van der Waals and hydrogen and are bundled into microfibrils, and these microfibrils are in turn bound to each other by hemicellulose and lignin. In lignin, which is a complex 3-D polymer, phenylpropane unit is present as the predominant building block. p-hydroxyphenyl propanol, (Coumaryl alcohol), guaiacyl propanol (coniferyl alcohol), and syringyl alcohol (sinapyl alcohol) are the major monomers present.

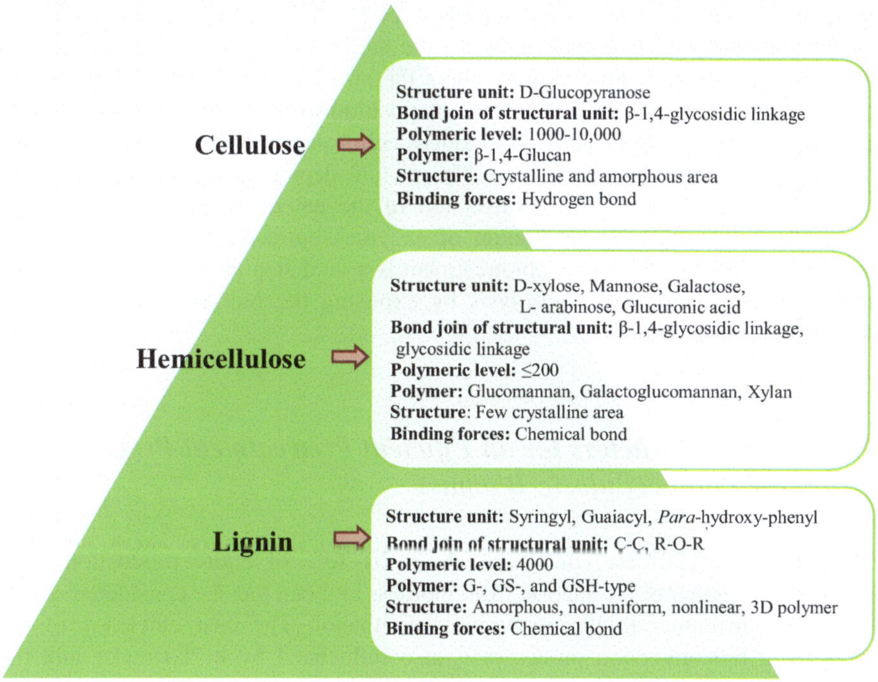

Fig. 17.1 Characteristic features of different components of lignocellulosic biomass

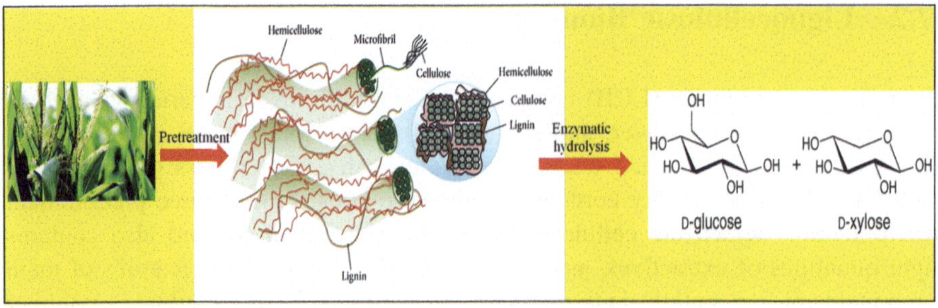

Fig. 17.2 Deconstruction of biomass into sugars

Due to its complex structure, lignin gives structural support, inflexibility and makes the cellulose fibrils extremely difficult to be attacked by both cellulases and microbes. The lignin also hampers enzymatic hydrolysis as the enzyme binds irreversibly onto lignin and thus not available to act on the cellulose chains (Kumar et al. 2009; Kapoor et al. 2017a; b).

17.3 Pretreatment

Pretreatment is the foremost step employed for ethanol production from LCB followed by hydrolysis of holocellulose to the sugars, fermentation of sugars, and purification of ethanol. Pretreatment alters the size, structure of the biomass and brings out the changes in its chemical composition to attain high sugar yields by easy carbohydrate hydrolysis. It also reduces cellulose crystallinity, augment pores in biomass, and lignin/hemicellulose. The picture depicting the deconstruction of biomass is given in Fig. 17.2. Conversion of the native biomass to free sugars would not only require huge amount of enzymes but would also result in low hydrolysis yields (<20%). Thus, pretreatment is a vital step to make easy access of cellulase during enzymatic hydrolysis by exposing the biomass (Brodeur et al. 2011; Kapoor et al. 2017a).

17.3.1 Crucial Factors for an Efficient Pretreatment Process of Lignocellulosic Biomass

The pretreatment process constitutes about 20–30% of entire production costs involved in the ethanol production from biomass and can thus be considered one of key cost contributors to the economics of ethanol. The cost efficiency of the downstream bioconversion processes is also influenced by it. To make sure the pretreatment is an effective process; several crucial factors need to be considered (Alvira et al. 2010).

1. Effective pretreatment method should have low capex which can be accomplished by avoiding the use of costly materials (catalyst, solvents, reagents, and biomass) during pretreatment and consequent neutralization (Bhutto et al. 2017).
2. Since pretreatment consumes substantial amount of energy in the ethanol process, it is essential to keep the demand of energy small while maintaining the process performance (Alvira et al. 2010).
3. Production of highly digestible pretreated solids which on enzymatic hydrolysis results in high sugar yields (>90%) in fewer than 3 days with as little as 10 FPU/g cellulase loadings (Yang and Wyman 2008).
4. Sugar recovery during pretreatment should be good, i.e., C5 sugar losses should be minimized.
5. Minimium amounts of inhibitors should be generated. High pretreatment severity can lead to incomplete degradation of hemicellulose and production of toxic compounds obtained from decomposition of sugar which could influence the enzymatic hydrolysis and fermentation steps. Hydroxymethyl furfural (HMF) results from the degradation of glucose and xylose, respectively. These on further degradation can form formic acid and levulinic acid. Low amount of inhibitor formation would result in direct use of the pretreated biomass for hydrolysis without employing washing/separation.
6. To establish adequate ethanol concentrations, manage recovery and other downstream costs, the sugars concentration from the collective pretreatment and enzymatic hydrolysis operations ought to be >10% (Bhutto et al. 2017).
7. Lignin should also be recoverable to be converted into important products to realize the biorefinery concept.

17.4 Pretreatment Methods

Various methods such as physical pretreatment (mechanical comminution, extrusion, ultrasound, and microwave), chemical pretreatment (dilute acid, alkali, organic solvents, ionic liquids, and ozonolysis), physicochemical pretreatment (steam pretreatment/autohydrolysis, ammonia, wet oxidation, and liquid hot water), and biological pretreatment have been explored for LCB pretreatment (Alvira et al. 2010; Kumar and Sharma 2017).

17.4.1 Physical Pretreatment

This is normally considered as the first step of pretreatment with an objective to decrease the biomass particle size. Moreover, it results in expanding the accessible surface area, reduces polymerization degree, and causes decrystallization of the biomass (Kumar et al. 2009).

17.4.1.1 Mechanical Comminution

Mechanical comminution which basically includes grinding, chipping, and/or milling techniques, alters the innate structure of biomass and crystallinity, and thereupon makes it susceptible to attack by cellulase (Alvira et al. 2010). Chipping lowers the particle size to 10–30 mm, whereas 0.2–2 mm size is achieved on milling or grinding. Thus, different physical method results in varying particle size. Milling and grinding dramatically trim down the size of biomass and crystallinity of cellulose owing to the shear forces induced at the time of milling, whereas chipping decreases the mass and heat transfer effects. Ball milling, hammer milling, two-roll milling, wet disk milling, and colloid milling are numerous categories of milling which result in improved enzymatic hydrolysis. The power required for size diminution is decided by biomass characteristics and its size. Oil palm frond fiber and empty fruit bunch resulted in glucose yields of 87 and 70%, respectively, whereas xylose yields obtained were 81.6 and 82.3%, respectively, when pretreated through ball mill (Mood et al. 2013; Zakaria et al. 2014).

Due to the low energy consumption of wet disk milling, it has been found to be more economical than ball milling; however, the glucose and xylose yields were lower than ball milling for most of the LCB. The major shortcoming of the milling process is that the amount of energy which is consumed during pretreatment is much more than the biomass theoretical energy content. The inefficiency of milling in removing the lignin which impedes the cellulase accessibility to cellulose by irreversibly binding on cellulose is also a disadvantage of this method. Thus, this method is expensive and cannot be used as a stand-alone method for pretreatment. This method is most implausible to be used at a commercial scale, unless coupled with other pretreatment methods.

17.4.1.2 Extrusion

In extrusion process, heating followed by mixing and shearing is applied on biomass which changes its physical and chemical structure when it makes a way through the extruder. The biomass structure is disrupted by the screw speed and barrel temperature as the fibrillar structure of biomass is broken down, and fibers are shortened and this in turn result in the increasing accessibility of cellulose to cellulases. Single and twin screw extruders have been extensively used to pretreat variety of LCB, ultimately resulting in high enzymatic hydrolysis (Alvira et al. 2010). Reducing sugar yields of 28.2% was obtained from the switch grass which was pretreated with the screw speed (200 rpm) and barrel temperature (75 °C), whereas big bluestem is pretreated at similar speed but temperature of 150 °C resulted in 66.2% and prairie cord grass produced 49.2% at 100 °C and 150 rpm (Karunanithy et al. 2013). Pretreatment of soybean hulls (in-barrel moisture, 40% wet basis) at screw speed (350 rpm), temperature (80 °C) resulted in 94.8% glucose conversion after enzymatic hydrolysis. This method has an advantage of huge shear, speedy stirring, mild barrel temperature, and brief residence time. This method does not result in HMF and furfural formation; hence, no washing/detoxification is required. Due to probability of continuous operation, versatility to process modification, and scale-up, extrusion can be easily customized for commercial-scale operation (Yoo et al. 2011; Zheng and Rehmann 2014).

17.4.1.3 Ultrasound

This pretreatment involves treating LCB with ultrasonic waves in the range of 10 kHz–20 MHz which generate both chemical and physical effects and change the anatomy of LCB. These effects augment the separation of bonds between the hemicellulose and lignin and degrade lignin-like compounds by attacking hydroxyl groups of the phenolic ring. Small cavitation bubbles which are formed during pretreatment also help in the separation and depolymerization of carbohydrates and hence deconstruct the biomass and increase availability of cellulase to cellulose. Thus, ultrasonic pretreatment enhances the saccharification of cellulose. Several crucial factors determine the success of this pretreatment: frequency, power, ultrasonic mode (continuous or pulse), temperature, solvent used, aeration, reactor design, reactor configuration, and biomass characteristics. The character and the harshness of the ultrasonic effects are determined by the frequency and power settings as frequency in the range of 10–100 kHz effectively causes cell rupture and holocellulose degradation (Bhutto et al. 2017; Kumar and Sharma 2017; Gogate et al. 2011).

Ultrasonic pretreatment is found to be more efficient for smaller particle size ranges as compared to larger particle size biomass. As different LCB responds differently to ultrasonic pretreatment; therefore, the conditions which have been optimized for one biomass may be somewhat different for others. Coupling of ultrasound with other technologies such as oxidative species addition, peroxide and ozone has the possibility of enhancing ultrasonic pretreatment and reducing the cost considerably. Even though ultrasound treatment is familiar as a feasible option on a laboratory scale, to consider it for the large scales, the operating parameters should be thoroughly considered and optimized for development of an effective pretreatment technology.

17.4.1.4 Microwave

In this method, biomass is immersed in chemical reagents (dilute) and biomass slurry is then exposed to microwave radiation for 5–20 min time duration (Alvira et al. 2010). Microwave result in the physical, chemical, and biological processes by producing heat and broad collisions due to dielectric polarization. Thus, this pretreatment changes the innate organization of biomass by breaking cellulose, degrading/partially eliminates lignin and hemicelluloses and ultimately leading to enhanced cellulose digestibility. Addition of dilute alkali reagent such as NaOH has been found to be very effective for this pretreatment. This pretreatment can be easily conducted and consumes very low energy as compared to conventional heating methods. Apart from these advantages, short processing time, minimum production of inhibitors, high uniformity, and selectivity make it a widely used method for pretreatment. It can be easily carried out in small equipment in short residence time and thus requires very low initial investment (Bhutto et al. 2017; Zhu et al. 2006). It is not feasible at commercial level.

17.4.2 *Chemical Pretreatment*

17.4.2.1 Organosolv Pretreatment

This method involves treating the biomass with an organic solvent. Inorganic/organic acid or base catalysts can be added as catalyst to improve the method. Acetone, ethanol, ethylene glycol, methanol, and tetrahydrofurfuryl alcohol are the most generally used organic solvents. The inorganic acids used include hydrochloric acid, sulfuric acid, and organic acids, i.e., acetylsalicylic acid, salicylic, and oxalic acid. In this process, if pretreatment is conducted without catalyst then temperature in the range of 180–210 °C is employed. However, addition of catalyst makes it feasible to perform reaction at lower temperature (below 175 °C). After pretreatment, pretreated solid is washed with ethanol or methanol and then subjected to water wash. After draining the organic part, organic solvent is evaporated and condensed. Further, recycling of the solvent is performed. The lignin is precipitated from the condensed black liquid by adding water and the filtrate thus obtained consists of hemicellulose sugars (mainly xylose). Thus, three independent fractions are obtained in organosolv pretreatment: arid lignin, an aqueous solution consisting of hemicellulose part, and purer cellulose (Mesa et al. 2011; Zhao et al. 2009).

This method results in solubilization of lignin and breaking of hemicellulose bonds and thus provides biomass material which is much more amenable to hydrolysis. This method is very effective for pretreatment of biomass containing higher percentage of lignin, e.g., softwoods. The following reactions get accomplished during organosolv pretreatment: (i) interior bonds present in lignins and those of lignin-hemicellulose part get hydrolyzed; (ii) hemicellulosic glycosidic bonds and some of these bonds in cellulose get hydrolyzed varying with the pretreatment conditions; (iii) degradation of the sugars into furfural and HMF (catalyzed by acid) succeeded by condensation reactions which occurs within lignin and degraded parts. There are numerous advantages of this method, e.g., (i) pure lignin generated during the process; (ii) easy recovery of solvent by distillation owing to low boiling nature of ethanol, methanol; (iii) fractionation into arid lignin, an aqueous hemicellulose part, and a comparatively purer cellulose (Bhutto et al. 2017). On treating sugarcane bagasse with ethanol (30%) at 195 °C for 60 min, 29.1% reducing sugars was produced (Mesa et al. 2011). Acetone–water (1:1 molar ratio) pretreatment for *Pinus radiata* conducted at pH 2.0, 195 °C, for 5 min gave ethanol yield of about 99.5% (Araque et al. 2008).

There are many disadvantages associated with this method, i.e., (i) high cost of organic solvents; (ii) requirement of specially equipped vessels to avoid leaks and prevent fires during use of volatile organic liquid at high temperature; (iii) cost involved in recycling solvents which could otherwise be toxic to enzymatic hydrolysis and fermenting organisms; (iv) to prevent dissolved lignin from precipitating, washing of the pretreated biomass with organic solvent is mandatory before water washing. Thus, this method is too expensive to be used for the pretreatment and commercial implication is not feasible unless technology is properly developed.

For organosolv pretreatment development in the future, focus ought to be on combined usage of biomass parts and reduction of pretreatment expenses (accomplished by decreasing the quantity of solvent utilized). Several other features, such as reduction of energy and consumption of chemicals, increase value of by-products can also be incorporated to improve the entire method (Zhao et al. 2009).

17.4.2.2 Acid Pretreatment

In this pretreatment, high sugar yields are achieved by solubilizing hemicelluloses, disorganizing lignin arrangement, decreasing cellulose degree of polymerization (DP), thereby increasing the cellulose accessibility to cellulase. Inorganic acids, i.e., H_2SO_4, HCl, HNO_3, H_3PO_4, and organic acid, i.e., HCOOH, CH_3COOH, and CH_3CH_2COOH are used (Pierre et al. 2015). Conventional method involves using concentrated H_2SO_4 and HCl (30–70%) for hydrolysis of LCB into monosaccharides. Although the pretreatment can be completed at very low temperature (T < 100 °C), the drawbacks, viz, (i) formation of high amount of inhibiting compounds (furfural, HMF, phenolic acids, and aldehydes) and (ii) metallurgy issues for the equipment due to corrosion issues necessitating further expenditure on operation and maintenance make this process less attractive.

These problems have been solved by using dilute acid (DA) which has been used for pretreatment of broad range of LCB, i.e., switchgrass, corn stover, poplar, rice straw, and *Acacia mangium* (Semwal et al. 2016; Xu et al. 2009; Du et al. 2010), etc. Dilute acid pretreatment commonly uses dilute sulfuric acid (H_2SO_4) at concentrations below 4% (w/w), high temperatures (120–210 °C), and pressures to achieve shorter reaction time (sec to min) and are thus appropriate for continuous operations. Plug flow, batch percolation, shrinking-bed, flow-through, and counter current reactors are various types of reactors which are in use for this method (Taherzadeh and Karimi 2008).

The mechanism is based on cleavages of β-1,4-glycosidic bonds initiated by the transmission of protons through biomass matrices and rapid protonation of glycosidic oxygen bonds between sugar monomers. Hemicelluloses hydrolysis, mainly xylan to xylose and other sugars which further give rise to monomers (furfural, HMF) and other (volatile) products are the major reaction that develops during this pretreatment. Hemicellulose is hydrolyzed by acid as particularly C–O bonds between sugar molecules get cleaved via the protonation of glycosidic bonds or pyranic oxygens. As hydrogen bonds in hemicellulose are not strong enough, hemicellulosic sugars are preferentially converted to 2-furaldehydes and then to 2,5-anhydride intermediates in the following order of reactivity: xylose > arabinose > mannose > galactose > glucose. Combined severity factor (CSF) predicts the severity of pretreatment process conditions and relates to the holocellulose and inhibitors formation. It is a function of temperature, reaction time, pH and calculated by the following formula (Eq. 1), where "t," "T," and "T_R" are the reaction time (min), operating temperature (°C), reference temperature(100 °C), respectively, and "pH" is that of the pretreatment hydrolyzate (Pedersen and Meyer 2010).

$$\text{Log } R_0 = \text{Log}_{10}\left(t * \exp\frac{(T - T_R)}{14.75}\right) - \text{pH} \quad (1)$$

The oligomers formed from hemicelluloses are inhibitory to cellulases and should also be hydrolyzed into monosaccharides by adjusting pretreatment conditions. During acid pretreatment, lignin is solubilized as lignosulphonates, condense, and precipitate in acidic environments (Semwal et al. 2016; Dutta et al. 2012).

Dilute acid pretreatment is suitable for continuous production due to rapidity and no need for recycling and thus is the most feasible, inexpensive, and effective for industrial scale. Furthermore, essential nutrients such as S and P are released on using <1% w/v H_2SO_4/H_3PO_4 that enhance fermentation step. It is also used to commercially produce furfural from cellulosic materials. Although consumption of low acid is a major advantage in terms of cost and process severity, the method needs relatively high temperature and pressure. This causes formation of degradation products which have negative influence on both hydrolysis and fermentation. Even though this problem can be circumvented by the use of washed pretreated biomass, it would not be economically practical owing to loss of the sugars and requirement of huge amount of water. Thus, it is desirable to take forward the whole slurry directly for enzymatic hydrolysis, but this would necessitate the need for neutralization using acid (Zhang et al. 2012).

Another important factor in the process of acid pretreatment is the high-solid loadings, which refers to the percent amount of biomass solids in the pretreatment mixture. By increasing the initial solid loading, it is possible to increase the sugar concentration to obtain increased ethanol yields while decreasing operating costs and energy utilization in the final product recovery steps. There are not enough studies on high-solids loading (>15% solids concentrations) regarding pretreatment, saccharification, and fermentation to determine the minimal concentration to produce economically feasible ethanol such as in terms of energy requirements for distillation. Dilute acid pretreated rice straw at 25% solid loadings yielded 83.3 and 31.9 g/L, glucose, and xylose post-enzymatic saccharification. The conditions, viz., 0.35wt% acid, 162 °C temperature for 10 min residence time were optimized in continuous pilot-scale horizontal reactor (Kapoor et al. 2017).

With the aim to reinforce the DA pretreatment, certain catalysts have proved quite useful. The mechanism involves targeting lignin thereby opening up the lignin–carbohydrate network. In this direction, sulfite pretreatment has earned notice as a method for wood species as free phenolic hydroxyl groups are created owing to sulfonation of lignin at C-α of the side chain which breaks the linkages (α-O-4 and β-O-4) between phenyl propane units. Thus, lignosulfonate portion and residual insoluble sulfonated lignin are formed which have enhanced hydrophilicity (Semwal et al. 2016).

Although high xylose production during pretreatment and high glucan hydrolysis during subsequent enzymatic hydrolysis are associated advantages of DA pretreatment but another inhibitor "pseudo-lignin" becomes a matter of concern. Under high severity condition, pseudo-lignin is formed during DA pretreatment which significantly inhibits enzymatic hydrolysis of cellulose. During DA pretreatment, polysaccharides (e.g., cellulose and xylan) hydrolysis results in the production of monosaccharides (e.g., glucose and xylose), which further get converted to HMF and furfural. Pseudo-lignin is formed by the further rearrangements of furfural and/or HMF (Hu et al. 2012).

Table 17.1 displays the process condition, advantages, and disadvantages of acidic catalysts. POET-DSM, Abengoa (USA) has set up commercial plants based on DA technology (Table 17.2).

Table 17.1 Description of acidic catalyst

	Concentrated H_2SO_4	**Dilute H_2SO_4**	**Organic acid**
Process Condition	*30% (w/v), ambient to moderate temperature for several hours	*0.5-5% (w/v), high temperature for a few minutes	*0.5-5%(w/v), high temperature for a few minutes
Advantages	* enzyme not required; *Low temperature and pressure	*High reaction rate; *Applicability of combined severity factor	*Less toxicity; *Less formation of inhibitory compounds
Limitations	*Catalyst toxicity; *Catalyst corrosiveness	*High temperature; *Inhibitor formation	*Expensive catalysts; *Pretreatment effectiveness

Table 17.2 Commercial 2G ethanol plant in world (http://www.etipbioenergy.eu/?option=com_content&view=article&id=273; http://www.praj.net/ethanol-plant.html)

Commercial ethanol plant	Location	Capacity MG/yr	Pretreatment technology	Feedstock used
DuPont	USA	30	Ammonia	Corn stover
Beta Renewables	Italy	20	Steam explosion	Wheat straw, rice straw, and Arundo donax
POET-DSM	USA	20	Acid	Corn stover and corn cob
Abengoa	USA.	25	Acid	Wheat straw
Raizen	Brazil	10	Steam explosion	Baggase and sugar cane straw
Granbio	Brazil	20	Steam explosion	Baggase and sugar cane straw

17.4.2.3 Alkaline Pretreatment

Alkali pretreatment targets lignin component of the biomass, thereby disrupting the ester and glycosidic side chains. KOH, NaOH, $Ca(OH)_2$, and NH_4OH, etc., are the frequently employed catalysts in this pretreatment. As a result of pretreatment, lignin structural alteration, acetyl removal, swelling, and partial decrystallization of cellulose with limited solubilization of hemicelluloses is achieved. Consequently, plant biomass exhibits higher cellulose accessible surface area to enzyme (Alvira et al. 2010).

The two important reaction mechanisms, i.e., solvation and saponification occur during alkaline pretreatment. Solvation causes the swelling of lignocellulosic cell wall, thus increases the internal surface area. Meanwhile, the alkali agent produces a charged carboxyl group as uronic ester linkages (4-O-methyl-D-glucuronic acids) which are connected to xylan backbone undergo saponification. It also cleaves lignin–hemicelluloses bonds and ultimately led to destruction of cellulose–hemicellulose–lignin matrix (Himmel et al. 1994). Therefore, the pore structures of lignocellulose are increased because of disappearance of the connecting bonds, thus making the lignocellulose components easily available for enzymatic and microbial degradation. This pretreatment has been used for various LCB, i.e., switch grass, rice straw, sugarcane bagasse, sugarcane tops, dendrocalamus, corn stover, and poplar wood (Sharma et al. 2013; Cheng et al. 2010; Sindhu et al. 2014), etc. The two main chemicals, i.e., lime, NaOH, and $Ca(OH)_2$ are mainly applied to alkaline pretreatment of LCB. Dilute NaOH pretreatment was established to be efficient for straws containing lignin contents in the range of 10–18% (Millet et al. 1976); however, it was found not to be efficient for softwoods having lignin content >26%.

Alkali pretreatment is usually executed at relatively lower pressure and temperature compared to acid and steam explosion pretreatment and usually degrades less sugar than acid pretreatment and thus consumes lower energy during pretreatment. The alkali pretreatment normally takes longer time if carried out at ambient conditions. Another limitation is alkali losses as salts get incorporated into the biomass and thus cannot be recovered. Substantial quantities of water are also required for washing step. It is essential to neutralize biomass before setting up enzymatic hydrolysis, and this makes downstream processing complicated. Further this would also increase the scale-up cost (Alvira et al. 2010).

17.4.2.4 Ozonolysis

This method involves the use of oxidant, i.e., ozone gas to lessen the biomass lignin content and thus results in the improved enzymatic hydrolysis. Biomass moisture content is the important factor which governs the pretreatment by ozone gas as moisture content is inversely related to lignin oxidization potential. This method has been investigated in many LCB, i.e., bagasse, wheat straw, cotton straw, peanut, pine, poplar sawdust. The attractive feature of this method is that it is generally conducted at ambient temperature and pressure. This method does not give rise to inhibitory compounds which can influence the later reactions, i.e., hydrolysis and fermentation; however, the application at industrial level is faced with the challenge of the requirement of high amount of ozone for pretreatment. To employ it commercially, research is in progress for production of industrially viable ozone concentrations (Mood et al. 2013; Kumar et al. 2009; Kumar and Sharma 2017).

17.4.2.5 Ionic Liquids (ILs)

The associated drawbacks of chemical/physical pretreatment methods have resulted in a surge to look for milder and environmental-friendly pretreatment agents. While, no pretreatment technology can be called a winner due to the different composition and constitution of LCBs, ILs have emerged as green solvents to address the issue of environmental safety and economic viability of pretreatment. Ionic liquids are endowed with the ability to dissolve cellulose due to their special composition involving organic cation and inorganic anion. As opposed to the regular volatile organic solvents, i.e., the ones used in organosolv process, ILs do not suffer from limitations related to biodegradability, toxicity, hydrophobicity, viscosity, electrochemical, and thermal stability. Moreover, nearly 100% recovery of used IL is feasible owing to various combinations of anion and cation. Besides, ILs posses high reaction rates, low volatility and are inflammable (Brandt et al. 2013).

Ionic liquids exert their dissolving effect by forming electron- donor–acceptor (EDA) complexes breaking the inter-chain hydrogen bonds in cellulose. The solubilizing power of ILs may be explained by the Kamlet–Taft (KT) parameters which are measured using UV–Vis spectra of certain dyes dissolved in an ionic liquid of interest. The solubilizing property of IL is significantly affected by the choice of its cation. Length of cation alkyl chain progressively reduces cellulose solubility. Presence of hydroxyl groups on the alkyl chains or on the anion functionalities decrease the solubility of cellulose owing to increase in IL hydrogen-bond acidity. Thus, IL which has comparatively small, non-coordinating cations, and small hydrogen-bonding anions is most appropriate for cellulose owing to its bi-functionality (amphiphilicity). The biomass type, particle size, time and temperature of dissolution are various other factors which affect the level of biomass solubilization (Sun et al. 2009).

After the dissolution of biomass, an antisolvent is added to precipitate the cellulose. The regenerated biomass thus obtained contains mostly cellulose with reduced lignin and hemicellulose contents. The recovered cellulose after IL pretreatment has the same degree of polymerization as the initial cellulose, but significantly reduced degree of crystallinity. Besides cellulose, chemical composition and supramolecular organization changes in lignin structure are also modified during the pretreatment. In general, as compared to herbaceous and hardwood lignocelluloses, softwood (pine) feedstocks undergo lesser delignification and hemicellulose removal. Compositional and structural changes in pretreated biomass during IL pretreatment improved the hydrolysis yield up to 50-fold (Dadi et al. 2006). Micro-structural modification of wheat straw and mustard stalk pretreated by [C_2mim][OAc] resulted in 97% glucose yield (Raj et al. 2016).

Ionic liquids have undoubtedly opened up new avenues for pretreatment, separation and fractionation but practical use of ILs for the proficient exploitation of LCB still poses many challenges. The high cost of ILs, recycling requirement, and lack of toxicological data are the critical issues. Besides, an insight into the mode of action on hemicellulose and lignin along with the addressal to inhibitor generation needs attention. Moreover, financial support for research on such issues is much required.

17.4.3 Physicochemical Pretreatment

17.4.3.1 Steam Explosion (SE)

One of the most extensively operated pretreatment for LCB involves subjecting biomass to pressurized steam in the range of 20–50 bar and 170–250 °C for several sec to few mins, and then abruptly depressurized. As the pressure is quickly lowered, detachment of fibers occurs due to the explosive decompression. Due to the high temperatures, autohydrolysis takes place which produces acetic acid resulting from hemicellulosic acetyl groups; moreover, water at high temperatures also operates as an acid. Thus, this pretreatment involves both chemical effects and mechanical forces. Apart from incomplete hydrolysis/solubilization of hemicellulose, the lignin is also removed slightly from the biomass. Hemicelluloses removal surges enzyme availability by exposing cellulose surface (Gaur et al. 2017).

This technology has been used for variety of agricultural residues/herbaceous biomass and hardwoods, i.e., olive residues, poplar, herbaceous residues (corn stover and wheat straw) (Brandt et al. 2013). The overall sugar recovery (pretreatment hydrolysate + enzymatic hydrolysate) has been found to be improved by the addition of acid (H_2SO_4), SO_2, etc. (Brodeur et al. 2011). Steam explosion of rice straw at 180 °C and 10 min resulted in total saccharification yield of 64%; however, prior soaking with 0.5% H_2SO_4 improved the yield to 87% (Sharma et al. 2015).

Careful optimization of steam explosion is required as its efficiency is influenced by particle size, temperature, and residence time. Similar to DA pretreatment, the term severity factor (Ro) has been used to describe the coupled consequence of both time and temperature and expressed as in Eq. 1. High temperatures can not only result in an increased hemicellulose removal which can enhance cellulose hydrolysis, but can also promote higher sugar degradation resulting in the formation of fermentation inhibitors. The major inhibitors produced are furan derivatives (HMF and furfural resulting from hexoses and pentoses degradation, respectively), phenolic compounds, and weak acids. Due to high temperature, lignin also undergoes chemical transformation, redistributed, and redeposited as lignin droplets, and this so called pseudo-lignin is detrimental to enzymatic hydrolysis. However, if temperature is low, partial destruction of lignin–carbohydrate matrix can cause soluble lignin components to condense/precipitate and thus produce less digestible biomass. Therefore, temperature is a crucial process parameter and need to be carefully optimized.

Optimization of SE process parameters can be carried out in a batch mode on a pilot scale as these reactors are cheap, versatile, need relatively small amounts of biomass and simple to manage. For demo scales, continuous plants are typically used as can work with high feeding rates. As compared to other pretreatment technologies, steam explosion process offers several interesting features and these are (i) lesser capital investment, lower energy consumption; (ii) less risky chemicals and conditions employed except for softwoods which require acid catalyst; (iii) almost complete sugar recovery; (iv) no recycling cost and lower environmental impact; (v) possibility of using large particle size for biomass (Alvira et al. 2010).

However, there are several limitations of SE such as (i) partial degradation of hemicelluloses (ii) side products formed could inhibit the subsequent steps; (iii) in softwoods, owing to little acetyl content in the hemicellulosic portion, acid catalyst (SO_2 or H_2SO_4) addition is essential for getting higher sugar yields, and this could lead to inhibitory compounds formation; (iv) biomass washing before enzymatic hydrolysis may result in the decrease of the overall hydrolysis yields by 20–25% due to elimination of soluble sugars; (v) detoxification methods required for the removal of inhibitory compounds (Agbor et al. 2011).

Due to its salient features, steam explosion has been successfully checked in both laboratory and pilot processes by several researchers and enforced at demonstration scale (Bacovsky et al. 2010). Commercial plants based upon steam explosion technology have been set up by Beta Renewables (Table 17.2).

17.4.3.2 Liquid Hot Water (LHP)

Liquid hot water, a hydrothermal treatment involves the use of water as a reaction medium at high temperatures (180–220 °C) and pressure (up to 5 MPa) to keep it in the liquid state (Mood et al. 2013). It is distinct from SE as rapid decompression is not required and does not involve any catalysts or chemicals, and this makes it more advantageous as compared to any physicochemical/chemical method. High hemicellulose sugar recovery and generation of hydrolysate containing low fermentation inhibitors also makes it more favorable to SE. The LHW results in the solubilization of hemicelluloses and partially lignin removal and thus makes the cellulose more accessible. Inhibitors formation is prevented by maintaining pH between 4 and 7 during pretreatment. This is done to avoid the formation of monomer as in this pH range; sugars are in form of oligomers. Similarly to SE, pretreatment process results in the slurry which consists of solid fraction (cellulose enriched biomass) and aqueous fraction (predominantly solubilized hemicelluloses), but unlike SE little or no inhibitors are formed in this.

This pretreatment has been effectively investigated for variety of biomass including corn stover, corn cob, sugarcane bagasse, rye straw, and wheat straw (Mood et al. 2013). LHW of wheat straw under optimized condition resulted in 91% glucose and 80% xylose yields (Perez et al. 2008). Essential advantages of LHP are (i) low cost of reactor due to complete exclusion of chemicals; (ii) no need for recycling of material; (iii) elimination of washing step owing to lower formation of inhibitory components; (iv) lower temperature operation minimizes the energy consumption; (v) lower concentration of degraded products due to higher water input.

In spite of it being high cost savings potential and environmental-friendly technique, it has not yet been employed beyond pilot-scale investigation due to its associated disadvantage of huge amount of energy requirement in downstream processing owing to involvement of huge quantities of water (Agbor et al. 2011).

17.4.3.3 Ammonia Pretreatment

Ammonia fiber/freeze explosion (AFEX), soaking aqueous ammonia (SAA), and ammonia recycle percolation (ARP) involve the use of liquid ammonia to pretreat biomass. AFEX is similar to SE, as a physicochemical process, but conducted at lower (<90 °C) temperatures. ARP on the other hand is carried out at higher temperatures. A customized version of AFEX, i.e., SAA utilizing aqueous ammonia has been put forward to treat biomass in order to reduce the liquid throughput during pretreatment. It is carried out in batch reactor at moderate (25–60 °C) to high temperatures (150–190 °C). In the AFEX and ARP processes, the LCB is treated with ammonia at a given temperature and high pressure. The resulting biomass with reduced lignin content and swollen cellulose has altered crystallinity. This results in the increased reactivity of the remaining carbohydrates rendering the pretreated biomass susceptible to hydrolysis. Compared to other pretreatment methods, ammonia treatment achieved close to theoretical glucose yields after enzymatic hydrolysis and that too at reduced enzyme loadings (Foster et al. 2001).

AFEX pretreatment involves contacting the biomass with anhydrous liquid ammonia in a closed vessel at ammonia to biomass loading as 1:1 or 1:2 for 10–60 min at pressures >3 MPa and temperature range 60–90 °C. The batch is held at the intended temperature for about 5 min followed by opening of vent valve leading to explosive decompression. Apparently, swelling of cellulose is brought about by the chemical effect of ammonia under pressure, thereby increasing the biomass accessible surface area. Along with the partial solubilization of hemicellulose as xylo-oligomers, transition of cellulose I to cellulose III takes place. Interestingly, the lignin distribution remains reasonably the same after AFEX; however, rigorous alteration in the structure of lignin is achieved. Thus, combined physical and chemical changes affect accessibilities of cellulose to the cellulase markedly increasing the susceptibility of the pretreated LCB to subsequent enzymatic hydrolysis (Galbe and Zacchi 2007). Since it is a mild process, it can minimize the production of sugar degradation products and fermentation inhibitors.

While, AFEX operates in batch mode, ARP employs ammonia in a packed bed reactor. In a flow-through column reactor maintained at high temperatures (150–180 °C), aqueous ammonia (5–15%) percolates through biomass at a flow rate (1–5 ml/min) for a residence time (10–90 min). After pretreatment is completed, ammonia is recycled or recovered (Kim et al. 2003). During this process, simultaneous removal of lignin and hydrolysis of hemicelluloses occurs along with a concomitant decrease in cellulose crystallinity. AFEX and ARP are effective methods for agricultural residues, herbaceous plants, and municipal solid waste, whereas ARP pretreatment can be applied for hardwoods too.

Advantages of AFEX/ARP include, (i) compatibility of pretreatment hydrolysate with fermentation organisms; (ii) increased feasibility at continuous industrial scale owing to easy recovery and recyclability; (iii) less formation of fermentation inhibitors owing to moderate temperatures (<100 °C), pH (<12.0), and short residence time compared to other physicochemical processes; (iv) high selectivity for lignin removal (70–85%); (v) reduction of water required for washing as superheated ammonia vapor can be used to shred the remaining ammonia in the pretreated biomass; (vi) availability of residual ammonia as a nitrogen source during fermentation (as recovery is close to 99%); (vii) economic viability due to price of ammonia being about one-fourth (on a molar basis) of sulfuric acid used in acid pretreatment (Kim et al. 2003).

Associated limitations of AFEX/ARP are: (i) less efficiency over high lignin containing lignocellulosic biomass (18–30% lignin) and softwoods; (ii) the ammonia cost; (iii) environmental concerns with the odor of ammonia at pilot as well as industrial-scale applications (Himmel et al. 1994).

The use of AFEX and ARP has not been reported beyond laboratory scale. Ammonia soaking technology has been successfully demonstrated by ICT (Mumbai, India) in a batch wise, at 4 ton/day feed at IGL, Kashipur which is in the process of being scaled-up to 10 tons per day, feed basis on a continuous scale.

Commercial plants based upon Ammonia technology have been set up by DuPont (Table 17.2).

17.4.3.4 Wet Oxidation (WO)

This method employs water and oxygen/air as catalyst and usually occurs at temperature >120 °C and pressures in the range of 0.5 and 2 MPa for about 30 min. Water behaves like an acid on raising temperature above 170 °C and thus catalyzes hydrolytic reactions. WO results in the fractionation of biomass as organic acid formed during the pretreatment process causes solubilization/hydrolysis of hemicelluloses, delignification, and oxidative reactions. Temperature, oxygen pressure, and reaction time are the three crucial factors which affect the efficiency of wet oxidation. Addition of chemical agents like alkaline peroxide and sodium carbonate has been found to reduce the reaction temperature, improves hemicellulose solubilization, and decreases inhibitory components formation (i.e., furfurals). Alkaline WO treatment (195 °C, 15 min) of sugarcane bagasse resulted in ~93% hemicelluloses solubilization and 50% lignin removal, thereby increasing the cellulose content to 70% and ultimately producing 75% sugar yield on enzymatic hydrolysis (Kumar and Sharma 2017; Martin et al. 2007).

Some of the interesting features of WO are (i) the use of inexpensive and readily oxidizing agents, e.g., available air or oxygen; (ii) decomposition of lignin to carbon dioxide, water, and carboxylic acids; (iii) formation of lower amounts of inhibitors, as compared to SE and LWH pretreatments. However, there are very fewer chances of this pretreatment reaching industrial scale due to immense expense of the hydrogen peroxide, requirement of expensive reactor material owing to high temperature and pressure and the flammable character of the pure oxygen (Bajpai 2016).

17.4.4 Biological Pretreatment

It comprises of microorganism specially fungi usage to increase the accessibility of LCB for enzymatic hydrolysis. Brown, white, and soft-rot fungi primarily degrade lignin, hemicellulose, and partially cellulose. White and soft rots basically attack both cellulose and lignin, whereas brown rots chiefly attack cellulose. Out of all fungi, white-rot fungi is the most useful and their action has been found to be due to presence of peroxidases and laccase. Diverse range of white-rot fungi such as *Pleurotus ostreatus, Ceriporia lacerata, Cyathus stercoreus, Ceriporiopsis subvermispora, Pycnoporus cinnabarinus,* and *Phanerochaete chrysosporium* have been investigated on various biomasses (Kumar and Wyman 2009). The main effects of different microorganism on various biomasses are given in Table 17.3. There are various parameters which influence this pretreatment such as biomass composition, particle size, moisture, and characteristics. Apart from this some other

Table 17.3 Main effects of biological pretreatment on various biomasses using different microorganism

Microorganism	Biomass	Main effects	References
P. ostreatus/ P. pulmonarius	Eucalyptus grandis saw dust	Hydrolysis increased by 20 times	Castoldi et al. (2014)
Fungal consortium	Straw	Hydrolysis increased by seven times	Taha et al. (2015)
Irpex lacteus	Corn stalks	Hydrolysis yield 82%	Du et al. (2011)
Ceriporiopsis subvermispora	Corn stover	Reducing sugar yield increased by 2–3 times	Wan and Li (2011)
Punctualaria sp. TUFC20056 s	Bamboo culms	50% delignification	Suhara et al. (2012)
Phanerochaete chrysosporium	Oil palm empty fruit bunch	Klason lignin content reduced 5.89%	Hamisan et al. (2009)

factors are: microorganism involved, concentration of inoculum, pH, aeration rate, time and temperature of incubation.

Advantages of this method are (i) environmental-friendly process; (ii) low energy demand; (iii) no discharge of deadly compounds to environment and no waste matter production throughout the process; (iv) fermentation inhibitors formation avoided during the process. Despite of its advantages, industrial application is normally ruled out as it requires continuous monitoring of microorganism growth. Long processing time and requirement of large space for functioning are other disadvantages of this method (Wyman et al. 2005). However, optimizing the pretreatment with the best strain and culture conditions is the right step for the future development of this process.

17.5 Conclusions

Lignocellulosic biomass (LCB) of non-food non-fodder value is both abundantly and cheaply available and thus finds application in the bioethanol production. However, the greatest complexity lies in the intimate structural organization of LCB components to harness fermentable sugars. Pretreatment of the LCB is of utmost importance to facilitate the release of sugars during enzymatic hydrolysis. Although pretreatment can be precious in terms of input such as time and energy; the advantages of upcoming strategies to surmount recalcitrance are massive. Different methods of pretreatment have been defined and widely studied for improving the digestibility of biomass. These bring out different kinds of alteration in the chemical and physical structure of the biomass which are outlined in Table 17.4. The main objective of an efficient pretreatment technology is to augment fermentable sugar recovery, reduce inhibitors generated during pretreatment, reduce consumption of chemicals and energy effort, generate valuable by-products, and reduce overall bioethanol cost. Owing to the difference in the composition of biomass, one method available for deconstruction of one biomass may not be applicable for another biomass. Table 17.5 presents application of different pretreatment technologies on same biomass resulting in different sugar yields. Same biomass responds differently to different pretreatments and the optimized conditions need to be derived for individual biomass. For energy crops like switch grass, all pretreatment methods look almost equally well efficient, while for agriculture residues like cotton stalk (much closer to hardwood species) require combined pretreatment strategies for overcoming recalcitrance.

Out of these many pretreatment techniques, physical and biological methods have a long way to go and intensive research would have to be conducted to make these commercially viable. Chemical and physicochemical methods appear to be the most efficient and hopeful technologies for industrial applications; however, chemical methods generate various by-products that inhibit subsequent biocatalytic conversion steps. Various inhibitors which are formed during chemical processes and different approaches employed to overcome these are given in Table 17.6.

Some of the major players to commercial cellulosic ethanol are Dupont (Nevada, Iowa), Beta Renewables (Tortona, Italy), POET-DSM (Emmetsburg, Iowa) in USA, Raizen/Iogen (Piracicabana) in Brazil, and Iogen Corporation in Canada, Granbio (Alagoas) [IEA bioenergy task 42 report, 2014], and most of these technologies are either based on DA pretreatment or steam explosion (Table 17.2). These technologies are, however, based on specific biomass and hence could not be preferred for the all kinds of biomass. IL pretreatment technology which can be applicable for wide range of biomass is very expensive, but development of cost effective recovery methods would offer a great potential for future.

Table 17.4 Alteration in the characteristic features of lignocelluloses biomass due to different pretreatment technologies (Alvira et al. 2010; Bhutto et al. 2017)

Characteristics	Mechanical	Acid	Alkaline	Lime	SE	AFEX	LHW	ARP
Cellulose decrysallization	***	n.d	n.a	n.d	n.a	***	n.d	***
Hemicellulose solubilization	n.a	***	*	**	***	**	***	**
Lignin removal	n.a	**	**	***	**	***	*	***
Increase in accessible surface area	***	***	***	***	***	***	***	***
Lignin structure alteration	n.a	***	***	***	***	***	**	***
Formation of toxic compound	n.a	***	*	**	***	*	*	**

*** High effect; ** Moderate effect; * Low effect; *n.d* Not determined; *n.a* Not applicable

Table 17.5 Various biomass feedstocks pretreated by different pretreatment methodologies

Biomass (feedstock type)	Pretreatment methods					Reference
	DA	Alkali	SE	IL	AFEX/SAA/APR	
Rice straw (agricultural residue)	0.5% acid; 121 °C/60 min; glucose yield = 359 mg/g 0.35% w/w acid, 162 °C and 10 min; glucan conversion = 72%	NaOH = 4%, w/v; 55 °C/(1–3 h); Glucan conversion = 39.2%	DA (1%), w/w/soaking at RT followed by SE at 180 °C/10 min; glucan conversion = 87%	[EMIM][OAc]; 45 °C/72 h; glucan conversion = 79.7%	Soaking in aq ammonia (20.93%); temp = 42.75 °C for 48 h; glucan conversion = 87%	Kapoor et al. (2017), Cheng et al. (2010), Sharma et al. (2015)
Mustard stalk (close to hardwood)	DA = 1% sulfuric acid (w/v); 160 °C/20 min; glucose yield: 780 mg/g	NaOH = 2%, w/v; 121 °C/2 h; glucose yield: 550 mg/g	100% steam at 200 °C (~15 bar)/10 min; glucose yield: 60 mg/g	[C2mim][OAc] = 90%, w/w; Temp = 130 °C; Time = 2 h; Glucan conversion = 97%	–	Raj et al. (2016), Kapoor et al. (2015)
Cotton stalk (close to hardwood)	Sulfuric acid = 3.5%, w/v; temp = 135 °C, time = 2 h, glucose yield = 64.36 mg/g; Sulfuric acid = 2%, w/v; temp = 121 °C/90 min; glucan conversion = 23.85%	NaOH = 3.5%, w/v; ultrasound waves at 420 W; time = 90 min, 25 °C; sugar yield = 167.03 mg/g NaOH = 2–3.0%, w/v; temp = 121 °C, 130 kPa, time = 40–90 min; sugar yield = 271.7 mg/g-60.8%	DA (1%, w/v) soaking at room temp. followed by SE at 200 °C for 10 min; glucan conversion = 68%	EMIMac = 90%, w/w; 150 °C 30 min, glucan conversion = 65%	–	Gaur et al. (2017), Wang et al. (2016)

(continued)

Table 17.5 (continued)

Biomass (feedstock type)	Pretreatment methods					Reference
	DA	Alkali	SE	IL	AFEX/SAA/APR	
Corn stover (agricultural residue)	Sulfuric acid = 0.75%, 160 ° C/0–5 min	NaOH: Na_2SO_3 = 1:1; 160 °C/30 min; glucan conversion = 90%		[EMIM] $[Ac]^-$ = 96%, w/w; 110 °C, 90 min; glucan conversion = 69%	Soaking with 15 wt% ammonia at 5 ml/min flow rate; 170 °C/ 90 min; glucan conversion = 92.5%	Mood et al. (2013), Kim et al. (2003)
Wheat straw (agricultural residue)	Sulfuric acid = 10%, w/w;100 °C, 30 min; glucan conversion: NR	NaOH = 5%, w/w, 100 °C, 30 min; glucan conversion: NR	100% steam; 180 °C, 20 min; glucan conversion: NR	[Ch][Tau] = 90;, w/ w; 80 °C/6 h; conversion = 79.7%		Eisenhuber et al. (2013), Ren et al. (2016)
Switchgrass (energy crop)	Sulfuric acid = %, w/ v; temp = 140 °C, time = 40 min, glucose yield = 83%	NaOH = 15.4%, w/w; temp = 130 °C for 30 min		[C2mim] [OAc] = 97%, w/w; 160 °C/3 h; g lucan conversion = 96%	Anhydrous NH_3 = 152 g/100gv; temp = 150 °C, time = 30 min, glucan conversion = 83% Aqueous NH_3 = 15%, w/v; temp = 160 °C, time = 60 min, glucan conversion = 92%	Garlock et al. (2011), Karp et al. (2015)

Table 17.6 Inhibitory compounds along with possible solutions to tackle inhibition problems (Jönsson and Martín 2016; Jönsson et al. 2013)

Inhibitors generated using various chemical methods	Techniques employed	Methods
1. Acid-base methods (Aliphatic carboxylic acids, phenolic compounds, furans, etc.) 2. Mild alkaline methods (Acetic acid, hydroxy acids, dicarboxylic acids, phenolic compounds) 3. Oxidative methods (Aldonic and aldaric acids, furoic acid, phenolic acids, acetic acid)	Physical	Liquid–liquid extraction: ethyl acetate, supercritical fluid extraction (such as supercritical CO_2), trialkylamine
		Liquid–solid extraction: activated carbon, ion exchange, and lignin
		Evaporation
		Heat treatment
	Chemical	Alkali ($Ca(OH)_2$, NaOH, NH_4OH)
		Reducing agents (i.e., dithiothreitol, dithionite, sulfite)
	Biological	Enzymes: laccase, peroxidase
		Microbes: *Coniochaeta ligniaria, Trichoderma reesei, Ureibacillus thermosphaericus*

Immense deal of research needs to be undertaken to understand the relationship between physical and chemical structure of biomass with the pretreatment. Therefore, it is highly required to have basic and applied research of each pretreatment process to craft its alliance with the rest of the processes to utilize the full prospective of the biomass for realizing our dream of bioethanol production.

Acknowledgements The authors would like to thank the Department of Biotechnology (DBT) and Indian Oil Corporation for supporting the writing work.

References

Agbor VB, Cicek N, Sparling R, Berlin A, Levin DB (2011) Biomass pretreatment: fundamentals toward application. Biotechnol Adv 29(6):675–685

Alvira P, Tomás-Pejó E, Ballesteros MJ, Negro MJ (2010) Pretreatment technologies for an efficient bioethanol production process based on enzymatic hydrolysis: a review. Bioresour Technol 101(13):4851–4861

Araque E, Parra C, Freer J, Contreras D, Rodríguez J, Mendonça R, Baeza J (2008) Evaluation of organosolv pretreatment for the conversion of Pinus radiata D. Don to ethanol. Enzyme Microb Technol 43(2):214–219

Bacovsky D, Dallos M, Wörgetter M, Task IB (2010) Status of 2nd generation biofuels demonstration facilities in June 2010. IEA bioenergy task 39: commercializing 1st and 2nd generation liquid biofuels from biomass 39:1–26

Bajpai P (2016) Pretreatment of lignocellulosic biomass for biofuel production. Springer, Singapore

Balat M (2007) Global bio-fuel processing and production trends. Energ Explor Exploit 25:195–218

Bhutto AW, Qureshi K, Harijan K, Abro R, Abbas T, Bazmi AA, Karim S, Yu G (2017) Insight into progress in pre-treatment of lignocellulosic biomass. Energy 122:724–745
Brandt A, Gräsvik J, Hallett JP, Welton T (2013) Deconstruction of lignocellulosic biomass with ionic liquids. Green Chem 15(3):550–583
Brodeur G, Yau E, Badal K, Collier J, Ramachandran KB, Ramakrishnan S (2011) Chemical and physicochemical pretreatment of lignocellulosic biomass: a review. Enzyme Res 2011
Castoldi R, Bracht A, de Morais GR, Baesso ML, Correa RCG, Peralta RA, Moreira RFPM, Polizeli MT, de Souz CGM, Peralta RM (2014) Biological pretreatment of Eucalyptus grandis sawdust with white-rot fungi: study of degradation patterns and saccharification kinetics. Chem Eng J 258:240–246
Cheng YS, Zheng Y, Yu CW, Dooley TM, Jenkins BM, VanderGheynst JS (2010) Evaluation of high solids alkaline pretreatment of rice straw. Appl Biochem Biotechnol 162:1768–1784
Dadi AP, Varanasi S, Schall CA (2006) Enhancement of cellulose saccharification kinetics using an ionic liquid pretreatment step. Biotechnol Bioeng 95(5):904–910
Dias De Oliveira ME, Vaughan BE, Rykiel EJ (2005) Ethanol as fuel: energy, carbon dioxide balances, and ecological footprint. AIBS Bulletin 55:593–602
Du B, Sharma LN, Becker C et al (2010) Effect of varying feedstock-pretreatment chemistry combinations on the formation and accumulation of potentially inhibitory degradation products in biomass hydrolysates. Biotechnol Bioeng 107:430–440
Du W, Yu H, Song L, Zhang J, Weng C, Ma F, Zhang X (2011) The promising effects of by-products from Irpex lacteus on subsequent enzymatic hydrolysis of bio-pretreated corn stalks. Biotechnol Biofuels 4:37
Dutta S, De S, Saha B, Alam MI (2012) Advances in conversion of hemicellulosic biomass to furfural and upgrading to biofuels. Catal Sci Technol 10:2025–2036
Eisenhuber K, Krennhuber K, Steinmüller V, Jäger A (2013) Comparison of different pre-treatment methods for separating hemicellulose from straw during lignocellulose bioethanol production. Energy Procedia 40:172–181
Foster BL, Dale BE, Peterson JBD (2001) Enzymatic hydrolysis of ammonia-treated sugar beet pulp. Appl Biochem Biotechnol 91–93:269–282
Galbe M, Zacchi G (2007) Pretreatment of lignocellulosic materials for efficient bioethanol production. Biofuels 108:41–65
Garlock RJ, Balan V, Dale BE, Pallapolu VR, Lee YY, Kim Y, Mosier NS, Ladisch MR, Holtzapple MT, Falls M, Sierra-Ramirez R (2011) Comparative material balances around pretreatment technologies for the conversion of switchgrass to soluble sugars. Bioresour Technol 102(24):11063–11071
Gaur R, Semwal S, Raj T, Lamba BY, Ramu E, Gupta RP, Kumar R, Puri SK (2017) Intensification of steam explosion and structural intricacies impacting sugar recovery. Bioresour Technol 241:692–700
Gogate PR, Sutkar VS, Pandit AB (2011) Sonochemical reactors: important design and scale up considerations with a special emphasis on heterogeneous systems. Chem Eng J 166:1066–1082
Gray KA, Zhao L (2006) Emptage M. Bioethanol Curr Opin Chem Biol 10:141–146
Hamisan AF, Abd-Aziz S, Kamaruddin K, Md. Shah UK, Shahab N, Hassan MA (2009) Delignification of oil palm empty fruit bunch using chemical and microbial pretreatment methods. Int J Agric Res 4:250–256
Himmel ME, Baker JO, Overend RP (1994) Enzymatic Conversion of Biomass for Fuels Production. American Chemical Society Washington, DC
Hu F, Jung S, Ragauskas A (2012) Pseudo-lignin formation and its impact on enzymatic hydrolysis. Bioresour Technol 117:7–12
Jönsson LJ, Martín C (2016) Pretreatment of lignocellulose: formation of inhibitory by-products and strategies for minimizing their effects. Bioresour Technol 199:103–112
Jönsson LJ, Alrikssonm B, Nilvebrant NO (2013) Bioconversion of lignocellulose: inhibitors and detoxification. Biotechnol Biofuels 6(1):16

Kapoor M, Raj T, Vijayaraj M, Chopra A, Gupta RP, Tuli DK, Kumar R (2015) Structural features of dilute acid, steam exploded, and alkali pretreated mustard stalk and their impact on enzymatic hydrolysis. Carbohy Polym 124:265–273

Kapoor M, Soam S, Agrawal R, Gupta RP, Tuli DK, Kumar R (2017a) Pilot scale dilute acid pretreatment of rice straw and fermentable sugar recovery at high solid loadings. Bioresour Technol 224:688–693

Kapoor M, Soam S, Semwal S, Gupta RP, Kumar R, Tuli DK (2017b) Impact of conditioning prior to dilute acid deconstruction of biomass for the production of fermentable sugars. ACS Sustain Chem Eng 5(5):4285–4292

Karp EM, Resch MG, Donohoe BS, Ciesielski PN, O'Brien MH, Nill JE, Mittal A, Biddy MJ, Beckham GT (2015) Alkaline pretreatment of switchgrass. ACS Sustain Chem Eng 3(7):1479–1491

Karunanithy C, Muthukumarappan K, Gibbons WR (2012) Effect of extruder screw speed, temperature, and enzyme levels on sugar recovery from different biomasses. ISRN Biotechnology 2013

Kim TH, Kim JS, Sunwoo C, Lee YY (2003) Pretreatment of corn stover by aqueous ammonia. Bioresour Technol 90:39–47

Kumar R, Wyman CE (2009) Effects of cellulase and xylanase enzymes on the deconstruction of solids from pretreatment of poplar by leading technologies. Biotechnol Prog 25(2):302–314

Kumar AK, Sharma S (2017) Recent updates on different methods of pretreatment of lignocellulosic feedstocks: a review. Bioresour Bioprocess 4(1):7

Kumar P, Barrett DM, Delwiche MJ, Stroeve P (2009) Methods for pretreatment of lignocellulosic biomass for efficient hydrolysis and biofuel production. Ind Eng Chem Res 48:3713–3729

http://www.etipbioenergy.eu/?option=com_content&view=article&id=273. Last Accessed 15 June 2017

http://www.praj.net/ethanol-plant.html. Last Accessed 15 June 2017

Laursen W (2006) Students take a green initiative. Chem Eng 774:32–44

Li DG, Zhen H, Xingcai L, Wu-gao Z, Jian-guang Y (2005) Physico-chemical properties of ethanol-diesel blend fuel and its effect on performance and emissions of diesel engines. Renew Energ 30:967–976

Malça J, Freire F (2006) Renewability and life-cycle energy efficiency of bioethanol and bio-ethyl tertiary butyl ether (bioETBE): assessing the implications of allocation. Energy 31:3362–3380

Martin C, Klinke HB, Thomsen AB (2007) Wet oxidation as a pretreatment method for enhancing the enzymatic convertibility of sugarcane bagasse. Enzyme Microb Technol 40:426–432

Mesa L, Gonzalez E, Cara C, González M, Castro E, Mussattoc SI (2011) The effect of organosolv pretreatment variables on enzymatic hydrolysis of sugarcane bagasse. Chem Eng J 168:1157–1162

Millet MA, Baker AJ, Scatter LD (1976) Physical and chemical pretreatment for enhancing cellulose saccharification. Biotech Bioeng Symp 6:125–153

Mood SH, Golfeshan AH, Tabatabaei M, Jouzani GS, Najafi GH, Gholami M, Ardjmand M (2013) Lignocellulosic biomass to bioethanol, a comprehensive review with a focus on pretreatment. Renew Sustain Energy Rev 27:77–93

Naik SN, Goud VV, Rout PK, Dalai AK (2010) Production of first and second generation biofuels: a comprehensive review. Renew Sustain Energy Rev 14(2):578–597

Pedersen M, Meyer AS (2010) Lignocellulose pretreatment severity-relating pH to biomatrix opening. New Biotechnol 27:739–750

Perez JA, Ballesteros I, Ballesteros M, Saez F, Negro MJ, Manzanares P (2008) Optimization liquid hot water pretreatment conditions to enhance sugar recovery from wheat straw for fuel-ethanol production. Fuel 87:3640–3647

Pierre JS, Duran L, Heiningen AV (2015) Fast pyrolysis of muconic acid and formic acid salt mixtures. J Anal Appl Pyrolysis 113:591–598

Raj T, Gaur R, Dixit P, Gupta RP, Kagdiyal V, Kumar R, Tuli DK (2016) Ionic liquid pretreatment of biomass for sugars production: driving factors with a plausible mechanism for higher enzymatic digestibility. Carbohy Polym 149:369–381

Ren H, Zong MH, Wu H, Li N (2016) Efficient pretreatment of wheat straw using novel renewable cholinium ionic liquids to improve enzymatic saccharification. Indu Eng Chem Res 55 (6):1788–1795

Semwal S, Gaur R, Mukherjee S, Chopra A, Gupta RP, Kumar R, Tuli DK (2016) Structural features of dilute acid pretreated *Acacia mangium* and impact of sodium sulfite supplementation on enzymatic hydrolysis. ACS Sustain Chem Eng 4:4635–4644

Sharma R, Palled V, Sharma-Shivappa RR, Osborne J (2013) Potential of potassium hydroxide pretreatment of switchgrass for fermentable sugar production. Appl Biochem Biotechnol 169:761–772

Sharma S, Kumar R, Gaur R, Agrawal R, Gupta RP, Tuli DK, Das B (2015) Pilot scale study on steam explosion and mass balance for higher sugar recovery from rice straw. Bioresource Technol 175:50–57

Silveira MH, Morais AR, da Costa Lopes AM, Olekszyszen DN, Bogel-Łukasik R, Andreaus J, Pereira Ramos L (2015) Current pretreatment technologies for the development of cellulosic ethanol and biorefineries. ChemSusChem 8:3366–3390

Sindhu R, Kuttiraja M, Binod P, Sukumaran RK, Pandey A (2014) Physicochemical characterization of alkali pretreated sugarcane tops and optimization of enzymatic saccharification using response surface methodology. Renew Energy 62:362–368

Soam S, Kapoor M, Kumar R, Borjesson P, Gupta RP, Tuli DK (2016) Global warming potential and energy analysis of second generation ethanol production from rice straw in India. Appl Energy 184:353–364

Suhara H, Kodama S, Kamei I, Maekawa N, Meguro S (2012) Screening of selective lignin-degrading basidiomycetes and biological pretreatment for enzymatic hydrolysis of bamboo culms. Int Biodeter Biodegr 75:176–180

Sun N, Rahman M, Qin Y, Maxim ML, Rodríguez H, Rogers RD (2009) Complete dissolution and partial delignification of wood in the ionic liquid 1-ethyl-3-methylimidazolium acetate. Green Chem 11(5):646–655

Taha M, Shahsavari E, Al-Hothaly K, Mouradov A, Smith AT, Ball AS, Adetutu EM (2015) Enhanced biological straw saccharification through co-culturing of lignocellulose degrading microorganisms. Appl Biochem Biotechnol 175:3709–3728

Taherzadeh MJ, Karimi K (2008) Pretreatment of lignocellulosic wastes to improve ethanol and biogas production: a review. Int J Mol Sci 9:1621–1651

Wan C, Li Y (2011) Effectiveness of microbial pretreatment by Ceriporiopsis subvermispora on different biomass feedstocks. Bioresour Technol 102(16):7507–7512

Wang M, Zhou D, Wang Y, Wei S, Yang W, Kuang M, Ma L, Fang D, Xu S, Du SK (2016) Bioethanol production from cotton stalk: a comparative study of various pretreatments. Fuel 184:527–532

Wyman CE, Dale BE, Elander RT, Holtzapple M, Ladisch MR, Lee YY (2005) Coordinated development of leading biomass pretreatment technologies. Bioresour Technol 96:1959–1966

Xu J, Thomsen MH, Thomsen AB (2009) Pretreatment on corn stover with low concentration of formic acid. J Microbio Biotechnol 19:845–850

Yang B, Wyman CE (2008) Pretreatment: the key to unlocking low-cost cellulosic ethanol. Biofuels Bioprod Bioref 2(1):26–40

Yoo J, Alavi S, Vadlani P, Amanor-Boadu V (2011) Thermo-mechanical extrusion pretreatment for conversion of soybean hulls to fermentable sugars. Bioresour Technol 102:7583–7590

Zakaria MR, Fujimoto S, Hirata S, Hassan MA (2014) Ball milling pretreatment of oil palm biomass for enhancing enzymatic hydrolysis. Appl Biochem Biotechnol 173:1778–1789

Zhang D, Ong YL, Li Z, Wu JC (2012) Optimization of dilute acid-catalyzed hydrolysis of oil palm empty fruit bunch for high yield production of xylose. Chem Eng J 181–182:636–642

Zhao X, Cheng K, Liu D (2009) Organosolv pretreatment of lignocellulosic biomass for enzymatic hydrolysis. App Microbiol Biotechnol 82(5):815

Zheng J, Rehmann L (2014) Extrusion pretreatment of lignocellulosic biomass: a review. Int J Mol Sci 15(10):18967–18984

Zhu S, Wu Y, Yu Z, Wang C, Yu F, Jin S, Ding Y, Chi R, Liao J, Zhang Y (2006) Comparison of three microwave/chemical pretreatment processes for enzymaric hydrolyisis of rice straw. Biosyst Eng 93:279–283

Chapter 18
Bioethanol Production from Sugarcane Green Harvest Residues Using Auxin-Assisted Pretreatment

Jegannathan Kenthorai Raman and Edgard Gnansounou

Abstract Bioethanol and biochemicals from agriculture residues are key alternative for transport fuels and fossil chemicals to environmental impact mitigation and rural economy. In biofuels and biochemicals production, pretreatment is the crucial and energy demanding process that enhances components hydrolysis. Any improvement in energy conservation in this process would be an advantage. In this study, auxin was used for sugarcane residues pretreatment and the advantage of their use over the conventional pretreatment was analyzed. The optimization of pretreatment process conditions using auxin followed by saccharification, fermentation, and surface area analysis shows that auxin could be a prospective agent for lignocellulose pretreatment at mild conditions (175 °C, 10.5 min and 2% auxin concentration). The optimized pretreatment conditions of auxin were less severe compared to optimized hot water pretreatment conditions that could lead to energy savings. Furthermore, the biodegradable nature of auxin could make the waste management easier in biorefineries.

Keywords Auxin · Bioethanol · Sugarcane green harvesting residues
Pretreatment · Saccharification · Fermentation

18.1 Introduction

World consumption of biofuels for transport such as bioethanol and biodiesel is gradually rising (REN21 2015), and several countries have announced policies to increase biofuel mandate in the transport sector. This indicates the keenness of society and the government to mitigate climate change and increase energy security. Most of the biofuels used now are first generation that compete with food due to

J. K. Raman (✉) · E. Gnansounou
Bioenergy and Energy Planning Research Group (BPE), Ecole Polytechnique Fédérale de Lausanne (EPFL), GC A3 444 (Bâtiment GC), ENAC IIC GR-GN, EPFL, Station 18, 1015 Lausanne, Switzerland
e-mail: jegannathan.kenthorairaman@epfl.ch

R. R. Singhania et al. (eds.), *Waste to Wealth*, Energy, Environment, and Sustainability, https://doi.org/10.1007/978-981-10-7431-8_18

land use and demand (Rulli et al. 2016). However, second-generation biofuels are being rigorously investigated and several commercial-scale operations were started in 2015 (Schwab et al. 2015). Second-generation biofuels mainly include bioethanol and biodiesel that are produced from non-food crop materials such as agriculture residues and jatropha. Among these, bioethanol has high potential to partially replace fossil fuel (gasoline) in transport sector due to the abundant availability of agriculture residues around the world (UNEP 2009).

The main constituents of agricultural residues are cellulose, hemicellulose, and lignin bound together to form a recalcitrant complex structure that demands several processes to access and convert the individual components. Pretreatment, enzyme saccharification, and fermentation are the processes involved in ethanol production from lignocellulosic biomass residues. There are several types of pretreatment techniques available such as acid, alkali, hot water, and organosolv to separate cellulose, hemicellulose, and lignin (Chaturvedi and Verma 2013; Sindhu et al. 2016). Each pretreatment technique has its own advantages and disadvantages. When dilute acid and hot water solution are used, the hemicellulose is removed and retained in the liquid fraction as C5 sugars. Whereas, the cellulose fraction remains in the solid fraction along with lignin. Later, the pretreated solid fraction is saccharified with cellulolytic enzymes to hydrolyze the cellulose to glucose. Lignin, resistant to pretreatment and saccharification, remains in the solid form that could be used as a burning fuel to provide energy to the processes.

In alkali and organosolv treatment, lignin is separated from biomass in the liquid fraction and both cellulose and hemicellulose remain in the pretreated solid fraction that are further enzymatically saccharified to C5 and C6 sugars. Both C5 and C6 sugars could be converted to ethanol or other desired biochemicals depending on the demand of the products and processes. Stand-alone second-generation production of ethanol is not economical in short term due to the competition from fluctuating fossil oil price. Therefore, the concept of producing multi-products along with ethanol called as biorefinery had received attention and several combinations of products are being investigated.

Pretreatment is the crucial step among the processes that determines the overall efficiency of the process. An ideal pretreatment should separate the constituents from biomass with minimum loss of the constituents and preferably at lower energy demand. Dilute acid and hot water pretreatment being more common, several chemicals are being researched in the pretreatment process such as aqueous ammonia (Kim et al. 2003), lime impregnation (Chang et al. 1998), ionic liquids (Liu et al. 2011), oxalic acid (Lee et al. 2011), maleic acid (Lee and Jeffries 2011), hydrogen peroxide–acetic acid (Wi et al. 2015), sulfite (Zhu et al. 2009), tetrahydrofuran (Cai et al. 2013), and phosphoric acid (Wang et al. 2016). As the trend for lignocellulose industry is toward production of bioethanol and biochemical products (such as furfural, succinic acid), the energy demand would be high in biorefinery and any improvement in energy conservation would be an advantage to the industry.

Auxins are growth regulators that stimulate cell elongation and cell division in plants (Sauer et al. 2013). They are used as herbicide (Grossmann 2007) and in plant cell cultures (Mori et al. 1994). The property of auxin to penetrate and stimulate cell elongation in biomass could be very useful in the pretreatment process that could favor reduction of the recalcitrance of biomass components at mild process conditions and thereby could reduce energy demand in biorefinery. In addition, auxins are biodegradable (OECD 1995; Boivin et al. 2005) which could ease management in wastewater treatment in biorefineries. The objective of this study is to use auxin as a novel chemical for pretreatment of sugarcane green harvest residues (GHR) to produce bioethanol, to optimize the pretreatment process parameters, and to compare auxin pretreatment with conventional pretreatment method. There are several auxins available in the market such as natural auxins (3-indoleacetic acid (IAA)) and synthetic auxins (2-4 dichlorophenoxyacetic acid (2,4-D) and 1-naphthaleneacetic acid (NAA)). However, 2,4-D that belongs to the phenoxyacetic acid family has been chosen in this.

18.2 Methods

18.2.1 Materials

GHR was provided by Stellenbosch University, South Africa, with size range 4–5 cm in a sealed polythene bag. The residues were dried in an oven at 40 °C overnight and further milled to pass through 80–20 mesh and stored in a zip lock bag until use at the Ecole Polytechnique Fédérale de Lausanne (EPFL). Sugars, ethanol, organic acid standards, cellulase (Cellulast® 1.5 L) and β-glucosidase (powder from almonds) enzymes and auxin (2,4-D) were procured from Sigma-Aldrich, Switzerland. Other chemicals used in this study were of analytical grade.

18.2.2 Compositional Analysis

Raw and pretreated GHR sample's composition was analyzed according to the National Renewable Energy Laboratory (NREL) standard procedures for biomass compositional analysis (Sluiter et al. 2010; NREL 2016). High-performance liquid chromatography (HPLC) (Agilent Technologies, Germany) equipped with an Aminex HPX-87P column and refractive index detector (RID) was used to analyze the sugars with Millipore water as the mobile phase at flow rate 0.5 ml/min, maintaining a column and detector temperature of 80 and 55 °C, respectively. Ethanol, furfural, hydroxymethylfurfural (HMF), and acetic acid were analyzed by same HPLC equipped with an Aminex HPX-87H column and RID detector with 5 mM sulfuric acid as the mobile phase at flow rate 0.6 ml/min maintaining a column and detector temperature of 65 and 55 °C, respectively. Moisture, ash, and insoluble lignin were analyzed by gravimetric method. Soluble lignin was analyzed using spectrophotometric method (Sluiter et al. 2011).

18.2.3 Auxin Pretreatment

The pretreatment process parameters optimization was based on response surface methodology (Table 18.1 and 18.2) using Design expert v 9 software with temperature range (175–250 °C) and residence time (1–20 min) as variable accounting to 13 sets of experiments. Auxin (2, 4-D) concentration of 2% w/v was used in all the pretreatment procedures. Experiments were carried out in a 300 ml pressure reactor (Parr, USA). Biomass: Auxin solution ratio (10% w/v) was taken in the reactor, and the mixture was soaked for 2 h in the reactor followed by heating at required temperature and residence time. After allowing the mixture to react at particular temperature and residence time, the mixture was cooled down to room temperature by circulating cold water into the cooling tube available in the reactor. Later, pH was measured that was in the range 3.3–3.8 (Table 18.1 and 18.2) and using vacuum filtration the liquid fraction and solid fraction in the mixture were separated. The solid fraction in the filter was washed with water to remove the auxin. The moist pretreated GHR and liquid fraction were stored in a refrigerator until further use. Sugars, organic acids, lignin, and ash composition of the pretreated solid fraction and liquid fraction were analyzed using the procedures mentioned above. For the liquid hot water pretreatment, experiments were carried out with the same procedure but with only water in place of auxin. To compare the effectiveness of two pretreatment processes, a combined severity factor (Overend and Chornet 1987) was used with the below formula considering the pretreatment temperature, time, and pH.

$$\mathrm{CSF} = \log[t * \mathrm{EXP}(t - 100/14.75)] - \mathrm{pH}$$

18.2.4 Saccharification and Fermentation

Filter paper assay and cellobiose assay were used for cellulase and β-glucosidase enzyme activity, respectively (Adney and Baker 2008; Zhang et al. 2009). Moist pretreated biomass (containing 0.6 g glucan) and citrate buffer (50 mmol, pH 4.8) were taken in a 50-ml conical flask with stopper cork. Cellulase enzyme (20 FPU/g glucan) and β-glucosidase enzyme (40 IU/g glucan) were added aseptically, and the conical flasks containing total saccharification mixture volume of 20 ml were incubated for 72 h at 50 °C and 150 rpm in a shaker incubator (Selig et al. 2008). After saccharification, the glucose-rich liquid was separated from lignin-rich solid fraction using vacuum filtration. The glucose present in the liquid fraction was analyzed to report the glucose yield (% theoretical maximum).

S. cerevisiae strain (Euroscarf BY4741) maintained in YPD agar slants was used for ethanol production from saccharified liquid in a batch anaerobic fermentation. The fermentation media was prepared using liquid fraction from saccharification (10 ml), yeast extract (1% w/v), and peptone (2% w/v) in a 50-ml conical flask with

Table 18.1 Pretreated biomass (solid fraction) composition after auxin treatment on various conditions based on the experimental design

Temp. (°C)	Time (min)	Combined severity factor	% Mass recovery	Glucan (% PT solid)	Xylan (% PT solid)	Lignin (% PT solid)	Glucan recovery (% initial)	Xylan recovery (% initial)	Lignin recovery (% initial)
175.0	1.0	−1.55	75.2	48.8	8.6	22.1	96.5	29.9	103.2
175.0	10.5	−0.42	72.5	51.6	6.1	23.0	97.4	20.3	102.7
212.5	1.0	−0.15	69.5	50.6	2.4	27.8	92.6	7.9	119.9
175.0	20.0	−0.12	71.8	51.1	5.1	23.9	97.5	17.3	107.6
193.8	5.8	−0.02	71.7	50.7	3.3	26.8	95.6	10.8	117.7
193.8	15.3	0.44	70.6	50.4	2.4	27.1	93.6	8.1	120.4
212.5	10.5	0.88	69.1	46.8	0.7	35.1	85.1	2.2	150.8
250.0	1.0	1.01	59.6	29.1	0.0	52.8	45.7	0.0	195.3
212.5	20.0	1.13	69.2	45.6	0.4	31.9	83.0	1.4	136.9
231.3	5.8	1.17	65.6	37.9	0.0	43.2	65.3	0.0	175.9
231.3	15.3	1.57	62.9	37.4	0.0	44.4	61.9	0.0	173.3
250.0	10.5	2.08	51.8	12.4	0.0	65.7	16.9	0.0	211.5
250.0	20.0	2.37	53.1	5.3	0.0	66.6	7.4	0.0	219.8

Table 18.2 Pretreated biomass (liquid fraction) composition after auxin treatment on various conditions based on the experimental design

Temp. (°C)	Time (min)	Combined severity factor	Glucose (mg/100 ml)	Xylose (mg/100 ml)	Acetic acid (mg/100 ml)	HMF (mg/100 ml)	Furfural (mg/100 ml)
175.0	1.0	−1.55	179.2	1550.6	201.4	4.1	182.7
175.0	10.5	−0.42	182.3	1504.7	238.1	4.9	267.2
212.5	1.0	−0.15	221.6	403.4	261.6	5.2	372.7
175.0	20.0	−0.12	211.6	1403.5	245.4	5.6	312.4
193.8	5.8	−0.02	214.6	930.9	249.7	5.7	401.5
193.8	15.3	0.44	176.7	752.4	259.1	6.6	483.0
212.5	10.5	0.88	206.5	15.7	275.2	10.2	490.6
250.0	1.0	1.01	123.1	1.2	244.6	17.3	199.0
212.5	20.0	1.13	162.1	3.6	279.7	11.8	430.1
231.3	5.8	1.17	95.6	2.1	286.8	14.6	411.7
231.3	15.3	1.57	83.5	2.2	292.5	15.8	363.6
250.0	10.5	2.08	48.7	1.5	260.1	12.7	194.5
250.0	20.0	2.37	27.9	2.8	206.9	21.6	137.4

stopper and sterilized at 121 °C for 20 min. *S. cerevisiae* strain (5 g/l) inoculum was added to the fermentation media aseptically, and the flasks were incubated anaerobically in a shaker incubator at 30 °C and 130 rpm for 48 h. After fermentation, the ethanol content was analyzed using HPLC to report the ethanol yield.

18.2.5 Optimization of Alternative Process

There are no common optimum conditions for different pretreatment processes. Therefore, to know the degree of advantage obtained from one pretreatment method, the optimum conditions for the individual method (in this case liquid hot water) have to be analyzed. Thus, a similar experimental design with the same process conditions range was also run for the liquid hot water pretreatment followed by saccharification and fermentation. The optimum process conditions of the two pretreatments were compared to know the advantage significance of one method over the other.

18.2.6 Surface Area Analysis

To get an insight into the effectiveness of pretreatment method on decrystallizing and breakdown of biomass in a particular process conditions, the surface area of dried untreated and pretreated biomass samples was analyzed using nitrogen (NOVA2200e, Quantachrome Instruments Co., USA) to know the impact of pretreatment on biomass.

18.2.7 Fourier Transform Infrared Spectroscopy (FTIR)

FTIR data would help to deduce the separation of biomass components before and after pretreatment. Therefore, FTIR spectra of dried untreated and pretreated biomass samples were recorded with FTIR spectrometer (Spectrum Two, PerkinElmer, Inc., USA). The dried samples were placed in the sample holder space ATR, and the spectra were recorded in the range of 4000–400 cm^{-1} with a resolution of 0.5 cm^{-1} (Li et al. 2014; Wu et al. 2016). All the experiments were performed in duplicate, and the average data are presented. Standard deviation of the duplicate measurements was less than 5%.

18.3 Results and Discussion

The compositional analysis results presented in Fig. 18.1 show that GHR is rich in glucan, xylan, and less in lignin. The ash content and acetyl group are in lower level. These results demonstrate that GHR could be a potential raw material for a biorefinery. Similar composition of GHR is reported in Krishnan et al. (2010) and Aboyade et al. (2013), whereas the GHR composition was different from several research studies reported in Canilha et al. (2012). The difference in composition of

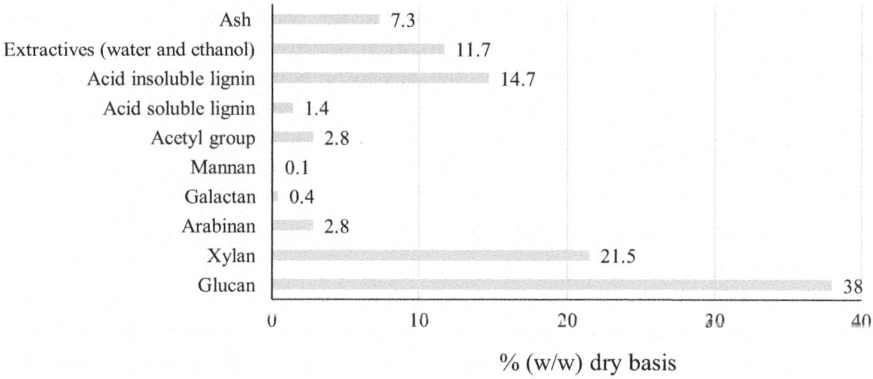

Fig. 18.1 Compositional analysis of sugarcane green harvest residues

GHR could be due to the geographical location, sugarcane variety, and harvesting season.

The biomass recovery after pretreatment plays a crucial part in determining the overall effectiveness of ethanol production process. As the pretreatment severity increases, the mass recovery decreases due to increase in degradation of the biomass. Therefore, it is crucial to obtain a balanced mass recovery after pretreatment which would not affect further process and at the same time increase the overall product yield. The mass recovery results of the auxin-assisted pretreatment in Table 18.1 show that the mass recovery and glucan recovery decrease upon increase in severity. A similar trend is followed for xylan as well; however, the lignin recovery follows the opposite trend. This could be explained by the humin formation upon increasing the process condition which shows a higher value for lignin in the measurement (Sannigrahi et al. 2011; Trajano et al. 2013). The composition of liquid fraction from the auxin pretreatment is shown in Table 18.2. Increase in harshness of the operating conditions increases the organic acids value meaning that more sugars were removed from the biomass. In the initial experiments the sugars are available in the liquid fraction, whereas in the later experiments the sugars are converted to HMF, acetic acid, and furfural. These results from the pretreated solid fraction and liquid fraction suggest that the optimum operating conditions are in the initial experiments conditions range that could be further explored by the saccharification and fermentation results.

The results from the saccharification of auxin-assisted pretreated biomass are shown in Table 18.3. Similar to pretreatment, the saccharification efficiency decreased with the biomass that was pretreated with harsh conditions. The maximum glucose yield (maximum theoretical) was obtained in the process condition (175 °C and 10.5 min). Low saccharification efficiency on increasing harsh conditions could be due to the humin in the pretreated biomass which is shown as lignin in the analysis (Table 18.1) that inhibits the accessibility of enzymes to convert cellulose. Further increase in enzyme concentration would increase the saccharification efficiency; however, the experiments were using these concentrations to be practical with the industrial situation due to the enzyme cost and the futuristic expected concentration level that has to be achieved. It is also worth to mention that these enzymes are in the development stage and several studies have used advanced enzymes that could achieve better saccharification efficiency.

The optimized pretreatment parameters could be only selected based on the ethanol yield after saccharification and fermentation. The amount of ethanol produced per gram raw biomass and the ethanol yield (theoretical yield) after fermentation of saccharified liquid fraction is shown in Table 18.3. The results confirm high yield at mild process condition. The low yield at high severity could be due to the low sugar content and the inhibitors present in the saccharified liquid. Considering all the processes from raw biomass to ethanol production and the sugars in pretreatment liquid fraction, the optimum process conditions based on the surface response plots (Fig. 18.2a–d) were found to be 175 °C, 10.5 min, and

Table 18.3 Results of saccharification and fermentation process relative to auxin treatments on various conditions based on the experimental design

Temp. (°C)	Time (min)	Combined severity factor	Glucose yield (% theoretical max.) upon saccaharification	Ethanol yield (% theoretical max.) upon fermentation	Glucose (g/g dry GHR) upon PT and saccaharification	Ethanol (g/g dry GHR) upon fermentation
175.0	1.0	−1.55	38.3	30.0	0.15	0.047
175.0	10.5	−0.42	48.4	45.5	0.19	0.086
212.5	1.0	−0.15	45.4	0.9	0.17	0.002
175.0	20.0	−0.12	46.8	40.2	0.19	0.076
193.8	5.8	−0.02	47.5	1.6	0.19	0.003
193.8	15.3	0.44	42.4	1.1	0.14	0.002
212.5	10.5	0.88	24.8	0.9	0.09	0.001
250.0	1.0	1.01	10.9	0	0.08	0
212.5	20.0	1.13	28.3	0	0.10	0
231.3	5.8	1.17	11.3	0.8	0.04	0
231.3	15.3	1.57	13.9	0.9	0.05	0
250.0	10.5	2.08	0	0	0.05	0
250.0	20.0	2.37	0	0	0.01	0

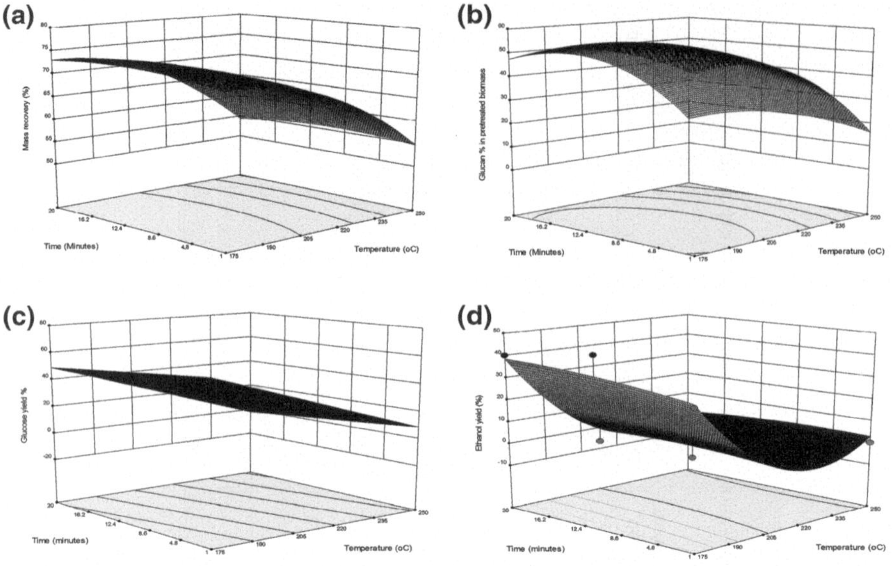

Fig. 18.2 Response surface for mass recovery, glucan, glucose yield, and ethanol yield for GHR auxin pretreatments

combined severity factor of −0.42. The effect of auxin concentration was tested for 1, 2, and 3% at the optimum temperature and time (175 °C, 10.5 min). The results were not significant compared to 2% concentration. A lower ethanol value of 0.078 g/g biomass and 0.068 g/g biomass was obtained at 1 and 3%, respectively, compared to that of 0.086 g/g biomass for 2%. This could be due to the change in combined severity factor (−0.45 for 1% and −0.34 for 3%) that influenced the pretreatment process and subsequent process.

The process optimization study results of liquid hot water treatment (Table 18.4, 18.5, and 18.6) and from response surface show that the maximum bioethanol production and xylose was obtained at 193.8 °C, 15.3 min, and combined severity factor of 0.26. This reveals that liquid hot water pretreatment needs higher treatment conditions and severity compared to that of auxin-assisted pretreatment for a similar bioethanol production from GHR leading to the conclusion that the addition of auxin has increased the pretreatment effectiveness compared to that of liquid hot water. The maximum bioethanol production data from auxin and liquid hot water pretreatment were in line with the literature data reported in the previous studies (Table 18.7) where lower reaction temperature, higher residence time, and different enzymes and dosages were used for pretreatment and saccharification.

To further support the effectiveness of auxin pretreatment, the surface area results of the pretreated biomass and raw biomass at 175 °C and 10.5 min shown in Fig. 18.3 reveal that the auxin pretreatment could increase the surface area by 130%

Table 18.4 Pretreated biomass (solid fraction) composition after liquid hot water treatment on various conditions based on the experimental design

Temp. (°C)	Time (min)	Combined severity factor	% Mass recovery	Glucan (% PT solid)	Xylan (% PT solid)	Lignin (% PT solid)	Glucan recovery (% initial)	Xylan recovery (% initial)	Lignin recovery (% initial)
175.0	1.0	−2.71	86.5	41.9	23.8	18.1	95.5	95.8	97.1
175.0	10.5	−1.27	77.1	45.6	21.7	20.4	92.7	77.9	98.0
175.0	20.0	−0.85	74.3	48.3	19.7	20.1	94.5	67.9	93.0
193.8	5.8	−0.46	64.8	55.0	11.3	24.0	93.9	34.0	96.6
212.5	1.0	−0.29	60.8	59.7	4.5	26.7	95.7	12.7	101.0
193.8	15.3	0.26	61.7	59.3	5.6	26.2	96.3	16.1	100.6
212.5	10.5	0.97	60.2	57.8	1.2	31.7	91.6	3.4	118.7
250.0	1.0	1.07	58.1	47.9	0.1	40.7	73.4	0.4	146.8
231.3	5.8	1.25	59.3	52.7	0.5	36.5	82.3	1.2	134.4
212.5	20.0	1.28	60.7	55.3	1.5	31.3	88.5	4.3	118.2
231.3	15.3	1.63	58.6	50.2	0.2	38.3	77.5	0.5	139.3
250.0	10.5	2.04	53.8	39.1	0.4	47.6	55.5	1.1	159.0
250.0	20.0	2.33	43.7	30.2	0.2	56.0	34.8	0.3	152.1

Table 18.5 Pretreated biomass (liquid fraction) composition after liquid hot water treatment on various conditions based on the experimental design

Temp. (°C)	Time (min)	Combined severity factor	Glucose (mg/100 ml)	Xylose (mg/100 ml)	Acetic acid (mg/100 ml)	HMF (mg/100 ml)	Furfural (mg/100 ml)
175	1	−2.71	80.0	79.3	17.5	2.1	15.5
175	10.5	−1.27	111.4	406.9	17.4	2.5	62.3
175	20	−0.85	128.2	647.5	16.2	2.4	89.7
193.75	5.75	−0.46	126.2	1119.7	17.1	2.2	162.5
212.5	1	−0.29	125.4	724.8	21.2	2.1	171.1
193.75	15.25	0.26	131.3	1013.5	24.4	2.2	235.2
212.5	10.5	0.97	128.0	47.0	40.9	2.0	124.6
231.25	5.75	1.25	74.9	3.7	99.7	3.4	131.6
212.5	20	1.28	99.7	7.2	80.7	2.9	181.5
231.25	15.25	1.63	45.3	4.0	118.9	4.7	130.2
250	1	1.07	52.8	1.0	122.4	5.2	124.2
250	10.5	2.04	54.3	4.2	132.0	4.4	106.1
250	20	2.33	34.1	2.5	76.5	2.5	41.1

compared to that of liquid hot water treatment. This could be explained by the penetration of auxin (Walters 1999) into the biomass in mild acidic conditions and the process conditions that supported efficient biomass hydrolysis leading to higher surface area compared to liquid hot water pretreatment despite their slow solubility in water. The soaking time of biomass in 2,4-D solution in this study (2 h) was selected based on the previous study by Barrier and Loomis 1957. However, investigating the impact of soaking time in the pretreatment process in future studies will provide additional information.

In addition, the spectra of FTIR analysis of raw biomass, auxin, and water-pretreated biomass at 175 °C and 10.5 min are shown in Fig. 18.4. The transmission band in the region 1730 cm^{-1} is due to the presence of ketone/aldehyde C=O stretch in hemicellulose (Sills and Gossett 2012; Xu et al. 2013). This band is visible in the raw biomass and fades off in other two samples meaning that the pretreatment was effective in hemicellulose removal. However, among the pretreated biomass samples, the absorbance in the auxin-pretreated sample is higher compared to the liquid hot water-treated sample at the same operating conditions. All the bands related to lignin (Sills and Gossett 2012; Xu et al. 2013; Kubo and Kadla 2005) could be observed in 1500, 2840, and 2937 cm^{-1} for raw biomass, auxin-assisted pretreated biomass, and liquid hot water-treated biomass meaning that the treatment retained lignin. The amorphous cellulose band in the region 790 cm^{-1} (Ang et al. 2012) is found to be higher for the auxin-pretreated biomass compared to the liquid hot water-pretreated and raw biomass. Whereas the crystalline cellulose band in region 1280 cm^{-1} (Sills and Gossett 2012) is lower for auxin pretreated biomass compared to other samples meaning the auxin assisted

Table 18.6 Results of saccharification and fermentation process relative to liquid hot water treatments on various conditions based on the experimental design

Temp. (°C)	Time (min)	Combined severity factor	Glucose yield (% theoretical max.) upon saccaharification	Ethanol yield (% theoretical max.) upon fermentation	Glucose (g/g dry GHR) upon PT and saccaharification	Ethanol (g/g dry GHR) upon fermentation
175.0	1.0	−2.71	20.8	2.3	0.09	0.002
175.0	10.5	−1.27	27.5	11.6	0.10	0.011
175.0	20.0	−0.85	32.0	23.1	0.13	0.028
193.8	5.8	−0.46	38.8	24.1	0.15	0.035
212.5	1.0	−0.29	55.4	42.8	0.23	0.093
193.8	15.3	0.26	49.3	44.0	0.20	0.085
212.5	10.5	0.97	57.3	49.4	0.23	0.102
231.3	5.8	1.25	54.9	44.2	0.19	0.069
212.5	20.0	1.28	55.9	37.4	0.21	0.070
231.3	15.3	1.63	50.0	37.3	0.17	0.047
250.0	1.0	1.07	48.2	40.9	0.15	0.046
250.0	10.5	2.04	44.2	31.6	0.10	0.018
250.0	20.0	2.33	25.2	20.3	0.03	0.002

Table 18.7 Comparison of present study results with the literature studies

Composition of GHR (glucan, xylan, and lignin)%	Pretreatment method and optimum conditions	Enzyme name and dose (FPU)	Sugar subject to fermentation	Ethanol yield g/g biomass	References
27.85, 19.41, 27.11	Dilute acid (121 °C, 60 min, 3% w/w)	Cellulase Zytex (60)	Glucose	0.049	Ang et al. 2012
38.5, 23, 15.6	Dilute acid (121 °C, 30 min, 1.5% w/w)	Accellerase 1000 (160)	Glucose	0.08	Sindhu et al. 2011
47.03, 6.89, 32.91	Dilute acid (121 °C, 30 min, 1.5% w/w)	Cellic CTec2 (30)	Glucose	0.0924	Jutakanoke et al. 2012
27.85, 19.41, 27.11	Glycerol-assisted transition metal and alkali (121 °C, 45 min, glycerol 6%, sodium hydroxide 5%, ferric chloride 1%)	Cellulase Zytex (80)	Glucose	0.089	Pereira et al. 2015
38, 21.5, 16.1	Auxin (175.0 10.5, 2% w/w)	Celluclast (30)	Glucose	0.086	This study
38, 21.5, 16.1	Hot water (193.8 °C, 15.3 min)	Celluclast (30)	Glucose	0.085	This study

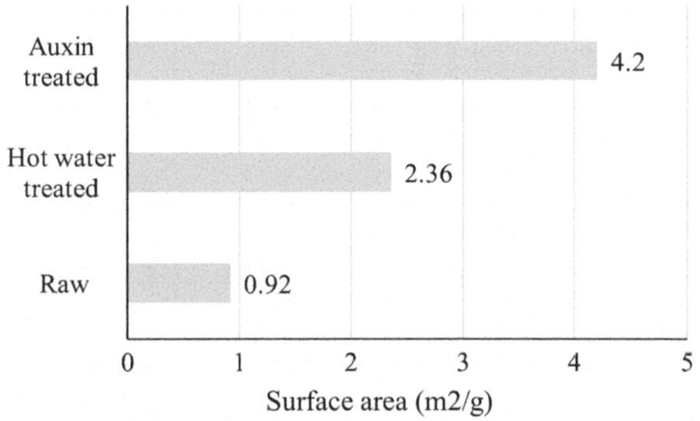

Fig. 18.3 Surface area comparison of raw and pretreated biomass

pretreatment is more effective in degrading the components. The FTIR results are in line with the chemical analysis results showing the effectiveness of auxin-assisted pretreatment. In addition, the biodegradable property of auxin and the possibility of converting the auxin wastewater to succinic acid and other biochemicals (Walters 1999; Kung and Wu 2012) could be advantageous that could make auxin a potential catalyst in biomass pretreatment.

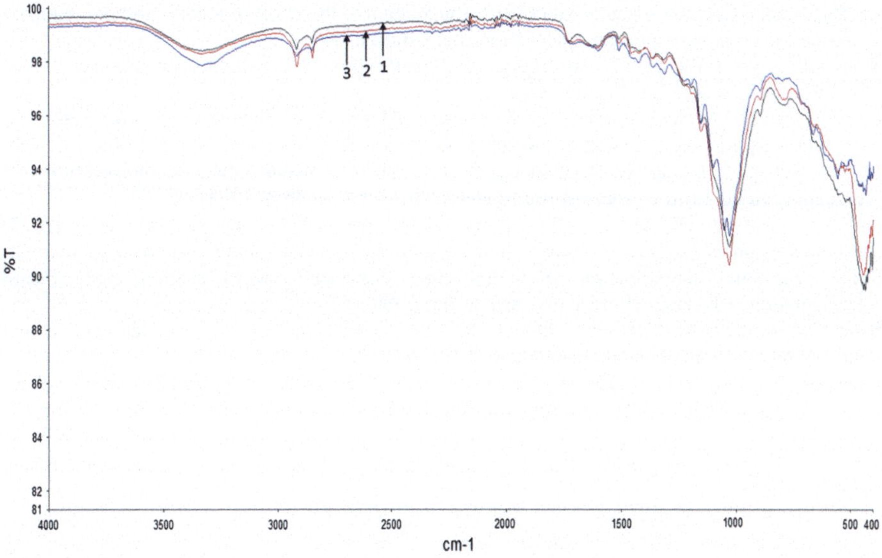

Fig. 18.4 FTIR spectra of biomass: **1** raw, **2** hot water pretreated, **3** auxin pretreated

18.4 Conclusion

Sugarcane residues were pretreated with auxin followed by saccharification and fermentation. The optimum pretreatment conditions with 2% auxin pretreatment were found to be 175 °C and 10.5 min. The process optimization, surface area, and FTIR results show that auxin pretreatment is effective at low severity compared to liquid hot water pretreatment. The biodegradability property of auxin would be beneficial in wastewater treatment and other biochemicals production that makes auxins potential candidates for biomass pretreatment. Nevertheless, further research on other auxins and biomass residues, economic and lifecycle assessments studies are necessary to deduce the full potential of auxins in lignocellulosic biorefinery.

Acknowledgements Authors acknowledge Prof. Johann Görgens (Stellenbosch University, South Africa) for his support with sugarcane green harvest residues samples and École Polytechnique Fédérale de Lausanne (EPFL) for financial support.

References

Aboyade AO, Görgens JF, Carrier M, Meyer Edson L, Knoetze JH (2013) Thermogravimetric study of the pyrolysis characteristics and kinetics of coal blends with corn and sugarcane residues. Fuel Process Technol 106:310–320

Adney B, Baker J (2008) Measurement of cellulase activities—laboratory analytical procedure (LAP), Technical Report. NREL/TP-510-42628. National Renewable Energy Laboratory ed. Golden, CO

Ang TN, Ngoh GC, Chua AS, Lee MG (2012) Elucidation of the effect of ionic liquid pretreatment on rice husk via structural analyses. Biotechnol Biofuel 7:5

Barrier GE, Loomis WE (1957) Absorption and translocation of 2, 4-dichlorophenoxyacetic acid and P32 by leaves. Plant Physiol 32(3):225

Boivin A, Amellal S, Schiavon M, Van Genuchten MT (2005) 2, 4-Dichlorophenoxyacetic acid (2, 4-D) sorption and degradation dynamics in three agricultural soils. Environ Poll 30:92–99

Cai CM, Zhang T, Kumar R, Wyman CE (2013) THF co-solvent enhances hydrocarbon fuel precursor yields from lignocellulosic biomass. Green Chem 15:3140–3145

Canilha L, Chandel AK, Milessi TSS, Antunes TAF, Freitas WLC, Felipe MCA et al (2012) Bioconversion of sugarcane biomass into ethanol: an overview about composition, pretreatment methods, detoxification of hydrolysates, enzymatic saccharification, and ethanol fermentation. J Biomed Biotechnol, Article ID, p 989572

Chang VS, Nagwani M, Holtzapple MT (1998) Lime pretreatment of crop residues bagasse and wheat straw. Appl Biochem Biotechnol 74:135–159

Chaturvedi V, Verma P (2013) An overview of key pretreatment processes employed for bioconversion of lignocellulosic biomass into biofuels and value added products. 3 Biotech 3:415–431

Grossmann K (2007) Auxin herbicide action: lifting the veil step by step. Plant Signal Behav 2:421–423

Jutakanoke R, Leepipatpiboon N, Tolieng V, Kitpreechavanich V, Srinorakutara T, Akaracharanya A (2012) Sugarcane leaves: pretreatment and ethanol fermentation by Saccharomyces cerevisiae. Biomass Bioenergy 39:283–289

Kim TH, Kim JS, Sunwoo C, Lee YY (2003) Pretreatment of corn stover by aquous ammonia. Bioresour Technol 90:39–47

Krishnan C, Sousa LC, Jin M, Chang L, Dale BE, Balan V (2010) Alkali-based AFEX pretreatment for the conversion of sugarcane bagasse and cane leaf residues to ethanol. Biotechnol Bioeng 107:441–450

Kubo S, Kadla JF (2005) Hydrogen bonding in lignin: a Fourier transform infrared model compound study. Biomacromol 6:2815–2821

Kung S, Wu R (2012) Transgenic plants: engineering and utilization. Academic Press, San Diego

Lee JW, Jeffries TW (2011) Efficiencies of acid catalysts in the hydrolysis of lignocellulosic biomass over a range of combined severity factors. Bioresour Technol 102:5884–5890

Lee JW, Houtman CJ, Kim HY, Choi IG, Jeffries TW (2011) Scale-up study of oxalic acid pretreatment of agricultural lignocellulosic biomass for the production of bioethanol. Bioresour Technol 102:7451–7456

Li X, Lu J, Zhao J, Qu Y (2014) Characteristics of corn stover pretreated with liquid hot water and fed-batch semi-simultaneous saccharification and fermentation for bioethanol production. PLoS ONE 9:e95455

Liu CZ, Wang F, Stiles AR, Guo C (2011) Ionic liquids for biofuel production: opportunities and challenges. Appl Energy 92:406–414

Mori T, Sakurai M, Seki M, Furusaki S (1994) Use of auxin and cytokinin to regulate anthocyanin production and composition in suspension cultures of strawberry cell. J Sci Food Agri 65: 271–276

NREL, 2016. Standard Procedures for Biomass Compositional Analysis. National Renewable Energy Laboratory. Golden, CO. http://www.nrel.gov/biomass/analytical_procedures.html#

OECD (1995) Guidelines for the Testing of Chemicals. Detailed review paper on biodegradability testing, Organization for Economic co-operation and development, Paris

Overend RP, Chornet E (1987) Fractionation of lignocellulosics by steam-aqueous pretreatments. Philos Trans R Soc. A. Math, Phys Eng Sci 321:523–536

Pereira SC, Maehara L, Machado CMM, Farinas CS (2015) 2G ethanol from the whole sugarcane lignocellulosic biomass. Biotechnol Biofuels 8:44

REN21 (2015) Renewables 2015 global status report. Renewable energy policy network for the 21st century: Paris, France

Rulli MC, Bellomi, D, Cazzoli, A, Carolis GD, D'Odorico P (2016) The water-land-food nexus of first-generation biofuels. Sci Rep 6:22521

Sannigrahi P, Kim DH, Jung S, Ragauskas A (2011) Pseudo-lignin and pretreatment chemistry. Energy Environ Sci 4:1306–1310

Sauer M, Robert S, Kleine-Vehn J (2013) Auxin: simply complicated. J Expt Bot 64:2565–2577

Schwab A, Warner E, Lewis J (2015) Survey of non-starch ethanol and renewable hydrocarbon biofuels producers. Technical Report NREL/TP- 6A10-65519. 2016. National Renewable Energy Laboratory ed. Golden, CO

Selig M, Weiss N, Ji Y (2008) Enzymatic saccharification of lignocellulosic biomass—laboratory analytical procedure (LAP), Technical Report National. Renewable Energy Laboratory NREL/TP-510-42629, ed. Golden, CO

Sills DL, Gossett JM (2012) Using FTIR to predict saccharification from enzymatic hydrolysis of alkali pretreated biomasses. Biotechnol Bioeng 109:353–632

Sindhu R, Kuttiraja M, Binod P, Janu KU, Sukumaran RK, Pandey A (2011) Dilute acid pretreatment and enzymatic saccharification of sugarcane tops for bioethanol production. Bioresour Technol 102:10915–10921

Sindhu R, Gnansounou E, Binod P, Pandey A (2016) Bioconversion of sugarcane crop residue for value added products—an overview. Renew Energy 98:203–215

Sluiter JB, Ruiz RO, Scarlata CJ, Sluiter AD, Templeton DW (2010) Compositional analysis of lignocellulosic feedstocks. 1. Review and description of methods. J Agric Food Chem 58:9043–9053

Sluiter A, Hames B, Ruiz R, Scarlata C, Sluiter J, Templeton D, Crocker D (2011) Determination of structural carbohydrates and lignin in biomass. Laboratory analytical procedure (LAP). Technical Report NREL/TP-510-42618. National Renewable Energy Laboratory ed. Golden, CO

Trajano HL, Engle NL, Foston M, Ragauskas AJ, Tschaplinski TJ, Wyman CE (2013) The fate of lignin during hydrothermal pretreatment. Biotechnol Biofuels 6:1

UNEP (2009) Converting waste agricultural biomass into a resource. Compendium of technologies, United Nations Environment Programme, Osaka

Walters J (1999) Environmental Fate of 2, 4-dichlorophenoxyacetic acid. Environmental monitoring and plant management. California Department of Pesticide Regulation, Sacramento, CA

Wang Q, Hu J, Shen F, Mei Z, Yang G, Zhang Y, Hu Y, Zhang J, Deng S (2016) Pretreating wheat straw by the concentrated phosphoric acid plus hydrogen peroxide (PHP): investigations on pretreatment conditions and structure changes. Bioresour Technol 199:245–257

Wi SG, Cho EJ, Lee DS, Lee SJ, Lee YJ, Bae HJ (2015) Lignocellulose conversion for biofuel: a new pretreatment greatly improves downstream biocatalytic hydrolysis of various lignocellulosic materials. Biotechnol Biofuel 8:228

Wu R, Zhao X, Liu D (2016) Structural features of formalin pretreated sugar cane bagasse and their impact on the enzymatic hydrolysis of cellulose. ACS Sustainable Chem Eng 4: 1255–1261

Xu F, Yu J, Tesso T, Dowell F, Wang D (2013) Qualitative and quantitative analysis of lignocellulosic biomass using infrared techniques: a mini-review. Appl Energy 104:801–809

Zhang YH, Hong J, Ye X (2009) Cellulase assays. In: Mielenz, JR (ed) Biofuels: methods and protocols methods in molecular biology. Humana Press, NJ, 58: p 213–231

Zhu JY, Pan XJ, Wang GS, Gleisner R (2009) Sulfite pretreatment (SPORL) for robust enzymatic saccharification of spruce and red pine. Bioresour Technol 100:2411–2418

Chapter 19
Cellulosic Biomass-Hydrolyzing Enzymes

Simranjeet Kaur Sandhu, Anshu Mathur, Ravi Gupta, Suresh K. Puri and Mukund Adsul

Abstract Biomass is a heterogeneous matrix of three interlinked polymers of cellulose–hemicellulose–lignin. Polymers such as cellulose and hemicellulose need to be fractioned and converted into monomers using enzyme-catalyzed process for the production of a value-added product. Cellulose is major polysaccharides within this biomass, and converting this cellulose into simple sugars requires synergic action of multiple cellulolytic enzymes. Enzymatic conversion of cellulose to simple sugars (for the production of platform chemicals) represents major costs in overall process of biomass to value-added product generation. So, extensive studies have been conducted to improve bacterial and fungal strains for maximum cellulase production along with accessory proteins that act synergistically and even in adverse conditions like in the presence of inhibitory components released during chemical and thermal pretreatment of biomass. This chapter covers biomass availability, microbial approaches for biomass-hydrolyzing enzyme production, different cellulases classification and characterization, and mechanism of enzymatic deconstruction of cellulose present in biomass.

Keywords Biomass · Cellulose · Cellulase · Mechanism

19.1 Cellulosic Biomass

Cellulose is a polymer of glucose abundantly available on earth. It is the structural portion of plants that cannot be used directly for food. Agricultural residues (sugarcane bagasse, corn stover, rice straw, cotton stalk, wheat straw, etc.), industrial waste (papermaking sludge), and forestry residues (branches leaves, twigs, saw dust, etc.) are the variety of sources of cellulosic biomass. Cellulosic biomass is mainly classified into forestry residues or woody fibers and agricultural residues or nonwoody fibers.

S. K. Sandhu · A. Mathur · R. Gupta · S. K. Puri · M. Adsul (✉)
DBT-IOC Centre for Advanced Bioenergy Research, Indian Oil Corporation Ltd
R & D Centre, Sector-13, Faridabad 121007, India
e-mail: madsul@gmail.com

R. R. Singhania et al. (eds.), *Waste to Wealth*, Energy, Environment, and Sustainability, https://doi.org/10.1007/978-981-10-7431-8_19

The wood fibers are structurally strong, physically larger, and denser than non-wood fibers. Despite of the fact that wood fibers are available throughout the year, no long-term storage is required and lesser inhibitors (furfurals) are formed after pre-treatment because of less pentose content; still non-wood fibers are playing bigger role in terms of their usage. Wood fibers also contain high lignin content, creating strong cellulose–hemicelluloses–lignin bond and act as physical barrier to enzyme digestibility (Tye et al. 2016). Moreover, approximately 200–600 Wh/kg of the energy is required to reduce the size of wood fibers, which is relatively higher in comparison to that of non-wood fibers (50 Wh/kg).

The principal framework of cellulose consists of glucose units linked by β-1,4 linkages forming straight chains. These chains are further linked together by strong hydrogen bonding and van der Waals forces to structure called microfibrils with high tensile strength, making it crystalline in nature. The diameter of microfibrils is about 5 nm and packed further into larger units called macrofibril (also called nanofibrillated cellulose) of diameters ranging from 20 to 50 nm. These macrofibrils are located in the secondary cell wall of the plant immersed into hemicelluloses and lignin matrix. Two-third part of the cellulose structure is crystalline in nature and rest is amorphous in nature, which is easier to degrade. Highly crystalline cellulose is resistant to enzymatic digestibility; thus, by pre-treating the cellulose by acid, base, or ionic liquids, etc., crystallinity can be decreased. Crystalline cellulose exists in two forms such as I_α and I_β depending upon their three-dimensional geometrical arrangement, i.e., triclinic and monoclinic unit cells, respectively. Source of cellulosic biomass affect the ratio of I_α and I_β within native cellulosic structures (Huber et al. 2012). Diversity of cellulosic biomass depends upon the chain length, position of hydrogen bond between chains, and degree of polymerization.

19.2 Availability of Cellulosic Biomass

Cellulose containing biomass types and their availability varies significantly, depending on the country or region. For example, major biomass sources are agriculture residues and forest residues in India and Canada, whereas cassava waste and other non-grain crops are available in China. Agricultural residues form readily available sources of biomass from current harvesting activities, and thus no additional land for cultivation is required. The major challenges to select cellulosic biomass are its uncertainty, surplus availability throughout the year, and cost to meet social and economic objectives. Cereal straw is collected after the grains have been removed, 4–5 inch from above ground part of the cereal part. In general, straw except wheat straw is considered as poor livestock feed because of slow rate of digestibility for ruminant animals. Therefore, it can be potentially utilized for other applications. According to Tye et al. (2016), there are approximately 51.3, 472.2, 376.8, 657.5, 12.0, and 1044.8 million tons of barley straw, wheat straw, corn straw, rice straw, sorghum straw, and sugarcane bagasse, respectively, produced annually in the world. Cellulose content in these residues ranges from 32 to 47%. It is estimated that from six major crops like barley, corn, oat, rice, sorghum, and wheat, 1580.2 million tons of cereal straw is produced annually in Europe, the USA, and China (Kim and Dale 2004). Sugarcane bagasse yield is estimated about 0.6 kg for 1 kg of sugarcane, which is being utilized in Brazil for ethanol production. Brazil alone produces 27% of world's sugarcane production. Among the agricultural residues, sugarcane bagasse and rice straw have attained the highest production yield.

19.3 Importance of Making Sugars for Different Applications

Cellulosic biomass being abundantly available and cheaper can be utilized to produce sugars and finally converted into other platform chemicals such as ethanol, butanol, methanol, dimethyl ether, green diesel. It is necessary to explore the alternative cheaper and eco-friendly biofuel resources as a strategy to reduce greenhouse effect and to meet the increasing demand and costs of fossil fuels (Iqbal et al. 2013; Asgher et al. 2013). Alternative sources include cellulosic biomass which normally requires a multistep processing to convert it into value-added products and these include (i) pretreatment either mechanical, chemical, or biological, (ii) hydrolysis with enzymes, and (iii) fermentation process. Biofuel produced from residues offers number of advantages. In the food vs fuel, policymakers often favor residue utilization over biofuels made from sugars and food crops like sugarcane and corn. Biofuels from residues also offer better greenhouse gas savings. Use of cellulosic biomass also encourages the removal of excess crop residues from the land which is otherwise burnt and causes environmental pollution and damages ecosystem, and its usage will generate revenues to the farmers.

There are numerous potential chemicals that can be derived from cellulosic sugars as shown in Fig. 19.1. The value of other platform chemicals obtained from cellulosic sugars far exceeds the feedstock costs and hence improves the profitability obtained presently only from cellulosic ethanol. The only limitation is that the quantity of these chemicals is low. Chemicals are at various stages of development, among which butanol production is close to commercialization. Butanol can be used as a diesel fuel additive. Through biological process, both pentose and hexose sugar can be utilized via phosphoenol pyruvate pathway to produce succinic acid, lactic acid, fumaric acid, and ethanol. Succinic and fumaric acid production requires carbon dioxide, while in the production of ethanol and lactic acid carbon dioxide is released. Succinic acid is a four-carbon dicarboxylic acid with the chemical formula $(CH_2)_2(CO_2H)_2$, used as a flavoring agent and pharmaceutical additive. Succinic acid can further act as base material for the production of polymers like 1,4-butanediol, gamma-butryrolactone, and tetrahydrofuran. Glutamic acid is produced using *Corynebacterium*, which acts as raw material for the production of monosodium glutamate (flavoring agent). Through chemical

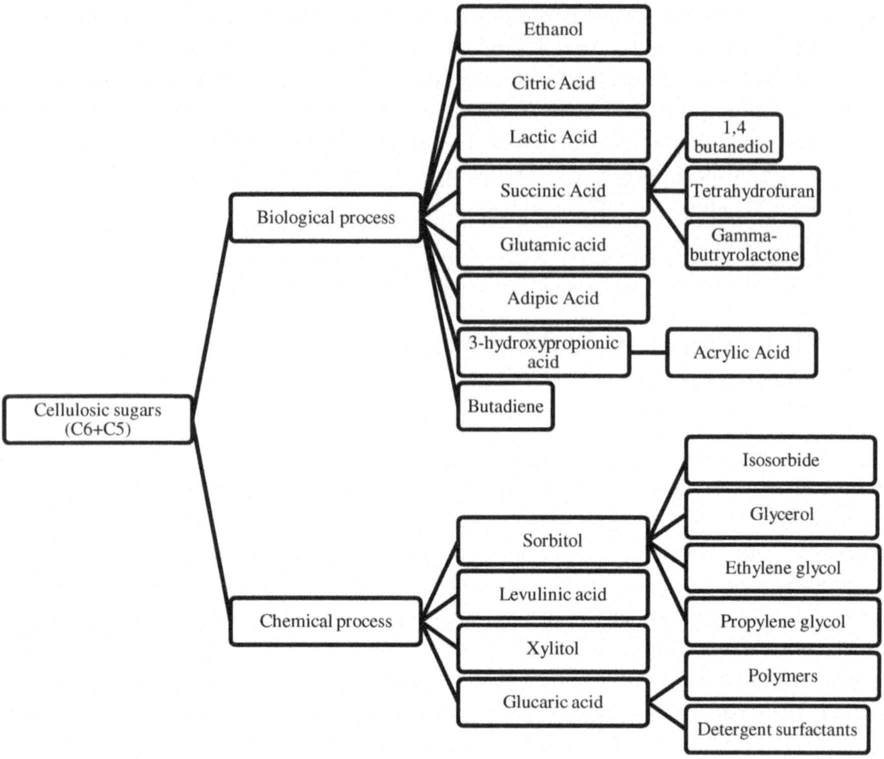

Fig. 19.1 Platform chemicals derived from cellulosic sugars

process, five-carbon sugars can be converted into levulinic acid. Levulinic acid is used as an intermediate to manufacture pharmaceuticals, synthetic fibers, and plastics. Sorbitol is produced by hydrogenation of glucose using Raney nickel as catalyst. It is a potential raw material of products like isosorbide, glycerol, ethylene glycol, and propylene glycol. Xylitol is produced by yeast or bacterial fermentation of xylose, with market potential similar to that of sorbitol. Similarly, other platform chemicals such as itaconic acid, 3-hydroxybutyrolactone, glycerol, aspartic acid, furfurals, gallic acid, and many more are obtained by chemical and biological treatment of pentose and hexose sugars.

19.4 Cellulases Enzymes

To get above chemicals, first it is needed to degrade cellulose into simple sugars. There are two ways to degrade cellulase, i.e., through chemical route and other is biological or enzymatic route. Biological degradation is simple and environmentally friendly way to degrade cellulose, and enzyme cellulases play an important role.

Cellulases refer to the class EC.3.2.1 of hydrolytic enzymes (glycoside hydrolases). It is a complex enzyme mixture consisting of three main classes of enzymes involved in hydrolysis of β-1,4 linkage in the cellulose polymer. Components of the cellulase system were classified below based upon their catalytic mode of action.

1. Endo-β-1,4-glucanase enzyme or 1,4-β-D-glucan 4-glucanohydrolases (EC 3.2.1.4): These are non-processive enzymes that attack randomly on intermolecular β-1,4-glucosidic linkages within the cellulose chains, thereby generating new chain ends. Carboxymethyl cellulose solution is commonly used for the assay of endoglucanases enzymes.
2. Exo-1,4-β-D-glucan cellobiohydrolases (CBH) or exo-1,4-β-D-glucanases (EC 3.2.1.91): These are processive enzymes involved in hydrolysis of reducing (CBH I) and nonreducing (CBH II) ends of cellulose modules to release glucose and cellobiose. Avicel and filter paper have been used for measuring exoglucanase activity among insoluble cellulosic substrates.
3. β-glucosidase enzyme (EC 3.2.1.21): This enzyme hydrolyzes soluble cellobiose or cellodextrins with a degree of polymerization up to six to produce glucose.

Glycoside hydrolases (GHs) are mainly produced by fungi and bacteria. In GHs classifications system, the different enzymes are categorized into structurally related GH families based on the distribution of hydrophobic amino acids in their sequences. According to the Carbohydrate-Active Enzyme Database, there are 145 families of GHs that have been identified so far and number will grow as more gene are sequenced (www.CAZy.org). The CBHs and endo-β-1,4-D-glucanases produced by *T. reesei* belong to GH5, GH6, GH7, GH12, and GH45, glycoside hydrolase families. GH7 is the only family that contains both CBHs and endo-β-1,4-D-glucanases (Druzhinia and Kubicek 2017). The β-glucosidases enzymes found in *T. reesei* belong to GH1 and GH3 families. The 36 families of GHs found in all cellulase producing filamentous fungi are GH1, 2, 3, 5, 6, 7, 10, 11, 13, 15, 16, 17, 18, 27, 31, 32, 36, 37, 38, 43, 47, 51, 53, 54, 55, 61, 63, 67, 72, 75, 79, 81, 92, 105, and 114 (Yang et al. 2011). Cellulases found in GH5 and GH7 families are most prevalent in biomass hydrolyzing cellulolytic fungi due its processive mode of catalysis. The β-glucosidases enzymes belong to GH1, and GH3 families' secretion systems are intracellular and extracellular enzymes, respectively (Guo et al. 2016).

Most fungal cellulases are organized in two structurally independent domains, cellulose-binding module (CBM) and catalytic core module. Largest part of the enzyme consists of catalytic core module, where the hydrolysis of the cellulose chain takes place. The catalytic core module and CBM are usually interconnected via glycosylated and a short flexible linker peptide. The length of the linker varies in size, from less than 20 to over 40 amino acids in the different enzymes. The role of the linker is probably to keep the two domains apart, and to restrict their movements with respect to one and another, so that the catalytic domain remains within close distance to the CBM, which binds on the surface of a cellulose fiber. The cellulose-binding module (CBM) is a small wedge-shaped domain consisting of approximately 35 amino acids. The function of the CBM is to bind on the surface of cellulose and serve as an 'anchor' for the enzyme, keeping it strongly adsorbed to the cellulose surface. Till today, based on amino acid sequence similarity in CAZy database, CBMs were classified into 84 families. Based on the structure of the ligand binding, alternate classification of CBM has also been proposed by Gilbert et al. (2013), which includes three types of enzymes: type A, B, and C that identifies the surface of polysaccharides, binds to the internal regions of glucan chains, and binds to the terminal region of glucans chains, respectively. β-glucosidases do not hold a CBM but hydrolyze soluble cellodextrins and cellobiose into glucose. There is no evidence that the fungal CBMs can penetrate into the cellulose fiber and disrupt the structure or have any catalytic activity.

Cellulases produced by anaerobic bacteria lack ability to penetrate cellulosic biomass and thus require another mechanism for cellulose degradation. This leads to development of complex cellulose system called cellulosome (Demain et al. 2005). Cellulosome complex consists of two parts, catalytic part which contains enzyme, CBM and dockerin and non-catalytic scaffolding part which contains cohesion part for dockerin attachment and cellulose-specific CBM. Scaffolding also carries C-terminal divergent dockerin that helps in attachment of cellulosome complex to cellulosic substrate for hydrolysis (Fontes and Gilbert 2010). The cohesins are modules made up of approximately 150 amino acid residues and usually present as tandem repeats in scaffoldings, while dockerins consist of ~70 amino acids containing two duplicated segments (~22 amino acid residues). The CBM includes of approximately 35 amino acid residues with highly glycosylated linker region consisting of serine, threonine, and proline amino acid residues (Gupta et al. 2013). The major difference between fungal, aerobic bacterial, and anaerobic cellulosomal enzymes is that the anaerobic cellulosomal enzymes carry a dockerin domain that carry the enzyme into the cellulose system, while aerobic bacterial and fungal enzymes hold a CBM for leading the catalytic domain to the cellulosic substrate for hydrolysis (Bayer et al. 2004).

19.5 Accessory Enzymes

Apart from above cellulases, other accessory enzymes such as oxidative and non-hydrolytic enzymes play an important role to speed up cellulose degradation. The enzymatic degradation of cellulosic substrate is an integrated process that depends upon the simultaneous synergistic action of different proteins. During the last few decades, effort has been made to recognize the role of non-catalytic and accessory proteins in enhancing the enzymatic hydrolysis of cellulose (Liu et al. 2015). The oxidative enzyme comes under the 'auxiliary activity' section.

(a) AA9/LPMO

According to Carbohydrate-Active Enzyme Database, there are 13 families of auxiliary enzymes discovered so far (www.CAZy.org). Out of which eight families are known to be involved in lignin degradation and four are directly active on polysaccharides, such as lytic polysaccharide monooxygenases (LPMOs) (Payne et al. 2015). The LPMO is currently grouped into sequence-based AA9, AA10, AA11, and AA13 families as auxiliary activities in CAZy database. So far two families of LPMOs have been characterized in detail, endoglucanases glycoside hydrolase (GH61) family that attack highly crystalline cellulose belonging to, second family contains bacterial and viral enzyme, formally classified as carbohydrate-binding module (CBM33). The CBM33 has also been reported in the fungus *Sporisorium reilianum* (Lombard et al. 2014). The LPMO degrades crystalline region of cellulose and creates more ends for cellobiohydrolases. The LPMOs are copper dependent and act as reducing agent for oxidative cleavage of either the C1 or C4 carbon of the glycosidic bond (Forsberg et al. 2014). The type 1 LPMOs involve in oxidation of C1 carbon (reducing end), creating an aldonolactone that further hydrolyzes to an aldonic acid. Type 2 LPMOs generate 4-ketoaldolase and the hydrated gem diol by oxidizing C4 carbon at nonreducing end, and type 3 LPMOs produce oxidized products at both the reducing and nonreducing ends (Beeson et al. 2012; Isaksen et al. 2014). The site of oxidation is determined by the position of the substrate on the catalytic surface. The copper activates molecular oxygen at the active site followed by its incorporation into the cellulose chain. Cellulose depolymerization by LPMO also involves cellobiose dehydrogenase (CDH, EC 1.1.1.99.18) which acts synergistically with LPMO, where CDH generates electron that is required for LPMO activity. The CDH has heme-binding cytochrome attached to flavin-dependent dehydrogenase enzyme via flexible linker. The CDH generates electron by cellobiose oxidation and shuttled this electron via heme-binding cytochrome to LPMO (Tan et al. 2015).

(b) Non-hydrolytic protein

Non-enzymatic expansins and other related proteins are familiar for acting in plant biomass degradation. The expansin proteins of fungi and bacteria are involved in contortion of cellulose chains and other cell wall polysaccharides by unwinding the cellulose macrofibrils (Saloheimo et al. 2002; Cosgrove 2000; Qin et al. 2004; Kim et al. 2009). Thus, an expansin protein leads to an increase in the efficiency and accessibility of the cellulolytic enzymes engrossed in hydrolysis of cellulose. According to Kende et al. (2004), expansins comprise of two distinct domains: an N-terminal catalytic domain (domain I) and second domain believed to be tryptophan-rich polysaccharide-binding C terminal domain (domain II) interconnected by a short linker. Domain I is structurally organized into double-psi beta-barrel fold, which is structurally similar to the catalytic domain of GH45 protein or fungal β-1, 4-D-endoglucanases (Kerff et al. 2008).

The swollenin (SWO) is well-known expansin-related protein in fungi (Saloheimo et al. 2002; Andberg et al. 2015). The SWO1 swollenin protein belongs to *T. reesei*, bears catalytic domain I and C-terminal cellulose-binding domain interconnected via flexible linker region, and thus functions in preparing cellulase for hydrolytic attack. Preparing cellulose for hydrolysis involves loosening, partial disruption, and swelling of plant cell walls. Fungal SWO differs significantly in size from the plant expansin as their amino acid residues are two times higher than the plant expansins and thus supports the hypothesis that the evolution of fungal and bacterial expansin-related proteins happened through independent domain fusion and horizontal gene transfer (Nikolaidis et al. 2014). Since recent studies have shown that the presence of swollenin improves enzyme cocktails designed for cellulose hydrolysis, new approaches for their production and purification attract the biofuels industry. Gram-negative and Gram-positive bacterial strains produce acid and basic expansin proteins, respectively. The expansins produced by bacteria bind to cellulose depending upon their electrostatic behavior. Another expansin-like protein isolated from the white-rot basidiomycete *Bjerkandera adusta*, BaLOOS1, lacks the carbohydrate-binding C-terminal domain II present in plant expansins, with both cell-disrupting and polysaccharide-binding activities bundled in domain I.

19.6 Mode of Action

There are two basically different strategies for the hydrolysis of cellulose (Payne et al. 2015) by the microorganisms: Anaerobic microorganisms follow the 'bound enzyme paradigm' (cellulosome) which involve intracellular enzymes (Bae et al. 2013), and aerobic microorganisms that secrete extracellular enzyme, i.e., the 'free enzyme paradigm' (Gupta et al. 2016). According to Koshland (1953), there are two major mechanisms of cellulose hydrolysis: either retaining or inversion. The retaining mechanism involves double displacement catalytic mechanism. Initially, the nucleophilic residue attacks the anomeric carbon with simultaneous protonation of the glycosidic oxygen leading to the formation of the glycosyl-enzyme intermediate and cleavage of the glycosidic bond. In the second step, a water molecule enters the active site and attacks the anomeric carbon simultaneously transferring a proton to the catalytic base, thus restoring the enzyme active site for subsequent catalysis. Inverting GHs employs a single displacement catalytic mechanism, wherein a water molecule conducts nucleophilic attack at the anomeric carbon. Catalytic base abstracts a proton from the attacking water molecule, while catalytic acid transfers a proton to the glycosidic oxygen to cleave the glycosidic linkage, resulting in an inversion of stereochemistry at the anomeric carbon (Fig. 19.2).

Regardless of whether the mode of action of the enzyme is inverting or retaining, the overall topologies of the active sites fall into only three general classes. The two topologies of active site found in cellulases, the cleft and tunnel (Davies and Henrissat 1995) as shown in Fig. 19.3. The cleft or groove is commonly found in endo-acting cellulases; it is an open structure that randomly binds to sugar units in cellulose substrate. The exo-acting cellulases such as exoglucanases or cellobiohydrolases possess tunnel-like active site. The tunnel-like topology allows these

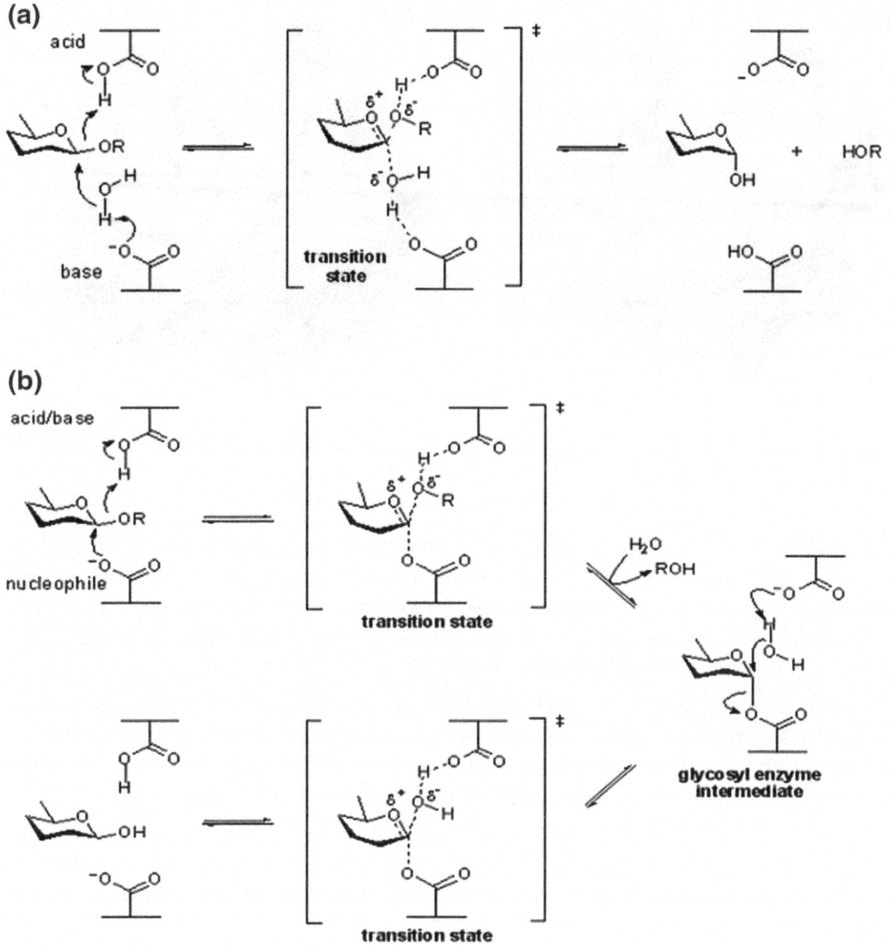

Fig. 19.2 Mechanism of cellulosic hydrolysis by cellulases, **a** inverting mechanism, and **b** retaining mechanism

enzymes to release the product while remaining firmly bounds to the polysaccharide chain. This explains the processivity concept of mode of action of CBH, i.e., sequential cleavage of cellulose by an enzyme.

19.7 Cellulase-Producing Microbial Strains

Many bacterial and fungal species are efficient producers of cellulases and hemicellulases enzymes. Preference has been given on to the use of fungi because of their ability to secrete complete cellulose complex into the medium and with a

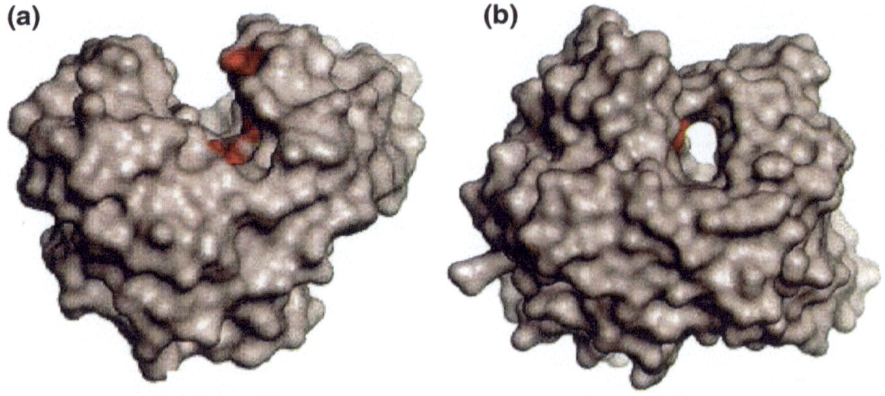

Fig. 19.3 Two topologies of active site found in cellulases glycosyl hydrolases **a** the cleft and **b** the tunnel form (Davies and Henrissat 1995)

higher titer. Commercially available enzymes for biomass hydrolysis are currently derived from fungi species of *Trichoderma*, *Penicillium, Humicola, Fusarium,* and *Aspergillus sp.* However, other cellulolytic systems of *Phanerochaete chrysosporium, Talaromyces emersonii, Melanocarpus albomyces,* and other anaerobic fungi belonging to genera *Neocallimastix, Cacomyces, Orpinomyces* have also been well characterized. The cellulases have also been isolated from thermophilic fungi such as *Sporotrichum thermophile*, *Thermoascus aurantiacus*, *Chaetomium thermophile, Humicola grisea* and *Myceliophthora thermophila*. These fungi are of more interest because of their capacities to produce thermostable cellulases. These enzymes have shown stability at highly acidic or alkaline pH as well as temperatures up to 90 °C. In order to overcome complexity of cellulosic substrates, fungi often produce a variety of cellulase components having different molecular weight, isoelectric point, amino acid composition, and sequence, etc.

Trichoderma reesei has been exposed to multiple rounds of strain improvement to escalate cellulase production. Cellulase production can be increased by the reduction of catabolite repression (Hood et al. 2007), reduction in protease activity (Nagendran et al. 2009), the development of methods to grow and produce cellulose by the microorganism on minimal media, and by using inexpensive cellulase inducers. *Trichoderma reesei* has ability to produce large amount of exocellulases enzyme, but it produces lesser amount of β-glucosidase enzyme. However, *Penicillium sp.* and *Aspergillus sp.* are well known to produce complete cellulases enzyme system but at low titer. This enzyme titer can be improved by strain improvement and media optimization techniques.

Bacterial species like *Bacilli, Pseudomonas, Cellulomonas* and few actinomycetes such as *Streptomyces* and *Actinomucor* are commonly used for lignocellulolytic enzyme production. Both anaerobic and aerobic bacteria are involved in the degradation of cellulose. The anaerobic bacteria are present in compost piles, wood chip piles, the soil, sewage, rumens, sludge, termite gut, on decaying plant waste, etc. Decaying agricultural/plant materials, soil is a natural habitat of these bacteria. The enrichment of these bacteria in nature is influenced by human activities (compost piles, wood processing plants, sewage plants, etc.). Rumen and gut (termites) are another natural habitat of these bacteria, where they degrade the plant material for host as well as their nutrition. Some examples of these cellulose degrading anaerobic bacteria are *Clostridium cellulofermentans, Clostridium aldrichii, C. cellulolyticum, C. cellulovorans, C. acetobutylicum, C. hungatei, C. cellobioparum, C. josui, C. herbivorans, C. papyrosolvens, Clostridium thermocellum, Bacteroides cellulosolvens, Acetivibrio cellulolyticus, Ruminococcus flavefaciens, Fibrobacter succinogenes, Ruminococcus albus, Butyrivibrio fibrisolvens,* etc. (Maki et al. 2009). Some examples of aerobic bacteria found in water, soil, feces, on plant waste are *Cellulomonas persica, C. fimi, C. gelida, C. flavigena, C. uda, C. iranensis, Pseudomonas fluorescens, Bacillus megaterium, Bacillus pumilus, Cellvibrio mixtus, Cellvibrio gilvus, Streptomyces cellulolyticus, Streptomyces lividans,* etc. (Roy 2008).

19.8 Cellulases Production

Isolation, screening, selection, and characterization have uncovered several novel cellulase-producing fungi and bacteria from heterogeneous environments. Diverse media compositions have been derived for cellulases enzyme production by bacterial and fungal strains. Media components include carbon source, nitrogen source, phosphorus source, and trace elements. However, source of carbon plays major role in defining the cost of enzyme production process. Industrial waste and agricultural residue can be ideal source of carbon with the following characteristics such as their availability in plenty at low cost, substrate accessibility, low viscosity, act as cellulose inducer, nontoxic, the degree of crystallinity, the degree of polymerization as well as the distribution and composition of lignin and have high nutrient value to the cellulose producers. Enzyme production using cellulosic industrial waste includes palm kernel cakes, coconut coir waste, vegetable wastes, rice husk, rice bran, coir waste, wheat bran, saw dust, rice straw, wheat straw, cotton stalk, corn cob, sugarcane bagasse, waste fiber sludge from pulp mill, etc., are shown in Table 19.1.

19.9 Comparison of Bacterial and Fungal Cellulases

Both bacteria and fungi produce cellulases, but almost all of the research on cellulase production on commercial level has been focused on fungi. Though bacteria have a high growth rate, produce cellulase with high specific activity, but the enzyme titers are very low. Moreover, the enzymes produced by bacteria are not easily accessible for cellulose hydrolysis because they are cell-bound enzymes.

Table 19.1 Cellulase production by bacterial and fungal strains using waste residues

	Microbial species	Carbon source	Nitrogen source	Activity	Reference
Bacteria	*Bacillus amyoliquefaciens* DL-3	Rice hull	Peptone	CMCase 153.0 U/ml	Lee et al. (2008)
	Geobacillus sp. HTA426	Sugarcane bagasse	Ammonuim sulfate	CMCase 103.67 U/ml	Potprommanee et al. (2017)
	Consortia of *Cellulomonas cartae, Pseudomonas fluorescence, Pseudomonas putida and Bacillus megaterium*	Banana pseudostem waste	NA	Filter paper activity 0.178 U/ml, β-glucosidase 0.602 U/ml, and CMCase 1.716 U/ml	Dabhi et al. (2014)
	Bacillus vallismortis RG07	• Sugarcane baggase • Rice husk • Rice bran	Ammonium sulfate	CMCase activity • 4105 U/ml • 3509 U/ml • 3110 U/ml	Gaur and Tiwari (2015)
	Bacillus halodurans CAS1	Rice Bran	Yeast extract	CMCase activity 3424 U/ml	Annamalai et al. (2013)
	Bacillus subtilis BY-3	Corn stover	peptone	CMCase activity 4.469 U/ml	Meng et al. (2014)
Fungi	*Trichoderma sp.*	palm kernel cake and vegetable waste	Ammonium sulfate and yeast extract	Filter paper 6.9 FPU/g and 50.1 FPU/g	Lah et al. (2016)
	Aspergillus japonicus C03	sugarcane bagasse	Peptone	Xylanase 102U/g and cellulase 13.25 U/g	Facchini et al. (2011)
	Aspergillus japonicas URM5620	Caster bean waste	Wheat bran	β-glucosidase 88.3 U/g, FPase 953.4 U/g and CMCase 191.6 U/g	Herculano et al. (2011)

(continued)

Table 19.1 (continued)

	Microbial species	Carbon source	Nitrogen source	Activity	Reference
	Aspergillus terreus	Rice straw	Ammonium sulfate, urea, and peptone	Filter paper activity 10.96 U/g	Narra et al. (2012)
	Aspergillus fumigatus	Sugarcane bagasse, wheat bran, orange peel and soybean bran	Ammonium sulfate, urea, and peptone	β-glucosidase 105.82 IU/g, FPase 5 FPU/g and CMCase 160 IU/g	Delabona et al. (2012b)
	Aspergillus ellipticus	Distillery spent wash with wheat straw	Distillery spent wash	β-glucosidase 26.68 U/g, FPase 13.38 U/g	Acharya et al. (2010)
	Trichoderma reesei SEMCC-3.217	Water hyacinth	corn steep liquor and ammonium nitrate	Filter paper activity 13.4 FPU/g	Zhao et al. (2011)
	Humicola insolens TAS-13	Sugarcane bagasse	Ammonium sulfate	CMCase 18.98 U/g/min, FPase 13.63 U/g/min and β-glucosidase 19.54 U/g/min	Ul-Haq et al. (2006)
	Trichoderma harzianum	Sugarcane bagasse	Ammonium sulfate and peptone	Cellulase 1.21 FPU/ml, xylanase 80 IU/ml and β-glucosidase 17.3 IU/ml	Delabona et al. (2012a)
	Myceliophthora heterothallica	Cardboard	Ammonium sulfate, urea, and peptone	Endoglucanase 2642 ± 561 U/g and β-glucosidase 244 ± 48 U/g	Teixeira et al. (2016)

Some bacterial species are appropriate for consolidated bioprocessing (CBP), which involves one-step microbial conversion of cellulosic biomass to ethanol, organic acids, butanol, and other platform chemicals. Anaerobic bacteria cannot be used for cellulose production because of slow growth rate. Enzyme secreted by fungi is extracellular and with high titers. So, bacterial cellulases with practical utility can be diversely expressed in fungi. Thus, biofuel and bioproduct industries are also focusing on improving fungal strains.

Acknowledgements All authors acknowledge Indian Oil Corporation and Department of Biotechnology (DBT), Govt. of India, for providing financial support.

References

Acharya BK, Mohana S, Jog R, Divecha J, Madamwar D (2010) Utilization of anaerobically treated distillery spent wash for production of cellulases under solid-state fermentation. J Environ Manage 91:2019–2027

Andberg M, Penttila M, Saloheimo M (2015) Swollenin from *Trichoderma reesei* exhibits hydrolytic activity against cellulosic substrates with features of both endoglucanases and cellobiohydrolases. Bioresour Technol 181:105–113

Annamalai N, Rajeswari MV, Elayaraja S, Balasubramanian T (2013) Thermostable, haloalkaline cellulase from *Bacillus halodurans* CAS 1 by conversion of lignocellulosic wastes. Carbohy Poly 94:409–415

Asgher M, Asgher Z, Ahmad HMN, Iqbal (2013) Alkali and enzymatic delignification of sugarcane bagasse to expose cellulose polymers for saccharification and bio-ethanol production. Indus Crops Prod 44:488–495

Bae J, Morisaka H, Kuroda K, Ueda M (2013) Cellulosome complexes: natural biocatalysts as arming micro compartments of enzymes. J Mol Microbiol Biotechnol 23:370–378

Bayer EA, Belaich JP, Shoham Y, Lamed R (2004) The cellulosomes: multienzyme machines for degradation of plant cell wall polysaccharides. Annu Rev Microbiol 58:521–554

Beeson WT, Phillips CM, Cate JH, Marletta MA (2012) Oxidative cleavage of cellulose by fungal copper-dependent polysaccharide monooxygenases. J Am Chem Soc 134(2):890–902

Cosgrove DJ (2000) Loosening of plant cell walls by expansins. Nature 407(6802):321–326

da Delabona SP, Farinas CS, da Silva MR, Azzoni SF, Pradella JG (2012) Use of a new *Trichoderma harzianum* strain isolated from the Amazon rainforest with pretreated sugar cane bagasse for on-site cellulase production. Bioresour Technol 107:517–521

Dabhi BK, Vyas RV, Shelat HN (2014) Use of banana waste for the production of cellulolytic enzymes under solid substrate fermentation using bacterial consortium. Int J Curr Microbiol App Sci 3(1):337–346

Davies G, Henrissat B (1995) Structure and mechanisms of glycosyl hydrolases. Structure 3:853–859

Delabona Pda S, Rosangela DP, Pirota B, Codima CA, Tremacoldi CR, Rodrigues A, Farinas CS (2012) Using Amazon forest fungi and agricultural residues as a strategy to produce cellulolytic enzymes. Biomass Bioenergy 37:243–250

Demain AL, Newcomb M, Wu JHD (2005) Cellulase, clostridia, and ethanol. Microbiol Mol Biol Rev 69(1):124–154

Druzhinia IS, Kubicek CP (2017) Genetic engineering of *Trichoderma reesei* cellulases and their production. Microb Biotechnol. https://doi.org/10.1111/1751-7915.12726

Facchini FD, Vici AC, Reis VR, Jorge JA, Terenzi HF, Reis RA, Moraes Polizeli L (2011) Production of Fibrolytic Enzymes by *Aspergillus japonicus CO3* Using Agro-Industrial

Residues with Potential Application as Additives in Animal Feed. Biopro Biosys Eng 34:347–355

Fontes CM, Gilbert HJ (2010) Cellulosomes: highly efficient nanomachines designed to deconstruct plant cell wall complex carbohydrates. Annu Rev Biochem 79:655–681

Forsberg Z, Mackenzie AK, Sørlie M, Røhr ÅK, Helland R, Arvai AS, Vaaje-Kolstad G, Eijsink VG (2014) Structural and functional characterization of a conserved pair of bacterial cellulose-oxidizing lytic polysaccharide monooxygenases. Proc Natl Acad Sci USA 111:8446–8451

Gaur R, Tiwari S (2015) Isolation, production, purification and characterization of an organic-solvent-thermostable alkalophilic cellulase from *Bacillus vallismortis* RG-07. BMC Biotechnol 15:19

Gilbert HJ, Knox JP, Boraston AB (2013) Advances in understanding the molecular basis of plant cell wall polysaccharide recognition by carbohydrate-binding modules. Curr Opin Struct Biol 23:669–677

Guo B, Sato N, Biely P, Amano Y, Nozaki K (2016) Comparison of catalytic properties of multiple β-glucosidases of Trichoderma reesei. Appl Microbiol Biotechnol 100:4959–4968

Gupta R, Mehta G, Deswal D, Sharma S, Jain KK, Kuhad RC et al (2013) Cellulases and their biotechnological applications. In: Kuhad RC, Singh A (eds) Biotechnology for environmental management and resource recovery. Springer, India, pp 89–106

Gupta VK, Kubicek CP, Berrin JG, Wilson DW, Couturier M, Berlin A et al (2016) Fungal enzymes for bio-products from sustainable and waste biomass. Trends Biochem Sci 41:633–645

Herculano PN, Porto TS, Moreira KA, Pinto GA, Souza-Motta CM, Porto AL (2011) Cellulase production by *Aspergillus japonicas urm5620* using waste from caster bean (*Ricinus communis l.*) under solid-state fermentation. Appl Biochem Biotechnol 165:1057–1067

Hood EE, Love R, Lane J, Bray J, Clough R, Pappu K, Drees C, Hood KR, Yoon S, Ahmad A, Howard JA (2007) Subcellular targeting is a key condition for high-level accumulation of cellulase protein in transgenic maize seed. Plant Biotechnol J 5:709–719

Huber T, Mussig J, Curnow O, Pang S, Bickerton S, Staiger MP (2012) A critical review of all-cellulose composites. J Mater Sci 47:1171–1186

Iqbal HMN, Kyazze G, Keshavarz T (2013) Advances in valorization of lignocellulosic materials by bio-technology: an overview. BioResources 8(2):3157–3176

Isaksen T, Westereng B, Aachmann FL, Agger JW, Kracher D, Kittl R, Ludwig R, Haltrich D, Eijsink VG, Horn SJ (2014) A C4-oxidizing lytic polysaccharide monooxygenase cleaving both cellulose and cello-oligosaccharides. J Biol Chem 289(5):2632–2642

Kende H, Bradford KJ, Brummell DA et al (2004) Nomenclature for members of the expansin superfamily of genes and proteins. Plant Mol Biol 55:311–314

Kerff F, Amoroso A, Herman R et al (2008) Crystal structure and activity of *Bacillus subtilis YoaJ* (EXLX1), a bacterial expansin that promotes root colonization. Proc Natl Acad Sci USA 105 (44):16876–16881

Kim S, Dale BE (2004) Global potential production from waste crops and crop residues. Biomass Bioenergy 26:361–375

Kim ES, Lee HJ, Bang WG, Choi IG, Kim KH (2009) Functional characterization of a bacterial expansin from *Bacillus subtilis* for enhanced enzymatic hydrolysis of cellulose. Biotechnol Bioeng 102(5):1342–1353. https://doi.org/10.1002/bit.22193

Koshland DE (1953) Stereochemistry and the mechanism of enzymatic reactions. Biol Rev 28:416–436. https://doi.org/10.1111/j.1469-185X.1953.tb01386.x

Lah TNT, Norulaini NAN, Shahadat M et al (2016) Utilization of Industrial Waste for the Production of Cellulase by the Cultivation of *Trichoderma* via Solid State Fermentation. Environ Process 3:803–814. https://doi.org/10.1007/s40710-016-0185-8

Lee YJ, Kim BK, Lee BH, Jo KI, Lee NK, Chung CH, Lee YC, Lee JW (2008) Purification and characterization of cellulase produced by *Bacillus amyoliquefaciens* DL-3 utilizing rice hull. Bioresour Technol 99:378–386

Liu X, Ma Y, Zhang M (2015) Research advances in expansins and expansion-like proteins involved in lignocellulose degradation. Biotechnol Lett 37(8):1541–1551. https://doi.org/10.1007/s10529-015-1842-0

Lombard V, Ramulu GH, Drula E, Coutinho PM, Henrissat B (2014) The carbohydrate-active enzymes database (CAZy) in 2013. Nucleic Acids Res 42(Database issue): D490–D495

Maki M, Leung KT, Qin W (2009) The prospects of cellulase-producing bacteria for the bioconversion of lignocellulosic biomass. Int J Biol Sci 5:500–516

Meng F, Ma L, Ji S, Yang W, Cao B (2014) Isolation and characterization of *Bacillus subtilis* strain BY-3, a thermophilic and efficient cellulase-producing bacterium on untreated plant biomass. Lett Appl Microbiol 59(3):306–312

Nagendran S, Hallen-Adams HE, Paper JM, Aslam N, Walton JD (2009) Reduced genomic potential for secreted plant cell walldegrading enzymes in the ectomycorrhizal fungus Amanita bisporigera, based on the secretome of *Trichoderma reesei*. Fungal Genet Biol 46:427–435

Narra M, Dixit G, Divecha J, Madamwar D, Shah AR (2012) Production of cellulases by solid state fermentation with *Aspergillus terreus* and enzymatic hydrolysis of mild alkali-treated rice straw. Bioresour Technol 121:355–361

Nikolaidis N, Doran N, Cosgrove DJ (2014) Plant expansins in bacteria and fungi: evolution by horizontal gene transfer and independent domain fusion. Mol Biol Evol 31(2):376–386

Payne CM, Knott BC, Mayes HB, Hansson H, Himmel ME, Sandgren M et al (2015) Fungal cellulases. Chem Rev 115:1308–1448

Potprommanee L, Wang XQ, Han YJ, Nyobe D, Peng YP, Huang Q, Liu JY, Liao YL, Chang KL (2017) Characterization of a thermophilic cellulase from Geobacillus sp. HTA426, an efficient cellulase-producer on alkali pretreated of lignocellulosic biomass. PLoS one 12(4)

Qin L, Kudla U, Roze EH, Goverse A, Popeijus H, Nieuwland J et al (2004) Plant degradation: a nematode expansin acting on plants. Nature 427(6969):30

Roy DH (2008) Cellulases of mesophilic microorganisms: cellulosome and noncellulosome producers. Annals New York Acad Sci 1125:267–279

Saloheimo M, Paloheimo M, Hakola S, Pere J, Swanson B, Nyyssonen E et al (2002) Swollenin, a *Trichoderma reesei* protein with sequence similarity to the plant expansins, exhibits disruption activity on cellulosic materials. Eur J Biochem 269(17):4202–4211

Tan TC, Kracher D, Gandini R, Sygmund C, Kittl R, Haltrich D, Martin BH, Ludwig R, Divne C (2015) Structural basis for cellobiose dehydrogenase action during oxidative cellulose degradation. Nat commun 6:7542. https://doi.org/10.1038/ncomms8542

Teixeira da Silva VDC, de Souza Coto AL, de Carvalho Souza R, Bertoldi Sanchez Neves M, Gomes E, Bonilla-Rodriguez GO (2016) Effect of pH, temperature, and chemicals on the endoglucanases and b-glucosidases from the thermophilic fungus *Myceliophthora heterothallica* F. 2.1. 4. obtained by solid-state and submerged cultivation. Biochem Res Int http://dx.doi.org/10.1155/2016/9781216

Tye YY, Lee KT, Abdullah WNW, Leh CP (2016) The world availability of non-wood lignocellulosic biomass for the production of cellulosic ethanol and potential pretreatments for the enhancement of enzymatic saccharification. Renew Sust Ener Rev 60:155–172

Ul-Haq I, Javed MM, Khan TS (2006) Sugar cane bagasse pretreatment: An attempt to enhance the production potential of cellulases by *Humicola insolens* TAS-13. Biokemistri 18(2):83–88

Yang B, Dai Z, Shi-You D, Wyman CE (2011) Enzymatic hydrolysis of cellulosic biomass. Biofuels 2(4):421–450

Zhao S, Liang X, Hua D, Ma T, Zhang H (2011) High-yield cellulase production in solid-state fermentation by Trichoderma reesei SEMCC-3. 217 using water hyacinth (Eichhornia crassipes). Afr J Biotechnol 10:10178–10187

Chapter 20
Consolidated Bioprocessing at High Temperature

Nisha Singh, Anshu S. Mathur, Ravi P. Gupta, Suresh K. Puri and Munish Puri

Abstract Replacing fossil fuels by biomass-derived ethanol (also known as second-generation ethanol or bioethanol) can provide the dual benefits of renewability and mitigation of the effects of global warming caused by the over-exploitation of petroleum-derived transportation fuels. However, the effective use of lignocellulosic biomass as a feedstock for the production of bioethanol is historically proven to be problematic and faces several technical challenges. A process configuration known as consolidated bioprocessing (CBP) has generated considerable research interests as the most cost-effective means of bioethanol production. However, insufficient production level of ethanol is the major roadblock, limiting their commercial importance. In this chapter, the research opportunities for developing thermoanaerobes for a high-temperature-based CBP and the associated technological challenges are discussed. The current industrial status of CBP is highlighted along with a detailed description of most promising candidate thermoanaerobes. Advanced technologies for improving the ethanol production level from these candidates are also discussed. A high-temperature-based biomass processing seems challenging; however, it could be the most rewarding approach for bioethanol production in the near future.

Keywords Bioethanol · Consolidated bioprocessing · Thermophiles
Anaerobic bacteria

N. Singh · A. S. Mathur · R. P. Gupta · S. K. Puri
DBT-IOC Centre for Advance Bioenergy Research, Research and Development Centre,
Indian Oil Corporation Limited, Sector-13, Faridabad 121007, India

N. Singh · M. Puri (✉)
Centre for Chemistry and Biotechnology, Deakin University,
Geelong, VIC 3217, Australia
e-mail: munish.puri@flinders.edu.au

M. Puri
Centre for Marine Bioproducts Development, Medical Biotechnology,
College of Medicine and Public Health, Flinders University,
Bedford Park, Adelaide 5042, Australia

R. R. Singhania et al. (eds.), *Waste to Wealth*, Energy, Environment,
and Sustainability, https://doi.org/10.1007/978-981-10-7431-8_20

20.1 Introduction

Ever-increasing demands of energy and environmental concerns have led to a growing interest in finding sustainable sources of energy. One such source is bioethanol, produced from renewable waste materials, having the potential to curb GHG emissions and become a carbon-neutral alternative of liquid transportation fuels (Salehi Jouzani and Taherzadeh 2015; Tan et al. 2016). Lignocellulosic biomass (such as agricultural wastes, forestry residues, and municipal solid wastes) represents an abundantly available cheap natural resource that contained large amount of sugars locked in the form of cellulose and hemicellulose which can be implied for the production of bioethanol (Zabed et al. 2016). However, due to its heterogeneous and highly recalcitrant nature, extracting the energy content of plant biomass is difficult and energy-intensive (Sánchez and Montoya 2013). To overcome this problem, conventionally a series of biologically mediated events, viz (i) biomass harvest and storage, (ii) pretreatment (physical/chemical treatment to expose cellulose and hemicellulose chains), (iii) enzyme production (cellulases and hemicellulases by appropriate microbial source), (iv) saccharification (enzymatic hydrolysis of cellulose and hemicellulose to its constituents), (v) fermentation of hexose (C6) sugars, and (vi) fermentation of pentose (C5) sugars, was followed by downstream processing to achieve the conversion of biomass to ethanol (Bhalla et al. 2013) (Fig. 20.1). However, such type of scheme requires multiple reactors and exogenous supply of costly cellulolytic enzymes which adds significant capital and maintenance cost (Lynd et al. 2005; Schuster and Chinn 2013; Hasunuma et al. 2013; Linger and Darzins 2013). Thus at industrial scale, the cost of bioprocessing would become more than the cost of feedstock, restricting the economical viability of bioethanol production. The total cost of feedstock, pretreatment, and enzyme treatment is estimated to account for two-thirds of the total production cost in which enzyme as the largest cost contributor (Himmel et al. 2010; Lynd et al. 2005).

Over the last 20 years, efforts have been made in the development of efficient and low-cost technologies to overcome these cost issues (Tan et al. 2016). This comprised mainly advancement in pretreatment technologies to achieve effective biomass conversion, enzyme engineering to reduce cellulase dosages, and genetic manipulation of fermenting microorganisms to gain sufficient production level of bioethanol. The detailed discussions on these process advancement efforts are beyond the scope of our focus and have been reviewed elsewhere (Karimi and Taherzadeh 2016; Zhang and Zhang 2013; Salehi Jouzani and Taherzadeh 2015).

In addition to above efforts, further process efficiency can also be gained using integrated process design (Chung et al. 2014). Process integration efforts generally entail in the consolidation of one or more of these steps to combat the high costs of bioethanol production such as combining enzymatic hydrolysis and fermentation of hexose sugars in simultaneous saccharification and fermentation [SSF] scheme or simultaneous saccharification and co-fermentation [SSCF] scheme that combines fermentation of both hexose and pentose sugars along with saccharification (Linger and Darzins 2013; Schuster and Chinn 2013; Sánchez and Montoya 2013).

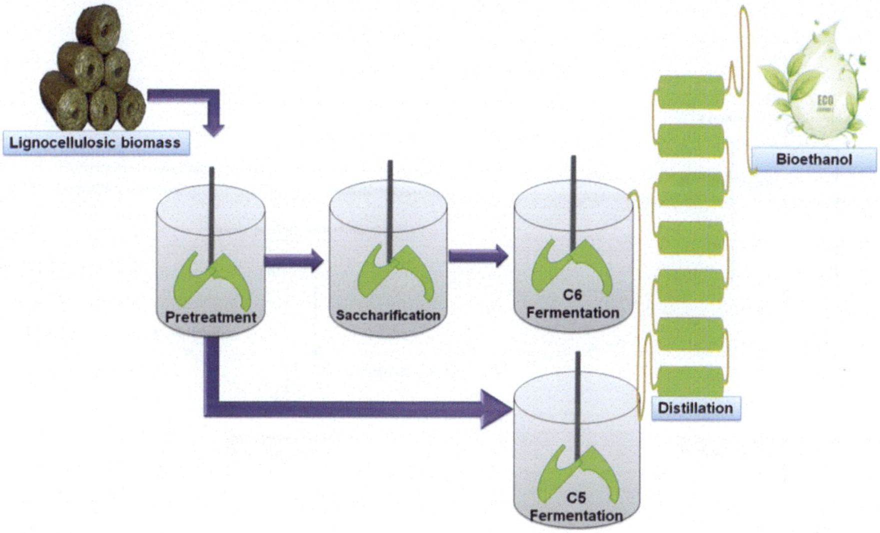

Fig. 20.1 Overview of bioethanol production from lignocellulosic biomass (various stages of ethanol production process are represented in the flow diagram)

However, the cost economy of supplying exogenous enzymes for both SSF and SSCF schemes cannot be ignored. Thus, the ultimate objective of low-cost biomass processing to bioethanol would be a single-stage process termed as "consolidated bioprocessing" [CBP] (Himmel et al. 2010; Lynd et al. 2005). During CBP, enzyme production, hydrolysis, and co-fermentation would occur in a single reactor employing a single microorganism or a consortium of microorganisms that are capable of reducing the recalcitrance of biomass with their robust enzymatic and metabolic capabilities (Lynd et al. 2005; Schuster and Chinn 2013; Hasunuma et al. 2013; Linger and Darzins 2013) (Fig. 20.2). Process integration in CBP offers huge economic benefits by avoiding high capital cost and addition of costly cellulolytic enzymes.

It is quite clear that an effective CBP-enabling microorganism is the heart of a CBP process (Salehi Jouzani and Taherzadeh 2015; Olson et al. 2012; Parisutham et al. 2014). While a number of mesophilic microorganisms including fungi, yeast, and bacteria are under development as a viable CBP candidate (Salehi Jouzani and Taherzadeh 2015; Amore and Faraco 2012; Haan et al. 2015), recently thermophilic and extremely thermophilic anaerobic bacteria (thermoanaerobes) have gained increased research interest (Blumer-Schuette 2014; Chang and Yao 2011).

The inherent benefits of using thermoanaerobes for CBP could potentially overcome many of the obstacles associated with the conventional biomass processing approaches. Despite the concept of bioethanol production via CBP at high temperature seems promising, the intricacies of single-step process are extremely challenging. In this chapter, we briefly discuss the current status and complexities of a commercially relevant high-temperature-based CBP and amenability of several

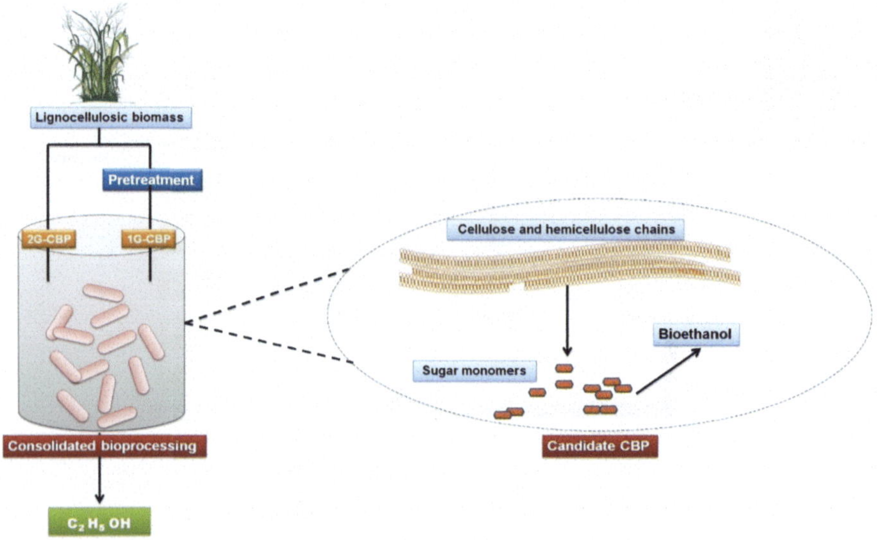

Fig. 20.2 Schematic representation of first-generation (1G-CBP) and second-generation (2G-CBP) consolidated bioprocessing scheme. The 2G-CBP is based on the microbial breakdown of untreated lignocellulosic biomass into bioethanol and promises maximum cost-effectiveness (Chung et al. 2014)

potential thermoanaerobes for CBP. We further discuss the various strain improvement strategies implied to attain efficient ethanol production by these candidates.

20.2 Consolidated Bioprocessing

CBP is a highly integrated process configuration where candidate CBP microorganism(s) performs "direct microbial conversion" of lignocellulosic biomass into bioethanol (Lynd et al. 2005; Schuster and Chinn 2013; Hasunuma et al. 2013; Linger and Darzins 2013). Essentially such candidate must be able to produce lignocellulolytic enzymes to depolymerize both cellulose and hemicellulose to their monomeric forms and subsequently ferment all the resulting sugars to ethanol. Owing to the simple processing configuration, fewer unit operations, and maximum cost-effectiveness, CBP is considered as a breakthrough for low-cost biomass processing (Lynd et al. 2005). Ideally a fully mature CBP technology would incorporate the microorganism(s) having following desirable traits:

- Efficient metabolic capabilities allowing co-fermentation of hexose and pentose sugars
- Equipped with a robust lignocellulose deconstruction machinery

- Ability to deconstruct untreated biomass to avoid additional capital investments associated with pretreatment
- Ability to withstand industrially relevant solid loadings and fermentation inhibitors
- Fermentation performance with yield, titer, and productivity as per industrial criteria (i.e., an ethanol yield of >90% of theoretical maximum, titer of >40 g L^{-1}, and productivity of >1 g L^{-1} h^{-1})
- High ethanol tolerance above a concentration of 5% (v/v)
- Minimal by-product formation

Unfortunately, no such single organism is available in nature or engineered so far that carries all these traits altogether (Parisutham et al. 2014). Therefore, research efforts are underway: (i) to isolate and characterize such organism (ii) or to develop existing cellulolytic or to ferment organisms for homoethanolic fermentation of lignocellulosic biomass (Olson et al. 2015).

Over the last decade, development of mesophilic cellulose-degrading and/or ethanol-producing organisms (including yeast, fungi, and bacteria) as platform CBP candidate has been a major focus of research by both industries and academies. Among the potential candidates for CBP, thermophilic anaerobes (T_{opt} > 60 °C) inherently possess several of the above-mentioned characteristics and are best suited for this highly integrated process configuration. This chapter will only focus on the prospects of utilizing thermoanaerobes for bioethanol production via CBP.

20.3 Thermophilic Anaerobic Bacteria for CBP: Challenges and Strain Improvement Strategies

Thermophilic anaerobes are interesting group of microorganisms that not only survive at high temperature but also can thrive to produce commodity products under such extreme environment (Canganella and Wiegel 2014). By definition, thermophilic anaerobes are described as the group of microorganisms that cannot use oxygen as the terminal electron acceptor and grow at elevated temperature of >50 °C (Canganella and Wiegel 2014). Thermophilic anaerobic bacteria are of particular biotechnological interest due to two main characteristics: (i) their ability to grow on major component of lignocellulosic biomass directly for the purpose of ethanol, hydrogen, or methane production (Taylor et al. 2009; Blumer-Schuette 2014) and (ii) presence of unique lignocellulose deconstruction machinery that can serve as an excellent source of thermostable enzymes for biorefinery applications (Bayer et al. 2008; Turner et al. 2007).

Due to these, in past years a renewed interest has been centered on the development of a high-temperature-based CBP to achieve the process benefits of elevated temperature saccharification. A higher process temperature, especially under anoxic conditions, would offer a number of potential benefits over mesophilic temperature process for CBP as follows:

(1) In case of a thermophilic fermentation system, cooling or heating of fermentor between different steps of bioprocessing would not be required, providing significant energy savings (Taylor et al. 2009; Lynd et al. 2005).
(2) As per Arrhenius equation, a 10 °C rise in temperature can double the rate of reaction; thus, a faster reaction rate could be achieved with thermophiles (Lin and Xu 2013; Turner et al. 2007).
(3) High-temperature process may advantageously increase the rate of lignocellulose deconstruction in a thermophilic fermentor and will require mild pretreatment (Kataeva et al. 2013).
(4) Microbial contamination (due to mesophilic bacteria and fungi) is a significant problem during industrial fermentation. High reaction temperature (>60 °C) and anoxic conditions can minimize these contaminants (Taylor et al. 2009; Lin and Xu 2013).
(5) Better solubility of reactants and products at higher temperature can provide efficient mixing, thus reduced energy input (Chang and Yao 2011).
(6) High temperature also facilitates easier recovery of product because ethanol vaporizes at temperature above 50 °C (Chang and Yao 2011).

Apart from these process benefits, thermoanaerobes offer some unique advantages during biomass bioprocessing. The enzymatic machinery of thermophilic and extremely thermophilic anaerobic bacteria is unique and highly efficient in many aspects compared to the conventional fungal enzymes (Xu et al. 2016; Resch et al. 2013; Brunecky et al. 2013; Kanafusa-Shinkai et al. 2013). Depending upon structural organization and mode of action, the biomass degradation schemes employed by thermoanaerobes can be classified into four systems, namely "free enzyme system," "cell anchored enzymes," "complex cellulosome system," and "multifunctional multimodular enzyme system." Such complex enzymes exhibit very high specific activity and have a natural ability to tolerate harsh bioprocessing conditions. However, for industrial production the major challenge is to achieve sufficient production level of these thermostable enzymes.

Although pretreatment is an absolute requirement to enhance the efficiency of saccharification, most of these physicochemical treatments are energy-intensive and feedstock-dependent (Karimi and Taherzadeh 2016). Different pretreatment methods are associated with sugar loss and production of toxic fermentation inhibitors. Thus, development of a bioprocess with mild or no pretreatment is always taken into consideration to promote cost savings and a toxin-free process. This is evident by the development of genetically modified plants that are less recalcitrant, such as switchgrass, alfalfa, aspen, and poplar (Bartley et al. 2014; Fu et al. 2011). Interestingly few extremely thermophilic anaerobic bacteria can perform "direct microbial conversion" of biomass without any pretreatment to produce commodity products like bioethanol and biohydrogen. In the past few years, a number of valuable studies have been conducted to evaluate the digestibility of untreated plant biomass either by thermophilic anaerobes itself or by the enzymes derived from them (Basen et al. 2014; Kataeva et al. 2013; Chung et al. 2014; Kanafusa-Shinkai et al. 2013; Brunecky et al. 2013). This can circumvent the 20% additional cost of pretreatment during biomass processing.

Thermoanaerobes are advantageous being capable to metabolize a broad range of substrates including both monomeric (hexoses and pentoses) and polymeric (cellulose and hemicellulose) carbohydrates generated after lignocellulose degradation (Demain et al. 2005; Blumer-Schuette et al. 2008; Chang and Yao 2011). This broad substrate utilization potential is one of the prime reasons for interest in these organisms for bioethanol production.

The utilization of hemicellulose along with cellulose is an essential aspect for cost-effective bioethanol production at industrial scale (Saha 2003). Majority of ethanologenic thermoanaerobes within the genera *Thermoanaerobacterium* and *Thermoanaerobacter* are effective hemicellulose-fermenting organisms and presented "preferential hemicellulose utilization." However, different thermophiles differ in their pattern of carbohydrate utilization, with more cellulose and C6 utilization observed for the thermophiles having T_{opt} 55–60 °C (Demain et al. 2005), while more hemicelluloses and pentose utilization was observed for the thermophiles having T_{opt} 70 °C or higher (Blumer-Schuette et al. 2008; Dam et al. 2011).

Majority of thermophilic anaerobic bacteria can perform co-utilization of C6 and C5 sugars derived from lignocellulosic biomass in an unbiased manner, which is an essential trait for any ethanol-producing candidate. This suggested the absence of carbon catabolite repression (CCR) (Gorke and Stulke 2008) in most of the extreme thermophiles (Vanfossen et al. 2009; Lin and Xu 2013; Shaw et al. 2008; Andersen et al. 2015; Georgieva et al. 2008).

The negative impact of inhibitors (mainly furfural, 5-hydroxy methyl furfural, weak acids, and phenolics) generated during biomass pretreatment on the fermentative microorganisms is a major obstacle and studied extensively (Akhtar et al. 2016; Karimi and Taherzadeh 2016; Zabed et al. 2016). To avoid this, often a cost-intensive detoxification step is performed prior to fermentation in conventional bioprocessing (Kundu et al. 2015). Adaptability to wide range of pH is another useful trait exhibiting thermoanaerobes as most suitable candidate for industrial bioprocessing. Biomass processing using thermoacidophilic and thermoalkaliphilic bacteria and their enzymes could avoid an additional neutralization step and thus promotes cost savings (Bhalla et al. 2013).

Despite these benefits, it is equally important to understand the limitations of these microorganisms before proposing their application for industrial production of bioethanol. One of the most important considerations for their implementation at an industrial scale is to achieve sufficient ethanol production level (Blumer-Schuette 2014; Taylor et al. 2009). None of the naturally occurring thermoanaerobe can convert lignocellulosic biomass into bioethanol at rates that meet current industrial targets. There are two main reasons for lower yield: (i) thermoanaerobes are heterofermentative and undergo mixed acid fermentation (Olson et al. 2015). Thus, production of ethanol is always accompanied by the formation of other products such as acetate, lactate, butyrate, propanol and (ii) most thermophiles with ethanol production ability are highly sensitive to increasing substrate concentrations. A high substrate concentration results in fermentation inhibition and thus lower product yield (Argyros et al. 2011; Lin and Xu 2013). Besides these limitations, the growth and cultivation of anaerobic bacteria require specific complex medium which also poses economic challenge (Turner et al. 2007). However, even with these limitations, the potential of these microbes cannot be neglected. In recent years, efforts in optimizing media for different thermophilic anaerobes and advancement in molecular biology techniques have paved the way to achieve improved performance in the near future.

20.4 Strategies to Improve Bioethanol Production by Thermophilic Anaerobic Bacteria

Considering the above-mentioned limitations, strain improvement via mutation and/or metabolic engineering remains the ultimate choice for the development of a potential microbial platform for a high-temperature-based CBP. In early the 1990s, classical approaches like UV and chemical mutagenesis were applied to create mutant with improved substrate conversion rate and metabolic properties. In past few years, a substantial progress has been observed in the metabolic engineering of thermophilic anaerobic bacteria to produce bioethanol (Scully and Orlygsson 2017; Lin and Xu 2013). This progress has been based on strategic advancement in genetic tools for thermophiles and parallel genome sequencing efforts. Some of the approaches are outlined for the development of a thermophilic engineered host as follows:

20.4.1 Improvement by Evolutionary Adaptation

Most of the wild-type thermoanaerobes possess lower tolerance (1–2% v/v) to ethanol, which is the major obstacle in their industrial application. For an economical biorefinery, ethanol concentration is required to be above 5% (v/v) (Burdette et al. 2002). Since the knowledge about thermophilic anaerobic bacteria is limited, the reason for their lower tolerance to ethanol is not clear. However, negative impact of ethanol accumulation on increased membrane fluidity and subsequently various cells functioning such as membrane transport and energy generation could be the possible cause of decreased cell viability and ultimately lower tolerance (Biswas et al. 2014; Shao et al. 2011; Georgieva et al. 2007).

Adaptation is a classical and a very prominent approach to improve ethanol yield and tolerance of thermoanaerobes. Evolutionary adaptation approaches are mainly based on (i) sequential transfer and continuous adaptation of strain on higher ethanol concentrations or (ii) cyclic transfer on lower and higher ethanol concentrations which may or may not be accompanied by mutagenesis (UV treatment or

Table 20.1 Evolutionary adaptation approaches for some thermophilic anaerobic bacteria

Bacteria	Adaptation strategy	Ethanol tolerance % (v/v)	References
C. thermocellum ATCC 27405	UV mutagenesis and sequential transfer in increasing ethanol concentration	4%	Tailliez et al. (1989)
C. thermocellum ATCC 27405	Sequential transfer in increasing ethanol concentration	5%	Timmons et al. (2009)
C. thermocellum SS21 and SS22	Sequential transfer in increasing ethanol concentration	3.2–4%	Rani and Seenayya (1999)
C. thermocellum ATCC 27405	Sequential transfer alternating zero and increasing ethanol concentration	5%	Shao et al. (2011)
C. thermocellum DSM 1313 [adhE*Δldh]	Sequential transfer in increasing ethanol concentration	5%	Biswas et al. (2014)
T. mathranii BG1L1	Immobilized continuous reactor	8%	Georgieva et al. (2007)
T. ethanolicus 39E-H8	Chemical mutagenesis	8%	Burdette et al. (2002)

C. Clostridium; *T. Thermoanaerobacter*; *Tm. Thermoanaerobacterium*

chemical mutagenesis). There are different adaptation strategies and their phenotypic effects have been discussed for some thermoanaerobes that are summarized in Table 20.1. With only a few cycles, evolutionary adaptation can yield strains with substantial improvements in ethanol tolerance (Table 20.1). Adaptation becomes more useful for newly discovered thermoanaerobes lacking important genetic information (Biswas et al. 2014).

20.4.2 Improvement by Metabolic Engineering

The screening of new thermoanaerobes has provided a foundation for industrially relevant CBP organisms, but efforts are needed for the targeted development that will enable homoethanolic fermentation of lignocellulosic biomass to bioethanol at desired production level (Scully and Orlygsson 2017; Olson et al. 2012, 2015). The development needs advancement in genetic tools that enable targeted elimination of genes leading to the formation of by-products other than ethanol (Taylor et al. 2009; Schuster and Chinn 2013).

Genetic engineering of thermophiles is still in infancy, and difficulties arise due to their unique physiological and genetic features such as (1) low G+C content, (2) need of thermostable markers and plasmids for transformation, (3) need of anaerobic conditions, (4) lack of proper medium for solidification at high temperature, and (5) limitation of established protocols for final selection of strains (Akinosho et al. 2014; Taylor et al. 2009; Lin and Xu 2013).

Nonetheless, in recent years the increasing availability of genome sequences of different thermophilic anaerobic bacteria and the development of omics-based technology have provided new insight into the microbial metabolism of thermophilic anaerobes that had been isolated decades ago (Olson et al. 2015). Current strategies involved in the genetic engineering of model thermophilic anaerobic bacteria are presented in Table 20.2.

Table 20.2 Ethanol yield by genetically engineered thermophilic anaerobic bacteria

Engineered strain	Target	Ethanol yield (g L^{-1})	Substrate concentration (g L^{-1})	References
C. thermocellum M0971 [ΔpyrF, Δpta::gapDHp-cat]	Elimination of acetate production	0.8	(5) Cellobiose	Tripathi et al. (2010)
C. thermocellum M1570 [Δhpt, Δldh, Δpta]	Elimination of lactate and acetate production	5.6	(19.5) Cellulose	Argyros et al. (2011)
C. thermocellum LL1210 evolved [ΔhptΔhydGΔldhΔpflΔpta-ack]	Elimination of lactate and acetate production	22.4	(60) Cellulose	Tian et al. (2016)
C. bescii JWCB032 [Δldh, adhE overexpression]	Eliminate acetate production	0.7	(10) Cellobiose	Chung et al. (2014)
T. saccharolyticum TD1[Δldh]	Eliminate lactate production	1.8	(5) Xylose	Desai et al. (2004)
T. saccharolyticum ALK2 evolved [Δldh, Δpta, Δack]	Elimination of lactate and acetate production	33	(70) Xylose	Shaw et al. (2008)
T. saccharolyticum M0355 [Δldh, Δpta, Δack]	Elimination of lactate and acetate production	25.3	(50) Cellobiose	Shaw et al. (2011)
T. saccharolyticum M1051 [Δldh, Δpta, Δack, express in gure ABCDEFG at ldh locus]	Elimination of lactate and acetate production and urea utilization	54	Cellobiose	Shaw et al. (2012)
T. mathranii BG1G1 [Δldh, PxylGldA]	Elimination of lactate production and heterologous expression of glycerol dehydrogenase	7.35	(5) Xylose and (2.5) glycerol	Yao and Mikkelsen (2010b)
T. mathranii BG1E1 [Δldh, adhE upregulated]	Elimination of lactate production and overexpression of adhE	0.49[a]	Xylose	Yao and Mikkelsen (2010a)

[a]Yield presented in g/g

C. Clostridium; *T. Thermoanaerobacter*; *Tm. Thermoanaerobacterium*; *C. Caldicellulosiruptor*

20.4.3 Improvement by Co-culturing

In the nature, microbes with different characteristics live in close association and exhibit highly symbiotic relationship to benefit each other, termed as microbial consortium. These consortiums often provide the opportunity to perform a complex task which is not possible by the individual organisms alone. Co-culture offers benefits over monoculture such as increased rate of substrate utilization, higher product formation, efficient sugar consumption, shortened lag time of fermentation, and possible reduction of substrate inhibition (Xu and Tschirner 2014).

One aspect of co-culture application is sequential culture in which two organisms with completely different growth requirements can be applied (Wen et al. 2014). A successful co-culture requires optimization of growth conditions so that both strains become compatible for one-pot cultivation (Blumer-Schuette 2014). Tripathi et al. (2010) have suggested the possible reason may be the secretion of some undetected metabolites by one organism and its utilization by the next partner, which will convert it into some product (Tripathi et al. 2010).

Fundamentals of CBP can also be achieved by employing co-culture of two or more thermoanaerobes such that one organism is saccharolytic while the other one is ethanologenic (Akinosho et al. 2014; Blumer-Schuette 2014).

The essential traits for a saccharolytic thermoanaerobes are the presence of complete set of lignocellulose deconstruction machinery, high specific activity, and stability of enzymes. Favorable features of ethanologenic partner included the following: higher ethanol yield, homoethanolic fermentation, improved ethanol tolerance, and multisugar fermentation. The ethanol production potential of various thermophilic co-cultures (both wild-type and genetically engineered) are summarized in Table 20.3.

20.4.4 Improvement by Immobilization

Immobilization of thermophilic anaerobic bacteria in a continuous system is a very promising approach to achieve high ethanol yield (Xu and Tschirner 2014). Immobilization offers recycling of microbes, which is cost advantageous, and cell immobilization entrapped them in a polymer or porous surface which protects strains from pretreatment inhibitors. There are very few reports of immobilization of thermoanaerobes that are available (Georgieva et al. 2008; Sittijunda et al. 2013; Xu and Tschirner 2014). Xu et al. have applied encapsulation of thermophilic co-culture for single-step ethanol production at 57 °C with sodium alginate as immobilization matrix (Xu and Tschirner 2014). The immobilized co-culture showed significant increase in ethanol yield from 1.36 to 4.67 g L^{-1} on glucose and 3.81 to 4.67 g L^{-1} on avicel. The immobilization of *Thermoanaerobacter* BG1L1 on a fluidized bed reactor resulted in sugar-to-ethanol conversion efficiency of 68–78% with an ethanol yield of 0.39–0.42 g/g during a continuous operation of 143 days at 70 °C using undetoxified wheat straw as substrate (Georgieva et al. 2008). In another study,

Table 20.3 Ethanol yield by thermophilic anaerobic co-culture system

Saccharolytic thermoanaerobe	Ethanologenic thermoanaerobe	Substrate	Yield (g L^{-1})	References
C. thermocellum ATCC 35609	*T. pseudethanolicus* ATCC 33223	Cellulose	4.5	Ng et al. (1981)
C. thermocellum ATCC 35609	*T. pseudethanolicus* ATCC 33223	Cellulose	6.6	He et al. (2011)
C. thermocellum ATCC 35609	*Thermoanaerobacter* sp. X514	Cellulose	12.2	He et al. (2011)
C. thermocellum DSM 1313	*T. saccharolyticum* JW/SL YS485	Cellulose	<3	Argyros et al. (2011)
C. thermocellum M1570	*T. saccharolyticum* ALK2	Cellulose	38.1	Argyros et al. (2011)
C. thermocellum ATCC 31924	*T. saccharolyticum* ATCC 31960	Cellulose	16	Saddler and Chan (1984)
C. thermocellum ATCC 31924	*T. saccharolyticum* ATCC 31925	Cellulose	9.7	Saddler and Chan (1984)
C. thermocellum ATCC 27405	*C. thermolacticum* ATCC 43739	Cellulose	4.67	Xu and Tschirner (2014)
C. thermocellum ATCC 27405	*C. thermolacticum* ATCC 43739	Treated aspen powder	2.78	Xu and Tschirner (2014)
C. thermocellum ATCC 27405	*C. thermolacticum* ATCC 43739	Untreated aspen powder	1.37	Xu and Tschirner (2014)

C. Clostridium; *T. Thermoanaerobacter*; *Tm. Thermoanaerobacterium*

Thermoanaerobacter pantosaceus was successfully immobilized in an up-flow anaerobic sludge blanket (UASB) reactor with glucose and xylose as substrates, and ethanol production was improved by 11% with a yield of 0.44 g/g sugar consumed in 24 h (Sittijunda et al. 2013). Immobilization studies so far suggest that it can improve pH tolerance and ethanol tolerance and can offer high substrate loading. Further advances can make immobilization a choice process for the continuous thermophilic ethanol production.

20.5 Candidate Thermoanaerobes for Consolidated Bioprocessing

The growing interest in the use of thermoanaerobes or enzymes derived from them for CBP is an emerging paradigm (Akinosho et al. 2014; Resch et al. 2013; Brunecky et al. 2013; Singh et al. 2017; Chung et al. 2014). Due to this, cellulolytic thermoanaerobes belonging to the genera *Clostridium* and *Caldicellulosiruptor* have received increased attention to be developed for CBP process via both mutation and genetic engineering approaches. Apart from this, xylanolytic members of the genera *Thermoanaerobacterium* and *Thermoanaerobacter* have also been subjected to the development for CBP via co-culture approach. Current progress in developing few potential candidate thermoanaerobes for CBP is presented below.

20.5.1 Clostridium thermocellum

C. thermocellum, a cellulolytic thermophilic anaerobic bacterium common in soils, self-heated rotting biomass and hot springs (Akinosho et al. 2014). Due to its maximal catalytic action on crystalline cellulose, *C. thermocellum* cellulosome has attracted significant interest, making it one of the most studied and reviewed cellulase enzyme systems of thermoanaerobes so far (Bayer et al. 2008; Blumer-Schuette 2014). Though the bacterium is equipped with highly efficient cellulolytic machinery and performs rapid fermentation of cellulose to ethanol, it suffers from some key detriments. Wild-type *C. thermocellum* strains typically give an ethanol yield of 10–35% of the theoretical maximum (Olson et al. 2015) and lower ethanol tolerance of 0.4–1.6% (v/v) (Brown et al. 2011). In addition to this, the inability of pentose sugars fermentation by this bacterium leaves hemicelluloses portion of the lignocellulosic biomass unutilized (Akinosho et al. 2014). More approaches have been applied to develop a high ethanol yielding strain, which includes: (i) targeted knockout of genes responsible for by-product formation (e.g., lactate, acetate, and formate) (Argyros et al. 2011; Biswas et al. 2014; Tian et al. 2016; Papanek et al. 2015) and (ii) directed adaptive evolution to increase ethanol tolerance (Brown et al. 2011; Shao et al. 2011).

The successful genetic engineering approaches so far resulted in *C. thermocellum* strain AG553 (*hpt, hydG, ldh, pfl*, and *pta-ack*), which is devoid of all the genes leading to the formation of products other than ethanol. The engineered strain presented an ethanol yield of 63.5% of the theoretical maximum albeit from lower cellulose concentration (5 g L^{-1}) (Papanek et al. 2015). More recently adaptive evolution of strain AG553 to achieve high biomass loadings was resulted in an evolved strain LL1210 (Tian et al. 2016). With about 95% cellulose conversion, this evolved strain LL1210 presented an impressive ethanol yield of 22.4 g L^{-1} (75% of the theoretical maximum) from 60 g L^{-1} initial cellulose loadings. These reports highly recommend the future research endeavors for the development of *C. thermocellum* as a potential CBP candidate thermoanaerobe.

20.5.2 Caldicellulosiruptor *sp.*

About twelve species of extremely thermophilic anaerobic bacteria within the genus *Caldicellulosiruptor* have been identified so far (Zurawski et al. 2015). All these members are capable to hydrolyze either polymeric or monomeric component of lignocellulosic biomass. *Caldicellulosiruptor* sp. has received increased interest particularly due to its unique enzymatic machinery involving highly efficient multifunctional enzymes in multimodular organization and broad substrate spectrum (Vanfossen et al. 2009).

One of the members of this genus, *Caldicellulosiruptor bescii* (formerly known as *Anaerocellum thermophilum*) is an extreme thermophile, possess the ability to grow up to 90 °C (Dam et al. 2011). The bacterium *C. bescii* DSM 6725 is specifically recognized for its ability to simultaneously solubilize insoluble lignin and other plant cell wall polysaccharides as carbon and energy sources (Kataeva et al. 2013). In particular, this bacterium displayed growth and utilization of different types of untreated biomass with low lignin (napier and bermuda) and high lignin (switchgrass) contents at 70 °C (Yang et al. 2009). Another study highlighted its ability to ferment industrially relevant high loads ($\sim$200 g L^{-1}) of untreated switchgrass and crystalline cellulose without growth inhibition (Basen et al. 2014). Realizing the direct biomass conversion potential, Chung et al. (2014) developed metabolically engineered ethanol-producing strains of *C. bescii*, accomplished by the deletion of genes involving lactic acid production and heterologous expression of a bifunctional acetaldehyde/alcohol dehydrogenase. This single-step ethanol production from untreated switchgrass at 33% of the maximum theoretical yield suggested a new paradigm of CBP. In addition to ethanol, one other strain, *C. saccharolyticus,* studied for its ability to produce hydrogen from switchgrass in single step without physicochemical or biological treatment (Talluri et al. 2013).

20.5.3 Thermoanaerobacterium saccharolyticum

T. saccharolyticum is an acid-tolerant (pH < 4.5), thermophilic, obligatory anaerobic bacteria having high hemicellulolytic potential and produces ethanol, acetate, lactate, H_2, and CO_2 as the by-products of xylan fermentation (Liu et al. 1996). *T. saccharolyticum* cannot be regarded as a true CBP candidate since it lacks cellulose degradation potential but can become an effective co-culture partner with other cellulolytic bacteria (Argyros et al. 2011) (see Table 20.3). In recent years, due to high amenability for genetic engineering, key enzymes involved in the catabolic pathway of *T. saccharolyticum* have been elucidated and highly effective metabolic engineering was performed to achieve homoethanolic fermentation by this bacterium. Both adaptive evolution and metabolic engineering approaches result in an engineered strain ALK2 which achieved an ethanol yield of 33 g L^{-1} (92% of the theoretical maximum) (Shaw et al. 2008).

20.5.4 Thermoanaerobacter mathranii

T. mathranii is a highly efficient pentose fermenting obligatory anaerobic thermophilic bacteria that gives ethanol yields of 62–90% of the theoretical maximum from glucose, xylose, and mannitol (Yao and Mikkelsen 2010b; Georgieva et al. 2008). Similar to *T. saccharolyticum*, *T. mathranii* is not a true CBP candidate but could be developed for CBP using co-culture approach. One of the patented strains of *T. mathranii* BG1 is under development by a Danish company BioGasol (www.Biogasol.com) for its industrial application in second-generation ethanol production. Apart from the above-mentioned candidates, few other members of the genus *Thermoanaerobacter* such as *T. aotearoense*, *T. ethanolicus*, and *T. italicus* are under development both for CBP co-culture approach and for direct fermentation of sugar-rich waste stream derived during conventional biomass processing.

20.6 Challenges in the Commercialization of Consolidated Bioprocessing

CBP is not a mature technology like conventional bioprocessing and is still under development. But realizing the potential of this highly integrated process, a few companies have adapted this technology for commercial scale production of cellulosic ethanol (Schuster and Chinn 2013; Salehi Jouzani and Taherzadeh 2015). Qteros was the first start-up based on CBP technology with a mesophilic anaerobic bacterium *C. phytofermentans* (trademarked as "Q microbe") which is their proprietary strain (http://www.qteros.com/). The target was to achieve a fuel cost of 1 dollar/gallon of ethanol. The company has also joined a commercial agreement with Praj Industries, India, in 2011 (Eco-Business 2011). Despite significant advancement in technology, the company had to close its demonstration facility in early 2012 due to financial problems (Freeman 2012). Mascoma, a US-based company established in 2005, has become a "proof of CBP concept" (http://www.mascoma.com/). Mascoma's platform is an engineered *Saccharomyces cerevisiae* strain (with heterologous expression of termite cellulases genes) and two engineered thermophilic anaerobic bacteria *T. saccharolyticum* and *C. thermocellum*.

Recently BluCon®-E DIREVO Industrial Biotechnology GmbH—A company based on novel CBP approach to low-cost lignocellulose fuel ethanol production at 70 °C by utilizing extremely thermophilic anaerobic bacteria. The cost price target is 2 dollars/gallon ethanol. The company has filed many patented strains out of which 7 are proprietary *Caldicellulosiruptor* strains and 8 are *Thermoanaerobacter* strains (http://www.direvo.com/). DEINOVE (http://deinove.com/en), a cleantech company has signed a collaboration agreement with ABENGOA, one of the world's leading ethanol producers, with the support of Bpifrance. The agreement was to focus for 36 months on the use of DEINOVE's Deinococcus bacteria for CBP to convert agricultural residues into ethanol at a competitive cost (Deinove 2014). A few companies have also started commercialization of potential sugar fermenting thermophilic anaerobic bacteria to pursue ethanol production. One such company is Estibio complemented by its sister company BioGasol ApS, founded in 2012 to commercialize its first- and second-generation ethanol production technologies (http://www.estibio.com/). The key finding of the technology is a proprietary "Pentocrobe" microbe (a genetically engineered thermophilic anaerobic bacteria; *T. italicus*) having co-sugar fermentation ability at 70 °C (http://www.estibio.com/products). The ongoing efforts and past success indicate that CBP-based biorefineries will be a reality in near future. Although for companies utilizing or attempting to utilize CBP technology, the major challenge would be to achieve maximum optimized condition in a single fermentor and faster processing time.

20.7 Conclusions and Future Prospects

This chapter presented multisubstrate fermenting thermoanaerobes as the most efficient candidates for the economic production of bioethanol via CBP. Thermoanaerobes are usually advantageous for CBP, however, lack necessary tolerance and production level of ethanol, pivotal for industrial conversion of lignocellulose. Companies focused on consolidated bioprocessing must overcome these challenges to enable efficient ethanol production. Different thermoanaerobes are in various stages of development to achieve higher productivity rates of bioethanol. Metabolic engineering efforts together with functional genomics aid in developing robust engineered thermophiles. With current pace of thermoanaerobe research, coming years should see rapid advancement in industrial application of these organisms as mono- and mixed cultures for single-stage bioethanol production.

Acknowledgements NS thank Deakin University, Australia and DBT-IOC for award of Ph.D. scholarship.

References

Akhtar N, Gupta K, Goyal D, Goyal A (2016) Recent advances in pretreatment technologies for efficient hydrolysis of lignocellulosic biomass. Environ Prog Sustain Energy 35:489–511

Akinosho H, Yee K, Close D, Ragauskas A (2014) The emergence of *Clostridium thermocellum* as a high utility candidate for consolidated bioprocessing applications. Front Chem 2:66

Amore A, Faraco V (2012) Potential of fungi as category I consolidated bioprocessing organisms for cellulosic ethanol production. Renew Sustain Energy Rev 16:3286–3301

Andersen RL, Jensen KM, Mikkelsen MJ (2015) Continuous ethanol fermentation of pretreated lignocellulosic biomasses, waste biomasses, molasses and syrup using the anaerobic, thermophilic bacterium *Thermoanaerobacter italicus* Pentocrobe 411. PLoS ONE 10:e0136060

Argyros DA, Tripathi SA, Barrett TF, Rogers SR, Feinberg LF, Olson DG, Foden JM, Miller BB, Lynd LR, Hogsett DA, Caiazza NC (2011) High ethanol titers from cellulose by using metabolically engineered thermophilic, anaerobic microbes. Appl Environ Microbiol 77: 8288–8294

Bartley L, Xu T, Zhang C, Nguyen H, Zhou J (2014) Switchgrass biomass content, synthesis and biochemical conversion to biofuels. CRC Press, Boca Raton, pp 109–169

Basen M, Rhaesa AM, Kataeva I, Prybol CJ, Scott IM, Poole FL, Adams MW (2014) Degradation of high loads of crystalline cellulose and of unpretreated plant biomass by the thermophilic bacterium *Caldicellulosiruptor bescii*. Bioresour Technol 152:384–392

Bayer EA, Lamed R, White BA, Flint HJ (2008) From cellulosomes to cellulosomics. Chem Rec 8:364–377

Bhalla A, Bansal N, Kumar S, Bischoff KM, Sani RK (2013) Improved lignocellulose conversion to biofuels with thermophilic bacteria and thermostable enzymes. Bioresour Technol 128: 751–759

Biswas R, Prabhu S, Lynd LR, Guss AM (2014) Increase in ethanol yield via elimination of lactate production in an ethanol-tolerant mutant of *Clostridium thermocellum*. PLoS ONE 9:e86389

Blumer-Schuette SE (2014) Thermophilic lignocellulose deconstruction. Fems Microbiol Rev 38:393–448

Blumer-Schuette SE, Kataeva I, Westpheling J, Adams MWW, Kelly RM (2008) Extremely thermophilic microorganisms for biomass conversion: status and prospects. Curr Opin Biotechnol 19:210–217

Brown SD, Guss AM, Karpinets TV, Parks JM, Smolin N, Yang S, Land ML, Klingeman DM, Bhandiwad A, Rodriguez M, Raman B, Shao X, Mielenz JR, Smith JC, Keller M, Lynd LR (2011) Mutant alcohol dehydrogenase leads to improved ethanol tolerance in *Clostridium thermocellum*. Proc Natl Acad Sci U S A 108:13752–13757

Brunecky R, Alahuhta M, Xu Q, Donohoe BS, Crowley MF, Kataeva IA, Yang S-J, Resch MG, Adams MWW, Lunin VV, Himmel ME, Bomble YJ (2013) Revealing nature's cellulase diversity: the digestion mechanism of *Caldicellulosiruptor bescii* CelA. Science 342: 1513–1516

Burdette D, Jung S-H, Shen G-J, Hollingsworth R, Zeikus J (2002) Physiological function of alcohol dehydrogenases and long-chain (C_{30}) fatty acids in alcohol tolerance of *Thermoanaerobacter ethanolicus*. Appl Environ Microbiol 68:1914–1918

Canganella F, Wiegel J (2014) Anaerobic thermophiles. Life 4:77–104

Chang T, Yao S (2011) Thermophilic, lignocellulolytic bacteria for ethanol production: current state and perspectives. Appl Microbiol Biotechnol 92:13–27

Chung D, Cha M, Guss AM, Westpheling J (2014) Direct conversion of plant biomass to ethanol by engineered *Caldicellulosiruptor bescii*. Proc Natl Acad Sci U S A 111:8931–8936

Dam P, Kataeva I, Yang S-J, Zhou F, Yin Y, Chou W, Poole FL, Westpheling J, Hettich R, Giannone R, Lewis DL, Kelly R, Gilbert HJ, Henrissat B, Xu Y, Adams MWW (2011) Insights into plant biomass conversion from the genome of the anaerobic thermophilic bacterium *Caldicellulosiruptor bescii* DSM 6725. Nucleic Acids Res 39:3240–3254

Deinove (2014) DEINOVE announces a collaboration agreement with a leading bioethanol producer. http://deinove.com/en/news/all-press-releases/deinove-announces-collaboration-agreement-leading-bioethanol-producer

Demain AL, Newcomb M, Wu JHD (2005) Cellulase, clostridia, and ethanol. Microbiol Mol Biol Rev 69:124–154

Desai SG, Guerinot ML, Lynd LR (2004) Cloning of L-lactate dehydrogenase and elimination of lactic acid production via gene knockout in *Thermoanaerobacterium saccharolyticum* JW/SL-YS485. Appl Microbiol Biotechnol 65:600–605

Eco-Business (2011) Soros-backed biofuels company Qteros partners with Indian ethanol plant builder. http://www.eco-business.com/news/soros-backed-biofuels-company-qteros-partners-indi/

Freeman S (2012) Qteros biofuels start-up closes Chicopee facility. http://www.masslive.com/news/index.ssf/2012/04/qteros_biofuels_start-up_close.html

Fu C, Mielenz JR, Xiao X, Ge Y, Hamilton CY, Rodriguez M, Chen F, Foston M, Ragauskas A, Bouton J (2011) Genetic manipulation of lignin reduces recalcitrance and improves ethanol production from switchgrass. Proc Natl Acad Sci U S A 108:3803–3808

Georgieva TI, Mikkelsen MJ, Ahring BK (2007) High ethanol tolerance of the thermophilic anaerobic ethanol producer *Thermoanaerobacter* BG1L1. Cent Eur J Biol 2:364–377

Georgieva TI, Mikkelsen MJ, Ahring BK (2008) Ethanol production from wet-exploded wheat straw hydrolysate by thermophilic anaerobic bacterium *Thermoanaerobacter* BG1L1 in a continuous immobilized reactor. Appl Biochem Biotechnol 145:99–110

Gorke B, Stulke J (2008) Carbon catabolite repression in bacteria: many ways to make the most out of nutrients. Nat Rev Micro 6:613–624

Haan R, Rensburg E, Rose SH, Gorgens JF, Zyl WH (2015) Progress and challenges in the engineering of non-cellulolytic microorganisms for consolidated bioprocessing. Curr Opin Biotechnol 33:32–38

Hasunuma T, Okazaki F, Okai N, Hara KY, Ishii J, Kondo A (2013) A review of enzymes and microbes for lignocellulosic biorefinery and the possibility of their application to consolidated bioprocessing technology. Bioresour Technol 135:513–522

He Q, Hemme CL, Jiang H, He Z, Zhou J (2011) Mechanisms of enhanced cellulosic bioethanol fermentation by co-cultivation of *Clostridium* and *Thermoanaerobacter* spp. Bioresour Technol 102:9586–9592

Himmel ME, Xu Q, Luo Y, S-y Ding, Lamed R, Bayer EA (2010) Microbial enzyme systems for biomass conversion: emerging paradigms. Biofuels 1:323–341

Kanafusa-Shinkai S, Wakayama J, Tsukamoto K, Hayashi N, Miyazaki Y, Ohmori H, Tajima K, Yokoyama H (2013) Degradation of microcrystalline cellulose and non-pretreated plant biomass by a cell-free extracellular cellulase/hemicellulase system from the extreme thermophilic bacterium *Caldicellulosiruptor bescii*. J Biosci Bioeng 115:64–70

Karimi K, Taherzadeh MJ (2016) A critical review on analysis in pretreatment of lignocelluloses: degree of polymerization, adsorption/desorption, and accessibility. Bioresour Technol 203:348–356

Kataeva I, Foston MB, Yang S-J, Pattathil S, Biswal AK, Poole Ii FL, Basen M, Rhaesa AM, Thomas TP, Azadi P, Olman V, Saffold TD, Mohler KE, Lewis DL, Doeppke C, Zeng Y, Tschaplinski TJ, York WS, Davis M, Mohnen D, Xu Y, Ragauskas AJ, Ding S-Y, Kelly RM, Hahn MG, Adams MWW (2013) Carbohydrate and lignin are simultaneously solubilized from unpretreated switchgrass by microbial action at high temperature. Energy Environ Sci 6:2186–2195

Kundu C, Trinh LTP, Lee H-J, Lee J-W (2015) Bioethanol production from oxalic acid-pretreated biomass and hemicellulose-rich hydrolysates via a combined detoxification process. Fuel 161:129–136

Lin L, Xu J (2013) Dissecting and engineering metabolic and regulatory networks of thermophilic bacteria for biofuel production. Biotechnol Adv 31:827–837

Linger JG, Darzins A (2013) Consolidated bioprocessing. In: Lee JW (ed) Advanced biofuels and bioproducts. Springer, New York, pp 267–280

Liu SY, Gherardini FC, Matuschek M, Bahl H, Wiegel J (1996) Cloning, sequencing, and expression of the gene encoding a large s layer associated endoxylanase from *Thermoanaerobacterium* sp. strain JW/SL-YS 485 in *Escherichia coli*. J Bacteriol 178:1539–1547

Lynd LR, van Zyl WH, McBride JE, Laser M (2005) Consolidated bioprocessing of cellulosic biomass: an update. Curr Opin Biotechnol 16:577–583

Ng TK, Ben-Bassat A, Zeikus JG (1981) Ethanol production by thermophilic bacteria: fermentation of cellulosic substrates by cocultures of *Clostridium thermocellum* and *Clostridium thermohydrosulfuricum*. Appl Environ Microbiol 41:1337–1343

Olson DG, McBride JE, Joe Shaw A, Lynd LR (2012) Recent progress in consolidated bioprocessing. Curr Opin Biotechnol 23:396–405

Olson DG, Sparling R, Lynd LR (2015) Ethanol production by engineered thermophiles. Curr Opin Biotechnol 33:130–141

Papanek B, Biswas R, Rydzak T, Guss AM (2015) Elimination of metabolic pathways to all traditional fermentation products increases ethanol yields in *Clostridium thermocellum*. Metab Eng 32:49–54

Parisutham V, Kim TH, Lee SK (2014) Feasibilities of consolidated bioprocessing microbes: from pretreatment to biofuel production. Bioresour Technol 161:431–440

Rani KS, Seenayya G (1999) High ethanol tolerance of new isolates of *Clostridium thermocellum* strains SS21 and SS22. World J Microbiol Biotechnol 15:173–178

Resch MG, Donohoe BS, Baker JO, Decker SR, Bayer EA, Beckham GT, Himmel ME (2013) Fungal cellulases and complexed cellulosomal enzymes exhibit synergistic mechanisms in cellulose deconstruction. Energy Environ Sci 6:1858–1867

Saddler JN, Chan MKH (1984) Conversion of pretreated lignocellulosic substrate to ethanol by *Clostridium thermocellum* in mono-culture and co-culture with *Clostridium thermosaccharolyticum* and *Clostridium thermohydrosulphuricum*. Can J Microbiol 30:212–230

Saha BC (2003) Hemicellulose bioconversion. J Ind Microbiol Biotechnol 30:279–291

Salehi Jouzani G, Taherzadeh MJ (2015) Advances in consolidated bioprocessing systems for bioethanol and butanol production from biomass: a comprehensive review. Biofuel Res J 2:152–195

Sánchez ÓJ, Montoya S (2013) Production of bioethanol from biomass: an overview. In: Gupta VK, Tuohy MG (eds) Biofuel technologies: recent developments. Springer, Berlin, pp 397–441

Schuster BG, Chinn MS (2013) Consolidated bioprocessing of lignocellulosic feedstocks for ethanol fuel production. Bioenergy Res 6:416–435

Scully SM, Orlygsson J (2017) Recent advances in genetic engineering of thermophilic ethanol producing bacteria. In: Gosset G (ed) Engineering of microorganisms for the production of chemicals and biofuels from renewable resources. Springer International Publishing, Cham, pp 1–29

Shao X, Raman B, Zhu M, Mielenz JR, Brown SD, Guss AM, Lynd LR (2011) Mutant selection and phenotypic and genetic characterization of ethanol tolerant strains of *Clostridium thermocellum*. Appl Microbiol Biotechnol 92:641–652

Shaw AJ, Podkaminer KK, Desai SG, Bardsley JS, Rogers SR, Thorne PG, Hogsett DA, Lynd LR (2008) Metabolic engineering of a thermophilic bacterium to produce ethanol at high yield. Proc Natl Acad Sci U S A 105:13769–13774

Shaw AJ, Covalla SF, Hogsett DA, Herring CD (2011) Marker removal system for *Thermoanaerobacterium saccharolyticum* and development of a markerless ethanologen. Appl Environ Microbiol 77:2534–2536

Shaw AJ, Covalla SF, Miller BB, Firliet BT, Hogsett DA, Herring CD (2012) Urease expression in a *Thermoanaerobacterium saccharolyticum* ethanologen allows high titer ethanol production. Metab Eng 14:528–532

Singh N, Mathur AS, Tuli DK, Gupta RP, Barrow CJ, Puri M (2017) Cellulosic ethanol production via consolidated bioprocessing by a novel thermophilic anaerobic bacterium isolated from a Himalayan hot spring. Biotechnol Biofuels 10:73

Sittijunda S, Tomás AF, Reungsang A, Sompong O, Angelidaki I (2013) Ethanol production from glucose and xylose by immobilized *Thermoanaerobacter pentosaceus* at 70 °C in an up-flow anaerobic sludge blanket (UASB) reactor. Bioresour Technol 143:598–607

Tailliez P, Girard H, Millet J, Beguin P (1989) Enhanced cellulose fermentation by an asporogenous and ethanol tolerant mutant of *Clostridium thermocellum*. Appl Environ Microbiol 55:207–211

Talluri S, Raj SM, Christopher LP (2013) Consolidated bioprocessing of untreated switchgrass to hydrogen by the extreme thermophile *Caldicellulosiruptor saccharolyticus* DSM 8903. Bioresour Technol 139:272–279

Tan H-T, Corbin KR, Fincher GB (2016) Emerging technologies for the production of renewable liquid transport fuels from biomass sources enriched in plant cell walls. Front Plant Sci 7:1854

Taylor MP, Eley KL, Martin S, Tuffin MI, Burton SG, Cowan DA (2009) Thermophilic ethanologenesis: future prospects for second generation bioethanol production. Trends Biotechnol 27:398–405

Tian L, Papanek B, Olson DG, Rydzak T, Holwerda EK, Zheng T, Zhou J, Maloney M, Jiang N, Giannone RJ, Hettich RL, Guss AM, Lynd LR (2016) Simultaneous achievement of high ethanol yield and titer in *Clostridium thermocellum*. Biotechnol Biofuels 9:116

Timmons MD, Knutson BL, Nokes SE, Strobel HJ, Lynn BC (2009) Analysis of composition and structure of *Clostridium thermocellum* membranes from wild-type and ethanol-adapted strains. Appl Microbiol Biotechnol 82:929–939

Tripathi SA, Olson DG, Argyros DA, Miller BB, Barrett TF, Murphy DM, McCool JD, Warner AK, Rajgarhia VB, Lynd LR, Hogsett DA, Caiazza NC (2010) Development of *pyrF* based genetic system for targeted gene deletion in *Clostridium thermocellum* and creation of a *pta* mutant. Appl Environ Microbiol 76:6591–6599

Turner P, Mamo G, Karlsson EN (2007) Potential and utilization of thermophiles and thermostable enzymes in biorefining. Microb Cell Fact 6:9

Vanfossen AL, Verhaart MR, Kengen SM, Kelly RM (2009) Carbohydrate utilization patterns for the extremely thermophilic bacterium *Caldicellulosiruptor saccharolyticus* reveal broad growth substrate preferences. Appl Environ Microbiol 75:7718–7724

Wen Z, Wu M, Lin Y, Yang L, Lin J, Cen P (2014) A novel strategy for sequential co-culture of *Clostridium thermocellum* and *Clostridium beijerinckii* to produce solvents from alkali extracted corn cobs. Process Biochem 49:1941–1949

Xu L, Tschirner U (2014) Immobilized anaerobic fermentation for biofuel production by *Clostridium* co-culture. Bioprocess Biosyst Eng 37:1551–1559

Xu Q, Resch MG, Podkaminer K, Yang S, Baker JO, Donohoe BS, Wilson C, Klingeman DM, Olson DG, Decker SR (2016) Dramatic performance of *Clostridium thermocellum* explained by its wide range of cellulase modalities. Sci Adv 2:e1501254

Yang S-J, Kataeva I, Hamilton-Brehm SD, Engle NL, Tschaplinski TJ, Doeppke C, Davis M, Westpheling J, Adams MWW (2009) Efficient degradation of lignocellulosic plant biomass, without pretreatment, by the thermophilic anaerobe "*Anaerocellum thermophilum*" DSM 6725. Appl Environ Microbiol 75:4762–4769

Yao S, Mikkelsen MJ (2010a) Identification and overexpression of a bifunctional aldehyde/alcohol dehydrogenase responsible for ethanol production in *Thermoanaerobacter mathranii.* J Mol Microbiol Biotechnol 19:123–133

Yao S, Mikkelsen MJ (2010b) Metabolic engineering to improve ethanol production in *Thermoanaerobacter mathranii.* Appl Microbiol Biotechnol 88:199–208

Zabed H, Sahu JN, Boyce AN, Faruq G (2016) Fuel ethanol production from lignocellulosic biomass: an overview on feedstocks and technological approaches. Renew Sustain Energy Rev 66:751–774

Zhang X-Z, Zhang Y-HP (2013) Cellulases: characteristics, sources, production, and applications. In: Yang ST, El-Enshasy HA, Thongchul N (eds) Bioprocessing technologies in biorefinery for sustainable production of fuels, chemicals, and polymers. Wiley, Hoboken, pp 131–146

Zurawski JV, Conway JM, Lee LL, Simpson HJ, Izquierdo JA, Blumer-Schuette S, Nookaew I, Adams MWW, Kelly RM (2015) Comparative analysis of extremely thermophilic *Caldicellulosiruptor* species reveals common and unique cellular strategies for plant biomass utilization. Appl Environ Microbiol 81:7159–7170

Chapter 21
Waste Valorization to Fuel and Chemicals Through Pyrolysis: Technology, Feedstock, Products, and Economic Analysis

Rupam Kataki, Neon J. Bordoloi, Ruprekha Saikia, Debashis Sut, Rumi Narzari, Lina Gogoi and Nilutpal Bhuyan

Abstract The decreasing fossil fuel reserves, rise in oil prices, and increasing awareness of environmental impact of continued fossil fuel use have made the quest for alternative energy sources significant throughout the world. In this regard, conversion of various types of wastes to biofuels and biomaterials offers a new paradigm of research in the changing world faced with these diverse problems. Lignocellulosic biomasses are the most predominant among different types of waste resources and are characterized by diverse nature and abundant supply. However, it also has numerous competitive uses which shrink the biomass resource base for energy production. There are numerous biomass materials which are produced as by-products, residues, or wastes from other processes, operations, or industries. The energy content of these materials can be usefully exploited and have the advantage of removing these materials from the landfill. This chapter presents an overview of the pyrolytic conversion of low-value biomass/bio-wastes, agricultural residues, bioenergy by-product, industrial agro-wastes, aquatic wastes, MSW, plastic solid wastes to bio-oil and biochar and their wide-ranging applications. Further, this chapter also reviews the pyrolysis technology and its economic analysis.

Keywords Pyrolysis · Biomass · MSW · Reactor · Biofuel · Biochar
Economic analysis

21.1 Introduction

The demand for energy is increasing sharply all over the world unlike its sources. But the present conventional sources of energy are inadequate to meet the increasing energy demands. According to International Energy Outlook 2016 (IEO

R. Kataki (✉) · N. J. Bordoloi · R. Saikia · D. Sut · R. Narzari · L. Gogoi · N. Bhuyan
Department of Energy, Tezpur University, Tezpur 784028, Assam, India
e-mail: rupamkataki@gmail.com

R. R. Singhania et al. (eds.), *Waste to Wealth*, Energy, Environment, and Sustainability, https://doi.org/10.1007/978-981-10-7431-8_21

2016), the world energy consumption will grow by more than 80% between 2010 and 2040. Total world energy use is forecasted to rise from 524 quadrillion British thermal units (Btus) in 2010 to 630 quadrillion Btus in 2020 and to 820 quadrillion Btus in 2040. In addition to such energy crisis, burning of fossil fuel is also responsible for increasing concentration of greenhouse gases in the atmosphere, subsequently resulting in global warming and associated adverse climatological impacts. Thus, rising energy demand, global energy crisis, and climate change have compelled almost all the nations of the world to search for renewable and alternative sources of energy. Renewable energy can contribute to socioeconomic development, energy access and security, energy safety, and diminishing the negative impacts on environment and health in addition to have a large prospective to mitigate climate change (IPCC 2017). The Renewable Energy Policy Network for the twenty-first Century (REN21) reported that renewable energy supplied 19% of global final energy consumption in 2012, of which modern renewable provided about 10% and the remaining 9% by traditional biomass. Furthermore, it is also projected that about 4.6 trillion kWh of renewable energy will be added to the grid by the end of 2035 (IEO 2011). Biomass, solar (e.g., photovoltaic solar cells and solar heat collectors), wind (e.g., wind turbines), water (e.g., hydropower, tidal energy), and geothermal resources are the sources of renewable energy. Biomass represents an abundant carbon-neutral renewable resource that can be converted to solid, liquid, and gaseous biofuels using appropriate conversion technologies. Biomass is clean as it contains a negligible amount of sulfur, nitrogen, and ash, which give lower emissions of SO_2, NO_x, and soot than conventional fossil fuels. Biofuels are considered among the most promising alternative options for increasing energy crisis and environmental degradation. This approach has triggered the scientific community to look forward toward the stepwise conversion of parts of the global economy into a sustainable bio-based economy with bioenergy, biofuels, and bio-based products as its main pillars which in turn ensure a gradual transition from petroleum-based economy to a diversified economy with renewable plant biomass as significant feedstock for both fuel and chemical production. Various processes can be used to convert biomass to different forms of energy. The biomass can be burned, transformed into a fuel gas through partial combustion, transformed into a biogas through fermentation, converted to bio-alcohol through biochemical processes, converted to biodiesel, pyrolyzed into a bio-oil, or transformed into a syngas from which chemicals and fuels can be synthesized. Thermochemical conversion technologies are advantageous over biochemical conversion technologies in their ability to utilize almost all types of biomass and recovery of both energy and the chemical value of the feedstock. There are four main thermochemical methods of converting biomass: pyrolysis, liquefaction, gasification, and combustion. Pyrolysis lies at the heart of all thermochemical fuel conversion processes, because in pyrolysis almost all types of feedstocks such as solid biomass as well as bio-wastes, which are difficult and costly to manage, can be converted to liquid, solid, and gaseous products depending upon various process parameters namely heating rate, final temperature, vapor residence time, sweeping gas flow rate, pyrolysis environment. Pyrolysis liquid often known as pyrolysis oil or bio-oil has advantages in transport, storage, combustion, retrofitting, and flexibility in production and marketing (Putun et al. 2002). Also, the biochar produced as a coproduct of pyrolysis can be a potential soil amendment with multiple benefits including increased soil fertility and C-sequestration (Choudhury et al. 2014; Saikia et al. 2015; Bordoloi et al. 2015). Pyrolysis produces energy fuels with high fuel-to-feed ratios, making it the most efficient process for biomass conversion and the method most capable of competing with and eventually replacing non-renewable fossil fuel resources (Demirbas 1998). The primary goal of pyrolysis is the optimization of high-value fuel products from biomass by thermal and catalytic means. In pyrolysis, the cellulose, lignin, and hemicelluloses components of wood pyrolyze largely to monomer and monomer-related fragments. Vapor phase cracking of the primary products proceeds through a stage of light hydrocarbons and oxygenates to the ultimate formation of aromatic tars and H_2, CO_2, CO, and H_2O (Evans and Milne 1987). Interest in the production of pyrolysis liquids from biomass and bio-wastes has grown rapidly in recent years, due to a number of possibilities (Harmsen and Powell 2011) such as decoupling liquid fuel production (scale, time, and location) from its utilization, minerals' separation on the site of liquid fuel production (to be recycled to the soil as a nutrient), producing a renewable fuel for boilers, engines, and turbines, power stations, and gasifiers, secondary conversion to motor fuels, additives, or special chemicals (biomass refinery), primary separation of the sugar and lignin fractions in biomass (biomass refinery). Industrialization, which is considered as the essence of modern development, however, also results in higher levels of energy consumption accompanied with enormous increase in the amount and diversity of waste generation which cannot only be resolved through disposal as it will create serious environmental and social problems. Alternative use of waste for energy production becomes increasingly interesting both from a waste management perspective to deal with increasing waste amounts while reducing the amount of waste deposited at landfills and from an energy system perspective to improve the flexibility of the energy system regarding feedstock diversity, increase the share of renewable energy, and reduce greenhouse gas emissions (Choudhury et al. 2014). These measures would reduce the quantity of wastes, generate a substantial quantity of energy from them, and greatly reduce pollution of water and air, thereby offering a number of social and economic benefits that cannot easily be quantified (IEO 2017). These bio-wastes can serve as an inexpensive raw material for conversion to value-added products including biofuels, chemicals, and improved animal feeds (Menon and Rao 2012; Choudhury et al. 2014). Production of biofuels and other value-added products from bio-wastes by thermochemical conversion also help to achieve the twin objectives of waste management, and energy and chemicals recovery.

21.2 Principles of Biomass Pyrolysis

Earlier, the term pyrolysis was generally associated with carbonization, in which the principal product was solid char. But nowadays, the term pyrolysis is often described as a process, in which oil is the preferred product. Since the past twenty years, many research works have been carried out to convert carbonaceous feedstock into liquids and gases, including valuable chemicals, intermediates, petrochemicals, and fuels by using the pyrolysis process. Therefore, the traditional slow pyrolysis can be used for the production of solid char, whereas the fast pyrolysis is used for producing higher-value fuel gas, fuel oil, or chemicals.

The changes that occur during pyrolysis process are shown below (Mohan et al. 2006):

(1) Transfer of heat from a source to increase the temperature inside the fuel;
(2) Primary pyrolysis reactions initiate at higher temperature and start to release volatiles and form char;
(3) The flow of hot volatiles toward cooler solids results during heat transfer between hot volatiles and cooler pyrolyzed fuel;
(4) Condensation of some of the volatiles also followed by secondary reactions to produce tar;
(5) Secondary pyrolysis reactions are autocatalytic, while primary pyrolytic reactions simultaneously occur; and
(6) During pyrolysis, thermal decomposition, reforming, water gas shift reaction, radical recombination, and dehydration reactions occur, which are a function of the process residence time/temperature/pressure profile.

Pyrolysis is a thermochemical decomposition of organic substance at higher temperatures (400–800 °C) in an inert atmosphere. Pyrolysis is a very complex process due to the occurrence of some of the simultaneous and successive reactions within the organic material. In this process, the organic constituents present in the biomass start to decompose as the temperature rises from 350 to 550 °C and then goes up to 700–800 °C in lack of air/oxygen. On thermal decomposition, the larger molecular weight compounds present in biomass break down into smaller molecules in the form of gases, condensable vapors (which in turn produce tars and oils on condensation), and solid residue as charcoal. Various process parameters such as reactor temperature, heating rate, pressure, reactor configuration, feedstock are accountable for rate and extent of decomposition of each component of biomass. The probable reaction with desirable end product of pyrolysis process can be depicted as (Balat et al. 2009; Jahirul et al. 2012):

$$\begin{aligned}\text{Biomass} \rightarrow \text{Biochar (FC, VM, Ash)} &+ \text{Liquid(Organic + Aqueous)} \\ &+ \text{Gas}(CO_2, CO, H_2, CH_4)\end{aligned}$$

(FC: Fixed carbon; VM: Volatile Matter)

Lignocellulosic biomass is a complex mixture of three major constituents, viz. hemicellulose, cellulose, and lignin. These constituents are unevenly distributed in the cell wall as a skeleton, linking material, and hard solids, respectively. Cellulose macromolecules regularly gather to form tough microfibers that function as the skeletal material of the cell wall, and the inner space is packed with amorphous hemicellulose and lignin linking material. Cellulose connects with hemicellulose or lignin molecules mainly through hydrogen bonds, while the connections between hemicellulose and lignin include both hydrogen and covalent bonds (Wang et al. 2017). Carbohydrates and lignin link tightly together in lignin-carbohydrate complexes, which results in residual carbohydrate or lignin fragments in extracted lignin or hemicellulose samples. The minor amounts of extractives present in the biomass decompose at different temperatures, and heating rate follows different mechanisms and pathways. The thermal degradation order of hemicelluloses, cellulose, and lignin can be summarized as follows (Balat 2008)

$$\text{Hemicelluloses} > \text{cellulose} > \text{lignin}$$

The complex reaction mechanisms of biomass pyrolysis can be defined in three main stages

$$\text{Biomass} \rightarrow \text{Water} + \text{Non - reacted residue} \quad (1)$$

$$\text{Non reacted residue} \rightarrow (\text{Volatile} + \text{Gases})_1 + (\text{Char})_1 \quad (2)$$

$$(\text{Char})_1 \rightarrow (\text{Volatile} + \text{Gases})_2 + (\text{Char})_2 \quad (3)$$

(Subscripts 1 and 2 indicate primary and secondary decomposition, respectively).

From the above reactions, it can be observed that pyrolysis proceeds in three steps. During the first stage of biomass decomposition, some rearrangement reactions such as water elimination, bond breakage, appearance of free radicals, formation of carbonyl, carboxyl, and hydroperoxide groups take place. In the second stage, decomposition of solid biomass occurs, which corresponds to the main stage of pyrolysis process. The decomposition proceeds to form the pyrolysis products with a high rate of reaction. Similarly, in the third stage, the char decomposes to form a carbon-rich residual solid at a very slow rate. The process of forming residual solid from the char is termed as secondary charring [Eq. (3)], which makes the char less reactive (Demirbas 2007). Further, biomass pyrolysis can also be categorized into either endothermic and/or exothermic reaction. Generally, cellulose pyrolysis is an endothermic process, whereas the lignin is an exothermic process. Similarly, wood pyrolysis is an exothermic process, where the main reason of heat generation is the secondary decomposition of volatiles that may be catalyzed by the remaining solid (Balat 2008).

21.3 Classification of Pyrolysis

Pyrolysis can be classified into six categories depending upon its operating conditions. Each category of pyrolysis has its own advantages and limitations. The main features and the operating conditions for each category are discussed below.

21.3.1 Slow Pyrolysis

Slow pyrolysis is traditionally used for the production of charcoal. In slow pyrolysis, biomass is pyrolyzed at a lower heating rate (0.1–0.8 °C/s) and longer residence time (5–30 min or even 25–35 h) at a temperature range of 300–550 °C. Charcoal is the main product of the slow pyrolysis process but liquid and gaseous products are also formed in a small quantity (Demirbas and Arin 2002).

In slow pyrolysis, lower heating rate and longer vapor residence time provide a suitable ambience and sufficient time for the secondary reactions to complete. Moreover, longer vapor residence time allows removing the vapors produced during the secondary reaction. This ultimately results in the formation of solid carbonaceous charcoal. With an increase in pyrolysis temperature, the charcoal yield decreases as the organic materials are combusted and cellulose and hemicellulose are destroyed at higher temperatures (Muradov et al. 2012; Demirbas 2004). The most commonly used reactors in slow pyrolysis are drum, rotatory kilns, and screw/auger (Roy and Dias 2017).

21.3.2 Fast Pyrolysis

The fast pyrolysis is characterized by higher heating rates (10–1000 °C/s) and very short residence times (0.5–2 s) up to a temperature range of 850–1250 °C. In fast pyrolysis, bio-oil yield is higher than the char and gaseous product yield. In general, fast pyrolysis produces 60–75% of liquid product, 15–25% of char, and 10–20% of non-condensable gaseous products. In fast pyrolysis, maximum amount of bio-oil yield (75%) can be obtained at around 500 °C. The biomass is heated up to a temperature at which thermal cracking can take place and the exposure time is minimized which favors the char formation. The higher heating rate favors conversion of the biomass to liquid product before it could react to form the undesired char. The bio-oil yield highly depends on feedstock properties and pyrolysis parameters. In general, bio-oil yields are the highest for woody biomass due to the presence of higher amount of cellulose and hemicellulose content compared to energy crops and agro-residues. After woody biomass, the highest bio-oil yields are reported for energy crops (reed), followed by agro-residues (flax straw, wheat straw, etc.). In fast pyrolysis, the product yields, i.e., bio-oil, char, and syngas, are affected by feedstock particle size. With the increase in particle size, the heat transfer rate decreases resulting increase in char yield and decrease in bio-oil and syngas yield. Moreover, smaller particle size mitigates internal heat transfer limitations and improves bio-oil yield. The bio-oil yield in fast pyrolysis can be maximized by optimizing both pyrolysis temperature and feedstock particle size (Roy and Dias 2017). The bio-oil produced by the fast pyrolysis is highly corrosive due to its low pH value. Higher heating value of bio-oil is approximately half of crude oil, which makes the upgrading of bio-oil necessary before using it (Tripathi et al. 2016).

21.3.3 Flash Pyrolysis

Flash pyrolysis is an improved and modified form of fast pyrolysis. The temperature reached in flash pyrolysis is in the range of 900–1200 °C with a very high heating rate of 1000 °C/s or even higher and a very short vapor residence time which is 0.1–1 s. Heat and mass transfer process along with chemical kinetics of the reactions and phase transition behavior of the biomass plays a crucial role in the product distribution in flash pyrolysis. The rapid heating rate combined with high temperature and low vapor residence time leads to high liquid yield but the char yield gets decreased. The biggest challenge to use flash pyrolysis on the industrial scale is to configure a reactor for flash pyrolysis, in which the input biomass particles reside for a very short time under the extremely high heating rate. The problem in flash pyrolysis reactors is the stability and quality of the bio-oil as it is strongly affected by the char/ash present in the product. Not only this, the char present in the bio-oil can catalyze the polymerization reaction inside the liquid product causing an increase in the viscosity of oil. Flash pyrolysis occurs with very fast heating rates and uses shorter solid residence time (0.5 s) than fast pyrolysis. The typical operating temperature for flash pyrolysis is 800–1000°C, and the biomass is supplied in the form of dust. This process gives a similar product distribution as fast pyrolysis (Roy and Dias 2017).

21.3.4 Vacuum Pyrolysis

Vacuum pyrolysis is the thermal degradation of biomass at low pressure in the absence of oxygen. The range of pressure usually taken is 0.05–0.20 MPa, and temperature range is kept between 450 and 600 °C. The rate of heat transfer in vacuum pyrolysis is similar to that of slow pyrolysis. In vacuum pyrolysis, low pressure/vacuum is used to remove the vapors. Also, low pressure tends the organic materials to be decomposed and devolatilized into its constituents at relatively low temperatures. The rapid removal of pyrolysis vapors produced during pyrolysis also decreases the vapor residence time considerably resulting in the reduction of secondary reactions and confirms the high liquid product yield. It is also observed that the vacuum pyrolysis of biomass also develops the porous structure of the product biochar. It was observed that the porosity of biochar produced depends upon composition of cellulose and lignin of the original biomass feedstock. It is observed that the biomass with high lignin content produces char having a macroporous structure and biomass with high cellulose content produces char having a microporous structure.

21.3.5 Intermediate Pyrolysis

Intermediate pyrolysis is usually employed in order to make a balance between liquid and solid products. High char and low liquid yield is the primary product of slow pyrolysis, while high liquid yield and comparatively low char are produced during fast pyrolysis. Intermediate pyrolysis is conducted in between slow and fast pyrolysis. The operating conditions during intermediate pyrolysis prevent the formation of high molecular weight tars and produce dry char, which can be used in agricultural fields and for other forms of energy generation along with good-quality bio-oil. The general operating condition for intermediate pyrolysis is with a pressure of 0.1 MPa. The temperature range selected for intermediate pyrolysis is 500–650 °C, heating rate range is 0.1–10 °C/min, and a residence time ranging from 300 to 1000 s. The product of intermediate pyrolysis generally contains 40–60% liquid, 20–30% non-condensable gases, and 15–25% char. The advantage of intermediate pyrolysis over the fast pyrolysis is that the liquid product does not contain high quantity of reactive tar and can be used directly in boilers and engines.

21.3.6 Hydro-pyrolysis

Hydro-pyrolysis is a comparatively new method for the conversion of biomass into high-quality bio-oil. The process involves hydrogen/hydrogen-based materials, which are fed to the reactor along with the biomass at a pressure higher than the atmospheric pressure ranging between 5 and 20 MPa. As heat transfer rate, residence time, and temperature are kept approximately similar to that of the fast pyrolysis, hydro-pyrolysis is often considered as fast pyrolysis under high pressure condition. As hydrogen is a reducing agent, the hydro-pyrolysis conditions reduce the oxygen content in the produced bio-oil and also obstruct the formation of char. In addition, hydro-pyrolysis enhances the hydrogen content of the liquid product. Often, catalyst is used during hydro-pyrolysis for the removal of O_2, water, and different CO_x from the liquid product. The use of catalyst also ensures the reduction in depolymerization and coking reactions. A very critical effect of the removal of oxygen and addition of hydrogen is the reduction in the requirement of recirculation of solid heat carriers as in catalytic hydro-pyrolysis both the pyrolysis stage and the catalytic stage are exothermic which is a significant advantage over other pyrolysis methods. The development of the catalyst for this purpose is still one of the most challenging parts of catalytic hydro-pyrolysis.

21.4 Feedstock for Pyrolysis

For establishing any bio-refinery, availability of rich quantities of lignocellulosic biomass is a necessary prerequisite. A sufficient quantity of renewable feedstocks like agricultural crops, residues, forestry materials, algal and other aquatic biomass are available worldwide for conversion into biofuel. Apart from these, energy crops like sweet sorghum and numerous grasses and municipal solid waste (MSW) are also drawing attention. However, all biomass resources cannot be considered as suitable for conversion into liquid fuels and the feedstock should be selected properly for thermochemical conversion and process optimization (Dhyani and Bhaskar 2017). The various feedstocks for pyrolysis are discussed below.

21.4.1 Biomass

Biomass is regarded as one of the important sources of renewable energy which can provide a solution to the problem of world energy crisis due to its potential availability as well as its positive impact on global warming and pollution (Hall 1997). Biomass is the only alternative energy source which can be converted to all the three forms of fuel, i.e., solid, liquid, and gas. Biomass pyrolysis has received a significant attention in the research arena during the last decade. Biomass is an intricate biological organic or inorganic solid material consisting of carbon, hydrogen, oxygen, and small amount of sulfur and nitrogen which can be derived from living or non-living organism and are naturally available. It comprises of different natural and derived resources like grass, woody and herbaceous species, aquatic plants, biosolids, agricultural residues, bagasse, wood wastes, sawdust. (Yaman 2004; Sharma et al. 2015). Wastes like animal manure, waste paper, sludge, and many industrial wastes are also considered as biomass because they are also a mixture of organic and inorganic compounds and can be processed to get energy (Tripathi et al. 2016). Biomass is mostly composed of three main components, cellulose (about 50%), hemicellulose [woods (10–30%) and herbaceous biomass (20–40%)], and lignin [woods (20–40%) and herbaceous biomass (10–40%)] on dry basis (Sharma et al. 2015). Other components present in biomass are extractives (protein, acids, salts) and minerals (alkali metals and chlorine in herbaceous biomass) (Zabaniotou 1999). These inorganic compounds remain in varying amount, in woods (less than 1%), herbaceous biomass and feedstock (15%), and agricultural and forestry residues (up to 25%) (Fitz et al. 1996).

21.4.2 Lignocellulosic Biomass

Among the biomass resources, lignocellulosic biomass is the most abundant non-edible biomass, largely composed of forestry and agricultural residues/wastes like woodchips and rice straw, wheat straw, corn stover, corn stalks, and sugarcane bagasse. Perennial bioenergy grasses, such as miscanthus, switchgrass, and giant cane, are also important grass lignocelluloses (Bentsen et al. 2014; Demirbas 2001).

21.4.3 Woody Biomass

The woody biomass can be categorized into two broad groups: softwood and hardwood (Pettersen 1984; Wang et al. 2017). Softwood originates from conifers and gymnosperm trees, including evergreen species, such as fir, pine, cedar, hemlock, spruce, and cypress. Softwood grows faster and is less dense than hardwood. Hardwood comes from angiosperm plants, most of which are deciduous. Willow, poplar, oak, cottonwood, and aspen are typical examples of hardwood. The cell structure of hardwood is more complex than that of softwood, containing large water-conducting pores or vessels that are surrounded by narrow fibrous cells (Brandt et al. 2013). Several varieties of woody biomass have been used as feedstock for pyrolysis. Different softwoods that have been studied include cedar, Douglas fir, pine, redwood, and spruce. The hardwood varieties that have been majorly studied are bald cypress, beech, birch, chestnut, eucalyptus, hybrid poplar, and oak. Moreover, wood wastes such as sawdust, particle board, waste from furniture factories, and other wood processing industries have also been used for pyrolysis (Dhyani and Bhaskar 2017).

21.4.4 Municipal Solid Waste (MSW)

The disposal as well as treatment and management of municipal solid waste (MSW) are a matter of major concern worldwide. The pyrolysis of MSW is considered as an innovative alternative for treatment of such wastes and to obtain different chemicals and fuels (Chen et al. 2014). MSW comprises of numerous materials which significantly differ in composition depending on community, consumer's incomes, and lifestyles, and the degree of industrialization, institutionalism, and commercialism. Generally, MSW includes leather, plastics, paper, rubber, metals, glass, ceramics, textiles, organic waste, soil materials, and other miscellaneous materials (Czajczynska et al. 2017).

The energy produced in MSW pyrolysis can be considered cleaner as the amounts of NO_x and SO_2 produced are lower due to inert atmosphere of the pyrolysis process. Moreover, the syngas produced can be washed before its combustion. The char produced from the pyrolysis of MSW is also of better quality (Saffarzadeh et al. 2006). Due to the added benefit of decentralization of pyrolysis system at a small scale which eventually reduces the cost of transportation, pyrolysis of MSW is receiving increasing attention in both small towns and big cities. Due to capital cost limitations, the distributed MSW treatment facilities fail to ensure environmental safety. A pyrolysis plant of proper capacity with energy product outputs can be considered as a suitable alternative when the products' quality is under fine control (Chen et al. 2014).

21.4.5 Plastic Solid Waste (PSW)

Plastic plays an important role in improving the living standard of human, and its demand has been ever-increasing. The continuous rise of plastic demand led to the increasing waste accumulation every year. This causes a serious environmental problem as plastic waste may take billions of years to degrade naturally. Plastics, derived from petrochemical sources which have high calorific value, can be converted to valuable energy. As plastic waste pyrolysis can be used for waste minimization, it has been gaining attention of late. Pyrolysis can produce a high amount of liquid oil (up to 80 wt%) at a moderate temperature of 500 °C (Sharuddin et al. 2016). Moreover, the parameters of pyrolysis process can be optimized to get the maximum product yield.

Thermoplastics and thermosets are the main types of plastics which are used as daily commodities. PSW mainly consists of high-density polyethylene (HDPE), low-density polyethylene (LDPE), polypropylene (PP), polystyrene (PS), polyvinyl chloride (PVC), and polyethylene terephthalate (PET). Polyethylene (PE) makes up about 40% of the PSW in the total waste stream making it a very common type of waste to accumulate in urban environments (Onwudili et al. 2009; Al-Salem et al. 2017). The two major polyolefin (PO) polymers used in the market, i.e., PE and PP, consist mainly of carbon and hydrogen atoms. It has been reported that PE has carbon content in the range of 83.9–86.1%, while the carbon content of PP ranges between 85.5 and 86.1% (Sorum et al. 2001). Due to the presence of high carbon content in the major polymers constituting PSW, pyrolysis has been a favored method for waste treatment. Other common types of plastics such as PS, PVC, and PVA have a high aromatic content and when pyrolyzed yield high fractions of aromatics (Al-Salem et al. 2017).

The pyrolysis of PSW can be considered as green technology as it does not cause water contamination like recycling. Moreover, the gaseous by-product of pyrolysis has a considerable calorific value and can compensate the overall energy requirement of the pyrolysis plant (Abnisa and Wan Daud 2014). As pyrolysis does not require an intense sorting process, it is less labor-intensive, and as such, the process handling is much easier and flexible when compared to recycling (Al-Salem et al. 2017).

21.4.6 Agricultural Waste

Agricultural waste which includes various agricultural crops like stalks, straw, shells of these crops is another important feedstock for biomass pyrolysis. Pyrolysis of agricultural waste is necessary, as a large part of the crop body is non-edible and goes as waste. For example, straw contributes to almost 50% of the yield of cereal crops and can produce more biochar in comparison with woody biomass (Zanzi et al. 1996). The agricultural wastes that have been used for pyrolysis include husks (olive, rice, etc.), shells (almond, cocoon, cotton, groundnut, hazelnut, etc.), stover (corn, stalk, cotton, rape, etc.), straws (corn, cotton, rape, wheat, etc.), and non-edible seeds (babool, grape, karanja, linseed, mahua, niger, rape, etc.). Apart from this, pyrolysis of other agricultural wastes, like apricot stones, bagasse, banana leaves, cherry stones, coir pith, cotton seed cake, garlic stem, grape residue, jatropha residue, jute sticks, olive waste, pepper stem, safflower seeds, sorghum bagasse, sunflower bagasse, tea waste, and tobacco waste, has also been reported. However, there is some limitation in using agricultural residues as feedstock for pyrolysis. The bulkiness of the agricultural residues limits their availability only in the local market as transportation costs are higher. The feedstock composition is also unpredictable and susceptible to dirt leading to equipment wear and increased maintenance cost (Dhyani and Bhaskar 2017).

21.4.7 Energy Crops

Energy crops can be grown on land which is not suitable for agricultural crops, thus reducing the competition for arable land. Bamboo, sweet sorghum, and many other varieties of grasses are the examples of crops with high energy values and have been studied for pyrolysis. Bamboo, with more than 1250 species, is a fast-growing, large woody grasses and abundant in Asia. In some bamboo species, the lignin content is as high as 29% (Scurlock et al. 2000). Sweet sorghum, a genus of grasses, is found in parts of Asia and Africa. The lignin content of raw sweet sorghum is 14–16% on a dry basis. Due to its drought tolerance, relatively low requirements of water and manure, and high yield under varied environmental conditions, sweet sorghum is gaining importance as an agricultural energy crop. The large sugar content in sweet sorghum can be converted to ethanol, while the bagasse can be used as a feedstock for thermochemical conversion processes (Zhao et al. 2009). In general, grasses have a high content of hemicellulose. The composition of a typical grass is cellulose 25–40%, hemicellulose 35–50%, and lignin 10–30% (Prasad et al. 2007). Various species of grasses like Bermuda grass, elephant grass, esparto grass, giant cane, switchgrass have been reported as a feedstock for pyrolysis (Dhyani and Bhaskar 2017). Apart from the feedstocks mentioned above, pyrolysis has also been done using aquatic biomass, cattle manure, household waste, oil shale, poultry litter, and sewage sludge as a feedstock.

21.5 Pyrolysis Reactor

The vital component of an efficient pyrolysis process is its reactor. The type of contact between particles in the reactor adds to the complexities of pyrolysis operation, and these depend on the configuration of the pyrolysis reactor (Jahirul et al. 2012). The parameters like size of the particles and vapor residence time have little effect on the amount of bio-oil, while the composition of bio-oil is greatly affected by these parameters (Wang et al. 2005). Many reactor technologies and designs have been researched over the last couple of years to get the optimum pyrolysis product as well as to produce high and suitable quality liquid yields. Nevertheless, various reactors have distinct features, ability to obtain different amount of product yields, advantages, and disadvantages. Some popular types of reactor designs are illustrated in the subsequent sections.

21.5.1 Fixed Bed Reactor

These are most widely used for large-scale commercial operations as well as for small-scale power and heat applications. Due to the high efficiency and ease and simplicity of operation, these reactors will be considered in large-scale commercial application. Although they have some challenges in scale-up and heat transfer limitation as well as elimination of tar, the current study in different conversion procedures, i.e., thermal and catalytic processes, has given feasible solution for tar removal (Kabir et al. 2017; Yang et al. 2006; Rao et al. 2004). Other important characteristics of fixed bed reactor are the high carbon conservation, long solid residence time, low gas velocity, and low ash carryover (Altafini et al. 2003; Leung et al. (2004). The technology used in these reactors is simple, suitable, and reliable for fine particle feedstock having comparatively uniform dimension (Chopra and Jain 2007). In fixed bed pyrolysis system, the reactor is fitted with a thermocouple to monitor the pyrolysis temperature and a gas entry pipe to supply N_2 gas into the reactor to create inert condition for the pyrolysis reactions. Fixed bed reactors can be made by using firebricks, steel, or concrete. Different parts of these reactors include a fuel feeding unit, an ash removal unit, a gas cooling and cleaning system, and a gas exit. In fixed bed reactor, solids move down through a vertical shaft and contact a countercurrent upward moving product gas stream. The presence of cyclone, wet scrubbers, and dry scrubber facilitates the cooling system and gas cleaning (Jahirul et al. 2012) (Fig. 21.1).

Fig. 21.1 Fixed bed reactor

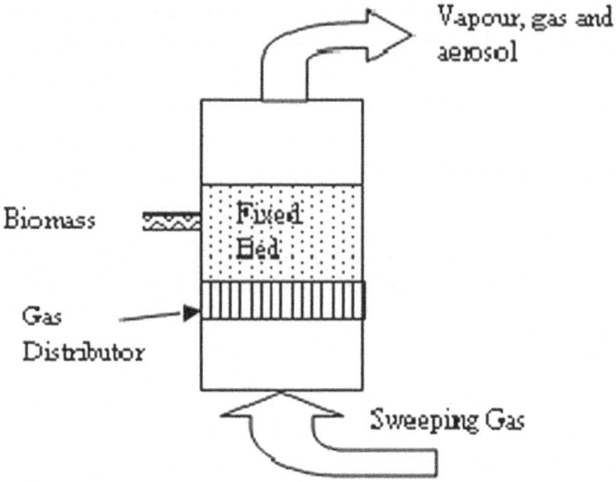

21.5.2 Fluidized Bed Reactors

Fluidized bed reactors are one of the most successful, popular, and the most explored pyrolysis reactors. On mixing the biomass particles at a high temperature in a stream of moving sand particles, biomass particles experience some rapid heating in these reactors. Due to the mixing of sand with biomass particles, high heat and mass transfer coefficients are achieved which is a favorable condition for fast pyrolysis. Heat to the bed is provided by the external combustion of the produced bio-oil or char, and transfer of this heat to the bed is occurred by both direct path (addition of hot solids to the bed) and indirect path (by passing hot gas/ steam through the tubes in the bed). Some other advantages of using fluidized bed reactor are as follows: It has a simple design, proficient control for pyrolysis reaction as well as over vapor residence time, wide range of contact area between the surface of liquefied portion and solid mass per unit size of bed, high relative velocity between both phases, and sufficient heat transfer within the system (Lv et al. 2004). However, it has some disadvantages like large reactor size, and the high construction and operation cost (Muradov and Veziroglu 2008). The fluidized bed reactors are mainly classified into three basic types: (a) bubbling, (b) entrained, and (c) circulating. These are explained below.

21.5.2.1 Bubbling Fluidized Bed Reactors

Bubbling fluidized bed reactors have a simpler design and are also simple to construct and operate. The high solid density in the bed of these reactors resulted in good control over temperature, efficient transfer of heat to biomass, solid-to-gas contact area, and storage ability (Jahirul et al. 2012; Scott et al. 1999; Mohan et al. 2006). Generally, sand is the solid phase of the bed and responsible for the rapid

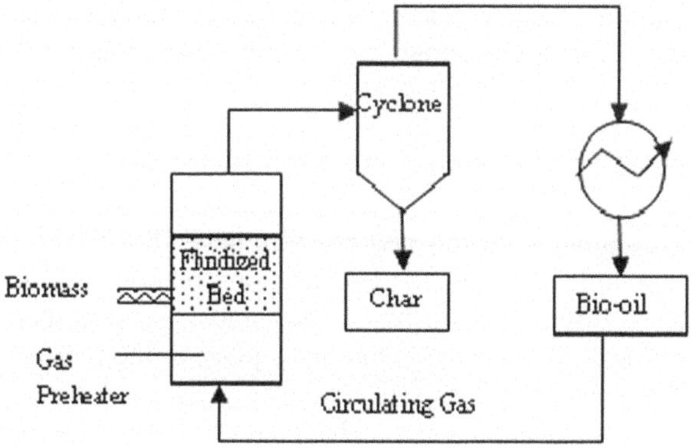

Fig. 21.2 Bubbling fluidized bed reactor (*Source* Dhyani and Bhaskar 2017)

heating of the biomass in an oxygen-free atmosphere, where it is decomposed into solid char, gas, vapors, and aerosols. Bubbling fluidized beds give the optimized yield of bio-oil products, which are good in quality. After exit from the reactor zone, the charcoal can be removed by a cyclone separator and stored. As char is rapidly separated, accumulation of char does not occur in the fluidized bed. Since the residence time of char is higher than that of vapor, it can act as an efficient catalyst in pyrolysis reaction temperatures for cracking of vapor, which necessitates the rapid and effective separation of char. Hence, it is important to carefully design the sand and biomass/char hydrodynamics. The remaining scrubbed gases, vapors, and aerosols then enter a quencher, where these are condensed to bio-oil. These bio-oils are then accumulated and stored, and the remaining vapors or the syngas may be recycled and reused to heat the reactor. The fluidizing gas flow rate controls the solid residence time and vapor. Another important feature of bubbling fluidized bed reactor is that in order to achieve a high biomass heating rate, it requires biomass of smaller particle sizes (less than 2–3 mm) (Dhyani and Bhaskar 2017) (Fig. 21.2).

21.5.2.2 Entrained Fluidized Bed Reactor

The entrained fluidized bed reactors are the reactors which operate at the co-current flow, and in this reactor, the biomass particles fall freely under the force of gravity or entrained under the influence of gas pulling force (Dhyani and Bhaskar 2017). The absolute features of this reactor are the data acquisition equipment, gas provider, fuel feeding system, gas preheating, and controlled extraction of gas and particles. Gravimetric screw feeder and an inclined vibrator are present in the fuel feeding system, which help in reducing the fluctuations. The grounded fuel reaches the reactor through central water-cooled air injection probe that passes through the gas preheater. At the top of the reactor, both fuel and preheated air are mixed. The internal dimension of the reactor is as follows: (i) 2 m in length and (ii) 79 mm in diameter which can withstand temperature up to 1500 °C. This reactor is suitable for studying heterogeneous reactions at high temperature and short residence times. The main disadvantage with the reactor is usage of huge amount of carrier gas.

21.5.2.3 Circulating Fluidized Bed Reactor

Circulating fluidized bed reactors have shorter residence time for char and vapor than bubbling fluidized bed reactors; otherwise, it is quite similar. The short residence time results in higher gas velocity and higher char content, and thus, faster transport of vapor and char takes place in the reactor. On the other hand, they have higher processing capacity, improved contact between the gas and solids, and better capacity to hold solids which are hard to convert to liquid. As the circulating fluidized bed reactors are mainly dependent on gas–solid convective transfer, the heat transfer rates of these reactors are not particularly high. In fluidized bed reactors, a separate reactor unit known as secondary char combustor is used as the typical heat supplier, which provides the heat for the process by burning the char for heating the sand and recirculation of the hot sand into the reactor. This leads to the possibility of ash transfer to the pyrolyzer (Dhyani and Bhaskar 2017; Boukis et al. 2001). Biomass ash is well known as a cracking catalyst for the organic molecules in the volatile pyrolysis products, which cause a loss of volatiles from the bio-oil yield (Mohan et al. 2006; Scott et al. 1999). The advantage of circulating fluidized bed reactor is that it is apt for large throughputs in spite of the complex hydrodynamics (Mohan et al. 2006) (Fig. 21.3).

21.5.3 *Ablative Reactor*

The reaction that occurs in ablative reactor is similar to melting of butter in a frying pan, in which the rate of melting can be considerably increased by pressing and

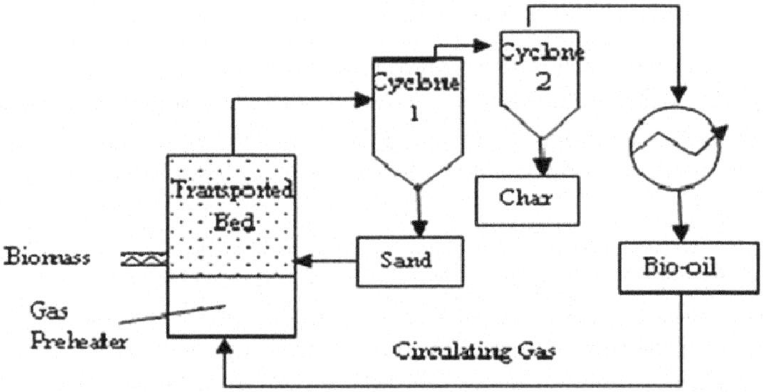

Fig. 21.3 Circulating fluidized bed reactor (*Source* Dhyani and Bhaskar 2017)

moving the butter over the hot pan surface. Ablative pyrolysis is primarily different from other pyrolysis processes as the reaction is limited by the rate of heat transfer taking place on sliding of biomass particles through a hot solid surface. On contact with the heated wall, materials start to melt, and after their exit, the remaining liquids start to evaporate as pyrolysis vapors (Jahirul et al. 2012; Jones et al. 2009). High pressure, high relative speed of the biomass particles on the surface where heat is exchanged, shear forces that diminish biomass particle size and increase surface area, and the temperature of reactor surface strongly influence the rate of the reaction of ablative pyrolysis reactor (Jahirul et al. 2012; Mohan et al. 2006; Peacocke et al. 1994a, b). These reactors can use particle of size up to 20 mm (Jahirul et al. 2012). Some disadvantages of ablative reactors are as follows (Jahirul et al. 2012; Dhyani and Bhaskar 2017): (a) Due to the low heat transfer coefficient and application of indirect heating, heat transfer from the hot surface of the reactor encounters some problems; (b) since the feed is to be worn away against the reactor wall, particle morphology, size, and free-flowing characteristics pose significant restriction; and (c) scaling is a linear function of heat transfer as it is controlled by surface area. Hence from the economic point of view, the ablative reactors do not provide the same advantage as the other reactors (Jahirul et al. 2012). The advantage of ablative reactors over other pyrolysis reactors is excessive grinding of feed material which is not required, and hence, larger biomass particles are allowed. The commonly used ablative reactors are ablative vortex and ablative rotating disk and are discussed below (Fig. 21.4).

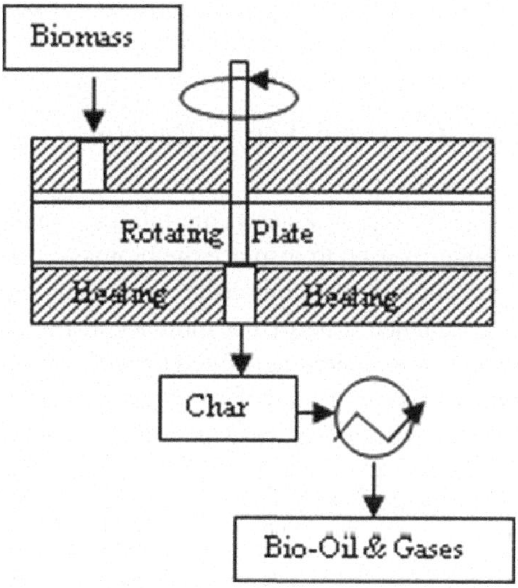

Fig. 21.4 Ablative pyrolysis reactor (*Source* Dhyani and Bhaskar 2017)

21.5.3.1 Ablative Vortex Reactor

For proper operation of an ablative vortex reactor, biomass particles should be carried in a hot stream of inert gas with a definite velocity and entered into the reactor tube tangentially. Due to this arrangement, the particles experience a high centrifugal force which allows them to slide on the reactor wall at high velocity and induce high ablation rates on the heated reactor wall (625 °C). As a result, the particles start to melt and leave a bio-oil film on the wall which vaporizes quickly. The particles which do not transform completely are recycled with a special solid recycle loop (Jahirul et al. 2012). Carrier gases remove the vapors generated on the reactor wall in 50–100 ms. This design meets the fast pyrolysis requirements and reported a high bio-oil yield (about 65%).

21.5.3.2 Rotating Disk Reactor

In the rotating disk or rotating plate pyrolysis reactors, biomass feedstocks are allowed to skim over a hot rotating disk. The main principle of the function of this reactor is that the biomass particles in touch with the heated plane could be decreased and vaporized due to the transportation of the heat from hot plane to the particles under pressure. This arrangement is useful as bigger particles can be pyrolyzed without grinding them. The most significant aspect of this type of reactor is that there is no requirement for an inert gas medium, thereby we get the processing apparatus which is small in size. Scaling is a matter of concern for the larger facilities as the process depends on surface area (Mohan et al. 2006).

21.5.4 Auger/Screw Reactor

Auger (or screw) reactors are tubular, continuous reactors, in which a rotating screw or auger is used to displace the biomass feedstock through an inert cylindrical heated tube. The screw accomplishes some functions like it mixes the feed and also controls the residence time of the biomass in the reactor. On passing through the tube, the feedstock temperature is lifted up to the required pyrolysis temperature (400–800 °C), which causes the devolatilization and gasification. Char is formed, and bio-oil is produced from condensation of gas, and non-condensable vapors are collected as biogas. In the configuration of auger reactors, the vapor residence time can be customized by altering the heated region through which vapor passing through prior to enter in the condenser (Jahirul et al. 2012; Mohan et al. 2006; Dhyani and Bhaskar 2017). Thus, this reactor configuration has some advantages such as (1) it is very compact, in some cases even portable, and it does not require carrier gas. These allow the reactor to be used on the site of biomass generation or where the biomass is abundantly available; (2) this design is energy efficient because it reduces energy costs as it operates at lower process temperatures (400 °C) and also saves the cost of

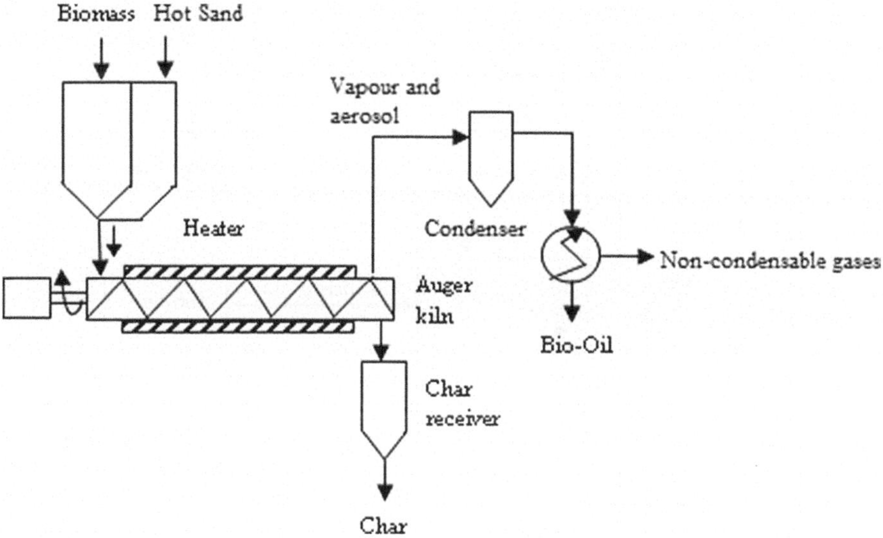

Fig. 21.5 Auger pyrolysis reactor

transport of feedstock to bio-refinery due to the on-site conversion of biomass which reduces the cost of operation (Dhyani and Bhaskar 2017); and (3) it operates as a continuous process (Fig. 21.5).

21.5.5 Rotating Cone Reactor

The most efficient approach of heat transfer to biomass in the pyrolysis process is the intense mixing of biomass and hot inert gas particles. However, mixing occurred in the fluidized bed processes requires too much inert gas, most of which are ineffective. In the rotating cone reactor, the biomass feedstock and hot sand are allowed to mix at the same time in the bottom of the cone. Thus, the mechanical mixing of biomass and hot sand takes place and they are transported upward by the rotation of the cone which causes centrifugal force to move the solids upward, instead of using inert gas to trigger the pyrolysis reaction. Upon exiting the cone, the vapors are directed to a condensation train. The solids such as char and sands are combusted in fluidized bed, where the sand gets re-heated before being introduced at the cone with new biomass feedstock. Although the rotating cone reactor has complex structure, it exhibits a high yield of bio-oil (Jahirul et al. 2012; Wagenaar et al. 2001). The pressure of outgoing materials is slightly above atmospheric level. The important features of this reactor include very fast heating and short residence time (Fig. 21.6).

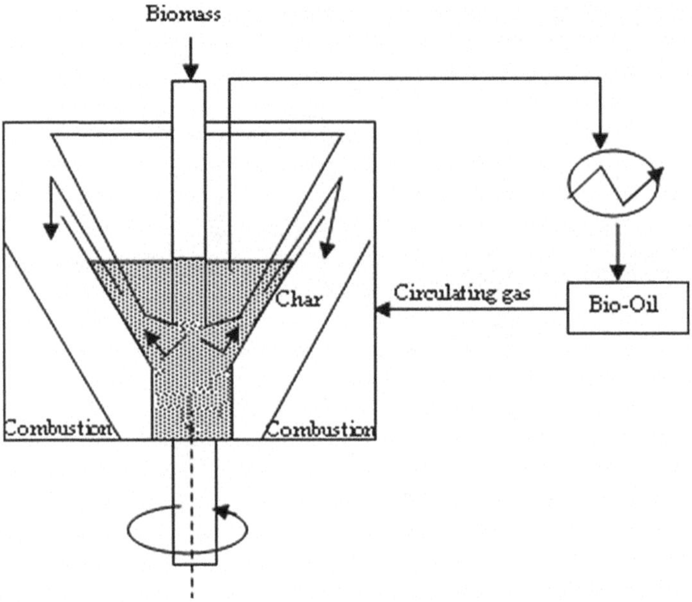

Fig. 21.6 Rotating cone reactor (*Source* Dhyani and Bhaskar 2017)

21.5.6 Vacuum Pyrolysis Reactor

Vacuum pyrolysis is actually a slow pyrolysis process, and it involves the slow heat transfer rate compared to other reactors (Mohan et al. 2006; Bridgwater and Peacocke 2000; Bridgwater et al. 2001). Due to lower heat transfer rate, bio-oil yields of 35–50% are achieved in vacuum pyrolysis reactor, which are lower than those of other reactors (Jahirul et al. 2012; Mohan et al. 2006). Although the vapor residence times are similar to those in fast pyrolysis, pyrolysis products evolve from the solid phase over a longer time frame. In this reactor type, a mobile metal belt carries biomass into the high-temperature vacuum chamber, where a burner and an induction heater are used to heat the biomass (Jahirul et al. 2012; Roy et al. 1997). A mechanical agitator stirs the biomass periodically on the belt. As these types of pyrolysis reactors are operating in a vacuum condition, they need unique solids feeding and expelling devices to keep the environment close all the time. Also, the vacuum pyrolysis occurs under reduced pressure (Mohan et al. 2006; Bridgwater and Peacocke 2000; Bridgwater et al. 2001). On heating the reactor, decomposition of the complex biomass particles takes place and the resultant products are then vaporized and vacuum rapidly takes it away from the reactor. They are then condensed into pyrolytic oils. The extent of secondary decomposition reactions is reduced due to the rapid volatilization occurred under vacuum. Therefore, the chemical composition of the pyrolysis products is similar to that of the complex biomolecules. This pyrolysis process is usually carried out at a temperature of

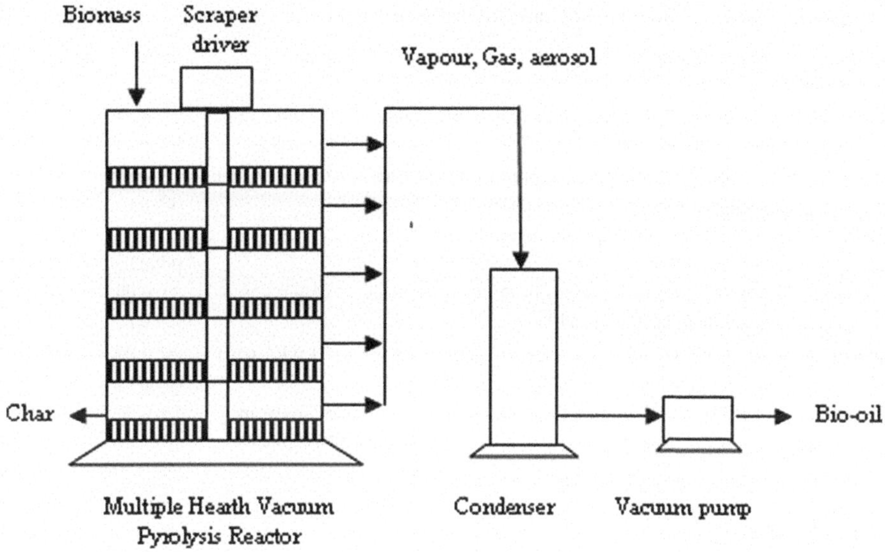

Fig. 21.7 Vacuum pyrolysis reactor (*Source* Ishak et al. 2012)

~450 °C and a total pressure of 15 kPa. The major benefit of vacuum pyrolysis reactors is that larger biomass particles of size 2–5 cm can be processed compared to fluidized bed reactors and they can easily achieve a short residence time for volatiles. On the other hand, disadvantages of the vacuum reactor are its complicated mechanical operation, poor heat and mass transfer rates, and high investment and maintenance cost. Also, installation of a larger capacity vacuum pump is important for the optimization of the reactor, which is susceptible to fouling, and it may increase the cost of the process (Jahirul et al. 2012; Mohan et al. 2006) (Fig. 21.7).

21.6 Pyrolysis Products

The three major products obtained from pyrolysis of biomass are solid residue called char, non-condensable gases, and condensable vapors that can be turned into a dark brown viscous liquid at suitable ambient temperature. Highest production of liquid product occurs within a temperature range of 350–500 °C (Fahmi et al. 2008). Yield of pyrolysis products mainly depends on the content of water present in the biomass, which contributes large quantities of condensate water in the liquid phase. This results in the extraction of water-soluble compounds from the gaseous and tar phases and thus the occurrence of larger decrease in gaseous and solid products (Demirbas 2000). Table 21.1 indicates the product distribution obtained from different modes of pyrolysis.

Table 21.1 Product distribution obtained from different modes of pyrolysis

Mode	Conditions	Liquid	Solid	Gas (wt%)
Slow	~500 °C, moderate hot vapor residence time 5–30 min	50 wt% in two phases	25 wt% (char)	25
Fast	~600 °C, short hot vapor residence time <2 s	75 wt% (oil)	12 wt% (char)	13
Flash	<650 °C, hot vapor residence time <1 s	75 wt%	12 wt% (char)	13
Vacuum	~600 °C, residence time 2–30 s, pressure applied 0.05–0.20 MPa	42 wt%	33 wt% (char)	25
Intermediate	~650 °C, long hot vapor residence time 300–1000 s	50 wt% (oil)	20 wt% (char)	30
Hydro	<500 °C, residence time <10 s	57 wt% in two phases	18 wt%	25

21.6.1 Bio-oil

Bio-oil is the liquid product obtained from the condensation of vapors produced from a pyrolysis reaction. Pyrolysis oils are usually referred to as bio-oil or bio-crude. Bio-oils are usually free-flowing liquids with a characteristic smoky odor. Bio-oil is a mixture of compounds with different size of molecules derived from depolymerization and fragmentation reactions of three building blocks of biomass. Bio-oil has potential to be used as fuel oil substitutes since it have heating values of 40–50% of that of hydrocarbon fuels. The main advantages of using bio-oil are summarized below (Balat et al. 2009; Chiaramonti et al. 2007):

- CO_2 is positively kept balanced in biomass fuel;
- Probability of using in small-scale power generation systems as well as use in large power stations;
- Storing and transportation capability of liquid fuel; and
- Have high energy density than biomass gasification fuel.

The distribution of the compounds in the bio-oil mostly depends on the type of biomass used and on the process parameters. Generally, it has been found that bio-oils are usually composed of oxygenated compounds (35–40 wt%) such as carbonyl, carboxyl, hydroxyaldehydes, hydroxyketones, sugars, carboxylic acids, and phenolic groups. The presence of these functional groups provides both potentials and challenges for the utilization of bio-oil. Most of the phenolic compounds are present as oligomers in bio-oil with a molecular weight ranging from 900 to 2500 (Meier et al. 1997). The presence of oxygenated compounds in the bio-oil is the primary reason for differences observed in the properties and behavior from hydrocarbon fuels. The high oxygen content results in a low energy density (heating value). Similarly, water is the most abundant component present in the bio-oil (Elliott 1994). It comes from the original moisture present in the feedstock and also produced as a product of dehydration reactions occurred during pyrolysis. Therefore, the presence of water content in bio-oil varies from 15 to 30 wt% depending on the feedstock and process conditions. Water component present in bio-oil is usually miscible with the oligomeric lignin-derived components because of the solubilizing effect of other polar hydrophilic compounds (low molecular weight acids, alcohols, hydroxyaldehydes, and ketones) which are originating from the carbohydrates decomposition. Hence, the bio-oil is immiscible in petroleum fuels (hydrophobic) and endures a lower heating value (Jahirul et al. 2012). The presence of water creates both the negative and positive effects on oil properties. It can lower the heating value and flame temperature of bio-oil. It also contributes to increase the ignition delay and decrease of combustion rate as compared to diesel fuels (Oasmaa et al. 2011). On the other hand, the presence of water slightly improves bio-oil flow characteristics by reducing the oil viscosity, which is beneficial for combustion (pumping and atomization). It also lowers the NO_x emissions. Bio-oil contains extensive amounts of organic acids, commonly acetic acid and formic acid which lower the pH of bio-oil (2–3). pH value only indicates the corrosiveness of oil, but it does not specify the concentration of acidic constituents which are mainly formed due to the degradation of hemicelluloses. Thus, the oils are corrosive to common construction materials such as carbon steel and aluminum and can affect some sealing materials but noncorrosive to stainless steels (Soltes and Lin 1984). Thus, the bio-oil is combustible in nature but inflammable because of the high content of nonvolatile components. Once the bio-oils are ignited, it burns with a steady self-sustaining flame. Further, the viscosity of bio-oil may become challenging as the bio-oil is stored over time, and some unfavorable reactions take place that makes the liquid too viscous to be a feasible fuel. Thus, the poor volatility, high viscosity, coking, and corrosiveness are the major difficulties of bio-oil to be used as a fuel which needs modification (Mullaney et al. 2002). Therefore, bio-oils are yet to reach viable standards because of the problems arising during its use as fuel in standard equipment such as boilers, engines, and gas turbines constructed for operation with petroleum-derived fuels.

21.6.1.1 Application of Bio-oil

Bio-oil can be used as a replacement of fossil fuel to generate heat, power, and chemicals. A general overview related to the area of combustion of bio-oil in boilers, diesel engines, gas turbines and in chemical point of view is depicted below.

Furnaces and Boilers

Furnaces and boilers are normally used for generating heat and power. They can produce less combustion as compared to turbines and engines. Various fuels ranging from natural gas, petroleum distillates to sawdust and coal/water slurries are worked as operating fuels for furnaces and boilers. Therefore, bio-oil could also be a viable option to be used as fuel for boiler applications since it encounters suitable emission levels, economic viability, and stable quality characteristics. Many studies have been reported till date on using bio-oil in boiler applications to replace heavy fuel oil (Freel et al. 1996; Gust 1997; Oasmaa et al. 2001). Bio-oils are known to have different combustion characteristics as compared to fossil fuels.

- Bio-oils exhibit poorer combustion characteristics while applying in boilers due to its high viscosity and water contents;
- Bio-oils obtained from different biomasses vary in combustion behavior and exhaust gas emissions;
- The flame from bio-oil combustion is longer compared to that of standard fossil oil;
- Harmful gas emissions from bio-oil in boiler applications are lower than those from burning heavy fuel oils except for particulate levels; and
- Some modifications of the burners and boilers are required for proper utilization of bio-oil in heat and power generation.

Diesel Engines

When diesel engines are coupled with oil combustion, the energy efficiency can be achieved up to 45% making itself a suitable candidate for the combined heat and power generation process (CHP). Medium- and slow-speed diesel engines can be used to utilize bio-oil due to its fuel flexibility and ability to perform in low-grade fuels. However, usage of bio-oil in diesel engine can possess some adverse effect such as deposition of carbon on pistons and other components of the engine combustion chamber; filter plugging; coking at injector; formation and deposition of heavy gum and wax, starting difficulty during cold weather; excessive wearing of engine; poor atomization piston ring sticking; and fuel pump failure of engine lubricating oil due to polymerization (Agarwal 2007; Labeckas and Slavinskas 2006; Ramadhas et al. 2005).

Over the last few decades due to the advantages of using bio-oils for power generation, it is being utilized in conventional diesel engine while few suggest that the application of pure pyrolysis oils should be limited to low-speed diesel engines with relatively high compression ratios, while blends of pyrolysis oil and methanol could be used in high-speed engines, especially with improved cetane number. Cetane number of fuels can be improved by using additives such as nitrated alcohol (Chiaramonti et al. 2003; Ormrod and Webster 2000; Moses 1994). With the use of improvised pyrolysis process and material for engine component, the difficulties associated with bio-oil may be overcome (Oasmaa et al. 2010).

Gas Turbines

Among the various applications, gas turbines can also be used to run power generators, industrial production processes, and providing power for aircraft. Currently, both liquid and gaseous fuels are being used for its operation. Biomass pyrolysis oil can be efficiently utilized by gas turbines with proper modification and redesigning in order to adapt with the unusual properties of biofuel. During modification, the effects of physical and chemical properties of bio-oils on atomization, efficiency of combustion, formation of soot, and emission of gaseous and particulate matter should be considered (Czernik and Bridgwater 2004). Materials used in fuel systems should be resistant to acid corrosion, and the blades have resistivity toward erosion, alkali hot corrosion, etc. During early 1980s at Teledyne CAE (USA), gas turbine was first tested on slow pyrolysis biomass liquids produced from forest and agricultural residues by Kasper and his coworkers (Kasper et al. 1983) by using a J69-T-29 combustor rig. The system consisted of an annular combustor and a centrifugal fuel injector rotating at shaft speed.

Chemicals

It is possible to extract useful chemicals from bio-oil by taking advantage of its most abundant functional groups such as carbonyl, carboxyl, phenols, and synthesized chemicals. For example, the formation of calcium salts and phenates occurs due to reaction between the carboxylic acids and phenols. Bio-oils have the potential to produce high-value chemicals for flavoring of food, phenols (adhesives for wood), fertilizer, acetic acid, sugars, and also chemicals for other industrial applications (Mohan et al. 2006; Jahirul et al. 2012). Fractionation of bio-oil could help to extract fine chemicals, petrochemicals, automotive fuels, and energy from bio-oil using standard refinery units. Thus, the ultimate goal of fractionation lies in the maximum valorization of the pyrolysis oil. The three main building blocks of biomass are the main base to extract specialty chemicals from bio-oil. The water phase of bio-oil also contains smaller organic components, e.g., acetic acid mainly produced from hemicellulose (https://www.btg-btl.com/en/applications/biochemicals).

21.6.2 *Char/Biochar*

Char is a stable form of carbon-rich compound that is produced when biomass is heated to temperature between 300 and 1000 °C, under low oxygen condition. It is composed of mainly carbon (C), hydrogen (H), oxygen (O), nitrogen (N), sulfur (S), and ash in different proportions. The char when intended to use as a soil amendment is termed as biochar. It is reported that application of biochar on soil can improve the soil quality, and therefore, a positive result is found on soil fertility and the growth of the plant. The incorporation of biochar into soil can alter soil physical properties such as structure, pore size distribution, density, soil aeration, and water-holding capacity, which positively affect the soil workability and therefore the plant growth (Downie et al. 2009). Biochar has a highly porous surface area with variable charge organic material, and therefore, it has the capacity to increase soil water-holding capacity, cation exchange capacity, surface sorption, and base saturation capacity when added to soil (Glaser et al. 2002; Bélanger et al. 2004; Keech et al. 2005; Liang et al. 2006). During conversion of biomass to biochar, 50% of the carbon that is present in biomass gets trapped in the new stable form and thereby reduces CO_2 emission from soil due to decomposition of biomass if otherwise applied to soil. This carbon-negative technology has the potential of net withdrawal of CO_2 from the atmosphere in a sustainable way. Therefore, use of biochar on soil creates a carbon sink in the one hand and recovers the fertility of soil to enhance ecosystem services on the other hand.

Due to the presence of condensed aromatic structure, biochar is found to be chemically stable. When biochar is produced at a low temperature, a considerable fraction of non-aromatic carbon is formed due to which biochars become susceptible to microbial attack and oxidation (Joseph et al. 2010). Biochars have a variable range of carbon half-life, with a range of 10,000–100,000 years. The stability of biochar in nature highly is dependent on a number of factors such as the type of biomass used for pyrolysis, the production conditions, soil properties, and climate (Lehmann et al. 2006).

21.6.2.1 Application of Biochar

Biochar has a huge potential in achieving environmental and agronomic benefits.

Climate Change Mitigation

When biomass is converted to biochar, the organic carbon that is entrapped in the biochar matrix becomes unavailable for microorganisms and other decomposers as substrate, which ultimately helps in long-term carbon storage in soils. Therefore, the conversion of biomass organic matter to biochar significantly increases the stability of the carbon in biomass leading to long-term carbon sequestration. On addition of biochar to agricultural fields, Kwapinski et al. (2010) reported a reduction in emission of GHGs including nitrous oxide and methane up to 50%. A significant decrease in N_2O emissions is observed from both incubation and field studies (Lehmann and Joseph 2009).

Enhancement of Soil Quality

Use of biochar as a soil amendment dates back to several thousand years in the region of Amazon, known as *terra preta*. Apart from being used as a soil conditioner, the biochar also possesses various beneficial qualities and therefore can be used for various purposes. Biochar is found to improve soil fertility by improving physical, chemical, and microbiological properties of soil. The observed effects on soil fertility have been explained mainly by a pH increase in acidic soils or improved nutrient retention through cation adsorption. Changes in microbial community composition or activity induced by biochar may affect not only nutrient cycles and plant growth, but also the cycling of soil organic matter. Biochar itself is enriched with nutrients, such as *N*, *P*, *K*, and can supply these to soil directly. Due to large surface area, its ability to absorb soluble organic matter, gases, and inorganic nutrients, and intricate internal structure, biochar provides a potential niche for microorganisms. Therefore, the population of soil microorganisms increases in soil added with biochar, though this effect of biochar disappears in the long run.

Biochar as a Sorbent for Contaminant Reduction in Soil and Water

Biochar has the capacity for remediation of contaminated soil and provides additional benefits to the environment. The science involved with it is still poorly understood although the use of biochar dates back to thousand years ago. Bioavailability and water solubility of pollutants are important characteristics to determine the significance of the pollutant. The concentration of Cd, Zn, and polycyclic aromatic hydrocarbon (PAH) in the pore waters decreases after the addition of biochar. Biochars have the potential to sorb several toxic elements, including arsenic (As), copper (Cu), cadmium (Cd), nickel (Ni), lead (Pb), and zinc (Zn). Due to high surface area and microporosity of biochar, sorption of organic contaminates from water onto biochar occurs (Goswami et al. 2016; Bordoloi et al. 2017).

21.6.3 Syngas (Synthesis Gas)

Syngas is mainly composed of carbon monoxide and hydrogen. Synthesis gas or, briefly, syngas is a mixture of carbon monoxide, carbon dioxide, and hydrogen. The name comes from its use as intermediates in the production of synthetic natural gas (SNG). Syngas or synthetic gas is formed through the gasification process. The formation of syngas is strongly endothermic and requires high temperatures. Gasification converts biomass into a mixture of gases called syngas containing CO, H_2, CO_2, CH_4, and hydrocarbon in the presence of a controlled oxidizing agent at a high temperature. The oxidizing agent used in the process includes oxygen, air, steam, or mixture of gases. Syngas having high heating value is produced when steam oxidation is done. Air gasification leads to the production of syngas with low heating value (Kumar et al. 2009). There are some other "non-gasification" processes which can produce synthesis gas (syngas) on large scale. These include partial oxidation, autothermal reforming (ATR), and steam methane reforming. In partial oxidation, feedstock material and the pure oxygen gas are mixed and reacted together which ultimately leads to produce syngas with a high concentration of hydrogen (H_2). In case of ATR, it is a mixture of thermal partial oxidation and low-sulfur natural gas in a single reactor. In this process, natural gas (methane) is used as feedstock to produce syngas with a high H_2 concentration that can be further used in refineries and other industrial processes. In steam methane reforming technology, a mixture of natural gas and steam is heated, which flows through pipes filled with a nickel-based catalyst. Various sources, including natural gas, coal, biomass, or virtually any hydrocarbon feedstock, can be used for the production of syngas and produced by reaction with steam or oxygen.

Syngas is used as fuel in internal combustion (IC) engines that are coupled to a synchronous alternator, which finally produces electricity. The energy generated in the form of combustion gases is used for the generation of thermal energy either in the form of hot water, steam, thermal oil, etc. Thermal energy that is produced by direct combustion of syngas can be used in a range of sectors such as the industrial sector, in chemicals, cement, food. Bioethanol and Bio-SNG (synthetic natural gas) are produced through the chemical transformation of syngas which can be later used for the production of automotive and transportation fuels, chemicals, and plastics.

Syngas can play an important role in future alternative energy production. It can be used as an alternative fuel for power generation applications and also have the potential for near-zero pollutant emissions, including greenhouse gases. The interest in power generation technology based on syngas is increasing in recent days.

21.7 Pyrolysis and Economic Analysis

Apart from other goals, the major objective behind the development of biofuels is to replace fossil fuel with renewable energy. Due to limited availability of land and other resources needed to produce biomass, the comparison of energy efficiency of the process designs from cost perspective is necessary. Since the cost of production of pyrolytic energy product is higher compared to the fossil fuel, therefore before commercialization of the pyrolysis process it has to prevail over its technical and non-technical drawbacks (Thornley and Wright 2001). The economic costs and benefits components considered are as follows:

- Collection/production of feedstocks;
- Transportation, storage, and pretreatment of feedstocks;
- Feedstock processing;
- Pyrolysis plant;
- Pyrolysis operation;
- Energy sales; and
- Product collection and storage.

Although the pyrolysis reactor contributes only 10–15% of the total capital cost, it is considered as the main component of a pyrolysis plant. The economic analyses are generally conducted to identify the preferred biomass sources and pyrolysis conditions on an energy value basis.

21.7.1 Pyrolysis Economics

To test the commercial viability of any pyrolysis process, it is very important to carry out an economic assessment of the process, where the production cost of pyrolysis has a major share. The cost involved in the production of pyrolysis plant can be broadly divided into two groups:

(i) Capital investment or fixed cost (includes pyrolysis module, basic equipment, feed handling and storage, and development of facilities) and
(ii) Operating or variable costs (include biomass harvesting or feedstock, maintenance, product transport, labor, utility, transport). Table 21.2 represents the different components which contribute to variable cost (Mullaney et al. 2002).

Apart from the above-mentioned factors, cost of biomass pyrolysis also depends on process technology, size of the plant, feedstock, year of construction, etc. By using the annuity method described by Islam and Ani (2000) and Polagye et al. (2007), the annual production costs can be estimated as follows:

Table 21.2 Approximate contributions from different components on variable cost (Mullaney et al. 2002)

Items	Percentage (%)
Biomass harvesting or feedstock	23–30
Maintenance	17–24
Utilities	22–25
Labor	12–19
Grinding	7–9
Transportation	5–7

$$\text{Annual cost} = \text{Operating cost} + (\text{annualized capital cost} - \text{annualized salvage value}) \tag{4}$$

whereas the annualized capital cost is determined by the following Eq. (5) (Islam and Ani 2000; Polagye et al. 2007):

$$\text{ACC} = \frac{(\text{total plant cost} + \text{construction cost})}{\{1 - (1+i)^{-N_p}\}} \times i_p \tag{5}$$

where ACC = annualized capital cost per year, i_p = interest rate, and N_p = plant lifetime.

The construction cost can be given by

$$\text{Construction cost} = \sum_{j=1}^{N_c} \frac{\text{total cost}}{N_c} j i_c \left(1 + i_p\right)^{N_c - j + 1} \tag{6}$$

where N_c = period of construction, i_c = construction cost/rate of interest, and i_p = project financing rate.

According to Islam and Ani (2000), the production cost can be calculated using the following parameters such as plant capacity, plant life, annual functioning period, maintenance labor, maintenance materials, overheads, insurance, other fixed operating costs, rate of interest, cost of feedstocks, labor charges, number of labors, cost of nitrogen gas, electricity cost, catalyst price, and catalytic validity. Table 21.3 summarizes the various parameters which are involved in the determination of unit production cost.

The profitability of an investment is evaluated by using the concept of net present value (NPV), which can be given by Thewys and Kuppens (2008), Voets and Kuppens (2011):

$$\text{NVP} = \sum_{n=1}^{T} \frac{CF_n}{(1+i)^n} - I_0 \tag{7}$$

where T = Time period of investment (years), CF_n = cash flow, i.e., the difference between revenues and expenditures after tax in year n, and I_0 = expenditure connected with the initial investment in year zero.

Further, CF_n can be calculated by using Eq. (7) (Jahirul et al. 2012):

$$CF_n = (1 - t) \times (R - E) + t \times D \tag{8}$$

where t = tax rate, R = revenue, E = expenditure, and D = depreciation.

Table 21.3 Assumptions and costs of the various parameters used to determine unit production costs for different size of pyrolysis plants

Plant size (tonne/day)	Feedstock	Capital investment (million $)	Annual operating Costs (M $)	Feed costs ($/ tonne)	Production costs ($/gal)	References	Type of pyrolysis process
2000	Corn Stover	200	12.3	83	0.26	Wright et al. (2010)	Fast pyrolysis
1650	Wood pellet	180	12	–	0.24	Polagye et al. (2007)	Fast pyrolysis
1000	Dry wood	68	10.6	44	0.41	Gregoire and Bain (1994)	Fast pyrolysis
1000	Wet wood	72	11.3	30	0.60	Cottam and Bridgwater (1994)	Flash pyrolysis
1000	Peat	76	10.2	20	0.61	Solantausta and Oasmaa (2003)	Fast pyrolysis
1000	Straw	82	10.2	42.5	0.64	Solantausta and Oasmaa (2003)	Fast pyrolysis
900	Wet wood	46	9.9	34	0.50	Luo et al. (2004)	Fast pyrolysis
550	Dry wood	48.2	9.6	45	0.71	Ringer et al. (2006)	Fast pyrolysis
400	Wet wood	14.3	8.80	36	1.02	Mullaney et al. (2002)	Fast pyrolysis
250	Dry wood	14	8.92	44	0.55	Gregoire (1992)	Fast pyrolysis
200	Wet wood	8.8	4.84	36	1.11	Mullaney et al. (2002)	Fast pyrolysis
100	Wet wood	6.6	2.84	36	1.48	Mullaney et al. (2002)	Fast pyrolysis
24	Rice husk	3.89	0.17	22	0.82	Islam and Ani (2000)	Fast pyrolysis
2.4	Rice husk	0.97	0.034	22	1.72	Islam and Ani (2000)	Fast pyrolysis

Source Jahirul et al. (2012)

21.8 Conclusion

Currently, human race is dealing with some grave challenges such as diminishing resources, changing climate due to human interference, energy crisis, and pollution, and it can be expected that in the near future, these problems will be more acute. The rapid growing population coupled with a sharp decline in natural resources in the past decade has forced world's policy makers to opt for sustainable development which can be achieved via finding alternative sources of energy and utilization of waste to produce energy. A plethora of waste (e.g., agricultural, industrial, MSW) is generated every day creating a significant problem in its management and disposal. Pyrolysis among the thermochemical conversion of wastes is one of the viable options that would help to achieve sustainable development. Apart from resolving issues associated with sustainable management of waste, pyrolysis also generates fuels, industrially important chemicals, materials, and valuable end products. Due to the potentiality of pyrolysis process to produce high yields of liquid fuels or chemical, the process can be commercially established as an emerging technology. In this regard, various reactor configurations and their modifications have been the target of research worldwide to achieve high efficiency. The potential of bio-oil is increasingly being acknowledged, leading to the identification of several ideal lignocellulosic species as feedstock for pyrolysis, including wood biomass, agricultural residues, and dedicated energy crops. Modification of pyrolysis technology deals with some challenges related to upgradation of the liquids and adaptation of its application due to unusual behavior and characteristics of the liquid product. Also, the presence of moisture in the biomass coupled with uneven distribution and varying nature of the feedstock has led to the production of bio-oil which is of inferior quality in comparison with the conventional fossil fuels. The presence of oxygen, acidic compounds, and instability in the bio-oil are some of the other major concerns in the establishment of pyrolysis on a substantial scale. Research in the field of bio-oil upgradation, production of high energy feedstock, and process intensification as a bio-refinery to lower the cost of production can help in overcoming these challenges. The successful and sustainable utilization of pyrolytic oils for substituting the fossil fuels will depend on the development of integral bio-refineries with a supply of quality feedstock, cheap and easy upgradation of bio-oil, reactors with high thermal efficiencies, and adapting energy market. The other coproduct of pyrolysis, i.e., biochar, has also wide applications in a number of fields. Production of biochar from various wastes and its subsequent use in the field as a soil amendment fulfill the twin goal of waste management and improvement of soil quality. Apart from agricultural benefits, biochar also possesses some environmental benefits like mitigation of GHG, remediation of polluted soil, and sequestration of carbon. Additional to these, biochar has a potential role as a catalyst in different reactions like syngas reforming, upgrading of bio-oil or biodiesel. In spite of positive results of biochar on soil and environment yet some long-term risks and challenges exist for the large-scale and long-term application of biochar. An evaluation of large-scale application and also the economic viability of biochar are needed to overcome these challenges. The techno-economic study explores the cost of converting biomass into energy via different types of pyrolysis technology. However, a detail economical assessment of industrial-scale pyrolysis plant is necessary.

Acknowledgements The authors would like to offer their sincere thanks and acknowledgement for the support from Tezpur University and University Grants Commission, New Delhi, India [Grant No. 42-723/2013 (SR)] received in the form of UGC-MRP. The authors RN, DS, and RS sincerely acknowledge the receipt of fellowships from UGC and CSIR, respectively.

References

Abnisa F, Wan Daud WMA (2014) A review on co-pyrolysis of biomass: an optional technique to obtain high-grade pyrolysis oil. Energ Convers Manage 87:71–85

Agarwal AK (2007) Biofuels (alcohols and biodiesel) applications as fuels for internal combustion engines. Prog Energy Combust Sci 33:233–271

Al-Salem SM, Antelava A, Constantinou A, Manos G, Dutta A (2017) A review on thermal and catalytic pyrolysis of plastic solid waste (PSW). J Environ Manage 197:177–198

Altafini CR, Wander PR, Barreto RM (2003) Prediction of the working parameters of a wood waste gasifier through an equilibrium model. Energy Convers Manage 44:2763–2777

Balat M (2008) Mechanisms of thermochemical biomass conversion processes. Part 1: Reactions of pyrolysis. Energy Sources Part A 30(7):620–635

Balat M, Balat M, Kirtay E, Balat H (2009) Main routes for the thermo-conversion of biomass into fuels and chemicals. Part 1: Pyrolysis systems. Energy Convers Manage 50:3147–3157

Bélanger NI, Côté B, Fyles JW, Chourchesne F, Hendershot WH (2004) Forest regrowth as the controlling factor of soil nutrient availability 75 years after fire in a deciduous forest of southern Quebec'. Plant Soil 262:363–372

Bentsen NS, Felby C, Thorsen BJ (2014) Agricultural residue production and potentials for energy and materials services. Prog Energy Combust Sci 40:59–73

Bordoloi N, Narzari R, Chutia RS, Bhaskar T, Kataki R (2015) Pyrolysis of *Mesua ferrea* and *Pongamia glabra* seed cover: characterization of bio-oil and its sub-fractions. Bioresour Technol 178:83–89

Bordoloi N, Goswami R, Kumar M, Kataki R (2017) Biosorption of Co (II) from aqueous solution using algal biochar: kinetics and isotherm studies. Bioresour Technol. https://doi.org/10.1016/j.biortech.2017.05.139

Boukis I, Gyftopoulou ME, Papamichael I (2001) Biomass fast pyrolysis in an air blown circulating fluidized bed reactor. In: Bridgwater AV (ed) Progress in thermochemical biomass conversion. Blackwell Science Ltd, Oxford, UK, pp 1259–1267. https://doi.org/10.1002/9780470694954.ch104

Brandt A, Grasvik J, Hallett JP, Welton T (2013) Deconstruction of lignocellulosic biomass with ionic liquids. Green Chem 15(3):550–583

Bridgwater AV, Peacocke GVC (2000) Fast pyrolysis processes for biomass. Renew Sustain Energy Rev 4(1):1–73

Bridgwater AV, Czernik S, Piskorz J (2001) An overview of fast pyrolysis. In: Bridgwater AV (ed) Progress in thermochemical biomass conversion, vol 2. Blackwell Science, London, pp 977–997

Chen D, Yin L, Wang H, He P (2014) Pyrolysis technologies for municipal solid waste: a review. Waste Manage 34(12):2466–2486

Chiaramonti D, Bonini M, Fratini E, Tondi G, Gartner K, Bridgwater AV, Grimm HP, Soldaini I, Webster A, Baglioni P (2003) Development of emulsions from biomass pyrolysis liquid and diesel and their use in engines. Biomass Bioenergy 25:101–111

Chiaramonti D, Oasmaa A, Solantausta Y (2007) Power generation using fast pyrolysis liquids from biomass. Renew Sustain Energy Rev 11:1056–1086

Chopra S, Jain A (2007) A review of fixed bed gasification systems for biomass. Agric Eng Int 5:1–23

Choudhury ND, Chutia RS, Bhaskar T, Kataki R (2014) Pyrolysis of jute dust: effect of reaction parameters and analysis of products. J Mater Cycles Waste Manage 16(3):449–459

Cottam ML, Bridgwater AV (1994) Techno-economic modeling of biomass flash pyrolysis and upgrading systems. Biomass Bioenerg 7:267–273

Czajczynska D, Anguilano L, Ghazal H, Krzyzynska R, Reynolds AJ, Spencer N, Jouhara H (2017) Potential of pyrolysis processes in the waste management sector. Therm Sci Eng Prog. https://doi.org/10.1016/j.tsep.2017.06.003

Czernik S, Bridgwater AV (2004) Overview of applications of biomass fast pyrolysis oil. Energy Fuels 18:590–598

Demirbas A (1998) Teaching practical chemical kinetics of pyrolysis reaction. Energy Educ Sci Technol 2:23–28

Demirbas A (2000) Mechanisms of liquefaction and pyrolysis reactions of biomass. Energ Convers Manage 41:633–646

Demirbas A (2001) Biomass resource facilities and biomass conversion processing for fuels and chemicals. Energ Convers Manage 42:1357–1378

Demirbas A (2004) Combustion characteristics of different biomass fuels. Prog Energy Combus Sci 30:219–230

Demirbas A (2007) Effect of temperature on pyrolysis products from biomass. Energy Sources Part A 29(4):329–336

Demirbas A, Arin G (2002) An overview of biomass pyrolysis. Energy Sources 24(5):471–482

Dhyani V, Bhaskar T (2017) A comprehensive review on the pyrolysis of lignocellulosic biomass. Renew Energy. https://doi.org/10.1016/j.renene.2017.04.035

Downie A, Crosky A, Munroe P (2009) Physical properties of biochar. In: Lehmann J, Joseph S (eds) Biochar for environmental management: science and technology. Earthscan, London, Dynamotive Energy Systems

Elliott D (1994) Water, alkali and char in flash pyrolysis oils. Biomass Bioenergy 7:179–185

Evans RJ, Milne TA (1987) Molecular characterization of the pyrolysis of biomass. Energy Fuels 1:123–137

Fahmi R, Bridgwater AV, Donnison I, Yates N, Jones JM (2008) The effect of lignin and inorganic species in biomass on pyrolysis oil yields, quality and stability. Fuel 87:1230–1240

Fitz HC, DeBellevue EB, Costanza R, Boumans R, Maxwell T, Wainger L, Sklar FH (1996) Development of a general ecosystem model for a range of scales and ecosystems. Ecol Modell 88:263–295

Freel BA, Graham RG, Huffman DR (1996) Commercial aspects of rapid thermal processing (RTMTM). In: Bio-oil production and utilization. CPL Press, Newbery, UK, pp 86–95

Glaser B, Lehmann J, Zechet W (2002) Ameliorating physical and chemical properties of highly weathered soils in the tropics with charcoal—a review. Biol Fertil Soils 35:219–230

Goswami R, Shimb J, Deka S, Kumari D, Kataki R, Kumar M (2016) Characterization of cadmium removal from aqueous solution by biochar produced from *Ipomoea fistulosa* at different pyrolytic temperatures. Ecol Eng 97:444–451

Gregoire CE (1992) Techno-economic analysis of the production of biocrude from wood; NREL/TP-430–5435. National Renewable Energy Laboratory, Golden, CO, USA

Gregoire CE, Bain RL (1994) Technoeconomic analysis of the production of biocrude from wood. Biomass Bioenerg 7:275–283

Gust S (1997) Combustion experiences of flash pyrolysis fuel in intermediate size boilers. In: Bridgwater AV, Boocock DG (eds) Developments in thermochemical biomass conversion. Blackie Academic & Professional, London, UK, pp 481–488

Hall DO (1997) Biomass energy in industrialized countries—a view of the future. For Ecol Manage 91:17–45

Harmsen J, Powell JB (2011) Sustainable development in the process industries: cases and impact. Wiley, Hoboken

http://www.eai.in/ref/ae/wte/concepts.html. Accessed 13 June 2017

https://www.btg-btl.com/en/applications/biochemicals. Accessed 20 May 2017

Intergovernmental Panel on Climate Change (2017) Forty-fifth session of the IPCC, Guadalajara, Mexico, 28–31 Mar 2017

International Energy Agency. World Energy Outlook 2016. http://www.worldenergyoutlook.org/factsheets/. Accessed 8 July 2017

International Energy Outlook (IEO 2011) https://www.iea.org/publications/freepublications/publication/WEO2011_WEB.pdf/. Accessed 3 Mar 2017

International Energy Outlook (IEO 2016) https://www.iea.org/media/publications/weo/WEO2016Chapter1.pdf. Accessed 22 Feb 2017

International Energy Outlook (IEO 2017) https://www.eia.gov/outlooks/ieo/pdf/0484(2017).pdf/. Accessed 12 Mar 2017

Ishak WNRW, Hisham MWM, Yarmo MA, Hin TY (2012) A review on biooil production from biomass by using pyrolysis method. Renew Sustain Energy Rev 16:5910–5923

Islam MN, Ani FN (2000) Techno-economics of rice husk pyrolysis, conversion with catalytic treatment to produce liquid fuel. Bioresour Technol 73:67–75

Jahirul MI, Rasul MG, Chowdhury AA, Ashwath N (2012) Biofuels production through biomass pyrolysis—a technological review. Energies 5:4952–5001

Jones SB, Holladay JE, Valkenburg C, Stevens DJ, Walton CW, Kinchin C, Elliott DC, Czernik S (2009) Production of gasoline and diesel from biomass via fast pyrolysis, hydrotreating and hydrocracking: a design case. Report No. PNNL-18284; U.S. Department of Energy, Springfield, VA, USA

Joseph SD, Camps-Arbestain M, Lin Y, Munroe P, Chia CH, Hook J, Zwieten L, Kimber S, Cowie A, Singh BP, Lehmann L, Foidl N, Smernik RJ, Amonette JE (2010) An investigation into the reactions of biochar in soil. Aust J Soil Res 48:501–515

Kabir G, Mohd Din AT, Hameed BH (2017) Pyrolysis of oil palm mesocarp fiber and palm frond in a slow-heating fixed-bed reactor: a comparative study. Bioresour Technol 241:563–572

Kasper JM, Jasas GB, Trauth RL (1983) Use of pyrolysis-derived fuel in a gas turbine engine. ASME Paper No. 83-GT-96

Keech O, Carcaillet C, Nilsson M (2005) Adsorption of allelopathic compounds by wood–derived charcoal: the role of wood porosity. Plant Soil 272:291–300

Kumar A, Jones D, Hanna M (2009) Thermochemical biomass gasification: a review of the current status of the technology. Energies 2(3):556–581

Kwapinski W, Byrne CMP, Kryachko E, Wolfram P, Adley C, Leahy JJ, Novotny EH, Hayes MHB (2010) Biochar from biomass and waste. Waste Biomass Valoriz 1(2):177–189

Labeckas G, Slavinskas S (2006) Performance of direct-injection off-road diesel engine on rapeseed oil. Renew Energy 31:849–863

Lehmann J, Joseph S (2009) Biochar for environmental management: science and technology. Earthscan, London, Dynamotive Energy Systems

Lehmann J, Gaunt J, Rondon M (2006) Bio-char sequestration in terrestrial ecosystems—a review. Mitig Adapt Strat Glob Change 11:395–419

Leung DYC, Yin XL, Wu CZ (2004) A review on the development and commercialization of biomass gasification technologies in China. Renew Sustain Energy Rev 8:565–580

Liang B, Lehman J, Solomon D, Kinyangi J, Grossman J, O'Neill B, Skjemstas JO, Thies J, Luizao FJ, Petersen J, Neves EG (2006) Black carbon increases cation exchange capacity in soils. Soil Sci Soc Am J 70(5):1719–1730

Luo Z, Wang S, Liao Y, Zhou J, Gu Y, Cen K (2004) Research on biomass fast pyrolysis for liquid fuel. Biomass Bioenergy 26:455–462

Lv PM, Xiong ZH, Chang J, Wu CZ, Chen Y, Zhu JX (2004) An experimental study on biomass air–steam gasification in a fluidized bed. Bioresour Technol 95:95–101

Meier D, Oasmaa A, Peacocke GVC (1997) Properties of fast pyrolysis liquids: status of test methods. Characterization of fast pyrolysis liquids. In: Bridgwater AV, Boocock DGB (eds) Developments in thermochemical biomass conversion. Blackie Academic & Professional, London, pp 391–408

Menon V, Rao M (2012) Recent trends in valorization of lignocellulose to biofuel. In: Satyanarayana T, Johri BN (eds) Microorganisms in sustainable agriculture and biotechnology. Springer, Dordrecht, pp 381–409

Mohan D, Pittman CU Jr, Steele P (2006) Pyrolysis of wood/biomass for bio-oil: a critical review. Energy Fuel 20:848–889

Moses C (1994) Fuel-specification considerations for biomass liquids. In: Proceedings of biomass pyrolysis oil properties and combustion meeting, 26–28 Sept, Estes Park, CO., NREL-CP-430-7215, pp 362–382

Mullaney H, Farag IH, LaClaire CL, Barrett CJ (2002) Technical, environmental and economic feasibility of bio-oil in New Hampshire's North Country. Final Report, New Hampshire Industrial Research Center (NHIRC), Durham City, NH, USA

Muradov NZ, Veziroglu TN (2008) "Green" path from fossil-based to hydrogen economy: an overview of carbon-neutral technologies. Int J Hydrogen Energy 33:6804–6839

Muradov N, Fidalgo B, Gujar A (2012) Production and characterization of duckweed bio-char and its catalytic application for biogas reforming. Biomass Energy 42:123–131

Oasmaa A, Kytö M, Sipilä K (2001) Pyrolysis oil combustion tests in an industrial boiler. In: Progress in thermochemical biomass conversion. Blackwell Science, Oxford, UK, pp 1468–1481

Oasmaa A, Elliott DC, Korhonen J (2010) Acidity of biomass fast pyrolysis bio-oils. Energy Fuel 24:6548–6554

Oasmaa A, Korhonen J, Kuoppala E (2011) An approach for stability measurement of wood-based fast pyrolysis bio-oils. Energy Fuels 25(7):3307–3313

Onwudili JA, Insura N, Williams PT (2009) Composition of products from the pyrolysis of polyethylene and polystyrene in a closed batch reactor: effects of temperature and residence time. J Anal Appl Pyrol 86:293–303

Ormrod D, Webster A (2000) Progress in utilization of bio-oil in diesel engines. In: PyNe News Letter. Aston University, Birmingham, UK, p 15

Peacocke GVC, Madrali ES, Li C-Z, Guell AJ, Kandiyoti R, Bridgwater AV (1994a) Effect of reactor configuration on the yields and structures of pine-wood derived pyrolysis liquids: a comparison between ablative and wire-mesh pyrolysis. Biomass Bioenergy 7(1–6):155–167

Peacocke GVC, Russell PA, Bridgwater AV (1994b) Ablative plate pyrolysis of biomass for liquids. Biomass Bioenergy 7:147–154

Pettersen RC (1984) The chemical composition of wood. Chem Solid Wood 207:57–126

Polagye LB, Hodgson KT, Malte PC (2007) An economic analysis of bio-energy options using thinnings from overstocked forests. Biomass Bioenergy 31:105–125

Prasad S, Singh A, Joshi HC (2007) Ethanol as an alternative fuel from agricultural, industrial and urban residues. Resour Conserv Recycl 50:1–39

Putun AE, Apaydin E, Putun E (2002) Bio-oil production from pyrolysis and steam pyrolysis of soybean-cake: product yields and composition. Energy 27(7):703–713

Ramadhas AS, Jayaraj S, Muraleedharan C (2005) Characterization and effect of using rubber seed oil as fuel in the compression ignition engines. Renew Energy 30:795–803

Rao MS, Singha SP, Sodhaa MS, Dubey AK, Shyam M (2004) Stoichiometric, mass, energy and exergy balance analysis of countercurrent fixed-bed gasification of post-consumer residues. Biomass Bioenergy 27:155–171

REN21-Renewable Energy Policy Network for the 21st Century. Renewables 2014 Global Status Report, 2014

Ringer M, Putsche V, Scahill J (2006) Large-scale pyrolysis oil production and economic analysis. Technical report NREL/TP-510–37779, National Renewable Energy Laboratory, Cole Boulevard, CO, USA

Roy P, Dias G (2017) Prospects for pyrolysis technologies in the bioenergy sector: a review. Renew Sustain Energy Rev 77:59–69

Roy C, Blanchette D, Korving L, Yang J, DeCaumia B (1997) Development of a novel vacuum pyrolysis reactor with improved heat transfer potential. In: Bridgewater AV, Boocock DGB (eds) Developments in thermochemical biomass conversion. Blackie Academic and Professional, London, UK, pp 351–367

Saffarzadeh A, Shimaoka T, Motomura Y, Watanabe K (2006) Chemical and mineralogical evaluation of slag products derived from the pyrolysis/melting treatment of MSW. Waste Manage 26:1443–1452

Saikia R, Chutia RS, Kataki R, Pant KK (2015) Perennial grass (*Arundo donax* L.) as a feedstock for thermo-chemical conversion to energy and materials. Bioresour Technol 188:265–272

Scott DS, Majerski P, Piskorz J, Radlein D (1999) A second look at fast pyrolysis of biomass—the RTI process. J Anal Appl Pyrol 51:23–37

Scurlock JMO, Dayton DC, Hames B (2000) Bamboo: an overlooked biomass resource? Biomass Bioenerg 19:229–244

Sharma A, Pareek V, Zhang D (2015) Biomass pyrolysis—a review of modelling, process parameters and catalytic studies. Renew Sustain Energy Rev 50:1081–1096

Sharuddin SDA, Abnisa F, Daud WMAW, Aroua MK (2016) A review on pyrolysis of plastic wastes. Energy Convers Manage 115:308–326

Solantausta Y, Oasmaa A (2003) Fast pyrolysis of forestry residues and sawdust, production and fuel oil quality. In Proceedings of international nordic bioenergy conference, Javaskyla, Finland, pp 1–3

Soltes EJ, Lin JCK (1984) Hydro processing of biomass tars for liquid engine fuels. In: Tillman DA, Jahn EC (eds) Progress in biomass conversion. Academic Press, New York, pp 1–69

Sorum L, Gronli MG, Hustad JE (2001) Pyrolysis characteristics and kinetics of municipal solid wastes. Fuel 80:1217–1227

Thewys T, Kuppens T (2008) Economics of willow pyrolysis after phytoextraction. Int J Phytorem 10:561–583

Tripathi M, Sahu JN, Ganesan P (2016) Effect of process parameters on production of biochar from biomass waste through pyrolysis: a review. Renew Sustain Energy Rev 55:467–481

Thornley P, Wright E (2001) Evaluation of Bio-Energy Projects. In PyNe Final Report to the EC; European Commission: Brussels, Belgium

Voets T, Kuppens T (2011) Economics of electricity and heat production by gasification or flash pyrolysis of short rotation coppice in Flanders (Belgium). Biomass Bioenergy 35:1912–1924

Wagenaar BM, Venderbosch RH, Carrasco J, Strenziok R, Van der Aa BJ (2001) Rotating cone bio-oil production and applications. In: Bridgewater AV (ed) Progress in thermochemical biomass conversion. Blackwell Science, Oxford, UK, pp 1268–1280

Wang X, Kersten SRA, Prins W, Van Swaaij WPM (2005) Biomass pyrolysis in a fluidized bed reactor. Part 2: Experimental validation of model results. Ind Eng Chem Res 44:8786–8795

Wang S, Dai G, Yang H, Luo Z (2017) Lignocellulosic biomass pyrolysis mechanism: a state-of-the-art review. Prog Energy Combust Sci 62:33–86

Wright MM, Daugaard DE, Satrio JA, Brown RC (2010) Techno-economic analysis of biomass fast pyrolys to transportation fuels. Fuel 89:S2–S10

Yaman S (2004) Pyrolysis of biomass to produce fuels and chemical feedstocks. Energy Convers Manage 45:651–671

Yang H, Yan R, Chen H, Lee DH, Liang DT, Zheng C (2006) Pyrolysis of palm oil wastes for enhanced production of hydrogen rich gases. Fuel Proc Technol 87:935–942

Zabaniotou AA (1999) Pyrolysis of forestry biomass by-products in Greece. Energy Sources Part A Recovery Util Environ Eff 21:395–403

Zanzi R, Sjostrom K, Bjornbom E (1996) Rapid high-temperature pyrolysis of biomass in a free-fall reactor. Fuel 75:545–550

Zhao YL, Dolat A, Steinberger Y, Wang X, Osman A, Xie GH (2009) Biomass yield and changes in chemical composition of sweet sorghum cultivars grown for biofuel. Fields Crop Res 111:55–64

The manufacturer's authorised representative in the EU is Springer Nature Customer Service Centre GmbH, Europaplatz 3, 69115 Heidelberg, Germany. If you have any concerns regarding our products, please contact ProductSafety@springernature.com

Printed and bound by CPI Group (UK) Ltd, Croydon, CR0 4YY
30/09/2026
02225446-0017